Defective Colour Vision

fundamentals, diagnosis and management

John Dalton (1766–1844)

Portrait courtesy of the Manchester Literary and Philosophical Society.

Defective Colour Vision

fundamentals, diagnosis and management

Robert Fletcher MSc(Tech), FSMC(Hons), FBOAHD, DOrth, DCLP, FAAO, FBCO

Professor of Optometry and Visual Science
The City University, London

Janet Voke BSc(Hons), PhD, MInstP, FAAO

Senior Lecturer in Visual Optics
Department of Applied Optics,
City and East London College

Adam Hilger Ltd, Bristol and Boston

British Library Cataloguing in Publication Data

Fletcher, R.
    Defective colour vision
    fundamentals, diagnosis and management
    1. Colour blindness
    I. Title   II. Voke, J.
    617.7′59   RE921

    ISBN 0-85274-395-5

Consultant Editor: **Professor W D Wright**

Published by Adam Hilger Ltd
Techno House, Redcliffe Way, Bristol BS1 6NX, England
PO Box 230, Accord, MA 02018, USA

Typeset in 10/12pt Times by Mathematical Composition Setters Ltd, Salisbury, UK
Printed in Great Britain by J W Arrowsmith Ltd, Bristol

This text is presented in memory of
Rowland Jackson Fletcher
who as an optometrist imparted his
fascination with the visual system to
his son and granddaughter, the authors.

Contents

Foreword

'Colour blindness' is an emotive term and a misnomer in most cases. Even the term 'colour defective' suggests that a certain colour norm should be the rule. Though no two people have the same colour sense it is those with more unusual colour perception and their interaction with society that is the concern of this text. Clearly, the setting by society of artificial norms for colour vision, for example for danger signals, exposes significant problems. Although a start has been made to replace outdated colour codes it is important that much more is done to introduce universally recognised colour coding schemes and patterns so that all variations of colour sighted people are included. If such a system were adopted there would be extreme cases only where unusual colour perception would exclude an individual from certain fields such as dye matching. In time too special lenses may well become available to assist even this small minority group.

In the past too many people have been excluded from jobs for which they are otherwise quite capable due to the use of outdated colour vision testing techniques. Colour vision testing at an early age and a greater public awareness would eliminate the trauma experienced, for example, by a young person suddenly unable to enter a chosen field of work.

This book deals also with the little-known acquired colour vision defects. The famous colourist Claude Monet, the French Impressionist, who later in life found his colour vision greatly improved by the removal of a cataract, is a good example of a sufferer from such a defect.

As an artist myself, with an unusual aspect to colour perception, I welcome this text. Colour vision is a complex matter and with numerous examples Fletcher and Voke have unravelled its mysteries, producing a valuable piece of essential reading easily accessible to all. Needless to say it will be especially welcomed by medical personnel, teachers, industrialists, careers officers and trade unionists who are in positions of responsibility with regard to colour vision and can directly influence society in this matter.

The Earl of Gosford
London, November 1984

Acknowledgments

The authors thank Muriel and Peter, who as tolerant wife and husband shared the burden involved in the preparation of this text. Special thanks are due to Nancy Cowley who typed the script with great care and Isobel Fletcher de Tellez who compiled the index.

Neville Goodman, Jim Revill and Bridget Johnson of Adam Hilger Ltd coordinated this work with enthusiasm and care. We recognise with gratitude the wise comments of Professor W D Wright (colour scientist) and Mr W O G Taylor (consultant ophthalmologist).

Introduction

The study of colour vision is an interdisciplinary subject which embraces aspects of physics, biochemistry, neurophysiology and psychology. Light is absorbed by pigments in the photoreceptor layer of the eye's retina initiating a photochemical reaction. By a transducer process, which is still largely a mystery, the various attributes of light energy are coded for transmission to the brain by neural signals; here the signals are later interpreted. The perception of colour is a psychophysical experience which is dependent on the physiological coding and processing that takes place in the eye and brain. The fault which gives rise to defective colour vision, lies in the retina and/or visual pathway.

Abnormal colour vision interests a wide range of people, including the millions who realise that their appreciation of colour is 'defective', their families, and many more who are responsible for the dangers and other consequences of the condition, including industrial and professional implications. Daily life depends upon colour to an enormous extent in education, packaging, medicine, sport, horticulture, transport and many industrial activities. It is the recognition of this widespread interest in how we see colours, the faults which underlie defective colour vision, or colour blindness, and their means of detection and diagnosis which has prompted this book.

Throughout a combined experience of over fifty years in this field both authors have on many occasions been requested to provide concise and directive guidance for colour vision examination and subsequent advice. A wide variety of interested groups or individuals have sought information and assistance, principally industrialists, trade union officials, medical and paramedical practitioners, parents, teachers and professional associations. This text provides an approach to the detection, interpretation and consequences of the conditions which come within the scope of the subject. Although specific guidelines have been offered in almost every chapter, one distinct aim has been the stimulation of personal decisions and choice, based on some understanding of the underlying theory, which has been presented as succinctly as possible. Although background material and explanation is included, there has been careful selection with the anticipated readership in view. Much information that could have been included has been excluded with reluctance. The extensive references have been chosen selectively, yet aim to provide valuable source material. Readers will approach the subject in different ways. It is recognised that some will be unfamiliar with elements of the underlying sciences which relate to colour vision; for this reason selected foundation material has been included.

The physical, psychophysical and physiological aspects of colour have been studied for many centuries; indeed colour vision has attracted the attention of some of the greatest scientific minds in Newton, Maxwell, Young and Helmholtz. The original concepts that were established provided an invaluable foundation, and led the way to the more recent contributions of Abney, Edridge-Green, Judd, Wright, Hurvich, Jameson, Rushton and many others. The last two decades have seen major and vital contributions by neurophysiologists too numerous to mention by name.

The present time is a most exciting period in vision research on account of the fruitful interdisciplinary cooperation between psychophysics, neurophysiology, clinical observation and application. The authors have attempted to portray each of these features and the relationships between them.

A chief objective in this text is the presentation of the development of colour vision testing, the procedures involved, and the advice and assistance that can be offered to those who are shown to have abnormal colour vision.

The term 'Daltonism' arose from John Dalton's account of his own defective colour vision in the eighteenth century, and is now implanted in romantic languages such as in Europe and Latin America. The authors consider it as a tribute to the British atomic chemist, as it is often the practice to name syndromes or conditions after the person who first described them with care. This book therefore frequently uses the word 'Daltonism' as synonymous with abnormal colour vision, or defective colour vision; it reduces the otherwise negative emphasis implied by the term 'colour blindness', which is in many circumstances an inappropriate description.

Colleagues from many countries in the fields of visual science and optometry, ophthalmology, physiology and psychology have brought scholarly influences to the authors' attitudes to the subject, through personal encounter at international conferences and their published research data. The views of industrialists (among them industrial medical advisers), community health specialists and teachers have also been invaluable. This opportunity is taken to express great indebtedness to hundreds of patients and observers; their own unique contributions, as a result of their presenting themselves to the authors in wide variety, have been anonymous but essential. Due acknowledgment to other writers has been attempted, within the bounds of a necessary restriction of references to manageable limits; unconscious omissions are regretted.

There remains no single ideal method of approach to colour vision testing and diagnosis; each case must be treated individually. The choice of approach is, to a large degree, a matter of opinion and the authors have indicated methods which they have come to favour. As there is no reason to doubt the application to others readers may wish to follow this advice, at least until personal preference can be developed through experience.

THE EFFECTIVE USE OF THIS BOOK

Readers will have different uses for this book according to their particular interests and requirements. It is recognised that only a minority will have immediate access to the literature which is recommended through the references cited; others will try to use this book almost in isolation from other source material. A few who seek assembled material for study of a particular aspect may find the references of greatest value. Perhaps a sizable proportion of readers will wish to concentrate on what they believe to be essential practical guidance; this is likely to reflect their responsibilities in daily life. The authors' attempts to show how to deal with the required topics expeditiously should assist. It is possible merely to turn to sections of immediate interest and use the book as a 'handbook'. For example, the authors' own methods of testing colour vision may be copied uncritically by those who seek definite and rapid guidance. Readers concerned specifically with the dangers of defective colour vision will find Chapters 12 and 13 reasonably self-contained.

The majority of readers will appreciate that the text has been designed as a coherent entity with the foundation material presented in conjunction with the applications recommended. The overlapping of some material should be noted as a feature of emphasis. Asterisks are used to identify the essential core of each chapter, sections which the authors strongly recommend as most valuable if some selection is necessary at a first reading. It

Main interest	Suggested chapters			
	Initial study	Orientation	Practical issues	Further study
Careers	Intro., 11	1 (end), 2, 7 (parts), 8	3, 10, 14	4, 6, 13
Education	Intro., 10, 11	1, 3, 4	5, 7, 8, 14	9, 12
Health	Intro., 3, 4	5, 10, 11	6, 7, 8, 14	1, 9, 12
Industry	Intro., 11, 12	1, 3, 4	2, 6, 7, 8, 14	5, 9
Scientific	Intro., 1, 2, 3	4, 10, 11	6, 7, 8, 14	5, 9, 12

is recognised that readers already well acquainted with visual or colour science will find introductory chapters on elements of these subjects redundant. Nevertheless the authors' experience has suggested that at least some readers will be grateful for these basic introductory concepts.

Maximum value will be gained by viewing the subject as a whole and attempting to assimilate the background material provided before seriously applying the methods described. The table on p. 3 provides some indication as to how different chapters are related to particular interests.

1 How colours are seen

The study of abnormal colour vision must be soundly based on established knowledge of normal colour perception. This chapter provides a review of the relevant anatomical and physiological processes involved in the coding of coloured stimuli by the visual system as far as is known to date. It must be stressed that detailed understanding of the way in which colour is coded and processed in the normal visual system is as yet far from complete. Furthermore virtually nothing is known at the present time of the mechanisms underlying the psychological/perceptual interpretation of colour.

Enquiry into the psychological basis of colour perception and colour vision defect was initiated by early investigators, principally Young and Helmholtz, but productive research in this field has, of necessity, been a more recent endeavour, assisted largely by advances in electronics, which have made possible recordings from single cells at each stage in the visual pathway. Electrophysiological recordings have led to elucidation of many of the functioning mechanisms of the visual system and in the past two decades intense activity has been concentrated on the investigation of colour coding. This review can provide only a glimpse into what is perhaps the most exciting field of colour vision research today.

Young (1802a) first recognised that trichromacy was a property of the eye, rather than of the light stimulus. Recognising that the retina could not contain sufficient 'particles each capable of vibrating in perfect unison' he proposed a grouping into the three principal colours, red, yellow and blue. In the same year (Young 1802b) he modified the choice of these colours to red, green and violet following Wollaston's correction. Five years later Young (1807) suggested an explanation of Dalton's colour vision, that he lacked the red-sensitive channel retaining only two types of receptor. The theory was not generally accepted until Helmholtz (1866) had adopted it and proposed overlapping sensitivity curves for the three mechanisms.

Present understanding of how colour information is transferred through the visual pathways and interpreted in the higher centres of the brain has come from three main sources—anatomical, physiological and photo-chemical/neurochemical research. These three fields are very much inter-related. Morphological examination of the retinal layers of the vertebrate eye with the aid of the light microscope was pioneered by Ramon y Cajal at the end of the last century, following early studies by Gottfried Treviranus in the 1830s. Important contributions in more recent times were made by Østerberg (1935) and Polyak (1941, 1957). More detailed histological studies using the electron microscope have been led by Dowling and Boycott (1966) and Kolb (1979). Light microscopy ($\times 10^3$ magnification) is sufficient to establish the basic neural elements but electron microscopy ($\times 10^4$ to $\times 10^5$ magnification) is needed to indicate the elaborate nature of the synaptic connections. In addition electrophysiological investigations involving the use of microelectrodes, to record the electrical impulses generated by single cells in response to specific stimuli, have proved most valuable in furthering understanding of the processing at each level of the eye–brain system. Hartline (1938) and Kuffler (1953) indicated different neural responses to light in vertebrates in relation to the presenta-

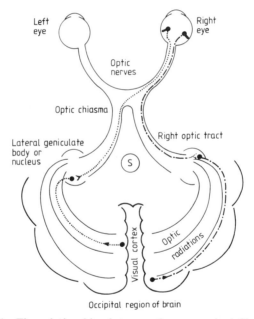

Figure 1.1 The relationships between the post-retinal fibres. Ganglion cell fibres travel in the optic nerves according to the region (right or left) of the retina involved, crossing (or not) at the chiasma, then reaching the lateral geniculate nucleus (LGN) via the optic tracts. After synapsing the final neurones from LGN cells pass to the visual cortex by the optic radiations. S represents the upper end of the fluid canal which forms an axis of the body.

tion mode and spatial location of the stimulus. These general findings were of fundamental value for later studies on colour processing.

It is generally believed that the trichromatic processing in the cone receptors is transformed to an opponent colour process in the horizontal cells early in the retinal processing (Tomita 1970). Analysis of the coding of brightness and colour in the retina, performed originally by Granit and Wrede (1937) on frog retinae and Svaetchin (1956) on fish retinae, has been possible more recently in the retinal ganglion cell layer of primates (de Monasterio and Gouras 1975).

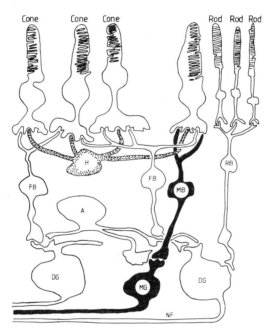

Figure 1.2 Representative primate retinal neuronal connections as established in the last two decades. H = horizontal cell; FB = flat bipolar cell with multiple connections; RB = rod bipolar cell with multiple connections; MB = midget biplar cell with single cone connected; A = amacrine cell; DG = diffuse ganglion cell; MG = midget ganglion cell; NF = nerve fibre (to optic nerve).

The difficulty of recording from single neurones in the retinae of primates, on account of the very small sizes of the structures packed in the neural matrix, has meant that most of the early studies involved fish retinae where units are more readily identifiable. Only in the last decade or so have such studies been possible on monkeys. More recently attention has been focused on the physiological mechanisms underlying colour vision in the cat and monkey, and since macaque and rhesus monkeys are known to have colour vision characteristics of a similar nature to normal human colour vision, these species have been widely used. At the level of the lateral geniculate nucleus (LGN) the single cell recording approach was pioneered by de Valois (1965) and Wiesel and Hubel (1966) in primates and at the visual cortex by Gouras (1970) and Zeki (1973) among others. The third approach has sought to elucidate the initial photochemical response and its interaction with the neural coding. Rushton (1955, 1958) built upon the pioneer studies of Wald (1945) and Dartnall (1957) in studies of the characteristics of photosensitive pigments in the receptors, while Baylor *et al* (1971) and Fuortes *et al* (1973) concentrated on the transfers from the photopigments and synaptic interaction and feedback processes in the next neural link. Neurohistological studies made by Marc and Sperling (1977), Marc (1980, 1982) and Marc and Lam (1981a,b) have more recently initiated some understanding of the chemical transfers at the synapses of retinal neurones. Each of these approaches will be discussed in this chapter.

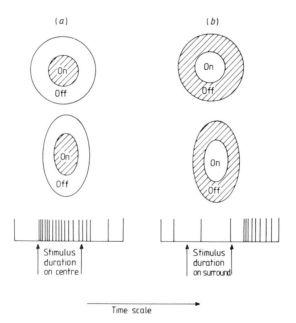

Figure 1.3 Receptive fields and their responses. The hatching represents the area stimulated. The fields related to retinal ganglion cells and LGN cells are approximately circular (top pair of diagrams) while those for the visual cortex are more elliptical (lower pair). In each case a field with an 'On' centre and an 'Off' peripheral area is repesented. In (*a*) the centre only is stimulated, giving the burst of 'action spikes' on the time scale shown beneath. In (*b*) the surround only is stimulated.

It is assumed that readers are familiar with the basic gross anatomy of the vertebrate retina and visual pathways (see figures 1.1 and 1.2). A rudimentary knowledge of the elements of electrophysiology is also assumed, such that messages are conveyed in the nervous system by changes in the number and frequency of spike electrical potentials. The catchment zone of a cell is termed the *receptive field*, and most cells at the various levels in the visual pathway have clearly organised receptive fields, such that some spatial locations give rise to an increase in firing of the cell when the stimulus is turned on ('on' cells) while other cells are stimulated into action when the stimulus is turned off ('off' cells). This results from the excitatory central zones and inhibitory surrounding areas. A further group of cells, termed 'on–off' cells, generate responses either when the stimulus is turned on or off. We know that neurones conveying visual information rarely act independently, or in isolation, so that neighbouring neurones influence the response, for example through lateral inhibition.

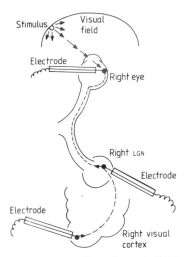

Figure 1.4 Microelectrode recordings from individual cells at different sites. A stimulus is located in part of the field of vision (for example a bright spot of light projected onto a curved screen). Electrodes are shown recording from a ganglion cell, from a cell in the LGN and from a cortical cell as the stimulus is moved to different parts of a (relatively small) receptive field.

The numerous stages in colour perception entail physical, physiological, and psychological processing. Davson (1980) or Barlow and Mollon (1982) should be consulted for introductory comment. The eye's refractive elements convey light to the retina. Significant modifications to the characteristics (colour) of the light reaching the retina, may occur as the rays pass through the ocular media.

Light traverses the neural layers of the retina to the photoreceptors which are in the innermost region (posterior) and is absorbed by the photopigments contained within the receptors; subsequently a neural discharge is initiated by which the signal is transferred to the brain. The retina has been shown to play a major role in separating out the coloured elements of the stimulus from brightness characteristics and coding these two attributes. Little is known about the mechanisms converting the photochemical change in the sensory receptors to an electrical potential. It involves a complex internal chemical reaction in which changes in the potassium and sodium balance play a major part. Pharmacological agents known as neurotransmitters act to 'carry' the coded signal from one neurone to the next as they are released from neurones at the synaptic junction. One important and novel characteristic of photoreceptors is their hyperpolarising action (production of a negative potential in the cell membrane) as a consequence of light absorption.

1.1*   RETINAL ANATOMY

The retina is a transparent ten-layered structure approximately 0.22 mm thick, which is highly sensitive to light. Rod- and cone-like receptor end-organs receive the light and four layers of neurones of different types are involved in coding, modifying and transferring the signals: bipolar cells, ganglion cells, horizontal cells and amacrine cells. All are packed in a highly organised manner with distinctive layers which can be identified by the light microscope. The outer pigment layer of the retina absorbs light, thus preventing reflections back through the retina which would degrade the image. The receptors are packed in a honeycomb-like mosaic of hexagonal arrangement. The spatial distribution of the rods and cones and their subsequent wiring pattern with other neurones determines the basic capabilities and characteristics of the entire visual system.

Synaptic junctions (near contacts) are made first between receptors (either rods or cones) and bipolar cells which then transfer the signals, via another set of synapses, through the layers to the ganglion cells (see figure 1.5). One receptor cell may activate many bipolars and one bipolar cell can activate many ganglion cells. Ganglion cell axons converge and leave the eye as the optic nerve, passing though the optic chiasma and tracts to the lateral geniculate nucleus. Horizontal cells connect receptor cells to each other laterally in a chain-like manner, and amacrine cells in the layer connect ganglion cells (figure 1.6). There are thus alternative pathways from the receptors to the bipolars, and two pathways from the bipolars to the ganglions. In addition a complex of supporting cells and fibres is found.

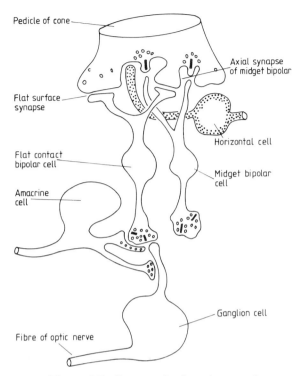

Pedicle of cone

Axial synapse of midget bipolar

Flat surface synapse

Horizontal cell

Flat contact bipolar cell

Midget bipolar cell

Amacrine cell

Ganglion cell

Fibre of optic nerve

Figure 1.5 Synapses in the primate retina.

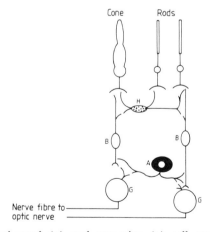

Cone

Rods

Nerve fibre to optic nerve

Figure 1.6 Horizontal (H) and amacrine (A) cell connections in the primate retina. B = bipolar cell; G = ganglion cell. Both horizontal and amacrine cells are connected to bipolars in addition to receptor or ganglion cells.

The rod and cone receptors (figure 1.7) have specific functions, controlling night (scotopic) and day (photopic) vision repectively. Both enclose visual pigment in membranes in flattened sacs in their outer segments. The rod pigment, rhodopsin, responds maximally to blue-green light at 505 nm (see figure 1.8) and, as no variation in the pigment for different rod receptors is found, the spectral sensitivity is common and does not facilitate colour vision. Rods govern vision at low levels of illumination on account of the fact that rhodopsin has a greater sensitivity than the pigments found within cone receptors. Rods are plentiful throughout the retina (except in one rod-free region called the fovea) and at the optic disc where no receptors are situated, but there are variations in density. There are approximately 120 million rod cells in the human retina. Individual rods are linked electrically with adjacent rods through horizontal cells; this allows a summation of response from a wider area than that served by an individual receptor.

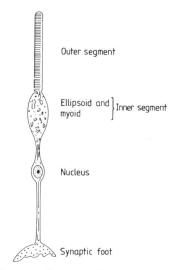

Figure 1.7   Schematic diagram of a cone receptor near the fovea.

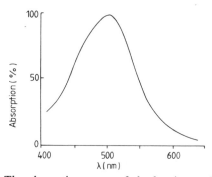

Figure 1.8   The absorption curve of rhodopsin or visual purple.

By contrast cone receptors contain a variety of visual pigments, each having a different spectral sensitivity and a specific wavelength at which they respond maximally, a fact which permits distinction between colours when cones are activated. Cones are distributed mainly in the central region of the retina, in a zone often called the *macula lutea*, which subtends some $4 \times 12°$ (approximately 1×3 mm), in which the foveola subtending approximately $1 \times 1°$ (0.24×0.3 mm) is most important. They number only four to seven million in total with approximately 25 000 cones in the fovea (see figures 1.9, 1.10 and 1.11). Østerberg (1935) made the first reliable count of receptors in the human retina. Foveal cones number 147 300 per mm^2 whereas parafoveal (peripheral) cone numbers are a mere 9500 per mm^2 at a position 2 mm from the fovea. Rods are first found 130 μm from the foveal centre; their maximum density rises to about 160 000 per mm^2 between 17 and 20° from the fovea. A marked drop then occurs towards the retinal periphery, although it is not as rapid as the peripheral fall-off of cones. Unlike rod receptors, which differ little in their structure throughout the retina, cones are of a slender formation (like rods) in the foveal region but fatter (more cone-like) in more peripheral regions. Individual cone receptors are joined functionally with each other through synaptic junctions with horizontal cells, except in the foveal region where each cone has virtually an independent link with the bipolar cell beyond it and this in turn with the ganglion cell beyond. This private pathway (the midget system) is the anatomical basis for the excellent acuity possible with foveal vision, since the signals are less combined or influenced by neighbouring receptors. Figure 1.12 shows the basic types of neural 'wiring' in the vertebrate retina.

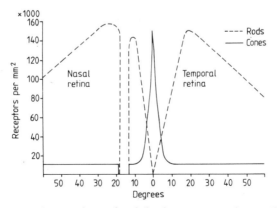

Figure 1.9 Relative numbers of rod (broken curve) and cone (full curve) receptors in different retinal areas. In the horizontal meridian through the optic disc (between 13° and 18° in the nasal retina) there are no receptors. The fovea is shown with a maximum of cones and a minimum of rods.

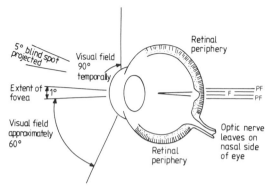

Figure 1.10 Cross section of an eye showing the major retinal areas. The exit of the optic nerve (projected into visual space, the 'blind spot') subtends about five degrees. An annulus around the fovea (F) is shown as the parafovea (PF) with the retinal periphery around this.

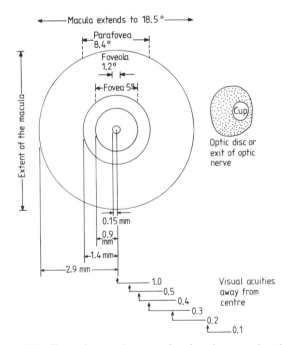

Figure 1.11 The dimensions and terms related to the central region of the retina. This diagram views the optic disc (of a right eye) from the direction of the pupil so that the fovea is seen on the left of the disc, to the temporal side of the disc. Linear and angular sizes approximate to those of the average eye. The visual acuities are shown as a proportion of normal foveolar acuity of 6/6; thus at an eccentricity of 1.5° a visual acuity of 6/12 or 0.5 might be expected.

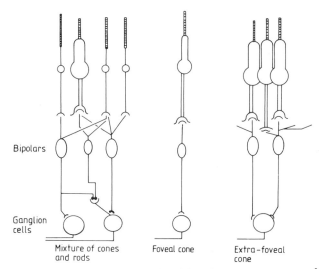

Figure 1.12 Basic types of neural 'wiring' in the vertebrate retina.

Colour vision is thus a function of cone vision and is confined to the retinal regions served by cones. Beyond 60°, vision is essentially mono-chromatic (Wooten and Wald 1973). The changeover from scotopic to photopic vision and vice versa is gradual, and at low levels of illumination both rods and cones function together in 'mesopic' vision. Anatomical, neurophysiological and psychophysical evidence supports this two-tier theory of vision.

1.1.1* Cone types

There has been much discussion and speculation as to how the cone recep-tors mediate colour vision, but with very little evidence until recently of ex-actly what form the organisation takes. Despite the very large number of cones it is unlikely that there should be a separate pathway for each colour or hue that the eye can discriminate. Simple colour mixing experiments conducted in the last century (notably by Maxwell, König and Helmholtz) indicated the trichromatic nature of colour which was clearly associated with the eye. However, it was unclear whether there were three types of cone receptors controlling vision, sensitive mostly to the long wavelengths (red light), the medium wavelengths (green light) or the short wavelengths (blue light), or whether, with no anatomical differentiation, there were separate neural channels. The favoured view has long been the existence of three types of visual pigments controlling colour vision. Physiologists working in the USA, Brown and Wald (1964) and Marks *et al* (1964), first indicated the existence, in the primate and human retinae, of three types of photo-pigment, responding maximally to red, green and blue light (see figure 1.13).

Experimenters in the UK used evidence of a different kind to substantiate this view (Rushton 1955, Rushton *et al* 1955, Ripps and Weale 1963). Work involved an analysis of the light reflected back from the retina in the living eye after a strong light had bleached the visual pigments (fundus reflec-tometry). This process is crudely analogous to the return of light in the ophthalmoscope beam. Recent histochemical studies by Marc and Sperling (1977) have clearly indicated the presence of three cone photopigments in primates, with the middle wavelength type being most numerous and the long wavelength type constituting overall about 33 per cent of the total. The short wavelength 'blue' coding receptors show greatest variation in density of population, with a maximum at about $1°$, being only 3–4 per cent of the total at $0.5°$. The proportion is increased to 20 per cent at $1°$ but by $5°$ has dropped to around 13 per cent. 'Red' and 'green' densities are therefore a maximum in the foveola, the region of maximum visual resolution. This is consistent with the marked superiority of the red and green channels for fine detail vision, compared with the blue system, measured psychophysically, for instance by Cavonius and Estevez (1975a).

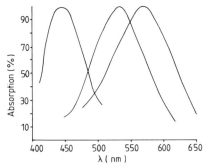

Figure 1.13   Absorption curves of cone pigment in the foveal region. The maxima and distributions follow data reported from various laboratories during the last two decades.

1.1.2   Neural connections of the retina

Although the transmission channels along which nervous impulses are conducted are well-established anatomically as separate units, knowledge of the way in which each of the retinal structures contributes to the total, and the mechanisms whereby signals are transferred from cell to cell, are less clear at present. Two main features of the organisation of the retina are central to colour processing—the functional linkage between cells (coupling or convergence) laterally through the receptive field, and the feedback mechanism which horizontal cells mediate either electronically or chemically to influence the final output of the cone receptors. The former is the anatomical basis for spatial summation and the latter for lateral inhibition, both of which are found frequently in sensory pathways. These features have a far greater consequence than mere economic channelling of responses, resulting in the cells losing their independence as single units; they greatly influence the sensitivity of the eye and the response to edges or demarcations of stimuli. On the deficit side the coupling means that spatial resolution is reduced.

1.1.3 Horizontal cells

The cell dendrites of horizontal cells make contact with cone receptors at their bases or pedicles (see figure 1.14). Two types of horizontal cells have been identified by Boycott and Dowling (1969) and these were originally thought to control the rod and cone signals respectively. However Kolb (1970) has shown that both types only make dendritic contact with cones.

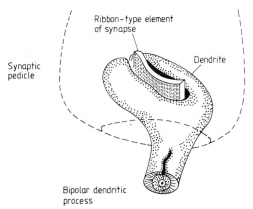

Figure 1.14   Simplified view of the dendritic process connecting with a synaptic pedicle. The bipolar dendritic process is seen entering so that the dendrite engulfs the ribbon-type element of the synapse. The vaculolar and probably multiple elements within this junction are not indicated except at the cut section of the bipolar dendritic process. After Pedler (1965) and Sjostrand (1965).

The difference between horizontal cell types is indicated by the numbers of cones they serve, one type typically contacting seven cones and others twelve cones. Near the fovea each cone is probably in contact with at least four horizontal cells and one foveal horizontal cell connects between seven and twelve cones, depending on its type. Towards the periphery the number of cones to horizontal cells increases to between thirty and forty; in this way they extend their lateral coverage, summating or gathering information over a wide area. Horizontal cells also link up with other horizontal cells, and they can also 'drive', or stimulate into action, bipolar cells and through them ganglion cells. The lateral interactions ensure that receptors outside the dendritic spread of a bipolar cell can communicate easily.

All neurones transfer their signals to adjacent neurones through pharmacological agents known as neurotransmitters contained within synaptic vesicles. Little is known as yet about the nature of the neurotransmitters which control neural activity in the retina (see §1.3).

1.1.4 Bipolar and amacrine cells

A number of different types of bipolar cells, the second neurone in the retinal chain, can be described on the basis of their morphology. In primates the bipolar pathways from receptors to ganglions are thought to to be exclusive for either rods or cones. Rod bipolar cells connect about fifty rods through base structures called spherules. Of particular importance for foveal vision are the midget bipolars which involve single pathways between individual cones and midget ganglions. Flat or brush bipolars involve diffuse pathways connecting five, six or seven cones, via horizontal cells or directly (see figure 1.15). Amacrine cells are intermediate lateral cells which receive inputs from bipolar cells and other serial amacrines, connecting with ganglion cells.

Ganglion cells show considerable variety morphologically, for example the diffuse types with wide coverage from bipolars and the midget ganglions which connect single midget bipolars to cones. Polyak (1941) suggested that midget ganglion cells, which he recognised as only present in primate retinae in foveal regions, could provide a private pathway to the brain for information from a simple cone via the midget bipolar. More recent morphological studies of primate ganglion cells have utilised the enzyme horseradish peroxidase to show retrograde transportation (see Leventhal *et al* 1981). These same authors have linked the cell types to their possible physiological roles.

1.1.5   Rod–cone interactions

The 'pooling' of responses from a wide area at various stages in the retinal network will be discussed in §1.2. Overall it appears that responses from rods and cones are dealt with by similar neurones within the retinal network. It is probable that differences in their response or transmission times prevent them from acting simultaneously. Nevertheless some interaction between rod and cone signals does occur, since according to Kolb (1970) the axon processes terminate on the bases of rods, called spherules, with each rod spherule receiving input from two horizontal cells. The horizontal cells in primates appear to mix the rod and cone messages so that rods may play some role in colour vision although the degree of interaction is, at present, uncertain.

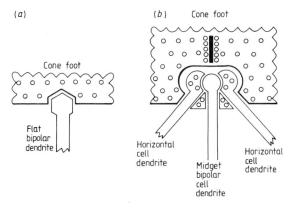

Figure 1.15 The cone foot synapse. (*a*) Simple connection. (*b*) Invaginated triad connection. Synpatic vesicles (containing transmitter substances) are shown in the cytoplasm. The serrated line represents the boundary of a small element of the cone foot shown.

1.2 RETINAL PHYSIOLOGY

1.2.1* Electrical responses—general

Light absorption initially causes a photochemical response and leads rapidly to electrical activity generated in the outer limb of the receptor; the first electrical response is termed the early receptor potential, and measures about 2 mV. In the human retina rods and cones contribute almost equally to this activity and the response depends on the wavelength of the stimulus even at this early stage. The transduction of light into neural activity is a unique process. Light causes a decrease in the cell membrane conductance, or an increase in its resistance, which leads to a hyperpolarisation; the membrane voltage becomes more negative than it was at its resting state. Hyperpolarisation of the cell is the opposite of what is normally expected of sensory receptors—depolarisation. The pulses (called action or spike potentials) rise to a maximum amplitude, which is always the same, and then gradually fall. Coding of the response is indicated by the number of responses and the frequency with which they occur rather than their amplitude. When stimulated by light the cell membrane maintains a reduced permeability to sodium ions, since in darkness there is a steady inward flow of these positive ions.

Measurements of the early receptor potential indicating electrical activity generated in the outer limb of cones were first made by Brown and Murakami (1964) by means of extracellular recordings. Later Murakami and Pak (1970) studied the early receptor potential using intracellular recordings. Recordings of this type indicate that at the stage of photoreception early electrical activity is influenced by the wavelength of the stimulating light. Typical values are of the order of 1 or 2 mV. An unusual characteristic of photoreceptors is their hyperpolarisation with respect to the membrane potential, on activation. Inhibitory or negative influences may result as opposed to the initial excitatory or positive response, as a result of feedback, thus modifying the response. An example occurs from horizontal cells to cone receptors (see §1.2.4).

Receptor cells and horizontal and bipolar neurones respond with a slow graded potential which varies in frequency with intensity of stimulus and/or wavelength. Initial coding of the stimulus is thus brought about. The first two cells in the chain give a hyperpolarisation which is generally associated with inhibitory action; the bipolar, amacrine and ganglion cells show mixed responses of hyperpolarisation or depolarisation. In the majority of cases stimulation of the final neuronal elements of the retina causes a brief depolarising potential. Amacrine cells are the first in the chain to show a brief spike potential rather than a continuous graded response. Ganglion cells also give spike responses, some in brief bursts in which the frequency of firing is directly proportional to the stimulus intensity (transient cells), and others giving a more sustained discharge. The first electrical recording from ganglion cells by Adrian and Mathews (1927) was followed by extensive experimentation in the 1930s and 1940s which gave a clear understanding of the mode of electrical response, but little knowledge of the mechanism by which it was generated.

Readers are referred to the excellent chapters on the anatomy and neurophysiology of the retina and central visual pathways by Ripps and Weale, Arden and Holden in Davson (1976); the discussion of the transduction process and electrical activity is particularly valuable.

1.2.2 Summated responses

It has already been indicated that retinal cells undergo varying degrees of convergence. The presence of up to seven million cones and 120 million rods with only one million optic nerve fibres (the axons of ganglion cells) going to the brain, shows that some ganglion cell responses reflect the activity of several hundred receptors. Electrophysiological recordings further demonstrate that a single ganglion cell may respond when any part of a large area of the retina is stimulated. The retinal area which when stimulated causes a cell to respond is called the *receptive field* of the cell, the spatial catchment area of the retina or visual field projection which is imaged on the particular cell which is responding. The retinal response is a summation of all the influences within the receptive field. Hartline *et al* (1956) first indicated that pooled responses are involved. Physiologists working towards the end of the last century (notably Mach and Hering) were aware of the mutually antagonistic portions within the retina and the concepts of excitation and inhibition. It is now possible to link these two central observations.

At the retinal level the opposing positive excitatory (on) and negative inhibitory (off) influences are organised in a clear, concentric spatial format (see figure 1.3). For a centre-on and surround-off receptive field, the response will be positive if the stimulus falls within the central zone, but will fire only when the stimulus is switched off, if it is moved to the surround zone. Other cells show opposite responses (central inhibition and surround excitation). For colour vision this opponent processing is of great importance since antagonistic colour responses are found at every level in the visual pathway. As an example, the central region may cause excitation when stimulated with red light and the surround inhibition when stimulated with green light; such a cell could be designated $+R/-G$ following the notation that the central response is indicated first, followed by the surround response. A positive sign indicates excitation and a negative sign inhibition. Kuffler (1953) first recorded the antagonistic centre–surround organisation of receptive fields in the cat at the retinal ganglion cell level. Considerable variation in receptive field sizes is shown throughout the retina; receptive fields are larger in peripheral retinal zones and for low levels of illumination.

1.2.3 Electrical responses—specific cell activity

The first recordings from single neurones of the retina were made by Svaetichin (1953) and MacNichol and Svaetichin (1958) who used fine microelectrodes to tap the signals from the inner nuclear layer of the fish retina. The responses have been named S potentials in Svaetichin's honour. Two distinct slow changes in potential were noted; a response which was independent of the wavelength of light used to stimulate the cell, and a second response which depended on the wavelength. The former had an action spectrum similar to the photopic sensitivity function, termed L responses, and are believed to code brightness information. The latter, designated C responses, were shown to vary in an antagonistic manner being positive to some wavelengths, for example red, and negative to others, for example green ($+R/-G$) (see figure 1.16). This was the first experimental evidence of colour opponency in the visual system, suggested by Hering almost a century earlier; furthermore it was identified in the retina. We now know that S potentials are the post-synaptic responses of horizontal cells which receive inputs from the photoreceptors and feedback onto them.

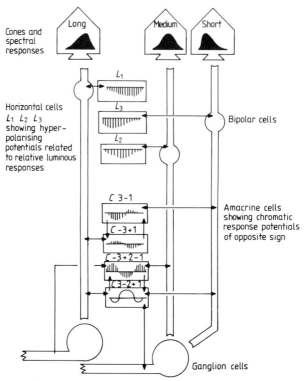

Figure 1.16 Simplified fish retina model after Svaetichin *et al* (1965) showing the levels at which nerve signals corresponding to specific stimuli are intercepted.

Few intracellular recordings have been made as yet from the bipolar cells or amacrine cells. Their distinctive concentrically organised receptive fields with large surrounds indicate that lateral interactions are involved. Kaneko (1971, 1973) has suggested that the receptive fields of colour-coded bipolar cells are produced by both the L-type and C-type horizontal cells. Some evidence of colour coding is shown by amacrine cells which depolarise on receipt of an input from bipolars. It is these latter cells which provide the large receptive field. Amacrines are linked in a series of synapses. In addition to providing an output response to ganglions they also link back with bipolars, though whether the feedback mechanism is inhibitory or excitatory is as yet unclear.

1.2.4 Inhibition and feedback in the retina

Hering first indicated the importance of a feedback mechanism in the visual process, realising that such a control mechanism would provide a wide operating range for the retinal neurones. Intracellular electrophysiological recordings from turtles suggest that the brightness-coding horizontal cells (L cells) modify the initial receptors' responses in a feedback mechanism giving an antagonistic inhibitory effect (figure 1.17). The hyperpolarisation activated in the receptors is partially cancelled on effective depolarisation through the horizontal cells (Baylor *et al* 1971). This action is not cone specific in turtles (Fuortes *et al* 1973) and it has also been suggested that in man the inhibitory modification probably also occurs between both like and unlike receptors since horizontal cells do not appear to differentiate between receptor classes (Kolb 1970).

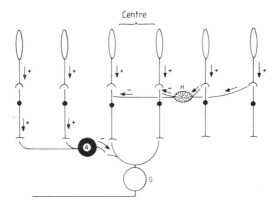

Figure 1.17 Possible inhibitory impulses contributing to different responses in different zones of ganglion cell receptive field. The central field contains the two middle receptors which stimulate the ganglion (G) cell (as an 'On' centre) normally. The surrounding groups produce an 'Off' response by inhibiting the ganglion cell. On the right horizontal (H) cell action is proposed and on the left similar effects are shown for amacrine (A) cells.

Local receptive field effects can be important at the receptor level. Fuortes *et al* (1973) found that light falling on a zone up to 40 μm from a small receptor group was responsible for increasing the response of a particular cell (the hyperpolarisation was enhanced). Beyond this distance, however, the cell was inhibited by giving a voltage response of opposite polarity (depolarisation); thus the wave response was reduced overall. It is thought that excitatory enhancement through nearby cones operates only between those having the same class of photopigment ('like' receptors in the colour-responsive sense) (Baylor and Hodgkin 1973 and Fuortes *et al* 1973).

Physiological evidence thus suggests that cones controlling the same wavelength range of the spectrum are coupled together, suggesting interactions between cones. More recent studies on the turtle, however, led Piccolino *et al* (1980) and Toyoda *et al* (1982) to suggest that the depolarising response of the R/G opponent-type horizontal cells to red light is mediated by feedback from the L cells to green responding cones. Piccolino *et al* (1980) furthermore suggested that the R/G cells can feedback onto blue responding cones. Their findings led them to suggest that 'red, blue and green' responding cones respond with a hyperpolarising effect to blue light from blue responding cones directly, but with depolarisation to green light on account of the fact that the hyperpolarisation of the R/G type horizontal cells to green light depolarises blue zones.

1.2.5*   Ganglion cell responses

The final neurones in the retinal chain, the ganglion cells, send true spiked messages to the brain via the optic nerve. Electrophysiological recordings from individual ganglion cells in the monkey retina were pioneered by Gouras (1968a,b,c), de Monasterio and Gouras (1975) and de Monasterio *et al* (1975a,b).

For some time the existence of two functional classes of ganglion cells has been established from both psychophysical and electrophysiological evidence, pioneered by Enroth-Cugell and Robson (1966). Transfer of visual data by means of these two distinct pathways is retained at subsequent levels of processing. A detailed discussion of their characteristics is not appropriate in this review; beyond a consideration of colour aspects it is sufficient to list their properties below.

Sustained discharge cells (also called tonic and X cells)	Transient discharge cells (also called phasic, Y and broad-band cells)
Longer maintained action potentials	Short burst of action potentials
Slow conduction velocity	High conduction velocity, implying larger axons and neurones than sustained cells
Mostly near fovea	Mostly in periphery
Carry colour information	Do not discriminate colour

Early studies by de Monasterio *et al* (1975 a,b,c) indicated at least 25 different types of retinal ganglion cell in the monkey, of which fifteen varieties code colour information in a colour-opponent manner, and six code spatial (form) information. This number has, more recently, been modified to eleven distinct functional classes (Gouras and Zrenner 1981a,b). Although the proportions of different types encountered varies with retinal position, as indicated by Zrenner and Gouras (1979), in general, all types are found in most retinal regions. Previous studies by Hubel and Wiesel (1960) had indicated colour opponency in the primate retinal ganglion cell layer.

The two main classes of receptive field, whereby excitatory centres (on-centre) are surrounded by inhibitory surrounds (off-surround) are maintained for both tonic sustained (X type) and phasic transient (Y type) cells (Gouras 1968a). Phasic cells, described as broad-band, are more common towards the peripheral retina with tonic colour-coding cells more numerous in foveal regions. Early recordings from monkey ganglion cells by de Monasterio and Gouras (1975) led them to classify opponent cells in either of two categories. The majority of opponent cells receive an input from the long wavelength and medium wavelength receptors but none from the short wavelength coding receptors ($+R/-G$ and $-G/+R$ cells); their spectral characteristics were similar to the Stiles π_4 and π_5 mechanisms, the fundamental colour mechanisms for red and green (see § 1.11.3). A second type was thought to receive an input from the short wavelength receptors with the other contribution coming from either the medium or long wavelength coding cells. These are simple opponent cells. Furthermore some foveal ganglion cells were seen to be arranged so that the central catchment area received an input from one spectral class of cones, with the surrounding area governed by the other two cone types together (trichromatic or double-opponent cells). Thus, where red and blue signals were seen to be processed in one region of the receptive field the term 'magenta' cells applied; similarly where inputs from both short and medium wavelength coding receptors were received together the cell was labelled as 'cyan'-coding. Similarly operating double-opponent cells have been noted in the visual cortex of monkeys by Gouras (1970) and Dow (1974).

Cells were also found which indicated colour-opponent properties (i.e. responding in an excitatory manner to some wavelengths and in an inhibitory manner to others, with a definite cross-over neutral point) only when one of the cone mechanisms was selectively depressed. These represented almost a third of the total (de Monasterio and Gouras 1975 and de Monasterio et al 1975a,b). Found most frequently towards the peripheral retina, these cells were called 'concealed colour-opponent' cells. A similar group of cells, which were originally thought to be non-colour coding cells, have been found in the lateral geniculate nucleus (Padmos and Norren 1975).

Non-colour-opponent (phasic) cells were seen to have an input from the long and medium wavelength coding receptors only. These broad-band cells were thought to code brightness information, as their action-spectra were similar to the photopic sensitivity function with a maximum at about 560 nm. Their action-spectra showed no variation for the centre and surround areas of the receptive field, indicating the absence of colour opponency. The cells share signals from R and G cones (Gouras 1968). Figure

1.18 shows a model of the primate retinal synapse circuits and figure 1.19 summarises the classification of cell types found throughout the foveal and parafoveal regions. From these figures it is evident that some yellow/blue-opponent cells ($+Y/-B$ and $-Y/+B$) are present, but the numbers are few, with the greatest proportion found outside the foveola. Recent research in independent laboratories has indicated that while cells subserving the R or G cone mechanism can be either of the on- or off-centre variety, those controlling the B cones are 'mostly and probably always' on-centre (Malpeli and Schiller 1978, Gouras and Zrenner 1979, 1981a,b). The special features of the short wavelength blue cone input will be discussed in §1.2.7. This characteristic of the blue-opponent cells has also been noted in the lateral geniculate nucleus of primates.

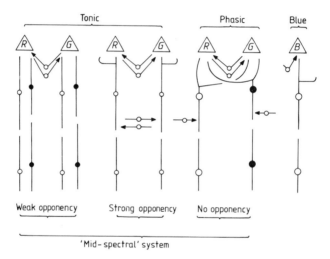

Figure 1.18   Model of primate retinal synapse circuits. Far left are R–G tonic cells, with horizontal cell feedback, giving weak opponent inter-actions between the R and G receptors. Middle left are R–G tonic cells with strong opponency involving amacrine cells, predominantly on-centre (or 0) bipolar and ganglion cells. The far right B cones are also con-nected to on-centre cells, while the middle right R and G receptors send direct signals to the underlying bipolar and ganglion cells. This model also involves cortical detectors for brightness and for hue contrast which are not shown here. After Gouras and Zrenner (1982).

A lack of blue off-centre cells at both the retinal ganglion cell layer and the lateral geniculate nucleus in the rhesus monkey has been reported by Malpeli and Schiller (1978); it appears that the majority of signals coding for short wavelength stimuli is by blue-opponent on cells.

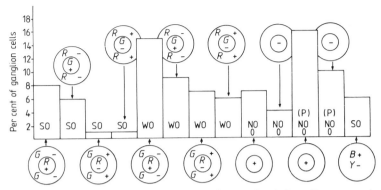

Figure 1.19 Varieties of primate ganglion cells (after Gouras and Zrenner 1982) from foveal and perifoveal regions of macaque retinae. $R +$ indicates 'red on', $G-$ 'green off' etc. SO indicates strong opponency of cones and WO shows weak opponency. Cells labelled (P) are phasic; all others are tonic.

1.2.6*   Red–green-opponent ganglion cells

The range of colour opponency shown by cells with R/G coding is much greater than for blue-coding cells (de Monasterio *et al* 1975a,b). In general, the G cone response integrates over the largest area of the cell's receptive field (Gouras and Zrenner 1981a,b). Some cells require indirect methods, such as chromatic adaptation, to reveal the colour-opponency characteristics. Gouras and Zrenner (1981a, b) have suggested anatomical correlates to these two colour-opponent forms, proposing that weak opponent cells result from horizontal cell interactions between all R and G cones and the strong opponent cells are mediated by amacrine cell interactions between selected bipolar cells which receive inputs from the same R and G cones.

Since broad-band, phasic cells could share the same R and G cones as the tonic cells, Gouras and Zrenner argued that there must be a separate set of bipolar and amacrine cells which produce this luminance/colour separation. They proposed a model of a possible pre-synaptic relationship between the two basic types of ganglions in relation to the cone input. In this model all ganglion cells controlling the mid-spectral cones (R/G) share identical horizontal cell circuits; separate sets of bipolars and amacrine cells thus determine their unique behaviour. Furthermore it is suggested that those R and G ganglions sharing weak cone opponency (i.e. required special conditions to indicate their opponent characteristics) would have a minimum of amacrine cell input and thus signal the cone responses. The R/G ganglions showing strong cone opponency are modelled as having a separate set of bipolars which receive an inhibitory input from a cone-specific amacrine cell. The short wavelength blue cone system is modelled using separate bipolar and ganglion cells and the rod system would use a further separate set of bipolars and amacrine cells. Gouras and Zrenner further related their physiological findings to morphological features of the retina by suggesting that the tonic ganglions sharing weak opponency could comprise the midget ganglion cell system (see figure 1.18).

1.2.7 Blue-opponent ganglion cells

Blue on-centre retinal ganglion cells are excited maximally by wavelengths between 440 and 450 nm and exhibit a positive response for all spectral wavelengths from 400–500 nm; longer wavelengths result in an inhibitory response (Gouras and Zrenner 1981a,b). It has further been suggested that the blue cones, like rods, might utilise only one set of bipolar cells, on account of the absence of blue responses in surround areas. This is substantiated further by the observation that blue cones usually feed only on-centre channels in monkey retinae.

Evidence from psychological experiments confirms that the short wavelength blue cone system has specific characteristics not shared by the other two cone mechanisms. The incidence of short wavelength signalling cones is sparse compared with the other two cone mechanisms, as indicated by histochemical studies and analysis of cone pigments (see §1.1.1). Furthermore colour vision defects affecting blue vision are rare in an inherited form but common in an acquired form (see Chapter 4). Among their peculiar properties, blue cones are known to contribute little to visual resolution and to brightness perception; such findings are consistent with the sparse incidence of blue-opponent cells in the primate.

Zrenner and Gouras (1981) indicated that the long latency at the onset and lack of inhibition at the offset of light they observed in the blue-opponent ganglion cells would limit the poor temporal resolution shown by psychophysical experiments. They previously showed a different flicker fusion frequency between the spectrally different ganglion cells (Zrenner and Gouras 1978).

A second study (Schuurmans and Zrenner 1981) indicated an electrophysiological correlate for the way in which the sensitivity of the blue cone system can be altered by yellow adapting lights as observed by Mollon and Polden (1977). This psychophysical phenomenon can be explained by interaction between the short and long wavelength sensitive cone mechanisms. Zrenner and Gouras (1978) found that a strong blue test light also stimulated the longer wavelength sensitive inhibiting cones which together would code yellow signals. This is a demonstration of inhibitory action between cones of unlike kind. Although not affecting blue cone sensitivity directly, the decrease in sensitivity of the short wavelength mechanism that results when a yellow adaptation light is withdrawn suggests that the blue cone's sensitivity is influenced by signals from longer wavelength sensitive cones. Schuurmans and Zrenner (1981) indicated how the neurotransmitter GABA (γ-aminobutyric acid) controls blue cone sensitivity through the long wavelength cones.

Zrenner and Gouras (1981) found that 6% out of the 52% of colour-opponent cells in their sample of 385 retinal ganglion cells coded blue signals; of these 5.7% were on-centre and only 0.3% off-centre to blue. The concentrically organised receptive field pattern typically found among R/G opponent cells was not present among B opponent cells, which displayed co-extensive fields indicating a poorer spatial antagonism for blue-sensitive ganglion cells than for R/G cells (see figure 1.16). The classical B/Y opponency is more suitably described in the opinion of these authors by a B/L nomenclature, where L represents antagonism from red (long) and/or green (long) cone mechanisms. Thus yellow is in effect signalled through the R/G opponent channels, probably combining later in the visual pathway to form the B/Y opponent channel. In the opinion of these researchers the signal for yellow is not antagonised by the blue-sensitive cones in the retina and probably not in the LGN.

The absence of a clear demarcation between off and on regions in the ganglion cells controlling blue vision suggests that the blue mechanism is not very well suited to detect brightness borders, while colour borders would be enhanced by such a system. Zrenner and Gouras (1981) considered that such an arrangement would allow the visual system to use small differences in colour for the detection of objects.

Colour-opponent cells which allow coding of various colours are believed by de Monasterio and co-workers to represent about 60–75% of retinal ganglion cells (see de Monasterio 1978 and de Monasterio and Gouras 1975). Similar numbers of cells in the dorsal LGN have been confirmed by Dreher *et al* (1976), Marrocco (1976) and Wiesel and Hubel (1966). Nevertheless no significant differences in the activity of colour-sensitive cells at the levels of retina, dorsal LGN, striate cortex and extrastriate cortex V4 could be found by de Monasterio and Schein (1982). A scarcity of B/Y opponent coding cells reported by other investigators was confirmed.

1.3 NEURAL TRANSMITTERS IN THE RETINA

Although the retina is part of the central nervous system which offers several advantages for neurochemical, histological and neuropharmacological studies, no substance has been established beyond doubt as a definite transmitter at the cellular level in the retina (Osborne 1982). Dopamine seems the most likely candidate, but a great many other chemicals have been proposed as possible neurotransmitter substances linking retinal neurones; of these the most promising are glutamate, aspartate, glycine, GABA and taurine. Preliminary evidence suggests that glutamate and/or aspartate may be the transmitters released from photoreceptors and bipolar cells. The inhibitory amino acid, GABA, is believed to be the transmitter released from horizontal and amacrine cells. Glycine and taurine may also have a similar function in conjunction with amacrine cells, and noradrenaline may also be released at this site. There is dispute over the question of whether acetylcholine is involved in the retina (Bonting 1976).

Knowledge of the transfer mechanism of visual signals in the retinal circuitry is at present largely confined to studies of the fish retina. As with other synapses the key clearly involves neuropharmacological agents released from neurone endings. Lam *et al* (in Westfall 1982) have indicated how uncertain is the nature of neurotransmitters in the vertebrate retina, although acetylcholine is a strong possibility in the case of cephalopod photoreceptors. Minute quantities of radiolabelled chemicals can be introduced to isolated retinal tissue to study the reactions of synaptic chains in the technique known as autoradiography. Most neurones complete the action of the neurotransmitters they release by an uptake mechanism of the chemical from the synaptic cleft or space back into the neurone. This mechanism provides the means of introducing the radiolabelled neurotransmitter. Results can be compared with carefully stained tissue examined to discover the detailed morphology.

This histochemical approach indicates separate chemical mediators for rod and cone linkages with subsequent neurones. Studies on the goldfish suggest that rods use glutamic acid or a similar chemical and at least some horizonatal cells utilise GABA. Marc *et al* (1978) and Marc (1980, 1982) should be consulted for reviews. GABA is thought to be the neurotransmitter mediating the feedback on red and green coding cones from L-type horizontal cells, as shown by Marc (1982). Amacrine cells with deep dendritic membranes also depend on GABA, while others with diffuse arbors tend to use glycine. In the fish retina two types of interplexiform cells are identified, one using the neurotransmitter dopamine and the other glycine.

Clear differences in the neural transmitters involved in colour coding neurones have been indicated by Marc *et al* (1978). For example, while long and medium coding cones are believed to use a common transmitter in the goldfish, the short wavelength coding cones use a different, as yet unidentified neurotransmitter. The antagonists of GABA, picrotoxin and bicuculline, tend to suppress the colour responses, probably by blocking the feedback mechanism from horizontal cells. Indirect evidence for the involvement of GABA in the blue cone mechanism of cat and monkey retinae is given by Schuurmans and Zrenner (1981), who found that the control mechanism of the blue cone's sensitivity, mediated by the longer wavelength sensitive cones, is blocked by the drug bicuculline. Dopamine is another synaptic transmitter substance which acts in an inhibitory manner in the retina, presumably in lateral interactions and feedback loops (see Bonting 1976). Figure 1.20 provides a schematic summary of neurotransmitter activity in the retina as far as is known to date.

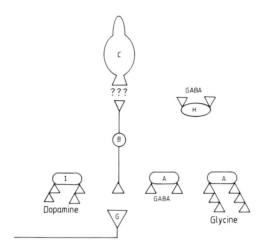

Figure 1.20 Retinal transmitter chemicals (based on studies of fish retinae). C = cone; B = bipolar cell; H = horizontal cell; A = amacrine cell; G = ganglion cell; I = interplexiform cell.

1.4 OPTIC NERVE AND TRACT

The retinal signals travel to the brain via axons of the ganglion cells forming one million optic nerve fibres organised in a systematic manner. Macular fibres from the fovea, signalling colour, occupy approximately one third of the optic nerve area on the temporal side, where they enter. They later move to a central location in the nerve, while towards the chiasma they shift medially.

At the optic chiasma 60 per cent of fibres decussate in man. Fibres from the nasal part of each eye cross in the chiasma, travelling centrally in the optic tract to the opposite side of the brain; those from the temporal part of the retina remain on the same side. Within the optic tract projected fibres from the retinal areas are separated. The fibres, carrying impulses coding colour, terminate at the lateral geniculate body or nucleus. Electrophysiological recordings from the tract indicate that the colour patterns remain unchanged from the retinal ganglion cells; thus the two classes of non-opponent cells coding brightness and four classes coding red–green and blue–yellow stimuli are indicated at this level in primates.

1.5*   LATERAL GENICULATE NUCLEUS

This 'relay-station' serves to receive the axons of the retinal ganglion cells and connects them by synapses with the higher centres of the brain. The stimulus characteristics of intensity, duration, colour and shape have been already encoded in the sequence of nerve impulses which leave the ganglion cell axons. Little if any recoding or functional reorganisation is thought to occur *en-route*. Although the coding pattern established in the retina is maintained throughout the visual pathway higher degrees of specificity are needed to evoke a response at the LGN. It is thought likely that the LGN performs a sensory spatial remapping operation to assist the preservation of the stability of objects in space. De Valois *et al* (1966, 1967) and Wiesel and Hubel (1966) first showed the characteristic colour response pattern of LGN neurones. Furthermore the two functional classes of ganglion cell response (sustained and transient, X or Y) are maintained with anatomical segregation at the LGN level to the striate cortex (Dreher *et al* 1976). Thus transient-type LGN cells receive fast-conducting afferents from the transient ganglions, projecting them to the visual cortex, similarly for sustained cells. The X-like (sustained or tonic) are mostly located in the paravocellular layer and the Y-like (transient or phasic) in the magnocellular layers of the LGN, so separation is complete.

1.5.1   LGN anatomical summary

Over two million neurones of the right and left parts of the LGN receive fibres from the retinal ganglion cells where the axons of the latter first synapse. Axons of the geniculate cells then form the geniculo-calcarine pathway or optic radiations, which later synapse in the visual cortex of the brain. Some 1350 cells project onto each square millimetre of the visual cortex.

The LGN has six layers of cells alternating grey matter (laminae of cells) and white matter (nerve fibres). The four dorsal columns are similar in type and the two ventral layers have larger cells and are called the magnocellular layers. Each layer receives tract fibres from one eye with layers 1, 4, 6 on the ventral dorsal side receiving inputs from the contralateral eye and layers 2, 3, 5 from the ipsilateral eye (figure 1.21). Each LGN layer contains its own retinotopic map of the contralateral visual field. The *macular fibres* are located in the upper posterior two thirds of the LGN. Some cells receive an input from both rods and cones while others deal with signals from cones alone. No evidence of cells which receive signals from rods alone has yet been established. The exact pattern of synapses is not yet clear; a single nerve fibre can apparently synapse with several LGN cells but it is not certain whether more than one fibre can converge onto a single cell. Nerve fibres entering divide into several branches, each which may terminate on a different neurone and despite this spread it appears that a definite mapping is maintained.

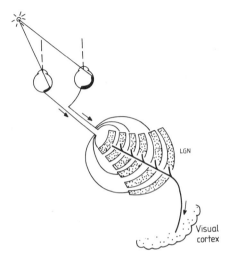

Figure 1.21   Connections in the LGN.

1.5.2   Physiology of the LGN

The *receptive fields* of cells in the LGN are of similar size to those at the retinal ganglion level, but on account of the multiple fibre convergence from the retina to each LGN neurone, individual LGN receptive fields can be thought of as representing a 'massed' retinal input. No clear evidence exists as to which retinal receptive field areas provide the input to the LGN receptive fields. Higher degress of spatial and colour-coding antagonism are shown than at the retina and the interactions can be highly variable and complex. For example, a colour-coded cell may have a receptive field centre which produces excitation to red at the onset and inhibition at the offset of the stimulus (a red 'on' cell). In the same spatial location the cell may show inhibition to blue at the onset of the stimulus and excitation when the stimulus is withdrawn (a blue 'off' cell). A second example is a red 'on' centre, blue 'off' centre, in one spatial zone, and the opposite response, red 'off' surround, blue 'on' surround in the surrounding region.

Spectrally non-opponent cells which respond maximally to mid-spectral wavelengths close to the peak of the photopic spectral sensitivity curve, are believed to code brightness information. Of 147 cells studied in *Macaca irus* by de Valois *et al* (1966) 44 (30%) were found to be spectrally non-opponent; 103 cells showed colour opponency, with a firing rate increased in one part of the spectrum and decreased in another and of these 54 were R/G coding cells and 49 Y/B coding. It was suggested that hue might be coded by the ratio of activity of the four classes of opponent cells and saturation represented by the ratio non-opponent to opponent activity. Wiesel and Hubel (1966) located fewer Y/B cells than R/G varieties. There is some dispute as to which cone mechanisms feed into the Y/B system. Wiesel and Hubel maintained an input from the blue and green mechanisms referring to the cells as coding B/G, on the basis of the input rather than the ouput.

Three groups of cells have been identified in the four dorsal layers, division being according to their receptive field organisation. Type I cells show colour opponency in an opponent centre–surround organisation, i.e. colour and spatial selectivity comprising 77% of the total, with one cone type being represented in the centre and another in the surround. This simple organisation is in contrast to the more complex types found at the retina by de Monasterio and Gouras and de Monasterio *et al* (1975a,b). The extent of the excitation centre was of the order of only two minutes of arc corresponding to only 10 μm on the retina. The proportions given are $+R/-G$ 35%, $-R/+G$ 18%, $+G/-R$ 16%, $-G/+R$ 6%, $+B/-G$ 1% (these latter are in effect $+B/-Y$). Some of these cells had an input from cones only, others from rods and cones. Type II cells have no centre–surround organisation, i.e. no spatial selectivity, but show colour opponency all over the receptive field. For near-foveal regions the receptive field centre was $\frac{1}{2}^\circ$. About half of these cells were of the type $+B/-G$ (i.e. $+B/-Y$) and represented 7% of the total with no rod input. Type III non-colour-opponent cells, have the same spectral sensitivity for both centre and surround; with several cone types contributing they represented 16% of the total. The non-opponent cells probably receive their input from only red and green cones since the blue mechanism is known to make a small overall contribution to brightness; their receptive fields vary in size from 1 to $\frac{1}{8}^\circ$ or less.

In the ventral layers two groups of cells were found. No information is available as to whether they receive cone or mixed rod and cone input but it is generally believed that both colour-opponent and luminance cells have both rod and cone inputs. One group behaves like Type III cells with centre diameters $\frac{1}{8}-\frac{1}{2}^\circ$, the others are described as Type IV cells and showed greatest response to red stimuli. It is thought that these at least have a pure cone input.

Trichromatic cells having an input from all three cone classes have been identified. These take the form of centre–surround with $+R/-(B+G)$, i.e. red excitatory centres and inhibitory surrounds composed of two cone types, which give a cyan response. Others $+G/-(B+R)$ give an inhibitory magenta response. Cells responding specifically to simultaneous and successive contrast were also identified.

Thus, in monkey LGN and, by inference, in man, most cells code both colour and spatial information but others deal specifically with either. This division of processing is carried on in the striate cortex, though the proportions appear to be markedly changed. Psychophysical evidence of channels conveying both colour and spatial frequency information, which supports the physiological findings, is numerous.

1.6*   STRIATE CORTEX

Nerve impulses signalling colour information are relayed from the LGN, via the visual radiations, to the main visual areas of the brain, the striate cortex (see figure 1.1). Motokawa *et al* (1962) noted how similar were the responses from spectrally opponent cells at the level of the optic radiations and from the striate cortex. Thus the optic radiations relay signals from the LGN to the cortex with little or no alteration to the coded impulses. The principal zones were designated Area 17 by Brodmann (1905) in his classical division, and then onwards to Area 18 (the occipital or parastriate cortex) and Area 19 (the praeoccipital cortex). Some non-colour fibres, concerned with eye movements, project to the superior colliculus and pretectal regions, as shown by Bernheimer (1899). The visual areas maintain strong foveal representation, having specific locations within the region for each part of the visual field. Retinotopic organisation or spatial mapping is thus preserved and fibres conveying signals from corresponding points of the two retinae remain close (Talbot and Marshall 1941). Foveal regions are spatially magnified in the LGN and striate cortex to ensure maximum resolution in this region. An estimated 1400 million neurones lie in this region, with impulses from the optic nerve fibres exciting over 5000 cell bodies. Approximately 100 cortical cells serve each foveal cone in the primate.

The synaptic connections, spread throughout seven horizontal layers, are exceedingly complex, and details of the pathways have yet to be established, as has any relationship between morphology and function. A valuable review of cell types and connections was given by Szentagothai (1973).

1.6.1   Physiology of the striate cortex

By means of electrophysiological recordings from simple cells the diversity of function of cortical cells has been revealed, each specific to particular characteristics of the stimulus, including colour. It is clear that significant reorganisation of the coded signals occurs here. Cells sharing the same function for analysis are grouped together in parallel coloumns which run from the pia to white matter of the brain. The neurones in different layers or lamellae have specific physiological functions, and these in turn project to different sites according to the requirements of the areas concerned.

The visual cortex appears to analyse the visual input by means of feature detectors which presumably operate in parallel channels. A significant relationship between colour and spatial processing is found at this level of the visual pathway and trichromatic and opponent characteristics are seen together. Receptive field organisation is more complex than elsewhere, suggesting that important recoding is probably occurring. The general functional division of cells into spectrally coding neurones (C type) and non-colour-opponent cells coding brightness (L type) is maintained.

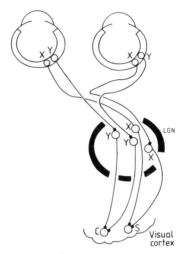

Figure 1.22 LGN relays showing X and Y channel connections. The former serve 'simple' cells in the cortex (S) while the Y cells send messages to complex cells (C).

Colour opponency, with trichromatic features, was first established at the striate cortex by Lennox-Buchtal (1962) and Motokawa *et al* (1962). In the former study more than 50% of neurones were shown to respond to a coloured input but later investigators found different proportions. Hubel and Wiesel (1968) located 6 out of 25 cells with a simple receptive field organisation to code colour, the majority responding positively to red and negatively to blue/green ($+R/-B/G$ cells). A further 12 out of 177 complex cells examined showed colour specificity. Dow and Gouras (1973) indicated the importance of carefully selected spatial features for the stimulus to evoke a response from colour-coding neurones in addition to colour. They noted how cortical colour-coding cells tend to be grouped in greatest concentration in the foveal projection layers and established 54% of all neurones examined to be colour coding; Yates (1974) confirmed these proportions. In general, brightness-coding neurones serve the peripheral retina predominantly (approximately 60% of the total compared with 31%

for the foveal cortex). Although responding positively to all wavelengths these demonstrate a maximum response around 550 nm, equivalent to the peak of the photopic sensitivity curve. Motokawa *et al* (1962) identified two groups of spectrally responsive cells. One group (C type) responded to all wavelengths (and also to white light) if the intensity level was sufficient, but had definite dominant peak responses in the red (600–640 nm), green (520–540 nm) and blue (460 nm) regions. The second (more numerous) group gave opponent on–off responses, responding with an increased firing to certain wavelengths and a decreased firing or non-firing to others, for both the introduction and the removal of a stimulus. These cells also fired to a white light stimulus, with on, off or on–off reactions depending on the size of the stimulus and the adaptation level involved. Michael (1978) found four classes of colour-coded cells in monkey striate cortex all with concentric receptive fields being of the R/G variety. All were most sensitive to the simultaneous presentation of two different colours and were thought to be involved in the perception of simultaneous colour-contrast; some cells appeared to receive contacts from red/green opponent geniculate fibres.

1.6.2   Receptive field patterns/trichromatic inputs

Hubel and Wiesel (1968) examined receptive field patterns in detail and noted how many cortical cells have an elliptical receptive field arrangement with orientation preference.

   One central consideration has been the way in which the cortex processes a predominantly colour-opponent input to achieve a trichromatic perception in vision, and analysis of receptive field organisation has assisted in such an understanding. A trichromatic input to LGN cells was put forward as a viable possibility by Hubel and Wiesel (1966). Using a chromatic adaptation technique, Gouras (1970) found inputs from all three cone mechanisms in half of the striate colour-coding cells, indicating a trichromatic interaction on a single neurone. A transient response was apparent among magenta/ green and cyan/red cells. Of all colour-coded cells 28% gave an excitatory input from only one cone mechanism (rarely the blue channel) showing only partial colour specificity, and 26% showed true colour opponency, responding to only very narrow wavebands; the latter had a trichromatic input. The characteristics of these cells led Gouras to suggest that striate cortical cells tend to be more wavelength discriminating than cells at lower stages of the primate visual system.

1.6.3   A model of colour coding

Gouras and Zrenner (1981a,b) have extended their physiological model of the retina to the striate cortex following observations that some cells in the striate cortex respond mainly to brightness, while others respond best to colour contrast, even though all share the same cones (Gouras and Kruger 1979). It is suggested that the R/G tonic system contributes to cortical cells coding both brightness and colour contrast, following physiological evidence that the R/G cone opponent system affects most brightness-sensitive cells in the foveal visual cortex (Kruger and Gouras 1980). The way in which this might be done was suggested by Gouras and Zrenner (1981a,b) as follows—R on-centre cells and G on-centre cells could lie on each side of a contour, both contributing to the excitation of the same cortical cell. When the ratio of R cone excitation to G cone excitation (or vice versa) was maximal, the cells would respond. A similar model is proposed for brightness and contrast-sensitive cells with the involvement of R on-centre and G on-centre tonic cells with weak and strong cone opponency on one side of a contour and tonic R off-centre and G off-centre cells on the other side of the contour, both exciting the same cell in order to respond maximally to brightness across a contour. Finally, since whiteness and blackness are also contrast phenomena which might be controlled by appropriate cortical detectors, it is suggested that whiteness may depend on the simultaneous excitation of retinal R, G and B on-centre cells on one side of a contour and off-centre cells on the other side (see also §1.12).

1.6.4   Relation between form and colour

Spatial selectivity is also an important feature of cortical cells, indicated by the complex opponent receptive organisation found at this level, e.g. double-opponent variety (complex and hypercomplex cells). Hubel and Wiesel (1968) found only 7% of complex and hypercomplex cells to show colour specificity although 25% of simple cells coded colour. The probability of a cell being colour selective was considered to decrease as its spatial selectivity increased, suggesting that the processing of form and colour might be carried out separately at the cortex. Gouras (1970) supported this hypothesis and showed that the converse also holds, i.e. as a cell's colour specificity increases its spatial selectivity decreases. From a total of 122 single cells, 42% were spatially specific, 24% colour specific and a small group (10% of total) had both properties; these were described as spatial–colour cells and were also highly orientation-specific. The spatially-specific cells had an input from both red and green channels, but rarely (if ever) from the blue. Spatial inhibition involved interaction within each cone channel rather than between different cone channels in these cells. The lack of blue involvement, and the narrow receptive fields ($1/16°$ across), indicate a likely involvement with high acuity processing. Colour-opponent cells (some showing trichromatic input) had more variable-sized receptive fields of the range $\frac{1}{4}$–$4°$ ($4°$ is the limit of monkey fovea). The largest group revealed fields showing a blue input, but with no spatial selectivity, i.e. no evidence of one colour channel in the centre and another in the surround. Boynton (1960) suggested that brightness discrimination is controlled by 50% of cells in the foveal projection of the striate cortex and indicates separate processing of hue and brightness at this level. The separate contributions of brightness and colour to the perception of illusions and depth, emphasise the parallel processing of colour and form.

1.7   PRESTRIATE CORTEX

Fibres from Area 17, the primary visual striate cortex, project onwards to the prestriate regions, Areas 18 and 19, and further areas classified by Zeki (1971). Degeneration studies have indicated connections between Areas 17 to Areas 18 and 19. Separate areas in the prestriate cortex each have their own independent set of afferent connections, and in each the visual fields are separately represented. The suggestion has thus been made that different areas of the prestriate cortex receive different types of visual data with a distinct division of labour. Separate areas for depth analysis and movement have been established. Zeki (1973) found only 8% of cells responding to colour in Area 18 but his later work (Zeki 1975 and 1977) indicated a region classified as V4 which, in the foveal projection areas, is rich in colour responsive cells. A restricted zone appears to be concerned with colour processing since the shape, orientation and direction of movement of the stimulus were not critical in eliciting a response. Cells in the prestriate cortex respond to narrow wavelength bands (Zeki 1973); those showing preference for a particular wavelength being grouped together into columns, producing minimal response to other wavelengths or to white light of varying intensities. Some cells responded to bands up to 50 nm wide with a sharp cut-off, others required greater wavelength specificity, others less. In general cells showed colour-opponent characteristics and variable-sized receptive fields. Those responding positively (excitatory) to a green centre and negatively (inhibitory) to a red surround were most common. A small group described as successive contrast cells were found, responding positively when the colour changed to its opponent pair.

In the superior temporal sulcus of the prestriate region two distinct regions have been identified by Zeki (1973). Cells in the medial area of the superior temporal sulcus showed direction selectivity only, while 87% of those in the lateral region were colour specific only, without orientation selectivity; all require binocular input to respond. Some showed narrow waveband responsiveness only, others responded to less specific wavelengths. Responses to red, blue and magenta (the latter responses arising when both red and blue stimuli were presented together) were most commonly encountered. Baizer *et al* (1977) found 16% of the total cells in Area 18 to be colour coded, with orientation and directional sensitive cells (responding maximally to a stimulus moving in a specified direction across the receptive field) more numerous. Monocular stimulation through either eye appeared to be as effective as binocular presentation for all but the orientation cells.

It appears, therefore, that the segregation of function by different cell groups persists from Area 17 to the higher visual centres. Colour is elegantly coded from an early stage in the visual pathway and individual cells at each level show ability to differentiate narrow wavebands. How these signals are decoded for perception remains a virtually unexplored field.

Evidence conflicting with that presented by Zeki (1977, 1978a,b,c, 1980) has been published by Schein et al (1982) who questioned the proposal that Area V4 cells in the primate extrastriate visual cortex do not specialise in colour analysis; their experiments revealed colour-selective cells in only 20% of the total and a few cells showed obvious colour-opponent responses.

1.7.1* Summary

The separation of luminance information for brightness perception and of wavelength information for colour perception begins early in the retina after the photoreceptor stage, following a (normally) trichromatic processing by three classes of cone pigments and activation of the rod pigment, rhodopsin. This separation is maintained throughout the visual pathway to the striate cortex.

Colour is processed in an opponent manner, involving a comparison signal from the cones controlling mid-spectral wavelengths (R/G opponent channel) and also involving comparison between cones controlling short wavelengths and the mid-spectral wavelengths (B/Y opponent channel). Luminance information is processed by means of signals from the medium and long wavelength ($R + G$) cones with no contribution from the short wavelength (B) cones. Such a mode of analysis and coding might well be expected from a bioelectrical circuit, since an isolated receptor such as is found in the retina can signal only the number of photons absorbed and provides no information concerning the wavelength of light.

Colour discrimination depends upon the presence of at least two sensors or mechanisms for coding. In the normal retina three classes of cone pigment mediate the initial response, acting as sensors. Limited colour vision experience is still possible when one of these cone classes is absent or faulty as in defective colour vision. For electrical transmission and transfer, difference or summation signals are preferred. Thus a comparison of the relative outputs of R/G and B/Y is a suitable means for coding colour in neural channels.

Spectrophotometric analysis of primate cone pigments and histochemical analysis of the primate retinal mosaic both indicate a sparsity of blue-coding neurones in the fovea, results that are consistent with the low proportion of blue-coding ganglion cells in the primate. Physiological and histochemical studies are in close agreement with human colour vision characteristics measured by psychophysical techniques. The sensitivity of the human blue mechanism, as measured by the increment-threshold technique of Stiles (1978) has a minimum at the foveal centre and a maximum at around 1 to 2 degrees from the fovea. Marc (1977) indicated how close this sensitivity is to the form predicted by the distribution of blue-sensitive cones, although he recognised that the magnitudes involved are poorly correlated.

Detailed morphological studies of the visual pathway will undoubtedly yield further understanding of mechanisms controlling neural colour-coding. This should elucidate neuronal connections in the retina at each level, linking anatomical observations to physiological data. When that has been accomplished we shall be in a position to consider the sites and operations involved in the various types of defective colour vision.

1.8* COLOUR MECHANISMS AND THE ASSOCIATED SENSATIONS

Physiological aspects of colour vision and the anatomical structures concerned must be related to the perceptual phenomena which some regard as the most important features of colour. The interface between physical stimuli, the processes and the percepts is complex and each element tends to demand separate study and explanation. Having assessed the normal structures and their workings it is necessary to move towards a more detailed consideration of the experience of colour when different stimuli are presented to the eye. Simpler and historic approaches can lead to the understanding of later and considerably modified views; the trichromatic approach is capable of substantial development.

Brindley (1970) and Sherman (1981) have indicated how the Russian geographer and glass manufacturer Michael Lomonosov speculated about the existence of the three retinal mechanisms in 1757. Palmer (1777, 1786) had similar ideas, which he extended to an explanation of defective colour vision. Thomas Young was a physicist, whose early experience as a physician undoubtedly influenced his colour vision studies. Young (1802) suggested that the retina used three sets of elements, processing the colours red, green and violet; earlier he had proposed red, yellow and blue. Young applied this approach to Dalton's defective colour vision (Young 1807), which he postulated to be due to a possible absence of the red-sensitive channel.

Once experimental proof of trichromacy had been established by Maxwell (1860, 1890d), Helmholtz (1866) adopted Youngs's hypothesis mod-

oncept by suggesting overlapping sensitivity curves for the three
s. Nevertheless Maxwell (1855a, 1860) was also aware that his
ixing experiments indicated trichromacy of vision. Maxwell
l three primary sensations which corresponded to the activity of
ent classes of receptor and attempted to derive the spectral sensitivity of the three mechanisms using the data of colour anomalous observers, being aware himself that the colour anomalous could be explained in terms of a missing sensation. In fact a derivation of the spectral sensitivities of the mechanisms is not necessarily a simple function of the physical stimuli required in colour mixing experiments. In his extensive account of Maxwell's contribution to colour science Sherman (1981) considers that Maxwell made a greater contribution to the revival of Young's trichromacy theory than did Helmholtz.

1.9 VISUAL PIGMENTS

1.9.1* Introduction

The trichromatic theory of vision rests upon the interaction of three groups of mechanisms which for many years have been thought to have as their basis three types of photopigments, maximally sensitive to different regions of the visible spectrum. Although the isolation of the postulated photopigments has not been achieved to date, objective evidence from three different approaches (microspectrophotometry, psychophysical measurements and fundus reflectometry) all suggest that three groups of photopigments are involved in normal colour vision. The peak sensitivities are around 435, 530 and 560 nm. Marks *et al* (1964) first demonstrated the three classes of visual pigment in the cones of man and monkey. Since the three cone photopigments in primates cluster around these spectral regions there has been a tradition, following early investigators, to refer to them as the blue, green or red pigments, although the wavelengths of maximum sensitivity in fact correspond to the violet, yellow–green and yellow spectral regions. A more exact designation, which is adopted here, is to refer to the three groups as the short, medium and long wavelength pigments. Jameson and Hurvich (1968) used the symbols α, β and γ, respectively.

The presence of at least two types of cone pigments, and a means for comparison of the consequences of their stimulation, is required for colour perception.

1.9.2 Rhodopsin and other pigments

Analysis of the content of the visual receptors was prompted following the histological studies of Schultz (1866) who noted morphological differences between the rod-like and cone-like retinal receptors and also their distinctive roles in dim and bright light. The key to the physiological distinction was seen to lie in the chemical characteristics of the pigments. Boll (1876) observed a fading effect of the dissected retina of a frog in daylight; the transition from an initial pink colour to transparency was at a rate which depended upon the intensity of the bleaching source. Since rhodopsin appears pink when viewed by transmitted light it absorbs predominantly mid-spectral wavelengths. The term 'bleaching' is used to describe the chemical changes which result from the capture of photons by rhodopsin. The cones appear not to contain any coloured pigment on observation. Kühne (1878) made futher investigations of the bleaching and subsequent regeneration of rhodopsin, which he named visual purple, and which provided König (1894) with the basis for an objective analysis of spectral sensitivity, recognising that wavelengths of equal energy for rod vision were equally absorbed by rhodopsin. Visual pigments are essentially selective absorbers of light with additive optical densities which vary across the visible spectrum with characteristic maxima, λ_{max} (see Dartnall 1957 and Wald 1965). A density spectrum was expressed as this variation of optical density, D_λ, as λ is altered where:

$$D_\lambda = \log \frac{\text{Light incident on specimen}}{\text{Light transmitted by specimen}} \; .$$

Modern usage favours (instead of optical density) 'absorbance' (A_λ) and the term 'absorptance' (or absorption factor) for the fraction

$$J = \frac{\text{Incident intensity} - \text{Transmitted intensity}}{\text{Incident intensity}}$$

which is related to transmittance (T), where

$$T = \frac{\text{Transmitted intensity}}{\text{Incident intensity}}$$

by the expressions $J = 1 - T$ and $A = \log 1/T$. Knowles and Dartnall (1977) gave a full explanation of these terms and their use. The absorbance (density) spectrum of a visual pigment has been shown by Dartnall to have the same shape at different concentrations. If the curves for different concentrations are plotted as *percentages* of maximum (density or absorbance) they are the same in all respects.

A standard method, used to determine absorbance spectra of pigments in their unbleached state, in solution, is to measure the absorbance under two conditions; first while unbleached and then after exposure to light. The difference, plotted as a 'difference spectrum' reveals the characteristics of the unbleached pigment—see §1.10 for details. The use of a frequency scale, in preference to a wavelength scale, for plotting absorbance spectra of visual pigments enabled Dartnall (1953) to demonstrate a similarity between the curves of different pigments, despite the different positions of the peaks (λ_{max}) of the curves. Crescitelli and Dartnall (1953) compared human vision with rhodopsin, noting a close agreement between the scotopic sensitivity for the human eye as measured psychophysically by Crawford (1949), and the absorption spectrum of rhodopsin, when due allowances were made for selective absorption in the ocular media. The 1951 CIE scotopic sensitivity curve shows a maximum at 507 nm; this compares well with a peak absorption (λ_{max}) of rhodopsin measured by Crescitelli and Dartnall (1953). Furthermore, when measured by electrophysiological methods the peak is at 493 nm, which is close to the λ_{max} of rhodopsin.

A central characteristic of visual pigments was established early in the study of photochemistry, and is known as the Principle of Univariance. This describes how the response of the visual receptors depends not on the wavelength of energy but the number of quanta caught by the pigment. The greater absorption by rhodopsin in the mid-spectral region merely indicates that the medium wavelengths produce a greater response of the photo-pigments than an equal quantity of light of short or long wavelength.

Nomograms, named after H J A Dartnall, are sometimes used in estimating the wavelength-related absorption features of visual pigments. Ebrey and Honig (1977) applied the method to interpreting data for normal and abnormal human colour vision published by Smith and Pokorny (1972).

The visual pigments of vertebrates have a common structure based on a protein (an opsin) carrying a chromophore of retinaldehyde, vitamin A aldehyde (previously called retinine). The opsin combined with the retinal explains the term rhodopsin. The carotenoid protein of rhodopsin is a complex molecule. Light is absorbed by rhodopsin or a cone pigment and the chromophore is isomerised. For detailed discussion of the photochemistry readers are referred to Davson (1976) or Dartnall (1972). The visual pigment is located in the disc-like membranous sacs located in the outer segments of the photoreceptors (see figure 1.7). The visual pigment selectively absorbs light of different wavelengths. A graph of the absorption at different wavelengths is described as the *absorption* spectrum. *Absorptance* indicates the relative amounts of light absorbed at the different wavelengths for a specific quantity of pigment. Measurement of the absorption spectrum by a spectrophotometer involves a comparison of the amount of light transmitted by the pigment and by a reference solution.

The consequence of light absorption by the retinal receptors is to initiate a neural signal by a transduction process (see Davson (1976) for details) which involves a hyperpolarisation of the photoreceptor membrane. Once bleached the pigment is temporally insensitive to the further reception of light. Each photon absorbed by the photopigment has the same consequence, although subsequent visual perfomance depends on the degree to which the pigments have been bleached.

Cone pigments and spectral sensitivity

Clearly a relationship exists between the characteristics of the cone photo-pigments of an individual eye and the corresponding spectral sensitivity features. To take a simple example, protanopes have been shown to lack the long wavelength sensitive pigment, erythrolabe, and demonstrate this deficit by an insensitivity to red. The relationship between the colour mechanisms of Stiles and cone pigment sensitivities is discussed in §1.11.

The distribution of rhodopsin, and the predominantly medium and long wavelength sensitive cone photopigments, appears to be relatively constant throughout the retina, at least in so far as their contributions to spectral sensitivity are concerned. Wooten and Wald (1973) and Stabell and Stabell (1981) both showed insignificant variations in the spectral absorption characteristics of the pigments between the fovea and peripheral retinal regions. The distribution of the short wavelength mechanism, however, changes considerably with retinal position (Stabell and Stabell 1981).

1.9.3   Objective measurements of visual pigments

Fundus reflectometry

The application of the principle of the ophthalmoscope to the study of visual pigments was begun in the early 1950s by independent British investigators (Brindley and Willmer 1952, Rushton 1952 and Weale 1953, 1955, 1959). This provided the first direct objective measurement of photopigments in their normal physiological state (*in situ*). Briefly, the technique involves the comparison of the beam reflected from the retina before and following bleaching at the fundus by a strong incident beam, thereby allowing an evaluation of the difference spectrum of the pigment, a measure of the difference in optical density of the pigment. Detailed discussion of the experimental technique is given below.

Such measurements allow a comparison of the photochemical changes (difference spectra) with psychophysical measurements (action spectra) on the same observers. Absorption properties and the kinetics of bleaching and regeneration can be studied by this technique. The approach was developed further to permit the measurement of human rhodopsin by Campbell and Rushton (1955) (see also Rushton 1956), followed by extensive measure-ments on cone pigments pioneered by Brindley and Rushton (1955).

Using his own technique Weale found λ_{max} (peak absorption) for human rhodopsin to be 507 nm. Dowling (1960) establised a direct relationship between the logarithm of the rod threshold and the amount of rhodopsin present in the rods.

Experimental details

Reflected light emerging from the back of the retina will have traversed the retinal receptors twice (firstly the incident pathway and secondly as the reflected beam). A proportion of the incident beam is absorbed by the photopigment, the value of which can be evaluated by allowing the light to fall on a photocell. If note is taken of the photocell value for the emergent beam before and after bleaching, the difference will represent absorption by the visual pigments; a measure of the density of the pigment is thus possible. Less pigment is available to absorb the emerging beam as it travels through and from the retina once bleaching has occurred, as indicated by the second photocell reading. Lights, each of a single variable wavelength, are used in turn to provide a measurement of the density spectrum of the pigment, since the reflected intensity is evaluated for a suitable spectral range. It is usual for a neutral density wedge to be inserted in the emerging beam to maintain a constant deflection of the photocell. Any decrease in pigment density by bleaching is then exactly compensated by the density increase required in the neural filter to maintain constancy. A quantitative measure is available by this method. The principal problems encountered were summarised by Boynton *et al* (1977), for instance the difficulty arising because of the fact that the light emerging from the eye is of low intensity.

1.10   MEASUREMENTS OF CONE PIGMENTS

1.10.1   Microspectrophotometry

Microspectrophotometry is a valuable technique which allows measurements of the absorption properties of cone pigments from individual photoreceptors. Developed by Caspersson (1940) it was initially used on the carp retina by Hanaoka and Fujimoto (1957) in Japan. More recently such measurements have been made on receptors in the monkey and in man, which are smaller.

A series of monochromatic wavelengths throughout the visible spectrum is imaged on the isolated outer segment of a cone receptor, the location of the visual pigment; a second beam is directed onto a clear region, for reference. The light intensities of the two beams are compared by photomultiplier, certain corrections are made, and the pigment transmission is determined.

1.10.2   Monkey cone pigments

Early axial measurements from primate foveal cones were made in the USA by Brown and Wald (1963, 1964) and Marks *et al* (1964) and by Liebman

Table 1.1 Summary of human and monkey photoreceptor visual pigments based on microspectrophotometric measurements. Adapted from Harosi (1982). Values given are wavelengths of maximum absorption λ_{max} (nm).

Reference	Rod rhodopsin	Cones			Source
		Short wavelength absorbing, 'B'	Medium wavelength absorbing, 'G'	Long wavelength absorbing, 'R'	
Brown and Wald (1963)	—	—	535	565	Human and Rhesus
			527	565	*Macaca mulatta*
Marks et al (1964)	—	445	535	570	Human and Rhesus
Brown and Wald (1964)	505	450	525	555	Human
MacNichol (1964)	—	447	540	577	Human and Rhesus
MacNichol et al (1982)	500 ± 0.8	430 ± 3.5	532	567	Macaque
Liebman (1972)	498	440	535	575 – 580	Human and Rhesus
Bowmaker et al (1978)	502 ± 2.5, 499 ± 1.0	—	536 ± 3.5	565 ± 2.5	Rhesus
Bowmaker et al (1980)	500	415	535	567	Cynomologus *Macaca fascicularis*
Bowmaker et al (1980)	497 ± 3.3	420.3 ± 4.7	533.8 ± 3.7	562 ± 4.7	Human
Dartnall et al (1982)	493.3 ± 2.3	419 ± 3.6	530.8 ± 3.5	558.4 ± 5.2	Human
Bowmaker et al (1982)	—	approx. 430	Considerable variation approx.	536 564 550	*Saimiri sciureus* Squirrel monkey
Harosi (1982)	498 ± 2	430 ± 5	532 ± 2	563 ± 2	Rhesus and Cynomol

(not published until 1972). More recent transverse scans have been made by Bowmaker *et al* (1978, 1980, 1982), Dartnall *et al* (1982), Harosi (1982) and MacNichol (1982). Table 1.1 presents a summary of the findings. These indicate the presence of three main classes of cone photoreceptor in most species, with sensitivity predominantly to short, medium or long wavelengths (violet, green and yellow–green spectral regions). There are marked similarities between human and many monkey cone pigments, paricularly those absorbing predominantly in the medium and long wavelength spectral regions (green and red cones).

1.10.3   Sparsity of 'blue' cones

A sparsity of short wavelength cones is noted in all primate studies to date. In the sample of 82 rhesus cones studied by Bowmaker *et al* (1978), 40 were long wavelength absorbing and 42 were medium wavelength absorbers. No short wavelength cones were found, although two (one from each of two animals) were examined in the Cynomologus monkey in the study by Bowmaker *et al* (1980) and three in the human retina (3/33) by Bowmaker *et al* (1980). Harosi (1982) located one short wavelength cone among 31 cones in total.

The studies of Bowmaker and co-workers show that the proportions of medium and long wavelength cones are similar. However Harosi (1982) found almost three times the number of medium wavelength absorbers than long wavelength cones. Bowmaker *et al* (1980) further noted how the short wavelength cones and rods kept their shape during measurements whereas the other two classes of cone did not. The sparsity of short wavelength absorbing receptors is confirmed by histochemical and psychophysical studies described elsewhere in this text.

In more than one study by Bowmaker and co-workers the wavelength at which maximum absorption occurred (λ_{max}) varied considerably within a range (for example in the 1980 study a range of 11 nm for medium wavelength cones and 15 nm for long wavelength cones). These differences were considered to be greater than would be expected as a result of experimental error. Furthermore λ_{max} for mid-spectral cone pigments were clustered rather than normally distributed. In the rhesus monkey clear clustering occurred at 534 and 541 nm for medium wavelength absorbers and at 563 and 567 nm for long wavelength absorbers. In man the clusters were at 528, 534 and 539 nm for medium wavelength cones and 555, 569 and 573 nm for long wavelength cones. No explanation has, as yet, been given for this finding but it is clear that considerable variability in the pigment characteristics is evident, even within one class (see Bowmaker and Mollon 1980).

Nunn and Baylor (1982a,b) report transduction studies in rods and foveal cones of *Macaca fascicularis,* a monkey thought to have photoreceptors similar to humans. Spectral sensitivity measurements indicate a rod peak at 491 ± 3 nm, and a function which is in good agreement with the standard scotopic curve. Peak sensitivities of 550–575 nm for long wavelength coding cones are similar to those recorded by Bowmaker and colleagues (see this section).

1.10.4*   Human cone pigments

Microspectrophotometric measurements from 7 human eyes involving 49 medium wavelength cones, 69 long wavelength cones and 11 short wavelength cones were made by Dartnall *et al* (1982). In two patients, supporting psychophysical measurements on a non-enucleated eye† were possible. The microspectrophotometric data are presented in table 1.1. It was considered likely that real variations in λ_{max} between observers, and even within a single retina, existed, particularly for the long wavelength cones. The possibility was raised of two classes of long wavelength cones, those with a λ_{max} at approximately 555 nm and those with a λ_{max} at approximately 563 nm. Additional support was provided by psychophysical measurements obtained for two of the patients, both colour normal as assessed by the anomaloscope and Farnsworth–Munsell (FM) 100 Hue test. One observer showed a λ_{max} at 553 nm for the long wavelength pigment, for the other, the λ_{max} was 561 nm. Close agreement was found between the log of the ratio of spectral sensitivity between 555 nm and 650 nm to the microspectrophotometric measurements in each observer.

Dartnall *et al* (1982) have also reported microspectrophotometric measurements for a human dichromatic eye, the first such measurements and most important ones. The healthy eye of the patient they examined was clinically tested the day before enucleation of his other eye. The FM 100 Hue test, the City University test and Ishihara and Okuma plates were consistent in classifying him as deutan. He had a neutral point at 503 nm. Microspectrophotometric records were obtained for 17 receptors; of these, 2 were short wavelength cones, 5 were rods and 10 were long wavelength cones. No medium wavelength cones were found; thus these results are consistent with the loss theory of dichromacy, as indicated by the fundus reflectometry studies of Rushton (1952) (see p. 45).

Good agreement is noted between the characteristics of the cones and the corresponding psychophysically determined π mechanisms of Stiles. Comparison between the rod pigment (rhodopsin) characteristics and the CIE relative luminous efficiency function for scotopic vision, V'_λ (see §1.9.2), similarly shows expected similarities, although the CIE curve is slightly narrower than the pigment measurements indicate.

†A non-enucleated eye is the eye left *in* when the other has been removed.

1.11   COLOUR MECHANISMS—COMPARISON OF EVIDENCE

The trichromatic theory has as its basis three rather loosely defined 'mechanisms' responding most prominently to long, medium and short wavelength regions of the visible spectrum; these are best known as the red, green and blue mechanisms. Physical, biochemical/physiological and psychophysical evidence combine to support this concept, each providing insight into the early processing of colours by the visual system. Such a three-pronged approach to colour vision has many advantages, not least the value of a combined research input from the three fields of physics, physiology and psychology, contributing data on colorimetry, neurophysiology and perception. The early development of the trichromatic theory had the advantage of such an interdisciplinary approach, for the majority of the founding fathers show evidence of being both physicist and physiologist in one.

Early colorimetry led to the determination of the colour matching functions in both normal and defective colour vision, the amount of red, green and blue required to match any given spectral wavelength. Cone pigment analysis has provided a glimpse into the likely means of mediation of trichromatic and dichromatic vision (see §§1.9 and 1.10). Psychophysical measurements of 'increment thresholds' using coloured backgrounds and test fields have established a vital link in the understanding of the characteristics of the colour mechanisms in both normal and defective colour vision (inherited and acquired). This combined approach has enabled informed speculation to be made about the nature of the spectral sensitivities of the three initial channels subserving normal colour vision, and the faults underlying defective colour vision. Caution is required in interpretation however. Despite good agreement between these three different approaches, the responses are not all generated as a result of the total activity of the visual system.

It is tempting to match colour matching functions with cone pigments by suggesting that the latter control the former. In the early discussion of his classical measurements of the colour matching functions Wright was careful to avoid such a claim or direct association (see Wright 1946). Stiles (1959), however, found that colour mixture characteristics could be nearly predicted by the colour mechanisms he described as the π functions (§1.11.3) and when Estevez and Cavonius (1977) made measurements of the colour matching and π functions on the same observers very close agreement was shown.

The psychophysical studies based on increment thresholds pioneered by Stiles (1946b, 1949a,b, 1953, 1959) (see Stiles 1978 for review), led to the establishment of three groups of functions which bear close resemblance to the absorption characteristics of the retinal cone pigments as studied by microspectrophotometry (see §1.10). Stiles, like Wright, was cautious in his interpretation, although he discussed the similarity of π_3 and π_4 to the spectra of known pigments (see Stiles 1959). Nevertheless the knowledge that has accumulated since these early measurements provides freedom and justification for some speculative observations. Rushton (1972) noted good agreement between the spectral sensitivity of three of Stiles' five mechanisms (π_3, π_4, π_5) and the absorption spectra of the cone pigments (see figure 1.23). Enoch (1972) summarised this association aptly with the statement that the psychophysically determined two-colour increment threshold method 'provides curves which certainly are substantially influenced by cone pigments', recognising that 'a larger number of intervening physical, chemical and neural interactive elements probably link conclusive integrations'. Wald (1964) was more pointed in his observation that the three functions he isolated psychophysically corresponded in essence to the cone pigment spectra. Puch and Siegel (1978) and Cavonius and Estevez (1978) have produced evidence that led them to suggest that Stiles' π_4 and π_5 mechanisms represent the spectral sensitivities of the medium and long wavelength receptors. The latter collaborators consider π_3 to be representative of the short wavelength fundamental mechanism with due allowances for macular pigmentation. The main reservation of Cavonius and Estevez was with π_5, which was discussed by Alpern and Pugh

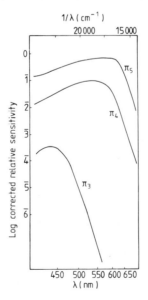

Figure 1.23 Sensitivity across the spectrum for three π mechanisms of Stiles, when modified to account for absorption by the cornea, lens and macular pigment. π_5 corresponds to the R cone pigment response, π_4 to that of the G cone pigment and π_3 to that of the B pigment. After Stiles (1953).

(1977) and Bowmaker *et al* (1978).

Among those who have undertaken recent microspectrophotometric analysis of cone pigments in primates and man Mollon has made additional extensive studies of colour mechanisms using the psychophysical technique of Stiles. This dual involvement has facilitated careful comparison between sets of data obtained by these differing approaches. Thus Bowmaker *et al* (1978) showed close agreement between the psychophysically isolated red mechanism (π_5) and the cone absorption spectra for the long wavelength (red) cones, when certain assumptions are made; for π_4 and π_5 see Bowmaker *et al* (1980). Boynton (1979) conceded that it is possible that π_5, π_4 and π_1 have spectral characteristics equivalent to those of the red, green and blue cones respectively.

1.11.1 Psychophysical measurement of colour mechanisms

The separation of rod and cone function is readily indicated by the two phases of the dark adaptation curve shown psychophysically (see figure 1.24). Following a period of preadaptation to a source of high luminance which serves to bleach the retinal pigments, the absolute threshold is measured repeatedly at intervals of one or two minutes in complete darkness for a total period of 20 to 30 minutes. Cone function dominates the initial stage before rod function takes over and determines threshold for the remaining period in darkness. Rod activity is furthermore revealed by the scotopic spectral sensitivity function.

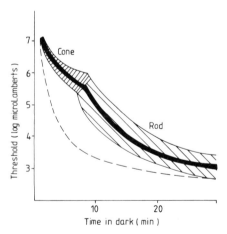

Figure 1.24 Dark adaptation. A broad band is shown within which the normal curve (after white light adaptation) should lie. This curve comprises cone and rod components which merge. The broken curve indicates the curve obtained following preadaptation to red light.

The photopic spectral sensitivity function (see §1.15) is considered in the colour normal observer to be a summation of the three colour mechanisms. Measurements of the foveal spectral sensitivity function by Stiles and Crawford (1933) showed significant variation according to the background field surrounding on which the test stimulus was superimposed. Evidence of three contributions was indicated by three peaks to the curve with maxima around 440, 540 and 590 nm; these were naturally thought to relate to the postulated three cone pigment varieties.

1.11.2 Increment threshold technique–experimental details and interpretation

In an attempt to measure the characteristics of the components of the foveal spectral sensitivity function Stiles made use of the increment threshold technique which involved the detection of a small coloured test field (foveal stimulus) in the presence of a larger coloured background. Increment thresholds determined for a series of wavelength pairs (test stimulus and adapting background stimulus) provided information of the character- istics of the separate colour mechanisms. The technique rests upon the effective removal of the influence of two cone mechanism groups from participation in the increment threshold measurement, using a selective adaptation source of suitably chosen colour and luminance, which serves to reduce their sensitivity. The mechanism which is least sensitive to the background wavelength is considered responsible for the detection of the test stimulus. The short wavelength (blue) mechanisms were isolated by a background stimulus of 570 nm a wavelength which stimulates both the medium (green) and long (red) wavelength mechanisms. To isolate the medium wavelength (green) mechanism a background of 628 nm was used and for the long wavelength (red) mechanism a background source of 535 nm (see figures 1.25, 1.26 and 1.60).

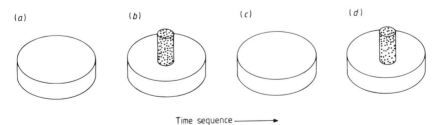

Time sequence ⟶

Figure 1.25 The increment threshold. In a sequence of several flashes of a spot of light some flashes are combined with smaller spots which form an additional stimulus at the centre. The central stimulus is adjusted in intensity until this incremental centre spot is just detected.

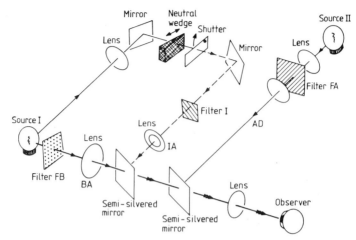

Figure 1.26 An increment threshold apparatus. Source I provides light for the background of the field of view via a filter and a lens which acts as an aperture (BA) to control the field size. Light also reaches the observer through two semi-silvered mirrors, a neutral wedge, a shutter and a filter (FI). This view is restricted by another small aperture (IA) at a lens. Source II provides an additional adapting field, filtered at FA, which can be superimposed at will.

This technique allows the spectral sensitivity of each of the three colour mechanisms in isolation to be determined in turn for a series of test stimuli of varying wavelength. The somewhat artificial situation of employing selective background adaptation enables detection of some wavelengths not normally detected by the isolated mechanism. The increment threshold measured was thus considered to depend upon the response of one colour mechanism alone. Although the use of the two-colour threshold technique of Stiles was unique for the purpose, in that it allowed separation of the contributions from the three mechanisms involved, previous experimenters were aware of the value of selective bleaching. Burch (1898) had shown how adapting the normal eye to coloured lights led to effects akin to a 'transient colour blindness'. Stiles used a 1° test stimulus presented in a flashed mode for between 100 and 300 ms every 3 s on a high luminance 10° adaptation field presented continuously. Both sources were of variable wavelength (monochromatic) provided by a colorimeter built for the purpose. The subject is usually dark-adapted prior to observations and required to maintain central fixation on the background adaptation field. Using a constant criterion the observer must indicate when the test stimulus is just detected, as a superimposed target to the background field. A variety of presentation methods can be used.

Thus the energy increment for detection, on the basic criterion of 50%

success, is measured for each experimental arrangement (choice of test stimulus, wavelength and background wavelength and luminance). The reciprocal value of radiance (a measure of threshold) provides a measure of sensitivity of the mechanism under test for the wavelengths used. The functions obtained by plotting threshold against background radiance (intensity) indicate two or more sections, with varying curves resulting from different combinations of test stimulus and background wavelength. Stiles interpreted the crossover point as the changeover between cone mechanisms in determining the threshold at different energy values for the adapting field.

1.11.3   Five colour functions

The five mechanisms established for foveal vision were named by Stiles the π mechanisms, using the Greek letter to abbreviate the word 'process'. They are shown in figure 1.27.

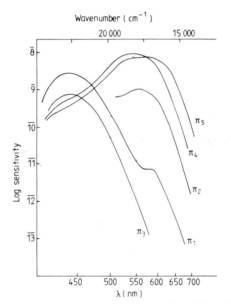

Figure 1.27   Five π mechanisms. After Stiles (1953).

π_1 = 'blue' mechanism—maximum sensitivity 440 nm. Similar spectral sensitivity to π_3 differing only in having a different background intensity.

π_2 = additional 'blue' mechanism (not always present).

π_3 = 'blue' mechanism—similar spectral sensitivity to π_1 differing only in having a different background intensity.

π_4 = 'green' mechanism—maximum sensitivity 540–550 nm.

π_5 = 'red' mechanism—maximum sensitivity 575–580 nm.

π_4' and π_5' mechanisms are modified green and red mechanisms shown at high intensity levels.

The sensitivity curves representing the colour mechanisms and indicated in figure 1.27 hold only when each mechanism acts alone in isolation. Stiles recognised the difficulty, and near impossibility, of isolating each mechanism fully, even with high luminance adaptation techniques to bleach out the contribution of the other two mechanisms. Later investigators highlighted this aspect further (see Pugh 1976 and Mollon and Polden 1977). The assumptions central to the development of the π concept are discussed in detail by Enoch (1972). The functions must therefore be considered as characteristic of the state operating when each mechanism in isolation is responsible for the detection of the test stimulus at threshold. For each set of conditions the mechanism with the highest sensitivity is considered to determine the threshold. Interactions undoubtedly occur as indicated by numerous investigators, among them Boynton *et al* (1964a) and Ikeda *et al* (1970).

1.11.4 Alternative approaches

Comparable functions to those of Stiles (though with some differences) were obtained by Wald (1964, 1966, 1967) using broad-band filter sources for the background chromatic adaptation field in preference to the narrow wavelength sources used by Stiles. This approach has found widespread application with many recent investigators of inherited and acquired colour vision defects, notably Marré (1969, 1973). Some examples of commercially available filters which provide a suitable background adaptation are given in table 1.2.

Table 1.2 A selection of commercially available filters used to isolate the colour mechanisms for increment threshold measurements.

Mechanism isolated	Filter specifications
Blue (short wavelength)	Yellow Corning No 3482 (Glass) at 7 log photopic trolands
Red (long wavelength)	Blue Kodak–Wratten 47B at 4.3 log photopic trolands
	Blue Kodak–Wratten 47 at 6.2 log photopic trolands
Green (medium wavelength)	Purple Kodak–Wratten 35 at 5.4 log photopic trolands

Using this technique Wald (1967) examined both normal and Daltonic observers and indicated the variations in mechanisms at different retinal positions, and the role of macular pigmentation at the fovea in particular in depressing the sensitivity of the blue mechanism. Peak sensitivities were noted at the following wavelengths for the three classes of mechanisms— 440 nm ('blue'), 548 nm ('green') and 580 nm ('red'). Although Wald claimed to isolate the photopigment sensitivities by this approach a more cautious view is taken by other authorities, for example Boynton (1979). Sperling and Harwerth (1971) have applied similar techniques to both human and monkey observers.

1.11.5*   Opponent colour theory

The opponent theory which owes its origin to Hering (1878) considers the spectral colours blue, yellow, green and red to be individual hues; these unique colours have a perceptual independence of their own. The concept arose from a consideration of how colours appeared from a psychological viewpoint, providing a contrasting approach to that developed by physicists of the nineteenth century, who based the trichromatic theory of colour vision on the matching of colours.

The rival theories of trichromatism and colour opponency competed until zone theories of colour vision were introduced early this century, principally by Müller (1924), which incorporated the two viewpoints.

Quantitative measurements of the opponent hue responses were made by Jameson and Hurvich (1955) using psychophysical methods. These measurements were later related to normal and abnormal colour vision under a variety of conditions, such as changes in adaptation and contrast (see Jameson and Hurvich 1956a,b, 1959, 1964). The technique involves the cancellation by superimposition of some spectral hues by others. Pairs of wavelengths which resulted in the perception of opposite hues (such as blue and yellow or red and green) were added and their energies varied until the opposite hues were not distinguishable in the mixture (Hurvich and Jameson 1955). Difference signals are considered to result in the opponent channels from the interaction of two receptor mechanism classes, for example the long and medium wavelength mechanisms. Data initially from fish retinae by Svaetichin (1956), showed clear evidence of colour opponency which was confirmed at various levels of the primate visual system by de Valois and co-workers (see §1.5.2).

The colour-opponent pairs, blue–yellow and red–green, involve the interaction of receptor mechanisms at the trichromatic level and neural opponent paired mechanisms. Each of these opponent systems is assumed to involve the acitivity of of all three cone types in normal colour vision (see Jameson, in Jameson and Hurvich 1972). Alteration of both photopigment and neural opponency is considered to provide a full analysis of defective colour vision (see Hurvich in Jameson and Hurvich 1972). The uniqueness of these hues holds at all luminance levels for red and green (Larimer *et al* 1974, 1975) and they can be added linearly, since a mixture of unique hues with any luminance ratio gives rise to either an achromatic perception or a unique combination. Hurvich (1977) considered such responses to be linearly related to the quantum catch of the three types of cone photopigments (see also Jameson and Hurvich 1968). The opponent model is considered by modern theorists to involve two stages of colour processing—a receptor trichromatic stage determined by cone photo-pigments and a neural colour-opponent coding initiated in the retina, possibly at the horizontal cell level (see §1.12.2).

1.12 COLOUR VISION MODELS

Proposed models of colour vision and its deficiencies can take different forms in attempts to visualise, communicate or modify existing theories. Frequently the term 'colour vison model' is applied to theoretical, cybernetic, often mathematical, descriptions of the physiological and psychophysical processes of colour perception. Occasionally the model involves a practical demonstration, such as that described by Walraven (1966) who portrayed his theoretical concept in an array of coloured lights with appropriate dimming, enhancement and switching.

This section, as a review of main approaches, is deliberately brief since interested readers are encouraged to consult the original sources. Degrees of speculation necessarily accompany the concepts protrayed by model builders, particularly for the vertebrate visual system. With the limited knowledge and understanding available, assumptions must be made; for this reason alone cautious analysis of the models is required.

Although the development of models can be a challenging and often valuable approach to the advancement of understanding of the brain's functions, compatability with current knowledge of neurophysiology and psychophysical observation is necessary for a plausible framework. The pioneering studies of the majority of those who sought to explain the perception of colour and the underlying mechanisms were really attempts at model building, using experimental data; such is the basis of scientific method and advance. Nevertheless it is likely that the most fruitful approaches will follow, rather than precede, further physiological understanding.

The difficulty of developing models of vertebrate colour vision is outlined by van Dijk and Spekreijse (1982) who made comparisons between the trichromatic carp and primate responses. The selective action of drugs on visual processes and the need to consider such modifications is further indicated by these authors. They indicated, for example, how non-linear interactions in carp are affected by ethambutol, with the result that a non-linear opponent system declines in action while underlying mechanisms are not affected. These features were related by these authors to ethambutol dyschromatopsia in humans.

The 'line-element' was an early theoretical approach. Helmholtz (1866) considered linear transformation of the components of three-colour mixtures in relation to the absorption properties of the receptors' photopigments, at the point of just-noticeable-differences.

1.12.1* Channel or zone models

The separation of colour vision processing into a series of zones or channels is thought to have originated from Donders (see Judd 1948 and Thomson 1952). Signals resulting from the reception of light by the photoreceptors are modified at each successive zone, often associated with a physiological level. Müller (1924) advocated such an approach, incorporating a specific luminance mechanism in the second and third 'zones' interpreting the earlier responses of the 'cone mechanisms'. Walraven (1962, 1973) built on this approach. The trichromatic response (R, G, B) at the cone receptors is considered to be converted to an antagonist opponent-coding, involving red–green and blue–yellow channels, early in the retinal processing, on the basis of electrophysiological evidence (see §1.2). A brightness-channel-coding luminance separate from colour was realised by Piéron (1939) as necessary to colour vision models. Considerable evidence now supports the view that the long and medium wavelength cone receptors, but not the short wavelength receptors, contribute to the luminance channel (Eisner and MacLeod 1980).

It is possible to envisage the integration of chromatic and luminance perception although early views inclined to separate the mechanisms (see §1.2). Walraven (1973) reviewed alternative models and included an extension in which he and Vos had considered groups of receptors and the involvement of contrast signals (see figures 1.28, 1.29, 1.30). The 'trita-signal' is concerned with the balance between red and green mechanisms; in normal trichromatism a 'deuta-signal' is present, related to yellow and blue responses. In this scheme protans have a red/green channel which misfunctions since there is red receptor loss.

Using the Walraven channel model and data such as dichromatic weighting Vos and Walraven (1970a) derived the likely response characteristics of three foveal receptor systems. A 'blue' response, antagonism to an $R + G$ combination, was combined with the signal obtained from the balance between R and G. Dichromatism was regarded as a channel-loss phenomenon as well as an unusual distribution of cone receptor densities.

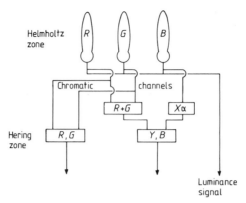

Figure 1.28   Schematic diagram of the retinal stages of the zones involved in colour vision processing. The factor α in the blue mechanism has not been determined. After Walraven (1962).

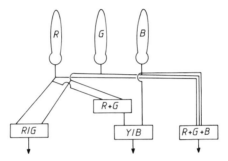

Figure 1.29   Model scheme in which the Young–Helmoltz trichromatic receptor zone transmits to a zone involving antagonistic processes. Here red and green are balanced separately from yellow and blue. A combined brightness channel involves contributions from all three of the trichromatic receptor zones. After Walraven (1973).

A description of neural signals sent to higher centres was added in a modification by Hård and Sivik (1981) of the Walraven approach (see figure 1.31).

Massof and Bird (1978) compared 'zone' and 'line-element' models, including a mathematical treatment for quantitative analysis. Readers are recommended to consult the full list of references in their paper. The postreceptoral 'zone' (associated with Hering-type opponency) deals with a further transformation of the signals (linear or non-linear) in order to produce perception of hue, brightness and purity. Between the physical events in the retinal receptors and the psychological events, physiological processes effect transformations portrayed in equations. A resultant colour can be identified as a vector, the magnitude of which expresses the brightness of the coloured sensation.

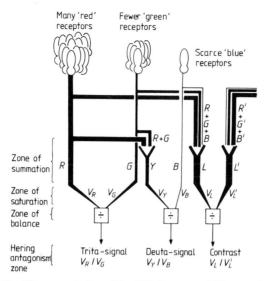

Figure 1.30 The extended model of Walraven and Vos. An indication of the relative numbers of receptors (responding selectively to different parts of the spectrum) is indicated at the top. After Walraven (1973).

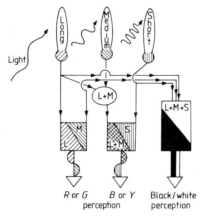

Figure 1.31 The initial signal model. After Hård and Sivik (1981) and Walraven and Vos (1981).

Colour television has been a useful model for several aspects of human colour vision, with special appeal for laymen. Using this approach, Hunt (1967) pointed out the effects upon discrimination of R and G receptors and their distributions. Unique yellow was seen to be the result of equality of R and G responses. Colour difference signals of $R - G$ and $R - B$ types were proposed; alternatively interactions such as $(\frac{3}{4}R + \frac{1}{4}B) - G$, simplified to $(\frac{3}{4}R - G)$ in the central fovea were introduced.

Advantages and limitations of physiological models were discussed by Vos (1982) in a review which covered many factors. He demonstrated alternative explanations for certain colour matches and made a case for approaching colour mechanism models quantitatively.

1.12.2 Opponent models

Various models of the postreceptoral activities involving antagonistic opponent coupling have been proposed.

Hurvich (1978) reviewed models developed over the preceding two decades in conjunction with Jameson, in which a reconciliation of physiology and psychophysics was presented within an 'opponent' framework (see, for example, Hurvich and Jameson (1957) and figure 1.32). Here the absorption spectra of the photopigments were portrayed as transformations of physical colour mixtures. A bell-shaped positive curve (roughly the V_λ response) can be superimposed on figure 1.33, representing the contribution to white sensation provided by all the receptors via the black/white opponent process. In such ways Hurvich (1978) used models to portray contrast effects and abnormal colour vision.

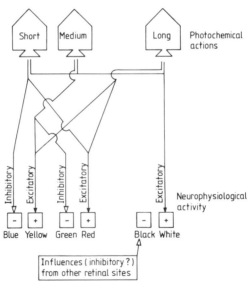

Figure 1.32 Opponent systems related to receptor actions. After Hurvich and Jameson (1957).

An extensive account of the history of 'zone theories' given by Wasserman (1978) placed emphasis on the contributions of Hurvich and Jameson. He also showed possible ideas of visual systems, for example dichromatism, where the opponent system is little affected by alternative varieties of photoreceptor distribution. Naka and Rushton (1966, 1967) envisaged L and M receptors influencing 'red/green units' by alterations of polarisation; multiple switch systems and resistances were used to explain data and relationships between recorded potentials and stimulus values. This treatment included a mathematical account, a hyperbolic function of stimulus intensity appearing as the 'S-potential' and membrane conductance being proportional to the stimulus. An opponent model which extended these ideas was devised by Lee and Virsu (1982) incorporating constants for cone mechanisms which enable the model to treat data obtained with coloured stimuli at the lateral geniculate level.

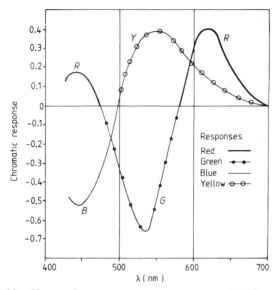

Figure 1.33 Chromatic responses at a postreceptoral level, according to an 'opponent' view. After Hurvich (1981)

1.12.3 Alternative models

Land (1965) proposed a very different model under the name of 'retinex'. This interpreted phenomena presented by mixtures of filtered lights in terms of lightness discrimination mechanisms.

One of the latest models of colour vision has been described by Hunt (1982) (see figure 1.34). Different zones are considered and the receptor sensitivities of Estevez (1979) are employed. In the linear response zone, which included the cones, it was suggested that the $(R + G)$ signal (the separate red and green components having been combined via horizontal cells) is generated as the basic 'achromatic response' responsible for such functions as V_λ, foveal viewing and hence good visual acuity. Opponent-cell combinations operate in the second zone, where non-linear (square-root) functions apply. The relationships $(R + G)^{1/2}$, $(R^{1/2} - C^{1/2})$ and $(G^{1/2} - B^{1/2})$ govern chromatic signals C_1, C_2 and C_3, these being interpreted in a third zone. A non-opponent signal $(R^{1/2} + G^{1/2} + B^{1/2})$ was described as C_0. The chromatic channels thus used were shown to be economical in their nerve fibre requirements.

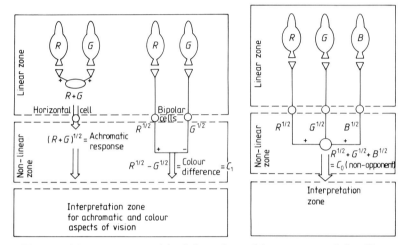

Figure 1.34 A recent model of the colour vision processes (after Hunt 1982). The left-hand diagram shows the possible interactions of R and G receptors. On the left the $(R + G)$ addition is modified to a square root form as this 'achromatic response' progresses through the non-linear zone. On the right bipolars operate a subtraction, in the non-linear zone, such as $C_1 = R^{1/2} - G^{1/2}$, $C_2 = G^{1/2} - B^{1/2}$ and $C_3 = B^{1/2} - R^{1/2}$. The right-hand diagram shows a non-opponent addition of three signals as C_0.

Gouras and Zrenner (1981a, b) made an extensive (neurophysiological) survey of colour perception, including models. Here the role of R/G opponent cells and the inhibition of a blue mechanism was stressed, with preference for a 'stage' model and a consideration of the discrimination of coloured borders.

A classical description of quantitative aspects of the Young–Helmholtz receptor concept was prepared by Hecht (1930). The ways in which colour mixture functions interact with V_λ functions are indicated in figure 1.35 according to Hecht's interpretation. Violet replaced blue and Hecht showed how early concepts of red, green and violet responses to the spectrum were developed by Helmholtz, Maxwell and Dieterici. Judd (1947) built upon work done earlier this century. His concepts are shown in figure 1.36. Excitation curves, adopted by the Optical Society of America in the 1920s, involved equal areas beneath the three curves to indicate 'equal contributions' to the sensation of white. Abney (1895) was responsible for providing the foundation of this concept and models of retinal activity were related to data obtained by Wright (1964). Excitation curves can be reconciled with opponent processes following the work of Judd (1947) and Svaetichin *et al* (1965). The Svaetichin concept (see figure 1.16) was based on electrode recordings from fish retinae using white light or narrow band stimulation. Such diagrams of models provide insight into the correlation between perception and the activity in retinal neurones. Excitation and inhibition in the retina correlated well with the opponent approach; inhibitory activity is often associated with blue or green stimuli.

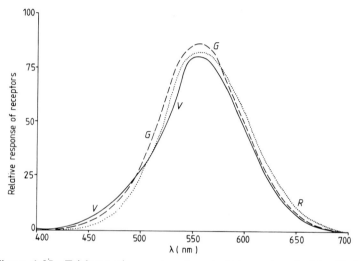

Figure 1.35 Trichromatic receptor characteristics as described by Hecht (1930). The sum of these gives the V_λ curve. Points at which the three curves interlace (around 500 nm) are related to neutral points of protanopes and deuteranopes since, if only two processes are operating, it is at such parts of the spectrum that the processes are in balance.

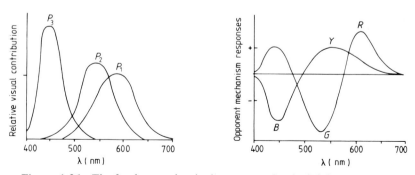

Figure 1.36 The fundamental retinal processes. On the left is a representation of the trichromatic contributions to vision across the spectrum (equalised). The opponent processes are shown on the right. After Judd (1947).

Between 1940 and 1950 there was considerable tension between 'opponency' and 'trichromatic' views, which Hartridge (1950) attempted to reconcile. His model, a polychromatic mixture of seven retinal mechanisms (with eight response curves), drew upon the modulators and other processes described by Granit (1947). Conditions of light level, size and retinal image position influenced the extent to which a 'trichromatic unit' operated, with or without other receptors, as indicated in figures 1.37 and 1.38. Elements were supposed to be lacking, or to combine, in different varieties of dichromatism and anomaly. However this manner of considering colour vision has receded into a virtually closed chapter in the history of the subject, despite the keen efforts of its protagonist.

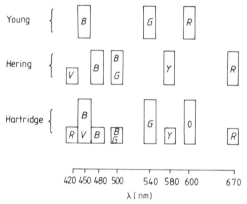

Figure 1.37 A possible reconciliation between the mechanisms proposed by Young and Hering. The polychromatic receptor system of Hartridge (1950) is seen at the bottom.

1.12.4 Colour appearance

Chromatic contrast has been considered by Ingling and Drum (1973) in the case of a boundary on the retina between two coloured images. Models of ganglion cell response were proposed. Data from LGN opponent cells and from human colour contrast experiments were related by these authors to colour vision models.

A theoretical model developed by Takahama and Sobagaki (1981) was specifically related to the phenomenon of chromatic adaptation. The long wavelength response mechanism, for example, has an initial element which is a 'modified von Kries transformation' expressing the effect of adaptation. A further stage is postulated in which the nonlinear transformation modifies the response before reaching the visual cortex. By this approach 'receptor noise' is included, as coloured field and background factors are combined in relationships designed to portray colour appearances.

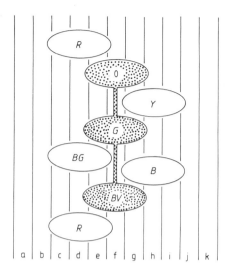

Figure 1.38 Schema based on the polychromatic theory of Hartridge (1950). The divisions a–k represent increasing light levels, showing how different sets of receptors come into operation. At low levels of stimulation (in divison b) colour vision partly begins, with a variety of dichromatism existing at c. Trichromatic vision is found at levels represented by d to h inclusive, based upon a basic 'trichromatic unit' of three receptors. At a and k vision is considered to be achromatic.

Hood and Finkelstein (1982) have reconciled physiological data with a model which 'tunes' a chromatic channel in different ways according to whether small or large areas of the retina are involved; small spots, which appear desaturated, may involve less opponent action and small, short duration pulses of colour could be detected by the achromatic system. The suggestion is that 'green plus/red minus' and 'red plus/green minus' cells could discount their minus features for small areas.

In a model where L and M response receptors only were represented (giving a neutral point at about 570 nm and resembling a tritanopic 'loss' situation) Tansley *et al* (1982) represented subjective estimates of the appearance of borders at the junction of equally bright retinal images. Here (L − M) opponent signals were combined with (L + M) non-opponent processes. Evoked pattern responses from the cortex of an observer could be predicted, as well as estimates of subjective border distinctness.

1.13* THE EFFECTS OF THE OCULAR MEDIA

A significant proportion of light is lost by absorption and scattering as it passes through the refractive elements of the eye; the attenuation is wavelength-selective in both cases. It is particularly important in the older eye on account of the yellowing of the crystalline lens and the possible increase in macular pigmentation with age because, along with natural pupil constriction with age, less light finally reaches the retina. Typical transmission curves, as shown in figure 1.39, show little wavelength selectivity between 550 and 1000 nm, most attenuation being for the blue and violet wavelengths in the visible region of the spectrum (Ludvigh and McCarthy 1938). The various ocular media (cornea, aqueous humour, crystalline lens and vitreous humour) have relatively similar transmission characteristics on account of the biological similarity of their largely water-based structures. Apart from a few absorption bands around 980, 1400 and 1950 nm the structures transmit a very high proportion of the incident light. The typical total transmittance through the entire young eye has been calculated as around 83.5% by Boettner and Wolter (1962). Harmful infrared and ultraviolet rays are not transmitted to the light-sensitive retina. Alpern *et al* (1965) give approximate estimates of spectral transmittance values for three living human eyes. A 'standard observer' ocular density spectrum has been calculated by van Norren and Vos (1974) using the CIE 1951 scotopic standard observer. Comparing the scotopic curve values with the rhodopsin absorption curve gives a total transmission for the ocular media. These authors also established that individual differences of ocular absorption are typically around 25% for the 350–420 nm range. These individual differences presumably account, at least in part, together with aging effects, for the variability in performance among so-called colour normal observers on clinical colour vision tests. Moreland (1972 a,b) for example has shown that a significant proportion of the population distributions of the green/red ratio of normals and anomalous trichromats on the Nagel anomaloscope can be attributed to pre-retinal pigment variability.

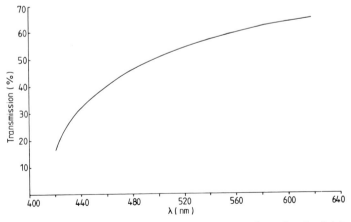

Figure 1.39   Total transmission of the ocular media. After Ludvigh and McCarthy (1938).

1.13.1 Scattered radiation

The pre-retinal media, particularly the lens and cornea, produce approx-imately the same amount of scattering as the fundus. Scattering in the lens and cornea is caused by the presence of inhomogeneities or small opacities. Blue light is scattered more than red light and scattering is in general greater in the older eye in the visible range (Boettner and Wolter 1962, Mellerio 1971).

1.13.2* Absorption by the lens

The crystalline lens absorbs more light than the rest of the ocular media together, and the gradual build-up of a yellow pigment with age results in greater absorption in the blue spectral region. It has long been known that the lens absorbs ultraviolet radiation, and observers who have had their lenses removed surgically for cataract (making them aphakic) are sensitive to the near ultraviolet wavelengths (Gaydon 1938). The spectral characteristics of lens absorption were calculated psychophysically by Wald (1949) using an aphakic observer, and Ruddock (1964) showed that the colour matches of an aphakic observer were consistent with the removal of a yellow filter from the visual system. The typical density spectrum of the lens is shown in figure 1.40.

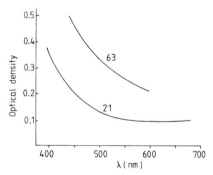

Figure 1.40 Spectral density curves of the human crystalline lens for ages
21 and 63 years. After Said and Weale (1959).

The variation of the ocular density of the lens with age has received considerable attention. Said and Weale (1959) showed that the increase between the ages of 25 and 45 years was almost the same as that between 45 and 63 years.

As to a possible function for the pre-retinal pigments Walls and Judd (1933) suggest that yellow nature of the lens tends to reduce the chromatic aberration of the eye and thus improve visual acuity because blue light, which is greatly dispersed, is most absorbed by a yellow pigment. A possible reduction in glare is also suggested.

1.13.3*   Role of macular pigment

A yellow, carotenoid, non-photosensitive pigment, xanthophyll, is present on the fundus and gives an elliptical zone extending between 5 and $10°$ in the foveal area a yellowish appearance called the yellow spot, *macula lutea*. The extent and optical density tends to vary slightly between individuals and significant variations in concentration or thickness of the pigment across the macula zone are reported by Wright (1946) and Ruddock (1963). It serves to absorb blue and blue–green wavelengths strongly thereby adding to the reduction of these wavelengths (on account also of the crystalline lens) which reach the retina. The suggestion that the pigment was a *post mortem* effect, as believed by Gullstrand (1906), was disproved when extensive psychophysical colour measurements at different retinal positions established its presence in the living eye. The higher pigment density shown among Egyptians by Ishak (1952) could be due to ethnic or environmental factors. A higher density of pigment is reported also for red-haired persons by Bone and Sparrock (1971). The suggestion that the pigment acted as a protective mechanism for the retina against harmful short wavelengths was made by Walls (1942).

The density spectrum of the macular pigment has been established by a number of techniques which give general agreement (Vos 1972) (figure 1.41). Favoured methods are to find the difference between cone thresholds at the fovea and retinal periphery (Wald 1959) or differences in colour matching and measurement of the chromaticities of white stimuli at different retinal positions (Ruddock 1963).

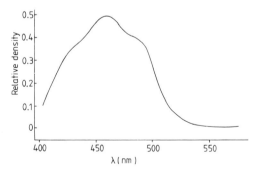

Figure 1.41 Human macular pigment densities at different parts of the visible spectrum. Approximate curve after Vos (1972). Reproduced by permission of the Institute of Perception, Soesterberg, Holland.

Utilising pairs of wavelengths having identical optical density values Moreland (1972a,b) chose such a pair of blue and green wavelengths whose ratio was independent of macular pigment and developed an interference filter anomaloscope for tritan defects which is unaffected by macular pigment variability. Having established the correction required to allow for changes in the blue/green ratio due to age variations in the crystalline lens, characteristics of the blue short wavelength pigment can be established. When making allowances for ocular media effects Weale (1963) considers the effects of macular pigment and pre-retinal components, principally the lens, should be separated. Nevertheless the overall effect is to produce a reduction in blue sensitivity.

It is possible that a greater density of macular pigment is found in the aging eye (Wright 1946, Stiles and Burch 1959). Opinions are divided on this issue; Kelly (1958), Ruddock (1965a,b) and Verriest (1974a) believe there to be no aging effect.

1.13.4 Age changes in colour discrimination

A gradual development of normal colour perception from birth towards youth and a gradual decline from around age 30 to 40 has been widely reported in the literature (Cook 1931, Tiffin and Kuhn 1942, Warburton 1954, Gilbert 1957, Lakowski 1958, 1962, Verriest et al 1962). Generally a red–green deterioration is believed to begin around age 55–60 years but a more marked loss in blue–green discrimination accompanies it from an earlier age, even as young as the third decade. Such changes can be detected by clinical colour vision tests, even pseudoisochromatic plates (Lakowski 1958).

The distinct tritan-like tendencies which become evident with advancing age in the normal eye have been fully studied by Verriest et al (1962, 1982), using the 100 Hue test. The 1982 study involved 109 males and 123 females between 10 and 80 years of age with at least 6/6 (20/20) acuity; they had no disease and normal colour vision was assessed by a perfect reading of the Ishihara plates. This confirmed the 1962 data among 480 normals ranging in age from 10 to 64 years. Young adults (20–29 years) make fewer errors, with error scores increasing in both younger and older age groups.

Although natural pathological changes at the retina and early visual pathway, typical of increasing age, together with a possible build-up of macular pigment, could contribute towards colour vision deterioration with age (Lakowski 1965a), the yellowing of the lens is considered to be the dominant factor (Ruddock 1965a,b). Aging effects also occur in both protan and deutan defectives (Lakowski 1974). Both deuteranopes and protanopes show an improvement in colour perception (as assessed by sensitive clinical test scores) with increasing age and a deterioration in later life. Barca and Vaccari (1978) found the average total error score on the

100 Hue test to increase by 4.47 per year although Verriest had earlier shown it to be 1.40 per year.

These consequences of aging affect the performance on a variety of colour vision tests (Jameson and Hurvich 1972) and the establishment of age norms for both normals and Daltonics at various clinical tests is desirable.

1.13.5   Transmission changes in the ocular media

No age changes are noticed in the transmission of visible wavelengths through the vitreous and aqueous humours (Boettner and Woltner 1962). The transmission of infrared radiation through the eye as a whole is also independent of age, although ultraviolet radiation transmission tends to decrease with age.

On the basis of a study among persons of all ages Pinckers (1980a) found that pseudoisochromatic plate tests are not sufficiently sensitive to show age variations in normal colour vision. He confirmed the finding of Boles-Carenini (1954) of a shift towards green in the Rayleigh match particularly in the age range 70–79 years.

The lens absorption increases as the yellow lens pigment accumulates with age, thus absorbing blue wavelengths predominantly and reducing the eye sensitivity to medium and short wavelengths overall. These age changes occur from 20 to 30 years (see Weale 1963). Significant changes in the lens pigment accumulation in the second and fourth decade were followed by less marked changes in the fourth and sixth decade (Said and Weale 1959).

1.14*   COLOUR MIXING BY NORMALS

Normal colour vision is said to be trichromatic; this is usually taken to mean that any spectral colour can be matched by an appropriate mixture of three reference stimuli or primaries (see Chapter 2 for discussion of colorimetry). As Graham (1965) points out it is more correct to indicate that the spectral light may be mixed with one of the 'primaries' to give a two-colour mixture that can match the mixture of the other two primaries. In this context we are concerned with the additive mixture of visible light of different spectral characteristics. Normal trichromats are said to require equal amounts of appropriately chosen red, green and blue stimuli to match an equal energy white. Colour matching curves indicate the spectral distribution for each of the three matching stimuli which when mixed together combine to give the perception of a colour equivalent to that of a reference wavelength. They are obtained by matching each wavelength in turn through the spectrum by a suitable mixture. Newton (1672) was aware of some order in colours and

BHS EAST HAM - THANK YOU
643 1 3760 0610159   020291

292340 26137 SLIPS 1   2.99
                  CASH   10.00
                CHANGE   7.01

                 TOTAL   2.99

PLEASE RETAIN YOUR RECEIPT
    AS PROOF OF PURCHASE

good service. We sometimes make mistakes. *Please let us know.* Thank you very much for your custom.

VAT Reg. No. 440 6445 66
BhS plc Marylebone House,
London. NW1 5QD

---

bhs

Our aim is to offer you quality and value with good service. We sometimes make mistakes. *Please let us know.* Thank you very much for your custom.

VAT Reg. No. 440 6445 66
BhS plc Marylebone House,
London. NW1 5QD

---

bhs

Our aim is to offer you quality and value with good service. We sometimes make mistakes. *Please let us know.* Thank you very much for your custom.

realised that a mixture of two colours in equal proportions could be represented by the mid-point of the individual components. Maxwell (1860, 1890a,b,c,d) conducted the first colour mixing experiments between 1852 and 1860, using his colour box to demonstrate the spectral colour mixture curves. Using this technique Maxwell measured the colour matching characteristics of several colour-normal observers, showing their individual variability; his red-blind observer (protanope) made all matches with only two matching stimuli. His results are now considered to be of qualitative interest only on account of the poor accuracy and experimental control. König and Dieterici (1893) used a Helmholtz colorimeter to establish reliable colour mixture curves for two colour normals and several dichromats (protanopes and deuteranopes). Other similar pioneer measurements were made by Abney (1913).

Guild (1931) provided new measurements and his results and those of Wright (1928, 1929) were used to define the colour matching functions of the CIE 1931 standard observer. Although the instruments used by Guild and Wright were quite different, the same viewing conditions of a 2° field of view were used confining the observations to foveal vision. Thus these corresponded to the viewing conditions used in the investigations on which the 1924 CIE V_λ curve was based; this V_λ curve was incorporated into the 1931 colour matching functions. Subsequently, it was realised that in some industrial applications of colorimetry, a set of colour matching functions was needed for much larger fields of view than 2°. A new large-field trichromator, built by Stiles at the National Physical Laboratory produced results on the basis of which a supplementary set of colour matching functions for 10° fields was defined by the CIE in 1964 (Stiles and Burch 1955).

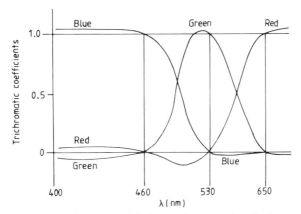

Figure 1.42 Wright's mixtures of three reference stimuli of wavelengths 460, 530 and 650 nm to match different parts of the spectrum. Shown as trichromatic coefficients for a normal observer. After Wright (1946).

Stiles' measurements of the colour matching functions for ten observers for 2° fields and his actual colour-matching data were in very close agreement with those of Guild and Wright. However, unlike the 1931 colour matching functions which were adapted to include the 1924 V_λ curve, Stiles was able to measure the complete functions for each individual observer, his results representing a homogeneous set of data. For this reason Estevez (1982) has suggested that the 2° data of Stiles and Burch (1955) are more suitable for visual research purposes than the 1931 observer.

Colour matches are additive remaining valid over a wide range of illuminations; nevertheless the matches do break down under high levels of adaptation (Wright 1946). Aspects of variability in colour matching, notably caused by ocular media absorption, have been the subject of comment by MacLeod and Webster (1982), Smith *et al* (1982) and Vienot (1982).

1.15   SPECTRAL LUMINOUS EFFICIENCY

The human visual system is not equally sensitive to all parts of the visible spectrum. Even one which is corrected to have an equal distribution of energy throughout is not seen with equal brightness throughout. The spectral senstivity or relative luminous efficiency function describes the relative sensitivity of the eye to different wavelengths of light. Although this describes essentially a luminosity response, the inseparable nature of colour and brightness must be borne in mind (Jameson and Hurvich 1953). In the past this was called a variation of 'visibility', hence the symbol V_λ was adopted; 'spectral luminosity' (another old term) has disappeared from use.

Values of V_λ depend on many of the variables involved in colorimetry, including characteristics of the stimulus, such as size, the state of adaptation of the eye, the retinal region involved, and the psychophysical technique. The sensitivity function of the eye shows a peak (λ_{max}) at 555 nm under photopic conditions (daylight, mediated by cone photoreceptors). There is a marked displacement (causing the 'Purkinje shift') towards 505 nm under scotopic conditions (dim light, mediated by rod photoreceptors). The visual system is thus more sensitive to shorter wavelengths at night, when sensitivity to long wavelengths in noticeably reduced. The measurement, one of sensitivity, involves essentially a comparison of absolute threshold responses throughout the visible spectrum for scotopic conditions.

1.15.1*   The classical V_λ curve

Comparative measurements of spectral brightness have been made since the early nineteenth century and the function is fundamental to both photometry and colorimetry. Historical reviews are presented by Gibson (1940), Sloan (1954) and Le Grand (1968). Classical measurements on 37 colour normals were made by Gibson and Tyndall (1923). Measurements of relative luminous efficiency under photopic conditions based on more than 200 observers were adopted by the CIE in 1924, to define a 'standard observer'. This hypothetical average set of data was understood not to be representative of every set of conditions but provided an approximation to 'normal' measurements. In 1931 the CIE V_λ curve was defined for a 2° field. A supplementary 'standard or average theoretical observer' describing photopic luminous efficiency for a 10° field can be derived. This differs from the 2° field in the short wavelength region, largely on account of the importance of macular pigmentation effects for the 2° fields, which are of less importance for 10° fields. The effect of rod receptors also has to be taken into account for large fields.

The V_λ function represents the total activity of the three cone mechanisms and separation of their individual contributions can be demonstrated by humps in the curve. Under some threshold circumstances the most sensitive cone mechanism (or the luminance mechanism) is represented.

1.15.2   Scotopic luminous efficiency

Under conditions of dark adaptation the senstivity of the eye to brightness at different wavelengths can be explored using large stimulus fields to allow measurement of rod function. Peripheral vision is typically used to facilitate the measurement of rod activity. Direct comparison techniques which are usually employed are easier than for cone vision, since no colour differences are present to complicate judgments. The clasical curves of Wald (1945) and Crawford (1949) show a peak around 507 nm. These data were incorporated into the scotopic standard young observer adopted by the CIE in 1951 although earlier notable measurements were made by König (1894) among others. The displacement from the photopic peak at 555 nm gives rise to the Purkinje shift which was described by the Czech physician Purkinje in 1825. When corrected for absorption characteristics of the ocular media (principally the lens since macular pigment is not involved in peripheral measurements) and when quantum considerations are used, the V_λ curve is similar to the action spectrum of rhodopsin, the rod visual pigment.

Sensitivity of the peripheral retina, studied using several approaches by Abramov and Gordon (1977), is greater for shorter wavelengths, an increase not explained completely by pigmentation at the macula. Spectral sensitivity curves for 'luminosity equality' of lights viewed peripherally were obtained by these authors. Practical vision, including many aspects of colour, is frequently dominated by the extensive periphery of the retina. This was emphasised by Weale (1960) in a summary of some of his studies of the V_λ and related performance of extrafoveal regions. Thus, in the important area of the visual field just within the restriction imposed by a fairly small spectacle frame, there is greatest relative sensitivity to blue lights.

1.15.3 Variations

The text by Graham (1965) provides a detailed account of normal variations in relative luminous efficiency. The major features of interest are shown in figures 1.43 and 1.44. Clearly the shape of the V_λ curve varies with a great many factors of observation, among them the size and luminance of the stimulus, the state of adaptation of the eye and the characteristics of the surrounding field. Different methods provide another source of variance and systematic differences are found between results obtained from flicker and step-by-step comparisons. However such discrepancies are small compared with inter-subject differences. Although careful calibration of the apparatus used is important, the *relative* performance of many observers can be compared provided the main source is maintained at a constant luminance. Seasonal variations are reported although ethnic variations are not. Le Grand (1968) should be consulted for general discussion. Variations are also described by Lee (1966); these can be considerable and in some cases data from colour normals can resemble those from cases of mild colour anomaly. Age variations, particularly for short wavelengths (where media effects are significant) have been discussed by Weale (1963) and Ruddock (1965a,b). Wright's (1946) data are valuable as reference material and allow the wide variations in the V_λ curve under different conditions to be seen for both colour-normal and anomalous observers. If the retinal area stimulated includes both cones and rods the luminosity curve may be taken to represent cone function at high intensities and rod function at low intensities.

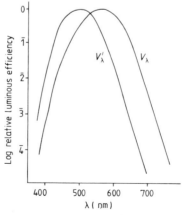

Figure 1.43 The relative luminous efficiency curves, photopic and scotopic.

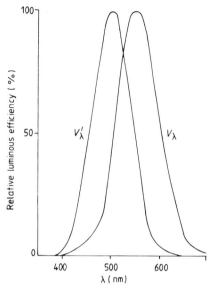

Figure 1.44 Relative luminous efficiency curves;

1.16* WAVELENGTH DISCRIMINATION

The detection of small differences in hue, purity of colour (saturation), and intensity, present in a coloured stimulus is of great significance industrially and is of considerable interest from a scientific viewpoint. The eye can detect an enormous range of colours in the spectrum, far more than the seven hues which Newton described.

A mathematical representation of the eye's sensitivity to colour was first discussed theoretically by Helmholtz (1866); he derived equations based on the 'line element', or just-perceptible-difference, using the trichromatic theory of colour vision. Later, Schrödinger (1920), Stiles (1946a) and Vos and Walraven (1970b) refined mathematical representations of colour discrimination.

Colour discrimination is more typically discussed using concepts of colorimetry adopted by international standardisation bodies such as the CIE (see Chapter 2). Colour-difference equations represent differences which are measurable by colorimetric instruments for industrial use, which may or may not relate to the responses of a human observer.

The sensitivity of the normal observer to differences in hue is discussed in this section; typical responses of those with abnormal colour vision are described in Chapter 3. Practical details relating to the use of such measures to investigate colour vision defects are found in Chapter 7. A measure of the just-noticeable-difference in wavelength which can be detected indicates the sensitivity of the human visual system to hue (a differential threshold).

This is usually achieved by presenting two juxtaposed fields of the same monochromatic wavelength, in Maxwellian view, subtending a 2° total field with the two fields separated by a sharp dividing line (bipartite field). The wavelength difference $\Delta\lambda$, which must exist between the two fields for a difference or change in the perceived colours to be detected, is determined. Equality of brightness is important throughout to ensure that hue and/or saturation changes only are measured; this is especially important for short wavelengths since these thresholds are very susceptible to such errors.

A series of wavelengths in the spectrum is taken for the test wavelengths, λ, (usually at 10 or 20 nm intervals) and the corresponding $\Delta\lambda$ is measured for each test wavelength to allow the hue discrimination curve for $\Delta\lambda$ against λ to be plotted (figure 1.45). The measuremnt of a psychophysical function such as hue discrimination involves the response of the entire system, and is thus influenced by psychological factors; the method of measurement must therefore be considered carefully. A subtle distinction is sometimes made between 'hue' and 'wavelength' discriminations, which is ignored here.

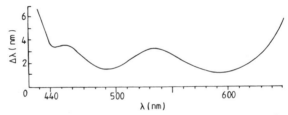

Figure 1.45   Wavelength discrimination across the spectrum for a normal observer. After Wright (1946) and Pitt (1944).

It is often simplest to increase or decrease the wavelength slowly and evenly in an ascending or descending manner until the observer reports a just-noticeable-difference. The mean is taken of several readings. Thus

$$\Delta\lambda = \tfrac{1}{2}[(\lambda + \Delta\lambda) + (\lambda - \Delta\lambda)].$$

The ability of the eye to detect differences (or rate of change) in hue varies considerably throughout the spectrum. At the two extremes, the deep reds and violets, hue discrimination is poor and large differences in wavelength are necessary to elicit a perceived change in hue. Near the centre of the spectrum, where the colours change rapidly (for example in the blue–green and yellow region), wavelength discrimination of 1 nm can be measured in many normal observers. Wright (1946) related these minima to the trichromatic response of the eye suggesting that at these spectral locations differences in the three receptor-mechanism responses are most marked. The nature of the hue discrimination curve varies significantly with numerous stimulus parameters (field size, luminance, retinal region stimulated etc).

Early measurements of wavelength discrimination were made by König and Dieterici (1884) and others. Wright (1946) should be consulted for historical background. Judd (1932) reproduced early measurements in a single hue discrimination curve and Wright and Pitt (1934) made important early contributions. In addition to the regions of inferior hue discrimination at the spectral ends, where large $\Delta\lambda$ values are found, other regions of poor discrimination are seen at approximately 460 and 530 nm. Secondary 'dips' have been reported in a number of classical studies, the most prominent of which is that found in the violet regions which Wright (1946) detected in four out of five normal observers. At 490 and 590 nm (and to a lesser extent around 450 nm) minima are noted in the curve, where hue discrimination is best. The hue discrimination threshold is smaller than the minimum wavelength change requried for colours to be named differently. Thus while differences in hue can be detected and the eye's differential sensitivity to hue is acute, distinct colour names cannot be given to each wavelength, for colour names are designated to wide ranges of wavelength.

1.16.1 Chromaticity discrimination

Measurements of colour discrimination taking into account the dual attributes of colour—hue and saturation—were first made by Wright (1941) and MacAdam (1942) using the CIE chromaticity diagram. Measurements of differences in chromaticity were made for both spectral wavelengths and desaturated colours. Lines of varying length on the chromaticity diagram represent a difference in colour between the two ends of the line (see p. 127). Discrimination 'ellipses' represent the sensitivity to chromaticity based on the standard deviation of colour matches in different directions, with respect to a point (see figure 2.15 and data by MacAdam (1942) and Brown (1951)). Equal differences in chromaticity do not correspond to equal perceptual differences. The size of the steps depends on both their location and the direction of change. A fully uniform colour space has yet to be developed but representation in the CIE UCS diagram provides a format which is most regular (see figure 2.16).

1.16.2 Hue discrimination by colour naming

The relationship between specific colour names and narrowly defined regions of the spectrum, which evoked the sensations associated with such names, was investigated early in the history of colour science. Judd (1940) has reviewed the data of 26 investigators. Thompson (1954) and Beare (1963) also made comparative studies, in the latter case preparing frequency distributions for basic colour names. Later a similar approach was used by Boynton et al (1964b) to derive a hue discrimination curve. Such techniques were applied to study hue discrimination in deuteranopes and protanopes by Boynton *et al* (1964b) and, in one case of inherited tritanopia and two cases of inherited tritanomaly, by Smith (1973). Graham *et al* (1976) derived wavelength discrimination curves from treatment of colour naming, closely agreeing with the more conventional data of others such as Wright (1946).

Viewing time can influence hue appearance. For example with long viewing durations yellow–reds and yellow–greens of moderate luminance appear yellower, while blue–reds and blue–greens are shifted towards shorter wavelengths, appearing bluer (Cohen 1975).

1.16.3   Factors influencing hue discrimination

Colour discrimination can be altered markedly by changing the conditions of observation from those considered to be standard, i.e. stationary stimuli with reasonably large fields at a suitable retinal illuminance to maintain photopic vision. A great many variables influence hue perception and discrimination; these include spatial factors such as field size, temporal factors, retinal illuminance, retinal position, degree of fixation, adaptation effects and individual normal variations on account of age. The psychophysical method chosen to investigate wavelength discrimination influences the size of just-noticeable-difference for most spectral positions. A summary of important considerations follows.

Experimental method

In a careful comparative study Siegel and Dimmick (1962) considered the presentation of a constant stimulus, requiring the observer to make a forced choice, to be more precise than exploring the range (method of limits) or allowing free adjustment to match by the subject. Five wavelengths to each side of the wavelength under study (comparison field) were presented in turn and the observer was required to state whether or not a hue match with the comparison field was possible with luminance adjustment. Wavelengths furthest from the comparison appeared distinctly different in hue. Repeated, random presentations for each wavelength allow standard deviations to be calulated. Wright (1946) compared the smaller values of $\Delta\lambda$ obtained by adjustment to a match with the just-noticeable-difference approach, indicating that a reasonably constant difference between data is obtained by the two methods. The choice of psychophysical method to be used will be influenced occasionally by the length of the time available for data collection; the constant stimuli approach can be the most time-consuming since it demands a large number of settings. The forced choice approach suffers from the additional disadvantage that it depends on the observer maintaining a consistent criterion in his responses.

Voluntary fixation tends to reduce colour discrimination and raise thresholds for wavelength differences below 500 nm by reducing saturation (McCree 1960a,b).

Stimulus exposure time

The study by Siegel (1965) on the dependence of colour discrimination on stimulus exposure time indicated how the wavelength difference threshold was halved for an exposure duration of 5 seconds when compared to one of 0.02 seconds. Uchikawa and Ikeda (1981) noted hue discrimination to be readily maintained for up to one minute with a gradual deterioration thereafter; this would suggest a short-term store for colour.

Field size

The size and position of the retinal image are potent factors in altering wavelength discrimination and the 'standard' procedure using a $2°$ field of fairly high luminance, surrounded by a dark field, gives 'typical' results only when central fixation is used. A $2°$ field is ideal as it attempts to avoid rod participation. Reduced wavelength discrimination results when fields are larger, or smaller. Sensitivity to wavelength differences is increased when the field size is increased to the optimum of between 1 and $3°$ and thereafter decreases. Judd (1932) recognised the relationship between accuracy of chromaticity settings and field size in early colorimetric studies.

Tritanopic tendencies are observed in normal observers when fields are reduced beyond $2°$. McCree (1960a) noted how the shape of the $\Delta\lambda$ versus λ function is altered. Higher $\Delta\lambda$ values occur for very small fields and such observations were confirmed by Ruddock (1966). Comparative values are shown in table 1.3 below (see figure 1.46 and §1.25).

Table 1.3 Typical just-noticeable-difference values for colour-normal observers for 75' and 15' fields (after McCree 1960).

λ(nm)	$\Delta\lambda$ for 75' field (nm)	$\Delta\lambda$ for 15' field (nm)
450	8	28
500	0.6	13
550	1.5	6
600	0.7	10

Luminance of stimulus

As luminance is lowered wavelength discrimination is impaired. Thomson and Trezona (1951) made observations using a $1°20'$ field with luminance levels ranging from near threshold to 10 000 times threshold. Observations by Bedford and Wyszecki (1958) and Walraven and Bouman (1966) should be consulted. A slight shift to shorter wavelengths for the spectral location where Δλ is minimum (sensitivity greatest) is also noted for low luminance levels of the stimulus. For the deuteranope this shift at low levels can be quite marked, and for some wavelengths an increase in sensitivity with decreased luminance is observed. Weale (1951a) and Walraven and Bouman (1972) noted the shift towards shorter wavelengths with decreasing luminance that is found for peripheral viewing in addition to foveal viewing. The effect of luminance is most significant over the wavelength regions 420–430, 480–500 and 570–585 nm and least for 530–550 nm and at the extreme red end. The yellow region is little affected by luminance changes, as shown by Siegel and Siegel (1972). Recent confirmation, in dichromats, that wavelength discrimination changes as lower intensities are used has come from Trick *et al* (1976); protanopes and deuteranopes showed elevated thresholds and their peak of sensitivity moved towards 460 nm.

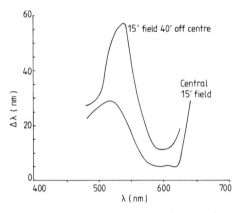

Figure 1.46   Discrimination of change in wavelength using small field size, with centre and off-centre fixation. After Thompson and Wright (1947).

Bipartite field versus grating

The conventional bipartite field used for most studies on wavelength discrimination is usually employed in clinical instruments designed for colorimetric presentation. Hilz *et al* (1974) noted the relative effects for both bipartite and grating presentations. Gratings targets are employed to measure wavelength discrimination objectively through the visual evoked cortical response (see, for example, Riggs and Sternheim 1969).

Juxtaposition of fields

It is usual that the two fields which are to be compared for hue difference are placed in juxtaposition, known as a bipartite field. Whereas separation of the two fields can influence blue–yellow discrimination a 'gap effect' is not noticed for red–green discrimination (Boynton *et al* 1977).

Retinal position

The special features of colour perception and discrimination in the periphery of both the normal retina and that of the colour anomalous are discussed in §1.23. Hue discrimination is poorest between 490 and 510 nm, with maximum sensitivity around 590 nm (Weale 1951a, 1960). The hue discrimination characteristics for peripheral viewing are similar to those obtained for foveal viewing with reduced luminances (Gilbert 1947, 1950). Red–green confusion sets in around 25°, as shown by Weale (1953b) (see table 1.4). By 45° vision is essentially dichromatic, being very poor in the red–yellow–green range but remaining relatively good in the blue–green regions. Such observations were generally confirmed by Boynton *et al* (1964b) for a 40° location using a colour-naming technique; marked desaturation of stimuli in the red and green regions was noted but yellow and blue spectral regions remained relatively unaffected. That the loss of wavelength discrimination follows the experience of desaturation in peripheral vision was demonstrated by Moreland and Cruz (1959). In their study, marked confusion in the range 495–540 nm (blue/green–green) was noted at 25° eccentricity, with extension to 420–590 nm (blue/ green–yellow) at 30° eccentricity.

Table 1.4 Typical just-noticeable-difference values for colour-normal observers for different retinal positions. After Weale (1953).

λ(nm)	$\Delta\lambda$ for foveal location (nm)	$\Delta\lambda$ for 10° extra-foveal location (nm)	$\Delta\lambda$ for 15° extra-foveal location (nm)
450	6	23	30
500	9	54	72
550	11	27	34
600	10	18	27

Surround luminance and colour

A light white surround tends to assist wavelength discrimination, as shown by Weale (1950) who used a 6° surround to a 50′ test stimulus. The ratio of stimulus luminance to surround luminance is a critical factor and Conners (1964) and Siegel and Siegel (1972) indicated how, once this ratio falls below approximately 1.0, discrimination deteriorates; increasing the ratio has little effect. The colour temperature of the surround does not appear to be critical (Pointer 1974).

Wavelength discrimination is significantly influenced by the presence of a coloured surround, particularly for large surrounds (see Brown 1952). Hurvich and Jameson (1961) noted how a yellow–red surround increases sensitivity at the long wavelengths, compared with a green-adapting source (see also Hilz *et al* 1974).

Preadaptation

Just as red–green matches made in an anomaloscope can be influenced by preadaptation (see §7.7), preadaptation can improve wavelength discrimination in some observers (Wright 1946). Colour adaptation was shown by Brindley (1953) to impair wavelength sensitivity in general.

Age

The increased absorption of all wavelengths (especially short wavelengths) by the ocular media with increasing age has already been discussed in §1.13. Age changes can be readily detected in measurements of wavelength discrimination using spectral sources (see Ruddock 1965a,b). Similar changes appear with clinical colour vision tests using surface colours such as the FM 100 Hue test.

1.16.4   Electrophysiological correlates

It is possible to train many experimental animals with colour vision to make subjective colour responses in such a way as to allow a hue discrimination curve to be plotted. A review is given by Jacobs (1982). Experiments with monkeys, which date from the work of Grether (1940), have shown that many species, notably macaque and rhesus, have hue discrimination ability which closely resembles that of man. Evidence also suggests that some, such as *Cebus* (capuchin) and the squirrel monkey, have anomalous colour vision closely resembling protanomalous trichromacy, although deuteranomalous characteristics may also be shown (Jacobs 1982).

Behavioural techniques involving monkeys with colour discrimination have been used to locate areas of the brain concerned with colour following deliberate lesions (see, for example, Weiskrantz 1963, Davidoff 1967a,b and Keating 1979). Answers to the questions as to how the primate visual system permits such good discrimination for hue have been partly forthcoming from electrophysiological studies. De Valois *et al* (1966, 1967) used the change in firing rate of the colour-opponent cells in response to a varying coloured stimulus presented to the animal, to measure an electrophysiological correlate of hue discrimination and to show how different cell types have varying degrees of discrimination throughout the spectrum. A rapid firing rate was taken to indicate a good wavelength discrimination capacity of the cells under examination. Care was taken to eliminate saturation and brightness differences. An increase or decrease in firing rate was noted, depending on the type of opponent cell under study and the direction of the wavelength change. Thus for a cell which showed excitation to long wavelengths and inhibition to short wavelengths ($+R/-G$, $+Y/-B$), a shift toward a wavelength shorter than the standard wavelength produced a decrease in firing rate and a shift toward a wavelength longer than the standard wavelength produced an increase in firing rate. The spectral response curve of $+R/-G$ cells to flashes of monochromatic light typically shows a wide range of wavelengths which produced excitation, then a sudden transition to a range of wavelengths which produced inhibition. At the changeover point where the wavelengths presented tended to shift the response from excitation to inhibition, excellent wavelength discrimination was found but good discrimination was also noted for regions of simple excitation or inhibition. De Valois and his co-workers considered these effects to result from chromatic adaptation to the receptor system absorbing the wavelengths.

On the basis of these experiments it has been suggested that the opponent cells in the primate visual system provide the underlying mechanism by which wavelength information is analysed and processed. Furthermore these workers suggest that the system which is most responsive to a change in wavelength at any point tends to control behaviour, since the psychophysical wavelength discrimination function corresponds to the lower bound of the sensitivity function of the individual cells rather than to their average.

1.17* SATURATION DISCRIMINATION

The range of colours between white and a particular part of the spectrum varies in purity or saturation, so that adding white to a spectral light desaturates the colour. This can be done progressively in either direction.

The threshold of saturation is usually expressed in terms of the 'first step' discernable from white, as a spectral light is added. Other steps exist until the completely pure spectral light is without admixture of white. It is clear from the shape of chromaticity diagrams that spectral yellow is relatively desaturated compared to green of about 520 nm wavelength (see Chapter 2).

Normal discrimination figures were published by Martin *et al* (1933) as well as by Chapanis (1944) and Wright (1946). Typical normal data, for photopic fields of about $2°$, are represented in figures 1.47 and 1.48.

It should be noted that different experimental conditions influence the results considerably. The judgment is a difficult one to make for most observers and the practical shortcomings of several methods were explained by Martin *et al* (1932).

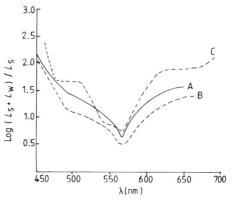

Figure 1.47 Saturation discrimination in normal subjects. Curve A shows mean data for a number of subjects. Curves B and C show the extent of normal variation between typical subjects. L_s and L_w are the luminosities of spectral and white light respectively. After Martin *et al* (1933). (1933).

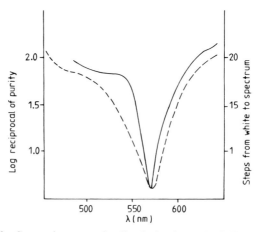

Figure 1.48 Saturation or purity discrimination. The full curve shows the number of just-noticeable steps between white and the spectral colour. The broken curve shows the log of the reciprocal of purity of the colour for the first just-noticeable step from white. Modified from several sources, including Wright (1946).

A standard white field has to be agreed and provided. The increment of colour must be superimposed on part of the field, while the balance of brightness with the white standard is maintained. Data are expressed as reciprocal threshold purity, plotted against a scale of wavelength in the form $(L_S + L_W)/L_S$, L_S being the 'luminosity' of the spectral light and L_W being the luminosity of the white. Wright (1946) expressed the colorimetric purity of a colour as $F_\lambda/(F_\lambda + F_W)$ where F_λ is the light flux of the monochromatic (spectral) light and F_W is the light flux of the white light. The state of light adaptation influenced results minimally. Weale (1960) expressed the fact that the number of steps between white and a certain part of the spectrum varies according to wavelength and also that at any part of the spectrum the steps vary in size, so that yellow has as few as ten steps compared to about a hundred elsewhere.

Onley *et al* (1963) used filtered lights to make magnitude estimates of saturation. Bartleson (1981) described the ways in which saturation is influenced by chromatic adaptation.

A term 'colourfulness', described by Hunt (1980), is related to saturation and is becoming widely used. Colourfulness is the attribute of vision whereby more or less chromatic colour is seen. Saturation is then described as colourfulness judged in proportion to brightness; it is a relative colour-fulness just as lightness is the brightness of an object as seen relative to the brightness of a white object illuminated in a similar manner. Bartleson (1980) has summarised the colorimetric context of saturation for CIE colour diagrams of 1960 and 1976 for those who wish to use such a system of calculation. The implications of saturation discrimination in tests for abnormal colour vision are discussed §7.10.4 while §3.4.4 includes data for abnormal subjects.

Abramov and Gordon (1977), using eccentrically viewed objects, detected a tritan tendency in the apparent desaturation which is more often associated with 'deutan' features of the (normal) peripheral retina. For stimuli of about $6.5°$ all colours appeared highly saturated and red is consistently well saturated at different locations at all sizes.

Peripheral field studies of saturation are of practical use, but since no standards exist, investigators must establish norms for their own apparatus. Milliken (1978) did this using the practical method of a bipartite test field of $2°$ made of Munsell chips (5GY, value 5) displaced $7°$ into quadrants of the field. He found significant quadrantic or hemi-field differences in peripheral saturation discrimination, attributable to the way in which cortical representation is organised.

1.18 LUMINANCE REQUIREMENTS FOR HUE PERCEPTION

The interaction of luminance and hue in colour perception is complex, and has been well investigated. The findings of one recent study are described as an example. As part of an analysis of the minimum luminance require-ments for hue perception, Conners (1968) studied red, green and blue targets (exposure duration 44 and 700 ms) at nine visual angles from $1°$ to $21''$ arc diameter. Small-field tritanopia influences perception for fields under $20'$ arc (see p. 113). With sufficient luminance, colours could be identified for stimuli as small as 0.35' arc diameter. Thresholds for perception of red were lower than for blue and green; thus red is identified more frequently than other hues of similar luminance.

Particular difficulty was noted with the green stimuli; at high luminance the 521 nm stimulus was often called white, so that increasing the luminance did not aid recognition beyond a certain level. A similar appearance was noted at low luminance.

1.18.1   The role of colour and luminance channels in detection

Shortly after propounding their theory of colour processing, Hurvich and Jameson (1959) showed that wavelength also interacts in luminance contrast conditions. The brightness of a coloured stimulus was affected more by an adjacent inducing field of similar wavelength than by one of equal luminance but different wavelength. This has led to the suggestion that the apparent brightness of a coloured stimulus results from both chromatic and achromatic activity.

Similarly, detection thresholds for a monochromatic stimulus can involve both chromatic and luminance channels. The psychophysical experiments of King-Smith (1975) and King-Smith and Carden (1976) confirm this view and even suggest that in some circumstances (for instance the detection of stimuli subtending small angles) the chromatic processing channel is in fact superior in sensitivity to the luminance system and plays the major role in detection. These authors demonstrated that coloured test stimuli, of all wavelengths except yellow, are as easily discriminated from white as they can be detected (i.e. discriminated from a blank), supporting the suggestion that the opponent-colour system is used for their detection. The discrimination of yellow from white was found to be less good than the detection of yellow, indicating that the yellow flashes were not detected by the opponent-colour system. Results from other studies confirm that the colours of spectral stimuli may be discriminated at their detection threshold. The view is consistent with a number of other observations—in particular the fact that temporal integration is greater in the opponent-colour system than in the luminance system, and the suggestion, based on psychophysical evidence, that spatial integration may also be greater in the colour channels.

King-Smith and Carden suggest that a stimulus will be detected if it exceeds the threshold of either the luminance or chromatic system, and the threshold will be determined by whichever system is most sensitive for the conditions that are operating. They believe that the colour channels provide the dominant contribution to detection for relatively long durations of stimuli (200 ms) presented on white backgrounds. By reducing either the stimulus duration (to 10 ms) or the stimulus size, or by eliminating light adaptation, the contribution of the luminance system becomes much more evident. It is suggested that in the red–green range brief spectral flashes of large or small diameter are detected mainly by the luminance channels.

The exact contribution of luminance and colour information to visual detection and recognition requires further study and caution is needed in relating the luminance and opponent-colour systems of psychophysics with the properties of ganglion and geniculate cells at the present time. Kerr (1974) confirmed the role of the colour mechanisms in detection by suggesting that if colour channels are operating at threshold, then the stimuli should be readily identifiable as coloured.

1.18.2   Hue changes with luminance

For more than a century it has been realised that the hues of most visual stimuli change with luminance (von Bezold 1873 and Brücke 1878). In general at low luminances wavelengths between 470 and 578 nm (blue–greens, greens and yellow–greens) appear greener, while hues around 578 nm (orange) appear redder; there is a shift toward shorter wavelengths. At high luminances hues below 500 nm (that is violet, blue and blue–green) appear bluer and hues above 500 nm (that is yellow–green, yellow, orange and red) appear yellower than those at medium intensities. Wavelengths 570, 508 and 476 nm, (yellow, green and blue respectively) do not undergo changes in hue appearance with intensity and are described as the invariant hues.

Various experimental techniques have confirmed these findings. One of the difficulties in controlling such measurements is the variation of perceived hue with stimulus duration, especially for reds of high luminance which, as shown by Cohen (1975), appear yellow for long duration exposures. Purdy (1931) used subjective hue matching, while Boynton and Gordon (1965) found a colour naming technique to be as sensitive as direct matching. Other studies of importance include those by Walraven (1961), Nagy and Zacks (1977) and van der Wildt and Bouman (1968). Attempts to explain the phenomenon have involved both the Young–Helmholtz theory and the opponent-colour theory. Walraven (1961) gives a review.

1.19   CONTRAST EFFECTS ON COLOUR PERCEPTION

The interaction of a coloured stimulus with its background (simultaneous contrast) which may well change with time (successive contrast) can alter colour appearance greatly. Simple theories of chromatic contrast effects suggest a selective change in sensitivity of visual mechanisms. Hence a physically achromatic area may appear green in the presence of a strong red surround owing to a reduction in the sensitivity of the red mechanism in the central as well as the surrounding area. These ideas are compatible with the known behaviour of visual pigments (their bleaching and regeneration) and are essentially based on lateral adaptation concepts. Opponent processing theory has now become a favoured explanation, offering the alternative view that stimulation of a specific retinal area induces an opponent or opposite response in adjacent retinal regions. This implies that the reduced response that occurs in conditions involving simultaneous brightness and colour contrast involves only the response of members of that opponent pair, and is independent of activity in other mechanisms.

It is well known that the apparent brightness and saturation of a coloured stimulus can be markedly affected, in addition to hue, by an adjacent coloured stimulus or surround.

Mount and Thomas (1968) showed that the brightness changes induced by a stimulus adjacent to a test target are independent of wavelength. In a study by Kinney (1967a) the importance of exposure time on simultaneous colour contrast effects was investigated. Only a very brief exposure of a blue surround was needed to induce a strongly saturated hue whereas the hue induced by a red surround increased in saturation as the exposure time increased. In an earlier study, Kinney (1962) found that the amount of colour induced increased (i) as the size of the inducing field is increased (ii) as the luminance ratio between the inducing and induced fields increased and (iii) to a small extent, as the purity of the inducing colours was increased.

1.20*   VISUAL ACUITY AND COLOUR

Visual acuity, which represents the ability of an observer to resolve fine detail (usually expressed as the angular resolving power of vision) is a complex function which varies considerably with the stimulus conditions and with the observer. The two most important variables are the illuminance level and the region of the retina stimulated, since good acuity is possible only with cone vision. Other important variables include the presentation time, spectral distribution, the shape and size of the stimulus, the task involved and adaptation effects.

Since the eye suffers from chromatic aberration (approximately 2.0 dioptres) it might be expected that the visual acuity with monochromatic light would be slightly better than with white light—since the blurring which results for some of the wavelengths of the white light will reduce the contrast (see, for example, Campbell and Gubisch 1967). Experiments with more than one colour of very different wavelengths are therefore often carried out with the subject looking through an achromatising lens. Such lenses (doublets or triplets involving different glasses) are overcorrected for chromatic aberration and have been described by Hartridge (1950). Reports in the literature on the variation of visual acuity with wavelength have been sparse and controversial, as the different luminance levels and states of accommodation used make comparisons between studies difficult. Luckiesh and Moss (1933) reported a 24% increase in acuity for single wavelengths, whereas Hartridge (1947) found no change in visual acuity between single and composite wavelengths. Von Bahr (1946) measured grating acuity for two colours, first separately and then superimposed; the acuity for one colour was found not to be affected by the second colour as long as the eye was in good focus for the first wavelength. Brindley (1954b) reported a lower grating acuity for violet light than for other wavelengths. Martin and Pearse (1947) measured the grating acuity for the accommodated eye in red and white light and reported a 4% superiority for red light of equal luminance, particularly at low illuminance levels. One of the most comprehensive studies involving different illuminance levels was by Shlaer et al (1942). They noted that the value of the maximum acuity reached at high intensities is little different for red and blue light; if anything a slight superiority is shown for blue light, contrary to Brindley's observation.

Brown et al (1957) suggested that the best acuity achievable in monochromatic light is, for all practical purposes, the same at all wavelengths. These authors found that for grating resolution varying the wavelength had no effect on the minimum amount of light required for vision.

1.20.1 Contrast

The value of measuring resolution at different contrast levels became fully evident in the early 1960s and led Campbell and Gubisch (1967) to note that differences in contrast sensitivity were predictable from optical theory for ·a grating acuity target in white and monochromatic light. Noting that slight improvement in acuity for monochromatic light was to be expected (on account of the chromatic aberration of the eye) they showed that this was in fact the case, with better spatial resolution for 578 nm that for white light. Their calculations demonstrated that the optical differences between white and monochromatic light should be most easily detected as a change in contrast rather than a change in straightforward visual acuity terms. Subsequent experiments with two observers showed an increased contrast sensitivity in monochromatic light over white light for all spatial frequencies. Since a large change in the contrast threshold of a grating may cause only a very small change in the conventional measure of visual acuity for that grating, it is not surprising that there has been a conflict of opinion on this matter over the years. Shlaer *et al* (1942), for instance, reported a change in acuity with wavelength only under conditions of reduced contrast.

Thus differences in visual acuity with wavelength are noticed only when a contrast change is present, and even then the change is small; for practical purposes the changes in acuity for different wavelengths are negligible.

The short wavelength cones contribute minimally to the luminance channel (Eisner and MacLeod 1980). Thus only the two longer wavelength sensitive cone mechanisms contribute to visual acuity.

1.21   STILES–CRAWFORD EFFECTS

Both the brightness and the colour of a beam of light, as perceived, depend on the part of the pupil through which it enters. Figure 1.49(*a*) illustrates the directional sensitivity, discovered by Stiles and Crawford (1933a), by which light pencils entering the pupil centrally are more effective in evoking a response than if they entered at the pupil margin. A small subjective change in colour (both in hue and saturation) is noticed when a monochromatic beam is incident at different angles on the same part of the fovea, as described by Stiles (1937) and known as the Stiles–Crawford effect of the second kind, to distinguish it from the variation in luminous efficiency of light entering first centrally and then peripherally. In practice this phenomenon shows itself by a slight but measurable shift in hue towards longer wavelengths (those greater than 560 nm). The hue becomes redder as the entry point moves towards the pupil periphery, giving an increase in apparent wavelength expressed as a positive shift. A reversal of the effect for wavelengths between 500 and 560 nm is seen, giving a shift towards shorter wavelengths, with a bluer appearance to hues and a negative shift. Various explanations have been proposed by Stiles (1937), by Walraven and Bouman (1960) and by Safir *et al* (1971), although these ideas are incomplete. Brindley (1953) noticed how adaptation to very bright lights caused a similar shift.

Experimental results for a small number of normal trichromats have been published (see Stiles 1937, Walraven and Bouman 1960 and Enoch and Stiles 1961). Such measurements are not easy since very careful positioning of the observer's pupil is necessary. Enoch and Stiles measured the complete colour change of hue and saturation and figure 1.49(*b*) indicates the hue changes for light entering at a point 3.5 mm from the pupil centre. Considerable variations between normal observers were studied by de Vries-de Mol and Walraven (1982), in addition to variations between deuteranomalous subjects.

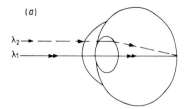

(a)

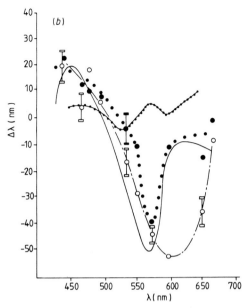

Figure 1.49 (a) Stiles–Crawford retinal direction effect of the second type. The two pencils of 'monochromatic' light enter the pupil at different positions

(b) Stiles-Crawford effect II. ⊷, Normals (3.5 mm off-centre), mean of 6 (Enoch and Stiles 1961). ——, Deuteranomalous (3.5 mm off-centre) (Walraven and Leebeck 1962). ○ and —·—·—, deuteranomalous (3 mm nasal) (Voke 1974). • and ••••, deuteranomalous (1.6 mm nasal) (Voke 1974). Bars show 2SD.

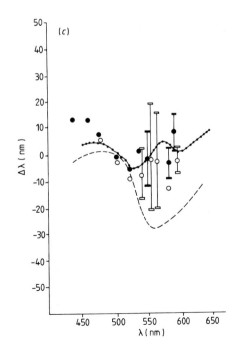

(c) ⊷, Normals (3.5 mm off-centre) mean of 6 (Enoch and Stiles 1961). ---, Protanomalous (3.5 mm off-centre) (Walraven and Leebeck 1962). ○, Protanomalous (3 mm nasal) (Voke 1974). •, Protanomalous (1.6 mm nasal) (Voke 1974). Bars show 2SD.

In Chapter 7 (p. 358) the application of these phenomena to abnormal colour vision will be considered, although dichromats have difficulty in making reliable colour matches of this type, in the short and long wavelength regions. Walraven and Leebeck (1962b) made approximate measurements of one protanomalous and one deuteranomalous person, noting how the shifts were greater than for normal trichromats. Voke (1974) used this method to examine anomalous trichromats and dichromats, at two eccentricities, for both nasal and temporal positioning of the beam. The negative shift in the 500 and 560 nm regions is more prominent in the colour defectives, as shown in figure 1.49(*b*) and (*c*).

1.22 SPATIAL AND TEMPORAL SENSITIVITY OF THE COLOUR MECHANISMS

1.22.1 Spatial sensitivity

The contrast sensitivity to stimuli varying in spatial frequency is measured following the example of Schade (1956); a typical function for the total visual system for white light is shown in figure 1.50. This allows a measure of visual resolution. Techniques whereby the individual colour mechanisms are isolated have been used to allow measurements of the spatial and temporal characteristics of the three basic classes of colour mechanisms. Green (1968) and Kelly (1973) used Stiles' chromatic adaptation methods, Cavonius and Estevez (1975a,b) used a spectral compensation method due to Estevez and Spekreijse (1974), whereby one class of cone receptors can be stimulated with independent control of adaptation. This was considered superior to the classical 'bleaching' of cone pigments, and significant discrepancies between the results of Cavonius and Estevez, Kelly and Green are considered by Cavonius and Estevez to be due to adaptation differences. Green (1968) concluded that the sensitivities of the red and green mechanisms were identical, whereas Kelly (1973) found a difference in both the shape of the sensitivities for the red and green mechanisms and in their absolute sensitivities; in particular supersensitivity of the green mechanism was noted. These differences led to contradictory conclusions about the physiology of cone mechanisms—Kelly's results suggesting that inhibition could occur both between and within mechanisms, whereas Green concluded that inhibition effects are confined to each cone system. Cavonius and Estevez (1975a) found the spatial frequency functions of red (π_5) and green (π_4) to be identical and similar in both shape and absolute sensitivity to the data of Green (1968). The shape of the functions suggested little, if any, inhibitory interaction between the colour mechanisms because they were at pains to remove hue clues at low frequencies. All agree that the blue mechanism has a lower absolute sensitivity (and lower resolving power) which is consistent with other psychophysical and electrophysiological data.

Studies by Johnson and Massof (1982a,b) to compare colour vision at different locations in the peripheral field with that of the central foveal region, with due correction of the optical factors involved, indicated different spatial properties of the colour mechanisms with varying eccentricity.

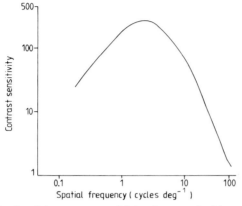

Figure 1.50   Spatial contrast sensitivity for a typical human observer.

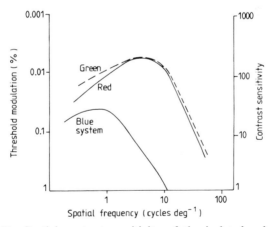

Figure 1.51   Spatial contrast sensitivity of the isolated red and green systems when hue information has been suppressed and of the isolated blue system for a normal observer. After Cavonius and Estevez (1976).

Colour-contrast sensitivity functions have been derived by Petry *et al* (1982) using VECPs (visually evoked cortical potentials) in response to coloured gratings. The VECP amplitude was shown to vary linearly with the log of the colour contrast.

1.22.2   Temporal sensitivity

Study of the visual system using intermittent light stimulation is often useful. The dependence of colour discrimination on stimulus duration is well established. It is usual to measure the flicker fusion threshold to sinusoidally modulated light as a function of temporal frequency to investigate the temporal sensitivity of the visual system in a similar manner to spatial sensitivity. The temporal function is frequently referred to as a 'de Lange curve' after de Lange's (1952, 1957) evaluation, which indicated greater sensitivity near 10 Hz and a drop at both lower and higher frequencies (figure 1.52). De Lange found the peak sensitivity to be nearly independent of wavelength. Green (1968) applied Stiles' chromatic adaptation isolation technique to determine the temporal frequency response for the separate colour mechanisms and Kelly (1974) considered the spatiotemporal interactions. Green found that the green mechanism has a sinusoidal flicker characteristic that is similar to that obtained with white light. The red mechanism has similar sensitivities at high temporal frequencies but the low frequency attenuation for the green was more marked than for the red mechanism. The marked differences in attenuation for the systems at low frequencies suggested different organisation of cone systems, indicating interactions between mechanisms.

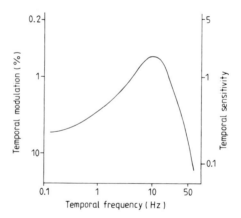

Figure 1.52   Temporal sensitivity of the red and green colour mechanisms together. This is a typical curve of the eye's total response to flicker. After de Lange (1954).

Kelly confirmed Green's data that at high temporal frequencies the green flicker sensitivity curve is similar to its achromatic counterpart, but reported a much flatter response for low frequencies. To a lesser degree this was also true for the red and blue curves. The maximum sensitivity for the red, green and achromatic curves coincided at approximately 8 Hz but the blue was much reduced, having a maximum between 1 and 4 Hz. Estevez and Cavonius (1974) used the spectral compensation method to study the response of the red and green colour channels in isolation and these were found to be identical. They maintain that it is impossible to know whether the results for red and green obtained by Green and Kelly were at the same adaptation level and this might account for the measured difference in sensitivities. As with equivalent measurements for the spatial domain, different conclusions have been drawn by those using different isolation techniques. When hue clues at low frequencies were removed, in the novel technique of Estevez and Cavonius, the results suggested that there are no temporal interactions between colour systems (figure 1.53). On the basis of binocular studies it was concluded that the detection of hue modulation occurs after convergence of information from the two eyes, i.e. at a fairly high level of the visual pathway. Green (1968) pointed out that similar contrast and temporal sensitivities for the red and green mechanisms would indicate identical visual resolving properties of the system. A lower overall sensitivity of the blue mechanism shown by Green (1968) and Cavonius and Estevez (1976) is likely to be a cause of the low flicker fusion frequency of the blue mechanism. Studies with Daltonics are considered in Chapter 7 (p. 359).

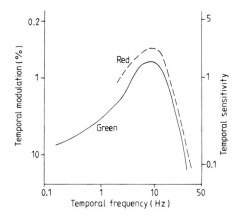

Figure 1.53 Temporal sensitivity of the normal green colour mechanism in isolation and the red colour mechanism in isolation. After Cavonius and Estevez (1976).

1.23*   PERIPHERAL COLOUR VISION

The peripheral retinal regions, taken to be those beyond the $2°$ foveal zone, are characterised by reduced visual function in photopic illumination on account of the preponderance of rod receptors. Visual acuity and colour discrimination are poor here compared with the fovea and local adaptation effects, in particular the fading of stimuli presented in the peripheral visual field (Troxler's effect (Troxler 1804)) make accurate observations difficult. Therefore, in experimental observations, non-foveal stimuli are often presented in a flashing mode to minimise these difficulties. Colour vision deteriorates, even for high intensities, towards the periphery. Most clinical colour vision tests are designed for foveal viewing and Verriest and Metsala (1963) do not advocate the use of pseudoisochromatic test charts or matching tests beyond 30°.

Early interest in the colour vision characteristics of the retinal periphery probably arose in connection with perimetric measurements of the visual field using coloured stimuli for the investigation of ocular disease. Von Hess (1889) laid important foundations by emphasising the need for controlled experimental conditions, and later investigators such as Abney (1913) showed the dependence of the size of the colour fields on the wavelength, luminance and angular subtense of the stimulus. Colours become desaturated, appearing less coloured at the extreme peripheral parts of the retina and in general reds and greens assume a yellow appearance while violets look bluer.

1.23.1   Peripheral colour experiences

For almost a hundred years the trichromatic nature of central vision in colour normals has been recognised, leaving aside the central foveal area with its tritanopic characteristics. The relative distribution of cone receptors throughout the primate retina and the intrusion of rod receptors outside the fovea greatly influences colour vision characteristics. Marc and Sperling (1977) showed the maximum density of the short wavelength 'blue' responding cones to be at about 1°, with the two long wavelength cone receptors being concentrated in the foveola. In addition the screening effect of the macular pigment must be considered.

Most authorities accept that, in general, colour vision progressively deteriorates towards dichromacy (red–green blindness) with an enhancement of sensitivity to blues beyond 30°. Exceptions to the observation must be noted (e.g. that of Sperling and Hsia 1957) which can be explained by the very short duration exposures used (see Moreland 1972c). Wooten and Wald (1973) also found the relative sensitivity of the short wavelength cone mechanism to decrease with perimetric angle in the far peripheral retina.

Beyond 60° vision can be considered to be monochromatic; such observations date from Purkinje (1825) and Aubert (1857). Modern investigations of importance include that of Weale (1953b), which reinforced the dichromatic nature of the external periphery, with enhanced sensitivity to blues extending from the parafovea outwards. Wald (1967) considered that all the classical forms of colour blindness are represented in colour normals in roughly concentric zones. Yet the three spectral mechanisms and cone classes appear to be represented in the peripheral regions at least to 80°, as indicated by Wooten and Wald (1973). Their suggestion that loss of colour sensitivity in the peripheral retina has a neural explanation is perhaps appropriate. Graham (1972) considered the site responsible for red–green defective vision at certain peripheral locations in the colour normal to be at or below the neurological level of the opponent system.

The contribution which the absence of the blue-absorbing macular pigment plays in this enhanced sensitivity to short wavelengths has been a matter of dispute. The studies of Weale (1951b, 1953b), Wooten et al (1975) and Stabell and Stabell (1980) all support the view that the absence of macular pigmentation and the greater participation of short wavelength coding receptors in the extrafoveal retina are jointly involved. Inui et al (1981) stressed that the transition zones are gradual. Moreland (1955) demonstrated trichromacy to 15°, dichromacy between 15 and 25° and near monochromacy as far as 45°. Later Moreland and Cruz (1959) noted that three matching stimuli are required up to 25°.

Colour matching techniques to investigate the perceived colour change with eccentricity have been undertaken by Dreher (1911) and Moreland and Cruz (1959). Monochromatic test stimuli presented at different retinal positions were matched with a foveal comparison field of three reference stimuli; red, green and blue. Rod receptors contribute significantly to these matches as studied by Stabell and Stabell (1975, 1976, 1977, 1979). A neural explanation involving inhibition in preference to a photochemical view is proposed by these authors.

A different approach which takes into account the effect of rod activity has been proposed by Trezona (1970, 1976). In order to equate the effect of rod activity in different parts of the colour field and to overcome the resulting distortion, a tetrachromatic colour match is achieved by a trichromatic match initially, followed by a brightness match at a level below the cone threshold using a cyan stimulus, which is chosen because it is near the peak of the rod absorption curve. Alternation between the two matches subsequently allows an acceptable colour match to be achieved.

The suggestion that rods may add a blue component to peripheral colour vision was made by Ambler (1974). Stabell and Stabell (1975, 1980) noted in particular the effect of rod intrusion on spectral sensitivity.

Variables such as field size and hue, saturation and luminance play a significant role in determining the colour sensations evoked by the retinal periphery. Thus Gordon and Abramov (1977) likened the hue characteristics for large peripheral targets to foveal ones for very small stimuli and Wooten and Wald (1973) concluded that 'in general the brighter, larger and more saturated a coloured stimulus (positioned peripherally) the wider the field in which it excites about the same hue sensation as centrally'.

The importance of stimulus size for peripherally located targets was indicated by Johnson and Massof (1982a,b) who measured colour thresholds out to $50°$.

Colour naming is unreliable in the retinal periphery for small targets except for reds, which were reliably named in the study of Gordon and Abramov (1977) even as far as $45°$. However Boynton *et al* (1964b) found difficulty with such measurements at this position. The findings of Burnham (1951) suggested that stimulating the retinal periphery can even improve foveal colour discrimination. Generally, however, peripherally located colours appear of uncertain hue, especially those of mid-dominant wavelengths. Sensitivity measurements of the isolated red mechanism (Stiles' π_5) at different eccentricities in the visual field made by Kitahara (1980) showed a marked decrease in sensitivity as the test field was moved from a foveal location to $20°$.

No major differences between foveal and parafoveal colour matching observations were observed in a relatively early study by Gilbert (1950) using high levels of illumination. A movement of the white point towards blue was noted, emphasising blue intensity; less red and more green was needed to match a yellow as the test field was displaced peripherally.

Some confusion surrounds an exact and concise description of the colour vision characteristics of the retinal periphery. On the basis of colour matching (mixture) measurements Moreland and Cruz (1959) favoured a tendency towards deuteranopia. The colour naming measurements of Boynton *et al* (1964b) reinforced this view. Early evidence of red–green 'blindness' came from Lythgoe (1931) who noted the persistence of only yellow and blue in the far periphery, and Wooten and Wald (1973) who indicated that peripherally situated targets of red and green appeared yellow. Nevertheless, evidence of tritan-like characteristics is also provided by Wooten and Wald's study where violet and blue appeared blue beyond $20–30°$ and the three colour vision mechanisms are reduced. Gordon and Abramov (1977) confirm such a tendency although they did not favour strict designation in terms of anomalous colour vision since the characteristics of peripheral colour vision vary widely with the stimulus size.

1.23.2   Wavelength discrimination studies

Peripheral deterioration in blue and green vision, particularly for low luminance stimuli positioned at 10 and $15°$ from the fovea, was shown by Weale (1951a) using hue discrimination measurements. Hue discrimination was poorest between 490 and 510 nm approximately and best around 590 nm. The hue discrimination curves for stimuli at these mid-peripheral locations are similar in general form to foveal measurements but with considerably elevated thresholds, particularly at low luminances. Gilbert (1947, 1950) had earlier noted similarities between parafoveal colour vision under photopic conditions and foveal colour vision under conditions of lower luminances, and dichromatic vision for small fields (15′) as far out as 40′ from the fovea was shown by Thomson and Wright (1947). Red–green confusion becomes evident at 25 and 45° and was described by Weale (1953b) as a tendency towards protanomaly. Conners and Kinney (1962) confirmed this by noting that the relative sensitivity to green, compared with red is highest in the 2–10° region before falling off rapidly.

1.23.3 Spectral sensitivity

A shift of the peak of the spectral sensitivity curve towards shorter wavelengths (the Purkinje shift) as the illumination level is reduced for foveal vision, is also experienced for peripherally placed visual targets as shown by Gilbert (1950) and by Weale (1953b). As already noted, an enhanced sensitivity to short wavelengths with increasing eccentricity is found; see for example Wald (1945), Weale (1953b), Sperling and Hsia (1957), Abramov and Gordon (1977). In one of the first measurements of the spectral sensitivity of the peripheral cones Wald (1945) noted greater sensitivity for green and blue spectral regions (below 578 nm) in the extrafoveal region at 8°, compared with foveal viewing. The biphasic characteristics of the V_λ function for cones at 8°, noted by Wald (1945), were confirmed by Wooten *et al* (1975) using stimuli presented at 30 and 70° in the peripheral retina; these investigators also confirmed the greater sensitivity of extrafoveal regions to short wavelengths.

The dependence of the spectral sensitivity curve on viewing conditions must be stressed; in particular the shape of the function measured peripherally frequently shows slight subsidiary humps which are especially pronounced at large eccentricities (Stiles and Crawford 1933a,b, Weale 1951b). Peripheral measurements are by no means easy on account of the poor visual acuity and local adaptation effects mentioned. Thus the precision of the observations recorded is poor.

Psychophysical studies of the relative contributions of the three cone photopigment classes to spectral sensitivity for different locations in the retina, such as by Stabell and Stabell (1981) support histochemical studies on the relative distribution of the three cone pigments. Whereas the relative spectral sensitivity and the weighted contributions of the medium and long wavelength cone photopigments remain invariant across the retina, the relative contribution of the short wavelength cone mechanism was shown to increase between the fovea and 17°, to remain essentially constant between 17 and 28° and to decrease between 28 and 65°. The absolute sensitivity of the medium and long wavelength cones was shown by Stabell and Stabell to decrease between the fovea and 65°.

1.23.4   Peripheral colour vision of colour defectives

The preceding comments deal with observations of colour perception when objects are projected onto the retinal periphery of colour normal observers. Moreland (1972c) suggested that Nagel's (1907d) comments on his own deuteranopia included trichromatic observations for a viewing field in excess of $10°$. Such observations were confirmed for *some* deuteranopes and protanopes by Jaeger and Kroker (1952). Two functional pigments were shown by Wooten and Wald (1973) to be present in the fovea and periphery of dichromats so that essentially the same changes as in normal observers are seen. The effect of retinal position on the neutral points of protanopes and deuteranopes studied by Massof and Guth (1976) indicated a shift towards shorter wavelengths for peripherally determined values relative to centrally positioned stimuli. The effect of rod intrusion in colour matches made by dichromats for foveal and parafoveal viewing has been discussed by Ruddock (1971) (see also §7.11). Boynton (1982) suggested that rod activity is responsible at least in some part for the tendency of dichromats to show trichromatic characteristics when using large fields (see also Rayleigh 1881 and Smith and Pokorny 1977).

1.24   THE PERIMETRIC APPROACH TO COLOUR THRESHOLDS

Perimetry was established as a clinical means for investigating the visual fields by von Grafe (1856) and assesses the functional capability of the visual system quantitatively by determining retinal sensitivity at different locations. Visual acuity and colour discrimination become progressively poorer in the peripheral zone of vision. Visual field measurement indicates the underlying physiological change from pure cone vision in the fovea, to a combination of rod and cone function in peripheral regions, and thus provides a valuable clinical assessment of the mechanisms controlling vision. Both kinetic and static modes of investigation have value in the detection and diagnosis of ocular disease.

Colorimetry has been traditionally carried out with a centrally fixated field of $2°$, often considered to correspond to the fovea centralis of the retina. Retinal topography in the central region is illustrated in figure 1.11, which summarises the terms applied to the different areas and indicates both linear size and angular subtense of each area. Assuming a bipartite field of $2°$ it is the central fovea, including the foveola, which is involved if fixation is reasonably central, although some deviation of the order of a few minutes of arc is likely, on account of small eye movements. The foveola is located approximately $15°$ to the temporal side of the optic disc (optic nerve exit) and the outward projection of this disc, as the blind spot, is found in the *temporal* field of vision. Fields of vision are measured monocularly.

Until 1945 moving targets were used almost exclusively (figure 1.54). Whilst maintaining foveal fixation the patient indicates when a test object of fixed size and intensity (usually a disc mounted on a small wand) is just visible against a dimly lit background as it is slowly moved across the visual field, from a point outside the visual domain, through the relatively insensitive periphery to the centre, from a 'non-seeing' area to a 'seeing' region. The line on a polar coordinate plot connecting just-visible points (difference threshold) in various meridians around fixation gives an *isopter*, an outline of the light-perception field. Insensitive zones where visual function is absent (scotomata) can be plotted. This kinetic mode is most suited for studying very small visual field defects and the profile of extent of the visual field. Traditional perimeters for kinetic plots consist of a semicircular arc, coloured black or grey. Targets are moved manually or projected as small spot stimuli.

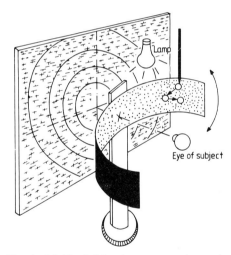

Figure 1.54 Classical field of vision instrument. A rotating arc is used for the periphery and a tangent screen for the central field.

Modern perimeter or visual field analysers (figure 1.55) have a hemispherical whitened bowl on which multiple static or mobile point stimuli can be projected at various angular distances along a particular meridian. There is control of the luminous intensity, size and wavelength of targets. Since the stimuli are stationary the method is referred to as *static perimetry*. This approach allows a detailed and accurate topographic determination of the visual threshold in various parts of the visual field. There is greater reproducibility should re-examination at a particular point be desirable, as is frequently appropriate, for quantification of defects. A fully automated read-out field plot in polar form is often provided and skilled technicians can operate the instrument. Because a plot of 'intensity required for seeing' against angular distance gives a profile view along a particular vertical section of the 'hill of vision' (a concept introduced by Traquair (1948)) this method is also known as *profile perimetry* (figures 1.56 and 1.57).

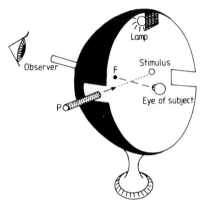

Figure 1.55   Modern perimeter or visual field analyser. The subject's eye is placed at the centre of the bowl. Stimuli are projected from the lens system P. The subject's fixation on point F can be observed through a telescope. Suitable background illumination comes from a lamp.

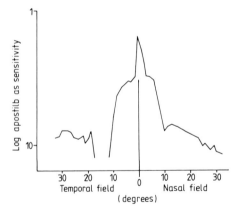

Figure 1.56   Tübingen perimeter static profile plots of the horizontal meridian of a normal left eye. A 10′ of arc stimulus is used at different intensities, superimposed on a 10 apostilb white background.

Sensitivity measurements based on luminance difference thresholds $\Delta L/L$ are involved. Patients may be dark adapted prior to measurements, sometimes followed by a period of light adaptation to the ambient level in modern instruments. The target (frequently circular) is presented below absolute threshold (sub-threshold) and its luminance is gradually increased by approximately 0.1 log units until it is just detected. The luminance is further increased (supra-threshold) and then gradually decreased until it is no longer perceived. A variety of psychophysical methods can be used (e.g. method of limits or constant stimuli) and the procedure is repeated several times. The luminance level required just to see the target for 50% of presentations is taken as the threshold. Luminance for the threshold is plotted against location on the field using units such as $cd\,m^{-2}$ or apostilb (1 $cd\,m^{-2}$ equals 3.14 apostilbs). If a dark background is used the measurement approximates to absolute threshold; when a white field is used a difference or increment threshold is recorded.

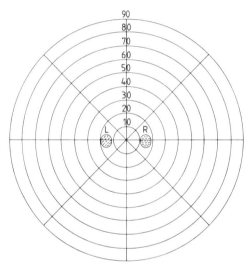

Figure 1.57 Visual field chart before completion showing expected positions of right and left blind spots.

The plotting of zones of sensitivity in the visual field is a powerful technique for investigating summation effects and can provide information about the receptive field organisation of the retina.

1.24.1* Colour perimetry

Most perimeters are supplied with both white and coloured test targets but, as indicated by Tate and Lynn (1977), the latter are rarely used. Perhaps examiners consider that coloured objects are an unnecessary complication if the patient is only required to detect the test object. Aubert (1857) first introduced them to investigate the variation of sensitivity for various colours with retinal position.

In the last two decades a renewed interest has been shown in colour field examination, partly as a consequence of the availability of monochromatic sources for test targets which can be incorporated easily into the automatic static instruments. This permits the testing of peripheral cone colour vision, separately from rod (colourless) vision. Clinical colour vision tests confine themselves to macular regions. For disorders of the visual system selectively affecting cone function outside the macular region colour thresholds are a valuable tool for diagnosis (see Sloan and Feiock 1972). Few clinical techniques are available to differentiate between individual receptor mechanisms although, as Hansen (1979) indicated, in many eye diseases selective impairment of the colour vision mechanism may be suspected. The determination of colour thresholds has provided researchers with valuable indications of the mechanisms of normal colour vision. It is pleasing to see techniques formerly confined to research laboratories now becoming widely used to provide a better understanding of the pathogenesis of eye disease and to aid clinical diagnosis.

Colour test objects can be considered as a mixture of a colour stimulus and a white (colourless) stimulus; the latter determines luminance, or 'brightness arousing capacity' according to Bailey (1980). Luminance can be varied without changing the chromaticity or colour attributes of a source, but changes in the spectral characteristics of a source affect chromaticity (hue and saturation). Great care must be exercised in colour vision examination to control the luminance element and in colour perimetry the distinction between intensity discrimination and colour discrimination must be made. As the test object is moved from the periphery to the central visual field its hue, saturation and brightness will appear to change.

Until very recently perimetry for coloured targets rarely involved more than the plotting of isopters. The shapes of the coloured isopters are traditionally accepted to be horizontal oval forms (Harrington 1971) with the red and green loci tightly positioned in the field and the blue isopter often more widely spaced. Dain et al (1980b) however, noted their circular form. Abney (1913) reported that green gave the smallest isopter, followed by red and then blue, but Wentworth reported the order from smallest to largest to be blue, green, red (see Dubois-Poulsen 1952). Wide variability in the colour specification, size and luminance of the targets probably accounts for these differences, for the field size tends to increase linearly with stimulus diameter. Ferree and Rand (1919) found that the isopters of all colours can be extended if large targets of high luminance are used. Traquair (1948) noted that colour targets are typically less 'intense' than white ones of the same size, and from this observation developed the belief that colour targets are similar to white targets of low intensity. More recent studies conducted for equally bright targets are in the order (smallest to largest) yellow, green, red, blue (Carlow et al 1976a,b). Moreland and Cruz (1959) extended earlier ideas of the colour isopters and a summary of the present view is that 'dichromatic' vision dominates the mid regions of the retina (20–40° from the fovea), that trichromatism exists more centrally and that the periphery tends to a form of monochromatism. Yet there are conditions where high intensity colour is recognised near the periphery of the field; thus Duke-Elder (1971) gave bitemporal 'colour hemiopia' as a possible result of pressure on the optic chiasma. There are probably variations between the nasal and temporal aspects of the field, as shown by Davidoff (1976a,b). Milliken (1978) found differences in the saturation discrimination of retinal quadrants.

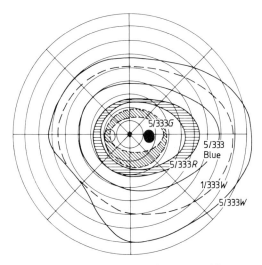

Figure 1.58 Right visual field plot showing normal isopters as obtained with a typical tangent screen. Isopters for 5 mm white and 1 mm white at $\frac{1}{3}$ m are shown. The blue 5/333 coloured isopter is less likely to depart from the values shown but the 5/333 red and green isopters may easily vary within the shaded areas, according to test conditions.

1.24.2 The need for standardisation

Great variability in the stimulus parameters and mode of operation of perimeters has been a major criticism of the diagnostic technique of visual field examination. This includes the chromaticity of the test object, its luminance and size and the intensity of the background on which it is presented. A research group on standardisation of the International Perimetric Society was formed in the 1970s and standard specifications for construction and modes of operation have now been published (ICO 1979) for use by manufacturers, clinicians and researchers. This helps to ensure the much needed uniformity of practice.

Automatic perimeters are built to the specifications of the individual manufacturer, but cooperation between manufacturers has not yet achieved a degree of standardisation. Standardisation is especially important where colour targets and backgrounds are used.

Although recommendations for standardisation of surface colour targets, for use in kinetic modes, have recently been made (Dain *et al* 1980b); this approach has largely been superseded by automatic methods involving static perimetry. The use of 'monochromatic' sources with automated perimeters such as the Goldmann and Tübingen, has provided a new impetus for the examination of 'colour fields'. Bailey (1980) notes that, according to Verriest, a provisory agreement has been made with Professor Goldmann (Goldmann perimeter) and Professor Harms (Tübingen perimeter) to fit, upon request, test and background colour filters of specified spectral composition and to provide for radiometric and photometric measurements.

1.24.3*   Increment thresholds on white backgrounds

When a colour target is used in perimetry it is important to distinguish between judgments involving intensity discrimination and those involving colour discrimination. In modern static perimetry, using coloured targets which are monochromatic or near-monochromatic presented on a white background, the target luminance is increased until the threshold is reached. Complications arise because the target *may* appear colourless at threshold (achromatic detection threshold) initially, with hue being perceived only when the luminance is further increased to the chromatic recognition threshold, or when a mobile target is moved more centrally. The colour threshold is often coincident with and sometimes even below the achromatic value. It marks the transition from target detection to recognition. The difference (photochromatic interval) depends on many factors, among them the wavelength of the target, the retinal area on which it is projected, target size and exposure and the state of adaptation of the eye. At the fovea the achromatic and chromatic thresholds are very close and the interval is especially small for long wavelengths. The variability in reaching the chromatic threshold is greater than for the achromatic threshold. Psychological factors also play a role, particularly expectation, for thresholds are generally lower if the observer is told to expect one of two hues rather than one of ten possible colours. Engelking and Eckstein (1920), Ferree and Rand (1924) and Ferree *et al* (1931) attempted to separate possible luminance involvements in the chromatic threshold. They flashed colour targets against a grey background of equal luminance. The observer in this case reported when the target colour differed from the background (grey) colour. This method was not universally adopted, perhaps because it was considered too complicated for routine perimetry (see Aulhorn and Harms 1972).

Zanen (1953) deserves credit for introducing both achromatic and chromatic static absolute foveal thresholds with monochromatic targets as a clinical method for distinguishing ocular pathology. He noted that chorioretinal diseases displayed themselves by an increase in the achromatic threshold, especially for reds. Optic nerve diseases frequently showed an increase in the photochromatic interval—the difference between the achromatic and chromatic thresholds. Verduyn-Lunel and Crone (1974) used monochromatic stimuli equated in brightness to a white background, so that visibility of the target was controlled by colour rather than by brightness differences.

1.24.4 Practical precautions

When comparing normal and pathological colour fields careful attention must be given to the stimulus conditions and the nature of the threshold—achromatic or chromatic. Typically, however, specification is confined to identification by the observer of the dominant hue (blue, green or red). Colour names should be avoided, where possible, because the subjective appearance of colour varies widely depending on its location in the visual field. Other precautions include calibration aspects and procedures, ably summarised by Verduyn-Lunel and Crone (1974). With our own additions and standardisaton recommendations drawn from ICO (1979) practical precautions are listed below.

(a) Stimuli should be exactly defined spectral sources from a monochromator or interference filters. A low pressure sodium lamp is a good example.

(b) Testing should be at high photopic intensity levels to activate cone receptors only.

(c) Test and surround luminance should be identical. This requires equalisation at each retinal position before threshold measurements are commenced because the effective luminance of colour targets is based on the photopic spectral sensitivity for cone vision and allowance must be made for rod involvement in the periphery. Hess (1889) was aware of this need. Ferree and Rand (1919) made their targets visible against the background only on the basis of colour and found them to be more sensitive than white test objects for many visual field defects. Variations in the V_λ response for different perimetric angles are noted (see Weale 1953b) so each position must be tested individually.

(d) Colour naming should be avoided because of the different subjective appearance of colours in the central and peripheral fields (Hedin 1979). Naming contributes significantly to variation in the size of colour fields among normal observers.

(e) The luminance of the coloured targets and surround or background should ideally be specified in physical rather than psychophysical units (Aulhorn and Harms 1972).

(f) Spectral transmission factors of neutral density filters used to control luminance should be known.

(g) The observer's task must be specified and made clear, whether it be a detection of luminance (achromatic threshold) or a colour recognition threshold, involving either a hue and/or saturation change. Verduyn-Lunel and Crone (1974) believe that the measurement of purity or saturation threshold (the first step from white) is the best colour threshold for comparison with achromatic perimetry.

(h) Both target and background field luminance should be specified to the centre of the entrance pupil of the eye (ICO 1979).

In the future it is anticipated that stricter recommendations in terms of CIE specifications for coloured targets and spectral specification in the plane of the entrance pupil of light sources and stimuli (including properties of filters), may be made (ICO 1979).

Dutch researchers have designed an instrument which incorporates many of the design and calibration requirements outlined above (see Verduyn-Lunel and Crone 1974). Dutch and Belgian workers have calibrated existing models, such as the Goldmann and Tübingen perimeters, for accurate colour perimetry (Francois *et al* 1964, Verriest and Israel 1965, Verriest *et al* 1974). These have been used to study observers with normal colour vision, inherited colour deficiency and acquired defects (Verriest 1960b, 1963, Verriest and Israel 1965, Verriest and Uvijls 1978). The red peak was shown to be missing in inherited protanopia, the green peak was broader in deuteranopia and the blue peak was missing in tritanopia. In blue–yellow acquired defects the sensitivity measured for blue targets is reduced and occasionally this is seen also for red targets. In some acquired red–green defects, blue targets are enhanced and red objects depressed in sensitivity.

With care, the variability in results for achromatic thresholds using coloured targets with the Goldmann static perimeter is no greater than with white test targets.

Verriest, writing in Pokorny and Smith (1979) considers the Tübingen perimeter more suitable for static increment thresholds than the Goldmann instrument. The Tübingen instrument normally has a background luminance of 3.18 cd m^{-2}.

The first study of detection thresholds using different monochromatic sources was by Wentworth (1930) and involved the projection of the target directly into the eye. Nolte (1962) first projected a monochromatic target on to the the surface of the perimeter. In both cases the centre of the retina (fovea) was shown to have a marked reduction in sensitivity for blue light compared to other wavelengths. The parafoveal zone was the most sensitive for short wavelengths. These findings are consistent with other psychophysical measurements (e.g. 'foveal' tritanopia) and with the histochemical studies of Marc and Sperling (1977) of the density of cones responding to short wavelengths in the fovea and parafovea.

1.24.5*   Clinical uses

As Bailey (1980) discussed, in a valuable review of the status of colour fields today, modern perimetric techniques assume that most ocular diseases impair both rod and cone function. However, the level of background illumination in most perimeters allows both rods and cones to function, so that a loss of cone function may not affect the peripheral visual field. Even at the highest background level available in the Tübingen perimeter, Sloan and Feiock (1972) found that rods are slightly more sensitive than cones to a white test object. Substituting a red target of equal luminance, however, effectively separates rod function from cones, because cones are more sensitive than rods to red light. Such a distinction allows pathological cone scotomata to be detected with the red target while visual fields are normal with a white target of equal luminance. Clearly coloured targets have a value for diagnostic purposes. Verriest and Uvijls (1978) measured the achromatic threshold in 120 pathological eyes using a modified static Tübingen perimeter. The technique was considered more sensitive than static perimetry for white test objects, and was even superior to visual acuity measurement and clinical colour vision tests for the examination of foveal vision.

Coloured perimetry using white backgrounds thus has very clear advantages over traditional colour vision testing using clinical tests based on pigments and confusion colours, as Verriest and Metsala (1963) showed that these more traditional techniques provide reliability only as far as 30°. Marmion (1977) found that in exudative diabetic retinopathy, static perimetry with coloured targets was more sensitive than even the FM 100 Hue test. Insensitivity to colour on account of age was measured by Maione and Carta (1972) using a modified Goldmann perimeter to allow achromatic isopters for coloured targets to be plotted. The marked loss of sensitivity to blues with age was particularly noticeable and clearly this method of examination is a sensitive one.

1.24.6   Use of perimeter-based instruments

The need for a simple portable apparatus for measuring selective disturbances of the colour mechanisms in both inherited and acquired defects was realised in the early 1970s. Although such methods are never likely to be routinely employed for evaluating the inherited defects, the potential for studying ocular disease was realised by European ophthalmologists, principally Verriest, Hansen and Crone. In many eye diseases selective defects

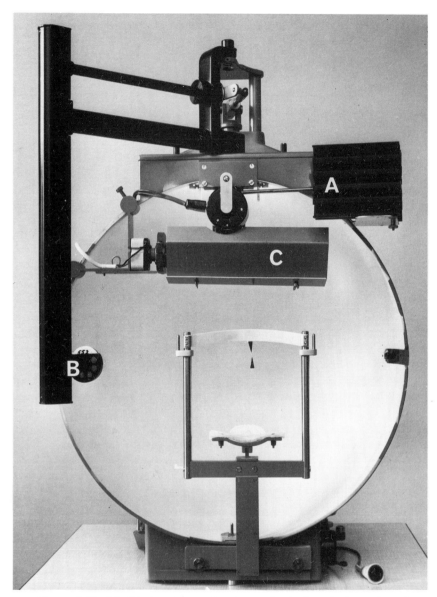

Figure 1.59 Modified perimeter showing method of producing coloured background (A and C) and filter wheel (B) used for stimulus. Photograph courtesy Dr E Hansen, Oslo

of the colour mechanisms are typical. Identification can lead to a greater knowledge of the underlying causes and a more efficient diagnosis. Using coloured targets it is possible to study three varieties of retinal damage (*a*) rod or cone dysfunction, (*b*) dysfunction of specific cone classes and (*c*) colour-opponent damage. The establishment of clinical thresholds to mark the distinction between normality and disease is clearly a desirable goal.

Isolation of the colour vision mechanisms requires relatively high luminances particularly for the coloured background adapting field. Calibration of the test and background wavelengths is required. Hansen (1974a) modified and calibrated the Goldmann perimeter to provide backgrounds of three broad-band colours and test stimuli of nine narrow wavelengths. Static achromatic thresholds were obtained at different positions in the visual field and spectral sensitivity curves indicating the Stiles π mechanisms were obtained (see §7.11).

1.25 BLUE DEFECTS IN NORMAL VISION

1.25.1 Small-field tritanopia

There is a tendency for normal colour vision to become tritanopic for foveal fields of 20' of arc or less, at low luminance levels and under special conditions of adaptation; to these the terms small-field, threshold and transient tritanopia, respectively, have been applied. König and Köttgen (1894) first described the relative insensitivity of the central fovea to short wavelengths, while Willmer (1944), unaware of their publication, 'rediscovered' the tritanopic confusions of small fields. Willmer and Wright (1945) made further observations, noting the dichromatic matches made by normal observers, which were similar to those of inherited tritanopia. Blue–green vision was seen to be impaired, with neutral points identified in the spectrum at 578 and near 410 nm (see the discussion on field size, p. 81).

Originally the phenomenon was thought to be confined to the central fovea but Hartridge (1945a,b) and Thomson and Wright (1947) proposed the concept of 'small field' tritanopia, noting that tritanopic colour matches for small fields could be made at retinal positions 20 and 40' of arc from the fovea, provided steady fixation was maintained. Since steady viewing is a necessary requirement, Brindley (1970) was led to suggest that the tritanopic effects in normal observers might be manifestations of the Troxler effect.

There has been much discussion as to whether the effect is confined to the central fovea and about its physiological basis. The relative scarcity of 'blue' receptors, particularly in the foveola (0.25 to 0.5°), established from histological studies by Marc and Sperling (1977), and the complete absence of blue-sensitive cones in a sample of 82 foveal cones examined by

Bowmaker *et al* (1978) using microspectrophotometric techniques, tend to support the view that low 'blue' cone density is a cause. However, Brindley (1954a) and Ruddock and Burton (1972) support a neural convergence theory.

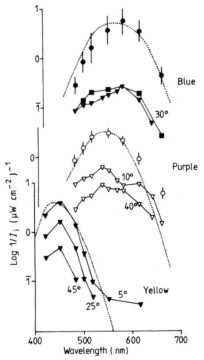

Figure 1.60 Spectral sensitivity of normal subjects measured against blue, purple and yellow backgrounds. Blue light (using a Wratten 47 filter) is of 165 lux. Purple light (using Wratten 34 A) is of 200 lux and yellow (from a low pressure sodium lamp) of 2300 lux. Mean values of seven normal males are indicated by circles with the variation (± 1 standard deviation) shown by the vertical bars. Broken lines indicate the red, green and yellow 'primaries' after Walraven (1974). The angular size of the target is 54'. Courtesy Dr E Hansen, Oslo.

Many studies have indicated that tritanopic effects can be experienced in parafoveal regions for small fields, but Wald (1967) maintained that the characteristic is confined to the central fovea, arguing that 'blue cones are well represented in non-foveal regions'. Weitzman and Kinney (1969) failed to establish total tritanopic matches for small fields in the parafoveal regions, except for confusions between green and blue which are typical of tritanopia. In fact 'deuteranomaly' described the effects they found more adequately than 'tritanopia'. They suggested a possible enhanced sensitivity to blue in the periphery, coupled with a decreased sensitivity to green which could produce such deuteranomaly. Moreland and Cruz (1955) were able to confirm small-field tritanomaly or tritanopia in the peripheral retina. Stiles (1949a) had earlier noted that the increment threshold of the π_1 (blue) mechanism is higher in the central fovea than $\frac{1}{2}°$ away, suggesting a foveal tritan anomaly. Since by the nature of its coarse spatial and temporal resolving characteristics, the short wavelength mechanism is unable to detect fine detail, tritan features in some parts of the retina are not surprising; for examples see Brindley (1954a) and for the inferior contrast sensitivity Green (1968, 1969), Kelly (1974) and Cavonius and Estevez (1975a).

A recent study by Williams *et al* (1981) established that foveal tritanopia is the result of the absence of functioning short wavelength cones. They conducted a series of experiments in which increment threshold measurements were carried out under conditions which isolated the medium and short wavelength mechanisms in turn, concluding that a central region 20–25′ of arc in diameter showed blue insensitivity; this could be explained neither by the Troxler fading effect nor by a increase in the density of 'macular' pigment in the fovea. The region involved was shown to be considerably larger than the central 7–8′ of arc foveal area described by Wald (1967) but these experiments do not rule out the possibility that such blue insensitivity is a characteristic of small fields elsewhere in the retina.

Changes in normal colour vision for small fields do not appear to be confined entirely to blue vision since several investigators have noted a change in the red–green responses for foveal fields (see Willmer and Wright 1945, Horner and Purslow 1947 and Ruddock 1969). Hartridge (1945a) examined the effect of target size, background luminance, target luminance, size and colour contrast on colour recognition, noting the point at which colour disappeared for different conditions. An increase in the target intensity allowed the visual angle at which loss of colour was noticed to be reduced. Luminance was less effective than visual angle in influencing hue perception, since a ten-fold increase in intensity was approximately equivalent to a doubling of the visual angle.

Many investigators have stressed the importance of steady fixation for the observation of these effects. McCree (1960a) and Ruddock and Burton (1972) considered small-field tritanopia to have different properties from those of the 'foveal' phenomena.

1.25.2 Transient tritanopia

When the eye is adapted to yellow light of high luminance the sensitivity to short wavelengths decreases markedly for a short period (5 to 10 seconds) following the removal of the adaptation source. This phenomenon is called transient tritanopia. On the basis of psychophysical experiments Mollon and Polden (1975, 1976, 1977) suggested as an explanation of the phenomenon that the blue cones are inhibited by the red–green cones.

1.25.3   Threshold tritanopia

Farnsworth (1955b) put forward the hypothesis that reducing the luminance of a field was equivalent to reducing the field size at constant luminance. However McCree (1960a) reported significant differences between threshold tritanopic effects, small-field and foveal effects. The results of variations of field size at three different luminance levels and the effect of varying luminance for five different field sizes were investigated with respect to hue discrimination.

2 Colour measurement and specification

Physical properties of coloured objects are related to psychological descriptions and physiological processes. Colour science, including standardised description and classification, is one of the concerns of the Commission International d'Eclairage (CIE). Numerous texts at various levels are available, but difficulty is often expressed by those who need to understand colour vision tests and the materials from which they are made; usually an elementary introduction to the terms and diagrams is sufficient to enable appropriate literature to be understood. Consequently the present outline is considerably simplified, in the light of experience with those who need an 'introduction', but it suggests sources of information and material which may be of more interest once the subject has been appreciated at an elementary level.

Newton published his first paper on the nature of coloured light in 1672 marking the beginning of the scientific era of the investigation of colour (see Newton's (1730) classic text *Opticks*). His demonstration of the dispersion of white light into the seven spectral colours (named by his assistant) with simple prisms indicated to Newton that the rays of light were not of themselves coloured but had a 'disposition to exhibit this or that particular colour'.

Newton was aware of 'primary' hues, which he called red, yellow, green, blue and violet-purple, with orange and indigo and 'an indefinite variety of intermediate gradations'. Simple colour mixing led him to the formation of white by the mixture of primaries, although he did not emphasise the concept of secondary colours. However, he equated the mixture of lights (additive) and pigments (subtractive).

Newton proposed the first colour diagram, a circle with colours surrounding white in the centre. Sherman (1981) mentions Newton as having made contributions to other aspects of colour such as saturation and desaturation and the varying apparent brightness of different hues.

2.1*   BASIC TERMS

An appraisal of assorted coloured objects soon reveals three ways in which they can be related or differentiated.

Assuming six coloured objects, respectively purple, green, yellow, red, orange and blue, we note that each differs in 'hue' from all the others. Since yellow appears 'more like' orange and green than the others, and so on, a logical colour circle (or arrangement of hues) can be made. Thus red, orange, yellow, green, blue and purple make a suitable circle, where purple meets red again (see figure 2.1 (*a*) and (*b*)). Figure 2.2 shows how purple does *not* appear in the solar spectrum formed by a prism; the circle is broken if 'spectral lights' are considered, but each has a distinct hue, or 'dominant wavelength', and a succession of names can be given to the major hues in a spectrum.

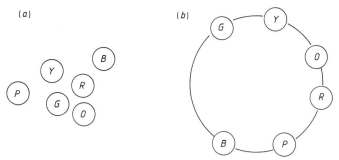

Figure 2.1 (*a*) Six typical colours in random positions. (*b*) The six colours in an order on a colour circle, allowing space for other colours, such as a blue–green.

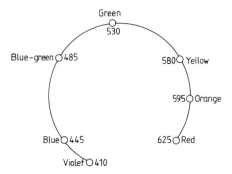

Figure 2.2 Seven spectral colours arranged on a spectral locus with appropriate (approximate) wavelengths in nanometres.

2.1.2 Purity or saturation (Munsell 'Chroma')

If red light from one projector is superimposed, on a white screen, upon green light from another projector a yellow mixture is produced, but this yellow is not as pure as its counterpart of the same dominant wavelength. A third, blue, projector can make the mixture 'white', provided the proportions are correct, which is the ultimate in 'desaturation' of the colour components. These mixtures appear in figure 2.3 which is a simple colour diagram, produced as a convenient triangle by Maxwell (1860) (see Judd 1961b). The purest natural colours, parts of the spectrum, form an outer boundary, any part of which can be desaturated or made less pure by superimposing white light. Coloured pigments when mixed as paints do not mix 'additively', as lights do, but white pigment mixed with a fairly pure red paint will produce a desaturated red. On the colour map these are located between the 'white centre' and the approriate 'pure' point on the perimeter. Accordingly the percentage of light of the dominant wavelength measures the extent of a colour's saturation.

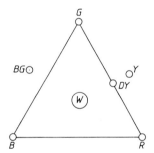

Figure 2.3 A simple colour triangle, R, G, B, Y and BG represent spectral lights, while DY is a desaturated version of Y.

2.1.3 Luminosity, brightness or lightness (Munsell 'Value')

White, grey and black are all equally 'neutral' and lacking in hue. They can all be located in the centre of the colour map but a two-dimensional map does not do justice to their differences in lightness. A printer may describe them as having different 'tones'. A three-dimensional solid now appears with white, grey and black each on an axis of their own, representing increasing 'luminosity' of three different neutrals. Figure 2.4 shows this with two arcs of 'spectral lights' corresponding to two different luminosities; an arc shrunk to nothing is located with black as an approximation for 'zero luminosity'. Two reds, each of the same hue and saturation, R and r are shown at different levels of luminosity.

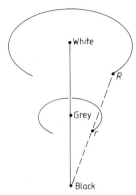

Figure 2.4   The three dimensions of colour emphasising luminosity or Munsell 'Value'.

2.1.4   Chromaticity

Frequently a colour is described in terms of both dominant wavelength and saturation, thus locating it on a two-dimensional colour diagram; the noun 'chromaticity' is used. This applies specifically to a CIE approved diagram, in which *xy* coordinates belong to each point, based upon physical measurements of the distribution of spectral energy and with reference to an assumed 'standard observer'.

2.1.5   Variations

A phenomenon named after Bezold and Brucke, who described it over a hundred years ago, is revealed as the intensity of most coloured lights departs from normal levels. Reds characteristically appear more yellow at higher intensities, yellower greens changing to yellow. Blue and green are 'invariable' as shown by Purdy (1931). Hartridge (1950) pointed out the involvement of a purple.

   Adaptation has many effects on colour appearance, many being classified by Burnham *et al* (1963) and juxtaposition of different combinations of colours produces perceptual variation. Le Grand (1968) and others have described, for instance, the 'Helmholtz–Kohlrausch' effect, involving different luminances.

2.2   THE MUNSELL SYSTEM, RELATED TO COLOUR VISION TESTS

A system widely used in industry and education was invented by Albert H Munsell between 1900 and 1918, as described by Nickerson (1940, 1969).

The Munsell system was developed by others, chiefly through industry and after 1942 by a corporation with trustees. Speciemens of painted graded paper, with either matt or gloss surface finish, form the backbone of large collections, classified in a distinctive manner. Matt papers (or 'chips', if about one or two centimetres square) form the basis of many colour vision tests, such as Farnsworth's. These papers are soiled relatively easily and are not cheap, but can be obtained with a statement of their spectro-photometric properties according to the CIE system to be outlined later (see Glenn and Killian 1940). Thus many of the familiar standards, such as 'paint cards' are described in Munsell terms. The Royal Horticultural Society's *Colour Chart*, which enables many plant colours to be matched, and British Standard BS 381C:1964 give suitable approximations in 'Munsell notation'. Granville *et al* (1943) gave data for about 1000 extra colours.

2.2.1 Munsell notation

The three attributes of colour are used in the order *hue, value* then *chroma*. Each has a numerical designation, according to the sample, a distinctive spacing and the insertion of an oblique stroke (/) between the 'value' and 'chroma' items. For example, a basic red colour is 5R 4/10, sometimes abbreviated to R 4/10, in which the hue is the one known as 5R at a value 4 and a chroma 10.

Greys or neutrals are called N, this letter being followed by a number indicating the Value of the neutral sample. Thus on a scale of 1–10, N5 represents a mid-grey.

Hues (each numbered on a scale of 1–10) are ranged round a circle and given distinctive letters. Principal hues are 5R, 5Y, 5G, 5B and 5P. Intermediate hues are 5YR, 5GY, 5BG, 5PB and 5RP. In this circle complementary hues (those which give neutral when additively mixed) lie opposite each other and neutral is central, as in figure 2.5. The positions of these ten hues are equally spaced, visually (with constant value and chroma), and all parts of the notation can be subdivided decimally.

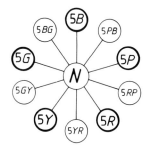

Figure 2.5   Five principal Munsell hues 5R, 5Y, 5G, 5B and 5P.

A three-dimensional aspect is added by including 'Value' which ranges from 1 to a practical limit of 9. Although many desaturated colours can be produced to fit into this Munsell solid, the limitations of colour chemistry restrict the number of higher value and/or higher chroma elements so that an asymmetrical envelope limits the solid, as seen in figure 2.6.

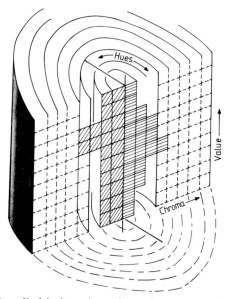

Figure 2.6 A cylindrical section of a colour space showing the three colour attributes of hue, value and chroma as used in the Munsell system.

The notation assumes that a standard, relatively blue phase of daylight is used (illuminant C, as described in Chapter 6), that the light falls at an angle of incidence of 45° to the surface and that the observer's line of sight is at right angles to the surface.

Given a Munsell notation for a colour, tables or graphs can be used to arrive at a specification in another colour system (e.g. CIE xyz) fairly accurately.

There is a reasonable relationship between Munsell Value, as subjectively appreciated, and percentage reflectance in physical terms (see Richards 1966).

2.3 COLOUR MIXING

2.3.1 The colour top

Maxwell's (1860) use of spinning discs of coloured paper was extended when Munsell developed his system. Coloured sectors are provided by interleaving discs with radial slits. These are spun fast to remove flicker. The angle of each coloured sector is proportional to the percentage contribution of that colour to the (additive) colour mixture (see figure 2.7).

Two discs, one black, one white, can be used when matching a neutral of unknown value and it is convenient to make a smaller central disc of the unknown sample, or hold it close to the mixture disc. Four discs can be used to match a sample colour and may include two with highly saturated colours, one neutral 'lighter' than the sample and one neutral 'darker'. The two coloured discs are ideally near to but on either side of the position of the sample on a colour diagram.

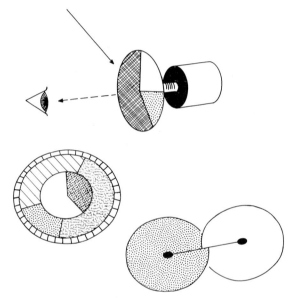

Figure 2.7 Maxwell's discs, spun on a motorised spindle. A scale indicates the percentage of each component colour in the mixture. The bottom right-hand diagram shows the discs being interleaved before rotation.

The disc method of colorimetry still has practical use, as described by Nickerson (1946), if carried out with discs of Munsell matt paper of known specification in CIE terms. By weighting the 'tristimulus' data according to the proportions of the discs used for a match it is possible to calculate the specifications of the sample.

2.3.2 Complementary colours

Two coloured stimuli mixed, as in a colorimeter, which then match white (or neutral) are called complementary colours. These pairs are of particular interest in colour vision testing. Examples of complementary colours are

> Blue, wavelength 460 nm complementary to yellow (570 nm)
> Blue–green, wavelength 490 nm complementary to red (610 nm)
> Green, wavelength 520 nm complementary to a mid-purple.

2.3.3* Colorimeters

Maxwell made quantitative mixtures of spectral lights with an ingenious prism device and descriptions of colorimeters are legion. Texts by Wright (1964), Le Grand (1968), Judd and Wyszecki (1975) and Grum and Bartleson (1980), among many others, can be consulted for details. Houston's (1932) relatively early method of mixing several 'reference stimuli' (sometimes called 'primaries', a term deprecated in BS 1611:1953) lends itself to applications both simple and complex. He used internal reflection in a relatively long wedge of glass, where separate lights entered at one end to emerge as a mixture at the other end. Burnham (1952) described an optical integrating bar in a simple design where four colours (light passing through four filters) could be mixed. Hunt (1954) designed a tricolorimeter in which filtered lights were mixed in a tunnel of four mirrors with diffusers at each end; this device provided 'desaturation' stimuli. The principle of a small colorimeter to provide a mixture suitable for matching many desaturated colours and based on the instruments mentioned, is outlined in figure 2.8. The similarity, in essence, to designs such as that by Donaldson (1935) is evident.

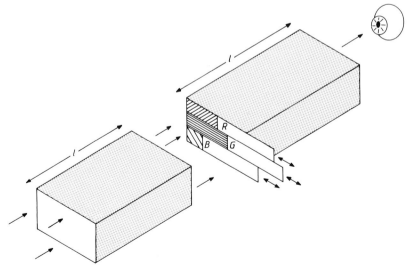

Figure 2.8 Simple mirror box tricolorimeter. Diffusing screens at each end of 'mirror tunnels' transmit light to the eye. Sliding shutters restrict the proportions of the three coloured filters in use.

Note that filters, placed successively in a single beam of light, colour the light in a 'subtractive' way. A mixture of paints also does this, by selective absorption. A visual tintometer described by Lovibond in 1887, Schofield (1939) and in a modern form by Chamberlin and Chamberlin (1980) uses filters to good effect.

Spectral stimuli have particular attraction. Three pure spectral lights can be mixed to match a good range of colours, as will be more evident later. Various designs using prisms have emerged, for example that of Wright (1946) and earlier by Helmholtz as well as diffraction gratings by Shirley (1966). The well tried method of spectroscope 'constant deviation' optics used by Fry (1981) deserves more widespread use, particularly as it can be adapted comparatively simply and cheaply.

Bouman *et al* (1956) constructed a very simple colorimeter specifically to study defective colour vision and this can be adapted in several ways. Mixing spheres illuminate two plates to give them a coloured appearance, one (uniform) plate being viewed through a series of very small holes in the other plate. At the viewing distance used the holes cannot be resolved but their colour contributes to desaturation of the perforated plate's colour.

Three laser monochromatic stimuli (441.6, 514.5 and 632.8 nm) have been used by Krauskopf *et al* (1981) in a device where the colours are mixed by a special diffusing suspension of tiny balls in a liquid.

2.3.4 Mixing three spectral lights

The literature of colorimetry describes how any colour can be matched by an additive mixture of three suitable coloured lights, within certain limits (for instance, two of the three lights may *not* combine to match the third). Also, the more saturated are all three, the larger is the range contained within their triangle on a colour diagram. Within this triangle there is no need to desaturate the sample. Spectral stimuli are therefore attractive and they can be obtained from dispersing prisms, by diffraction, or by interference filters. The working of monochromators (ideally double monochromators so stray light contamination is eliminated) is explained by Crawford *et al* (1968) and in many optical textbooks. Such monochromators can be combined in optical systems. More simply, interference filters (described by the same authors) offer cheap and approximately monochromatic sources, filtering tungsten or arc lamps. Regan and Tyler (1971) adapted a graded interference filter as a wide bandpass *double* monochromator and Wright (1972) showed how an interference filter can be inclined at different angles in a plane polarised light beam; this gives a source of near monochromatic lights of different peak wavelengths and is another simple approach.

Tristimulus values

A tristimulus set of 'reference' stimuli consists of red, green and blue monochromatic spectral lights. The set used by Wright (1946), which comprised wavelengths 460, 530 and 650 nm, is typical. A different mixture of the three can be used to match each part of the spectrum, in turn, giving the type of result shown in figure 2.9. A 'negative' value represents the amount of one of the reference stimuli (RS_1) which is often needed as an addition to the spectral light being used as a sample (SL) to be matched, in order to provide a mixture (RS_1) + (SL) sufficiently desaturated for a visual match to a combination of the other two reference stimuli (RS_2 and RS_3). This can be expressed by the equation

$$RS_1 + SL \;\rightleftharpoons\; \text{(Matches) } RS_2 + RS_3$$

although this ignores quantities. Bringing all the reference stimuli to one side it is rewritten

$$SL \;\rightleftharpoons\; RS_2 + RS_3 - RS_1$$

which expresses the 'negative' element, actually a real amount added to the measured sample, optically, in the colorimeter. Each of the three spectral lights can be matched by 100% of itself and zero contribution from the others; the sum of the three is always 100% and obviously their relative proportions are important. In figure 2.10 this is shown by a simple colour triangle formed by three reference stimuli R, G and B, where the part of the locus of spectral colours between blue and green is well outside the triangle. If the colorimeter used filters they, being very desaturated, would form a smaller triangle and always would need one negative element when matching spectral lights. A typical blue–green part of the spectrum, say with wavelength 500 nm, requires a mixture of some red before it can be matched to appropriate proportions of such blue and green stimuli.

It is possible to use completely theoretical reference stimuli (for example X, Y and Z as used by the CIE) each assumed to be more saturated than a real stimulus light so that the spectral locus is contained within their triangle and no negative quantities are needed for calculation. Any set of filters or spectral lights can be related to this $X Y Z$ system, mathematically.

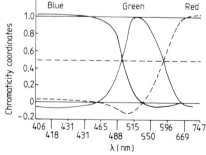

Figure 2.9 Spectral coordinates of the contributions of three sources, respectively of wavelengths 460, 530 and 650 nm (after Wright 1964).

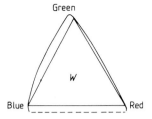

Figure 2.10   Colour triangle as developed by Maxwell, giving the locus of spectral colours. The broken line shows purple colours.

The *relative* amounts of the three reference stimuli in any mixture are expressed as

$$x = \frac{X}{X + Y + Z}$$

$$y = \frac{Y}{X + Y + Z}$$

$$z = \frac{Z}{X + Y + Z} \; .$$

A chromaticity diagram

In figure 2.11 the usual *xy* coordinates are used and *z* can be found for any point within the triangle by noting that $x + y + z = 1$ and $z = 1 - (x + y)$ so that only *x* and *y* need be used to locate a colour.

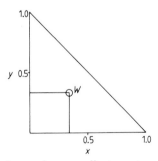

Figure 2.11   A basic form of *xy* coordinate system colour triangle, upon which the spectral locus can be plotted as in figures 2.13 and 2.15.

Descriptions of this CIE chromaticity diagram, in different degrees of detail, have been provided by Murray (1952), Hunt (1957), Bergmans (1960), Le Grand (1968), Wright (1969), Bouma (1971), Judd and Wyszecki (1975) and many others. Modifications of details of the system are published by the CIE from time to time.

To calculate the *xy* chromaticity coordinates of a colour sample, such as a piece of Munsell paper, a series of steps take place which take into account at different parts of the spectrum (*a*) the energy distribution of the light illuminating the paper, (*b*) measurements of the energy reflected from the paper, (*c*) the sensitivity of a standard observer's eye to the spectrum and (*d*) colour matching functions, assumed to apply to a standard observer's eye. These relative tristimulus values of the spectral components of an equal-energy spectrum (shown in figure 2.12) are called $\bar{x}$, $\bar{y}$ and $\bar{z}$, for the theoretical stimuli *X*, *Y* and *Z*. They are used when the visual angle involved is approximately $2°$ and are standards called 'distribution coefficients'. The $\bar{y}$ also represents the 1924 CIE $2°$ curve.

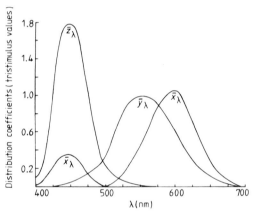

Figure 2.12   Tristimulus values, related to an equal energy spectrum, for the CIE standard observer's responses, as specified for a $2°$ field in 1931.

As Richards (1966) pointed out, there are likely to be linear relationships between the $\bar{x}$, $\bar{y}$ and $\bar{z}$ colour mixture functions and the spectral sensitivity of retinal cone mechanisms.

Standard data for an individual colorimeter are combined with the results of the match made with the paper sample to compute the *x* and *y* coordinates needed. These can be transformed into any other system by using a formula or tables. At this stage it is assumed that we are dealing with colours of a particular level of luminosity.

Munsell colours and the xy *diagram*

Using data for many Munsell samples at a given 'Value' such as Value 5, it is possible to locate them according to Hue and Chroma on the *xy* diagram; useful diagrams have been published by Glenn and Killian (1940) as well as by Granville *et al* (1943) and by the Munsell Corporation, but enlarged versions are necessary for accurate work. The data are tabulated, often with respect to alternative light sources which have different spectral distributions of energy. Figures 2.13 and 2.14 show such diagrams.

With reference to illuminant C, some typical Munsell papers have the characteristics set out in the table below.

Nominal notation	X	Y	Z	x	y	Measured notation
2.5R 4/6	16.11	12.05	10.89	0.4125	0.3085	2.5R 4.01/5.95
5Y 4/3	11.46	11.82	06.91	0.3795	0.3915	5.1Y 3.97/2.85
10G 3/4	04.58	06.55	07.25	0.2491	0.3563	10G 3.00/4.2
5BG 5/6	14.14	19.72	26.63	0.2338	0.3260	5.1BG 4.99/6.0
7.5RP 5/12	32.28	19.98	22.55	0.4314	0.2670	7.5RP 5.02/12.0

Certain 'lines of constant hue' are shown in figure 2.13, radiating from the point representing neutral. These curved lines, joining points which share the 'dominant hue' at which each line reaches the spectrum locus, are slightly different according to the 'Value' which is being considered.

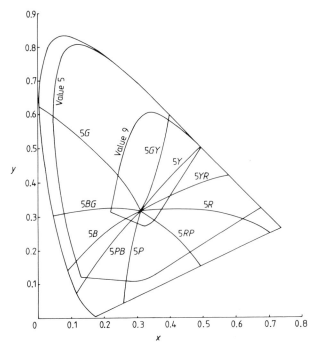

Figure 2.13   Munsell main hues plotted on CIE *xy* system indicating two 'Value' levels by loci within which it is possible to obtain Munsell chips at those Values. Lines of 'constant hue' (for a Value of approximately 5) radiate from the neutral point.

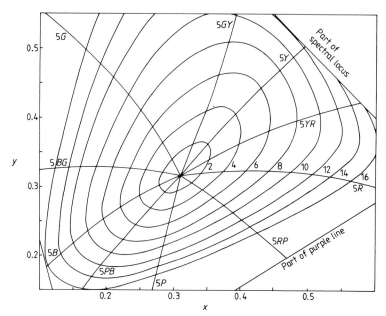

Figure 2.14   Loci of constant chroma (between Chroma 2 and Chroma 16) and the ten major constant hue loci, all at Value 5. The ellipse for Chroma 4 gives the approximate locus for the sixteen 'caps' used in the Farnsworth D.15 test and the 85 'caps' used in the FM 100 Hue test.

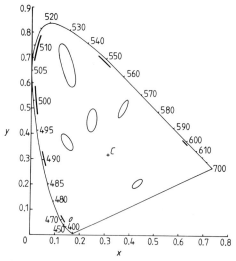

Figure 2.15   Six lines are shown which approximately represent equal steps of subjective discrimination (magnified ×3) as shown by Wright (1946). Five MacAdam ellipses (magnified ×10) are shown. Some idea of the variation of the just-noticeable-difference (jnd) in parts of the diagram is gained by assuming that an ellipse axis represents about 3 jnd.

A uniform chromaticity diagram

Dissatisfaction with the *xy* diagram is readily expressed when it is seen to display colour difference thresholds in a non-uniform way. The parts of the spectrum are arranged along the boundary of the diagram in a non-linear scale of wavelength. MacAdam (1966) has shown by a series of elliptical areas, how to express the visual tolerance limits surrounding a particular chromaticity; we do well to remind ourselves that these are, in fact, *visual* limits. Wright (1946) provided selected data for limits of discrimination and figure 2.15 indicates how equally noticeable differences assume different proportions according to the starting point. An ideal uniform chromaticity diagram would reduce the MacAdam ellipses to circles and despite limited success the work of Judd (1935) and others enabled a UCS (Uniform Chromaticity Scale) diagram to be adopted by the CIE. This projective transformation of the *xy* coordinates uses *u'* and *v'* coordinates and is shown in figure 2.16.

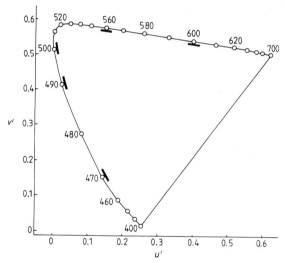

Figure 2.16   The 1976 CIE UCS ($u'v'$) chromaticity diagram. This modi-fication of the (uv) 1960 diagram shows five spectral discrimination lines, which appeared in figure 2.15. In the more recent diagram these lines are more equal in extent. Ellipses which were shown in figure 2.15 approx-imate more closely to circles on this ($u'v'$) diagram.

2.4   COLOURS IN COLOUR SPACE

Three features by which a colour can be described, for instance in terms of Munsell Hue, Value and Chroma, have been mentioned. Authors such as Bouma (1971) have shown how the results of mixing two colours can be determined by three-dimensional vectors, locating the mixture in colour space. The CIE system using X, Y and Z reference stimuli attributed all the luminosity of a colour to the Y axis, so that the X and Z axes lie on the 'alychne' a plane of stimuli of zero luminosity. At any section of the colour space contained within these three axes a chromaticity diagram of appropriate 'luminance' can be drawn and within this space the three 'real' stimuli adopted by the CIE can be traced as subsidiary axes; these intersect the spectral locus of each successive chromaticity diagram. Figure 2.17 shows a perspective view of this space.

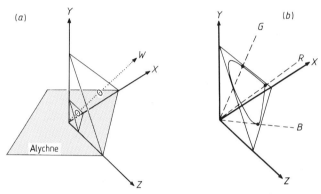

Figure 2.17 Colour space showing the projection of the CIE chromaticity diagram at different luminosities. In (a) the plane of zero luminosity is seen to contain the X and Z axes, all the luminosity being expressed by the Y axis. Two planes (each of constant luminance) are drawn as equilateral triangles, the white position axis transfixing their centres. In (b) the chromaticity diagram is seen in perspective, its spectral locus cut at the positions of three 'cardinal' stimuli having wavelengths 700, 546.1 and 435.8 nm.

Judd (1945) used the confusions and other responses of dichromats to derive functions called W_d, W_p and K to replace the X, Y and Z functions of normals; he took W_d plus K as a pair representing the vision of a deuteranope, while W_p plus K described a protanope. W described the 'warmer' and K the 'colder' elements of the dichromat's vision. Later, Judd (1948) applied the data to the question 'what colours do colour blind people confuse?' He showed how many Munsell colours could be given 'deuteranopic' or 'protanopic' descriptions. In this way a nominal 10R 5/4 would have a renotation of 5Y 5.1/2.3 for a deuteranope with a 'deuteranopic reflectance' W_d ($= Y$) of 0.2028 and a deuteranopic chromaticity coordinate, w_d, of 0.5774. The same sample would have, for a protanope, W_p of 0.1749 and w_p 0.5477, with the protanopic renotation 5Y 4.7/1.7. The data indicate the lower response to this red stimulus on the part of the protanope.

An adequate specification of the three-dimensional plot for any colour is therefore a pair of chromaticity coordinates such as x and y, as well as the extent of Y.

2.5 SAMPLES AND SETS OF COLOURS

Paint manufacturers have popularised cards on which areas demonstrate the colours of their products but there is often doubt whether these give acccurate matches of the finished painted surfaces. Viewing conditions obviously influence appearances and even photoelectric colorimeters require control of the type of illumination and the directions of incident light and its collection. Highly graded filters are available for matching coloured lights and sets of standardised surface colours are used for many purposes, including ceramic tiles. Magnesium oxide coating has been used as a standard white for many years and a clean surface of a (controlled quality) calcium carbonate block is suitable. A well known collection of samples in the *Munsell Atlas* (or *Book*) *of Colour,* available in matt or glossy finish, serves as a useful comparison chart for an unknown sample. British Standard 381C:1964, on a smaller scale, gives a set of samples designed to meet requirements of specific industries, such as paint manufacture and Supplement No 1 (1966) gives colorimetric values of BS 31C samples for different illuminants in terms of x, y and luminance factor $B\%$. The *Colour Chart* of the Royal Horticultural Society, although designed for plant colours, offers a wide range of samples on cardboard; CIE specifications and Munsell notations are usually provided. A Swedish Standard NCS (Natural Color System) atlas is available, containing a wide range of useful colour samples (see Hård and Sivik 1981).

The *Methuen Book of Colour*, by the London publishers, gives a selection of printed specimens and *Wilson's Colour Chart* (British Colour Council) included highly saturated colours. Foreign examples include Kornerup's (1968) *AIMS* collection of ink blending samples in book form. There are German and Japanese systems in various forms. Those dealing with colour vision deficiency sometimes require a few samples of coloured paper, including graded neutrals. Although the cheapest source may be an artists' supplier, or personal mixing of paints, it is best to order small quantities of Munsell or similar papers. Sets of such papers, for example those suitable for teaching colour principles, often provide most of what is needed. In some cases a spectrophotometric analysis of the reflecting properties and a CIE specification can be purchased with a special sample and this shows the way in which batches of colour may vary.

Stanley Gibbons' (1974) *Stamp Colour Key* provides 200 colours 2.5 cm × 1.5 cm each with a central 6 mm aperture; most are dark and fairly impure colours. Coates and other printing ink manufacturers make sets of colours to use as samples for letterpress work.

3 Inherited colour vision deficiencies

3.1 HISTORICAL INTRODUCTION

Faulty colour vision obviously arose long ago as a result of a gene mutation. Although many ancient scholars and philosophers wrote about colour, few recognised that differences in colour vision, amounting to 'colour blindness' could occur. As a result little has been reliably traced to indicate the very early origins of abnormal colour sensations. Plato (429–347BC) was probably one of the earlier speculators on colour vision, for although Aristotle (384–322BC) taught that colours arose from the intermingling of lightness and darkness Plato is believed to have realised that individual variations are likely (cited by Bell 1926). It is unlikely that colour blindness was recognised by the ancient and mediaeval writers on optics and vision since colour as a specific attribute of vision was not discussed.

Even the great optical writers of the seventeenth century, Kepler, Gassendi, Huygens and Newton made no reference to abnormal colour vision. The first published work on 'colour blindness' was probably by Tuberville (1684) involving a young girl from Banbury. This was probably an acquired colour disturbance, vision being confined to black and white. Boyle (1688), the renowned British chemist, described one person with unusual colour vision (as cited by Walls 1956); this individual saw 'some colours ... amiss' but again it is unclear whether this was an inherited or acquired condition.

The communication difficulty between the colour normal and colour defective perhaps explains why little attention was given to unusual colour vision in the early literature. Such observations are essentially subjective and often of a subtle nature and many Daltonics are not aware that their colour perception is abnormal.

The first clear case of inherited colour deficiency to be recorded came almost a century later, described by Huddart (1777) and involving two brothers by the surname Harris. Huddart wrote that one of the brothers 'had reason to believe that other persons saw something in objects which he could not see; that their language seemed to mark qualities with confidence and precision, which he could only guess at with hesitation, and frequently with error'. Harris's observations were careful, for he indicated his dependence on other clues such as size and shape in recognising cherries from leaves on a tree. Huddart included the comment that Harris was an intelligent man, implying that perhaps the underlying fault was probably of a physiological nature rather than involving inferior mental capacity. A statement indicating Harris's normal visual acuity and the shared colour problems of his brother, suggests that these were the first inherited cases of colour deficiency to be recorded. Huddart made the first *ad hoc* examination of at least one of the brothers, requesting colour names to various coloured ribbons. Light green was named yellow, pale red as blue and orange was named green.

Palmer (1777), a mysterious author who was the subject of a fascinating investigation by Walls (1956), believed in trichromacy and he postulated an absence of a set of receptors as a cause of colour blindness.

A year after the case of the Harris brothers came to light Lort (1778) outlined the case of J Scott who had three generations of family members similarly affected, including one of his sisters. This was the first recorded example of the defect in a woman. Scott's own colour experiences are recorded thus: 'I do not know any green in the world. A pink colour and a pale blue are alike. A full red and full green are the same . . . a good match; but yellows (light, dark and middle) and all degrees of blue, except those very pale, I know perfectly well . . . . A full purple and deep blue sometimes baffle me'. It is likely that Scott and his similarly affected family members were protanopes, since he mistook claret (red) garments for black. His observations include the comment that both his father and maternal uncle and one of his two sisters had similar colour vision difficulties, although his mother did not. The two sons of this sister were affected but his own son and daughter were not.

A few scattered examples appeared in the last two decades of the eighteenth century of other individuals so affected (see Sherman 1981).

It was the observations made in 1794 by John Dalton and published four years later (Dalton 1798) which provided the first serious study of the condition. The term 'Daltonism' which has persisted in much of Europe to this day as a general description of red–green inherited defects, is used frequently in this text. Dalton is best known for his contributions to the atomic theory. Readers interested in biographical details are referred to the work of Greenaway (1966). Here it is sufficient to state that he was born in 1766 in Eaglesfield, Cumberland, in the north of England. Despite his poor scientific training Dalton detected and observed characteristics of his own colour defect and that of others and further attempted to explain (in fact incorrectly) this unusual defect of vision, providing us with sufficient data to be reasonably sure that he was a protanope. By his own account Dalton was fairly well acquainted with the theory of light and colour before becoming aware of his own peculiarity in colour vision. During the early 1790s he became interested in botany and he felt some colours to be 'injudiciously named'. 'The term pink in reference to the flower of that name, seemed proper enough; but when the term red was substituted for pink, I thought it highly improper; it should have been blue, in my apprehension, as pink and blue appear to me very nearly allied; whilst pink and red have scarcely any relation .... With respect to colours that were white, yellow or green, I readily assented to the appropriate term. Blue, purple, pink and crimson appeared rather less distinguishable; being, according to my idea, all referable to blue'. In 1792 at the age of 26, he noted the change in colour appreciation of a geranium in daylight and candlelight. By candlelight it looked 'pink' 'but it appeared almost an exact sky-blue by day'. The visible spectrum appeared to Dalton as varying shades of blue and yellow with purple alone a possible third colour. Reds and greens were frequently confused by him; he described blood as being not unlike 'bottle green colour' and a green leaf and a stick of red sealing wax looked identical. 'Woollen yarn dyed crimson or dark blue is the same to me and green woollen cloth such as is used to cover tables, appears to me a dull dark brownish-red colour—it resembles a red soil just turned up by the plough. Stockings spotted with blood and dirt would scarcely be distinguishable'.

Dalton deliberately sought to discover other individuals, in addition to his brother, with similar colour sense and used a series of coloured ribbons as a crude test. He was aware of Huddart's account of Harris of Maryport, not far from his own home town. He corresponded with an acquaintance in Maryport who questioned one of the surviving colour deficient family members, four of the six sons of the family having anomalous vision. After receiving a report of colour naming using 20 ribbons sent to Maryport, Dalton was convinced of the similarity of vision between that of the four Harris brothers and his own. He had found two out of 25 persons and on another occasion one of 25, like himself. Yet he found that no parents, nor children, of anomalously sighted persons had anomalous colour vision, and he found no females affected at all.

Although this account is not the first report of colour anomalies, it was the first of a scientific nature with detailed comments; furthermore Dalton provided the basis for the inheritance mode based on familial analysis. His theory to explain the origin of the defect was less well-founded. He supposed the humour of the eye to be coloured blue, thus absorbing red, with the exception of a few red rays which 'may serve to give the colour that faded appearance'. Young (1807) extended his three-colour theory of colour vision to account for Dalton's abnormal colour vision, with the suggestion that there might be 'an absence or paralysis of those fibres of the retina which are calculated to perceive red'. Sherman (1981) indicates that this explanation was in fact rejected by Dalton. The question which was central to the discussion was whether the characteristic involved blindness (i.e. an absence of vision) or a difference in vision. In 1833 Dalton discussed the matter in correspondence with John Herschel, a leading optical expert. Herschel considered that Dalton 'and all others so affected, perceive as *light* every ray which others do. ...' He considered the normal-eyed to have three 'primary sensations' whereas those like Dalton 'have only two'. Mixtures, he said, involved the mixing of the three primary hues, but in the case of the anomalous 'the tints are referable to two'.

A *post mortem* examination of Dalton's eye media showed them to be normally transparent, dispelling his own theory of the explanation for his unusual colour perception. Following Dalton's account over twenty years lapsed before any further detailed discussion on the phenomenon emerged. Goethe (1810) indicated the difficulty in communication, and a possible first citing in a medical text can be attributed to Wardrop (1808).

The occupational consequences of faulty colour vision appear to have first come to light when Harvey (1826) reported the difficulty of a tailor who repaired a dark blue coat with a patch of crimson and a black dress with crimson, indicating protanopic tendencies in present day classification. The case of a weaver who was forced to leave his trade on account of constant errors in selecting threads was cited by Colquhoun (1829). Brewster (1826a, b) further brought the attention of the scientific community to the perception of the colour deficient. According to Sherman (1981), Herschel (1845) was the first person to utilise spectral colours in the investigation of colour vision defects in 1827; two coloured circles were displayed and colour naming was required. He employed a 'very thorough series of tests'. Previous investigators had used coloured cloth, glass or paints. Nevertheless for thirty years after Dalton's work this remained 'the best of all published reports'.

3.2*   THE TYPES OF DEFECT AND THEIR NAMES

Normal colour vision is not easy to define, since a range of slight variations is acceptable as normal; furthermore these variations change with age. Variations in performance at colour vision tests, considered in Chapter 7, demonstrate that different criteria of 'normality' are inevitable and that subtle differences appear. Assuming that averaged results of the majority of observers indicate 'normal trichromatism' it is appropriate to define departures with some reference to matches of white on a tricolorimeter.

Normal trichromatism: matching 'white' with R, G and B 'normally'.

Anomalous trichromatism: matching 'white' with R, G and B, with a bias to a greater proportion of one of the stimuli depending on the type of anomalous trichromatism.

Dichromatism: matching 'white' with two stimuli only, the choice depending on the type of dichromatism.

Monochromatism: matching 'white' with any other stimulus.

Subdivisions are used and 'abnormalities' do not appear in equal numbers. Many different terms are used to describe these states collectively; 'colour blindness', 'defective colour vision' or 'Daltonism' are each used.

A normal trichromat and a dichromat will find difficulty in appreciating each other's colour perception. The colour normal could look at the world through a coloured filter or adapt his retina to intense monochromatic light, though neither method would give him a complete idea of the experiences of those with abnormal colour perception.

Seebeck (1837) was aware that unreliable results frequently arose by asking persons with abnormal colour vision to name colours. He recognised at least two classes of defective colour vision, those confusing red and green and those showing difficulty with blues, and demonstrated both qualitative and quantitative variations. Rayleigh's experimental discovery of anomalous trichromacy in 1881 led to further classification of defective colour vision. Of the 23 males he examined with his colour box requiring the mixture of monochromatic red and green to match yellow, 16 required equal amounts of the two stimuli, while 5 used excess green and 2 excess red. Nagel (1905) developed the principle of colour mixture for a clinical test, the anomaloscope (see Chapter 7), and demonstrated significant individual variations among anomalous trichromats. Von Kries (1924) introduced the terms 'protanopia, deuteranopia and tritanopia' and was aware of the extreme anomalous subject. Franceshetti (1928) also proposed a distinct group intermediate between the dichromat and those anomalous observers with relatively good colour discrimination.

3.2.1 The generic terms

The foundation for a set of terms was provided by Seebeck (1837) as described by Edridge-Green (1891) and von Kries (1924). Almost the same descriptions are in general use today but there are inconsistencies, which sometimes originate in a view of normal colour vision, so that a standard for universal use has not been accepted. Ball (1974) proposed that the recording of different varieties of defects should be by means of symbols.

Throughout this text 'Daltonism' is used as a synonym for the inherited varieties of 'colour vision defect.' A separate chapter will describe the 'acquired' defects, and the present chapter outlines the different varieties of inherited defect, which are often termed 'congenital'; this is despite the association of the word with the moment of birth. There is a need for an adequate generic title, although it is likely that any word will blur the distinctions between acquired and non-acquired conditions. Most common English descriptions of the condition involve words such as 'blindness', 'defect' or 'deficiency' all of which have negative connotations. To some extent this could be considered to be derogatory to both the individual and the family involved. In languages derived from Romantic sources it is common to use the term 'Daltonism', with as little stigma as when 'Malapropism' is used by an Englishman. The word 'Daltonism' was in use in Switzerland about 1800, was mentioned by Prévost in 1827 and has appeared widely in the literature (see Abney 1895, Duke-Elder 1932, Wasserman 1978 and also Pokorny and Smith 1979). Brewster (1844) was among those who considered 'Daltonism' to be derogatory to a famous Englishman, and some dislike its use today. Yet it is popular to associate many syndromes and even 'undesirable' conditions with the names of those who described them, without offensiveness. Any who feel that Dalton himself (being dead) suffers by association with a 'defect' should consider the needs of many living persons. We should use a generic term, or one of the more specific descriptions, according to the circumstances, advisedly and with a good conscience.

A modern trend seeks to avoid the 'labelling' of a condition, preferring that an individual's difficulties or handicaps should be stated, instead. This reintroduces the difficulties of a description such as 'colour blindness' (*Farbenblindheit* when used in 1867 by Helmholtz), which could be resisted on several accounts, as the following pages will show; perhaps total monochromatism is the only condition which it describes. It is difficult to accept that 'red–green confusion' is any better.

3.2.2* Conventional categories

Dichromatisms of the 'red–green' type were the first conditions to be recognised. The fascinating history of the emergence of distinctive terms has been described by Helmholtz (1924, English edition) with important contributions to the text by von Kries. Wilson (1855), Judd (1943) and Wasserman (1978) extended this literature significantly. As long ago as 1811 Wardrop distinguished the two 'red–green' types of deficiency and Seebeck (1837) showed how those relatively insensitive to red lights were in a minority. Von Kries proposed the name 'protanopes' for those who 'lack the first component ... of the normal'. His terms 'deuteranopia' and 'tritanopia' followed the trichromatic approach, which Maxwell (1855a) supported.

Dalton's suggestion that an unusual coloration of his ocular media had been the cause of his condition had been disproved at autopsy. However his explanation of a colour vision deficiency by 'absorption' (some wavelengths being absorbed in the ocular media before reaching a normal retina) should be noted for it explains some acquired defects. An alternative cause, proposed by von Kries (1924) arose from dissatisfaction with the the term 'red-blindness'; he suggested that a 'reduction of the normal retinal mechanisms could account for the visual deficiency'. The 'loss' of one mechanism is compatible with modern concepts of dichromatism. This fact emerged as an understanding of the cone pigments developed, as will now be described.

The extraction of a cone pigment named iodopsin, from the predominantly cone retina of the chicken, by Wald (1937) and later from fowls by Wald et al (1955) was followed by the isolation of two different cone pigments from the fovea of the rhesus monkey Macaca mulatta by Murray (1968).

The suggestion that variations in colour vision might result from individual differences in receptor sensitivity was made by König and Dieterici (1893). König (1894) proposed the absence of one photopigment class in the fovea of the dichromat to explain defective colour vision in association with the other unaltered pigments. This followed Young (1807) who first propounded that the red mechanism might be lacking in Dalton's eye as an explanation of his protanopia. A century and a half passed before Rushton (1955) provided objective evidence with the absence of the erythrolabe (red-catching) pigment in protanopia, by measuring the remaining pigment, chlorolabe, in situ. Photopigment characteristics were related to spectral sensitivity by König and Kottgen (1894) and König (1894) who indicated how the protanopic sensitivity curve corresponded to the green sensation of normal vision. Dartnall (1953) and Bowmaker (1973) added to our understanding of this association while Wald et al (1955) proposed that the sensitivity ratio of the cones related to the absorption ratio of the visual pigment.

Another possibility is that a 'collapse' of function may be involved, perhaps by the fusion of otherwise separate normal processes. This could be at a post-receptoral level, an idea which was adopted by Pitt (1935).

It is generally agreed that a 'reduction' system is indicated by the fact that protanopes and deuteranopes accept the colour matches made by a normal person.

Persons with highly atypical colour vision are reported from time to time, generally from acquired causes, but in some instances such cases do appear to be genetically determined—for example the abnormal dichromat reported by Ruddock and Bender (1972).

Dichromatism

Dichromatism is usually divided into three categories all of which involve severe difficulties with colours.

(i) Protanopia. This condition is particularly associated with deficiency of appreciation of red. It is likely to arise from absence of the 'longwave' sensitive cone pigment, erythrolabe (see §3.3.2).

(ii) Deuteranopia. While being associated with deficiency of green vision, this type of dichromatism involves characteristic confusions of colours.

(iii) Tritanopia. This is a very rare condition, where sensitivity to blue is impaired and where there are confusions between blue and green. Walls (1964) at one time doubted its existence.

There is a fourth category 'tetartanopia' more obscure than tritanopia, and somewhat resembling it; this was described by Müller, as reported by Judd (1943) (see §3.4.1).

Anomalous trichromatism

The third Lord Rayleigh (1881) conducted colour mixing experiments involving members of his family. As a result he identified slight colour vision anomalies, later termed 'protanomalous' and 'deuteranomalous' by Nagel (1904). These were forms of trichromatism and less 'severe' than dichromatism but with a tendency to some characteristics of dichromatism. Donders is reported to have confirmed the existence of anomalous trichromatism soon afterwards and Nagel developed the anomaloscope to demonstrate the conditions qualitatively and quantitatively. It was Rosencrantz (1926) who showed the wide variations in the matches of protanomals and Francheschetti (1928) was able to identify the extreme forms of protanomaly and deuteranomaly, to be considered in §7.7 when the anomaloscope is described. Additional descriptions of cases followed, by Watson (1914) and by Pitt (1935), while Nelson (1938) collected data on a series of subjects of both types. McKeon and Wright (1940) should also be consulted for protanomalous data.

Tritanomaly was identified as a separate but rare condition by Engelking (1925) and Hartung (1926). It has been reviewed by Schmidt (1970) and 'incomplete tritanopia' was considered as a possible defect by Cole and Watkins (1967). Statistics show that deuteranomaly is the most prevalent of these disorders of trichromatic vision and it is understandable that it is therefore likely to attract most attention. However, protanomaly is accompanied by significant dangers on account of difficulties with red perception. Table 3.1 compares the frequency of each condition.

Table 3.1 Main types of inherited colour vision defects with approximate proportions of appearance in the population.

	Per cent in UK	
Condition	Male	Female
Protanopia	1	0.02
Protanomaly	1.5	0.03
Deuteranopia	1	0.01
Deuteranomaly	5	0.4
Tritanopia and tritanomaly	Very small	

Anomalous trichromacy is not only divided into 'protan', 'deutan' and 'tritan' categories. There are degrees of anomaly (or grades) which were demonstrated by the experiments and data of Nelson (1938), McKeon and Wright (1940) and Wright (1946), who indicated the possibility of a continuum of subjects of varied departure from normality. A series of categories is given below, which is in accord with the practical criteria explained in §7.7 and with the views of specialists such as Walls and Mathews (1952) and Schmidt (1955b).

Protanopia   (Dichromatism)
Extreme protanomaly
(Ordinary) protanomaly   }   (Anomalous trichromatims)
Deuteranopia   (Dichromatism)
Extreme deuteranomaly
(Ordinary) deuteranomaly   }   (Anomalous trichromatism)
Tritanopia   (Dichromatism)
Tritanomaly   (Anomalous trichromatism)

Achromatopsia

Complete absence of colour perception, achromatopsia ('without colour') or monochromatism ('one colour') is rare, especially in an inherited form (see Pitt 1944b). Most cases encountered arise as a consequence of pathology, specifically retinal cone dystrophy or degeneration, the term 'cone dysfunction syndrome' typically applying. A variety of acquired cases have appeared in the literature, usually of a progressive nature and frequently giving rise to monocular loss, unlike the inherited forms. One or two cases of hysterical total absence of colour vision, which may be temporary, are cited, and others as a consequence of cranial trauma. Weale (1953a) quotes a case resulting from gunshot wounds; such acquired cases are discussed in Chapter 4. In the present chapter the achromatopsias which are inherited, permanent and non-progressive are discussed. Hering (1891) and König (1897) made early investigations of the psychophysical characteristics of the condition, although the inherited anomaly was known as early as the seventeenth century, as indicated by Waardenburg (1963). Suspect cases are occasionally brought to the attention of colour vision experts, but when examined often prove to be simple dichromats; care and caution are thus needed in diagnosis (see also §§3.5 and 3.7.3).

3.3   THE NATURE OF INHERITED DEFECTS

This section will describe how defects of colour perception could be attributed to (a) alteration, reduction or absence of photopigments or of neural elements or (b) selective absorption before light reaches the retinal receptors.

3.3.1   Dichromatism (cause)

Protanopia arises on account of a 'loss' of the red responsive ('long wavelength' sensitive) mechanism in the retina. The state is one of dichromatism since two mechanisms are presumed to remain. Thus it is also a 'reduction' of the normal state since one of the normal cone pigments (erythrolabe) is absent. Rushton (1955) demonstrated the absence of erythrolabe in the protanope, thus confirming concepts of protanopia which go back over two centuries, since Palmer's views were comparable despite his ignorance of cone function.

Deuteranopia has often been associated with 'fusion' or 'collapse' of normal functions, as described by Hurvich (1972). It has been suggested that the signals from the normal red and green mechanisms are abnormally combined, or that all the receptors governing red and green vision contain both of the 'long wavelength' absorbing pigments. However, Boynton (1979) considered, on the basis of neutral point data, that fusion dichromacy is unlikely. The fact that the V_λ function in deuteranopia is similar to that of the normal gives little support to the idea of 'loss' of a green mechanism, but Rushton (1970) clearly believed that retinal densitometry proved that one pigment is missing in deuteranopia. Psychophysical estimates by Miller (1972) of visual pigment densities in dichromats further enhanced the views that the protanope lacks the normal 'red' pigment and the deuteranope the 'green' sensitive pigment.

Walls and Mathews (1952) assumed the cause of dichromatism to be receptoral, not cortical, since no case of a red−green defect with hemianopic features had been reported. However Fincham (1953a,b) found reasons to believe that 'cone monochromats' have cortical abnormalities, while the most common disorder of the dichromat is retinal. Explanations of tritanopia remain difficult on account of the dearth of observers; the assumption is usually that of a 'lost' mechanism. In the opinion of Hurvich (1972) the receptor and neural mechanisms might both be abnormal. One feature of normal retinae is related, notably the sparsity of 'blue-sensitive' cones. No blue-sensitive pigment has been revealed *in situ* in normal human retinae and no *lack* of such a pigment has yet been demonstrated in tritanopia. Walls and Mathews (1952) considered tritanopia to be 'a collapse of blueness responsitivity upon the distribution of greenness responsitivity' and tetartanopia to be a 'lack of the blueness receptors'. Increment threshold studies in tritanopia enabled Cole and Watkins (1967) to suggest that the condition could be a reduction system caused by merging of the blue process with another. Although responses to different parts of the spectrum through normal receptor mechanisms might be normal the possibility exists for faulty transmission of these signals through the opponent responses which commence in the retina. Such a 'neural loss' has been postulated by Hurvich and Jameson (1955) and accords with opponent views of colour vision. Such concepts are difficult to reconcile with evidence of a 'pigment loss' at the receptoral level. It is possible that the parafoveal colour vision responses of dichromats have some resemblance to trichromatism, with a rhodopsin-type pigment operating in addition to the pigments in the dichromatic eye (see Smith and Pokorny 1977).

A simple model of the physical, physiological and perceptual stages of colour vision can be used as a means of indicating the typical features of dichromatism. An approach by Hecht (figure 3.1) provides a simple explanation of a neutral point at 490 nm if the R mechanism is assumed to be absent; the remaining pair are in balance, or cross, at the appropriate point. At this 'blue–green' region the short and medium responses have been excited equally, as if a suitably neutral stimulus had stimulated them. On either side, spectral lights excite one or the other of the two mechanisms to a greater extent and a series of 'blues' or 'greens' appear.

In figure 3.1(a), a normal set of cone pigments respond to white light and the post-receptoral responses initiate the resultant, described as the perception of a white or neutral. In figure 3.1(b) blue–green light leaves out the 'long wavelength' parts of the sensorium because the L pigment is not affected. In figure 3.1(c) the S pigment is unaffected and the M pigment absorbs relatively little. In figure 3.1(d) yellow light is absorbed equally by the M and L pigments so the resulting perception of yellow comes from a balance between two sensory messages.

The remaining parts of the figure portray a protanope's responses but the characteristic V_λ features of the protanope influence the description. Figure 3.1(e) suggests that the L pigment is replaced with M', considered to be the same as M for practical purposes. The 'long wavelength' sensory message

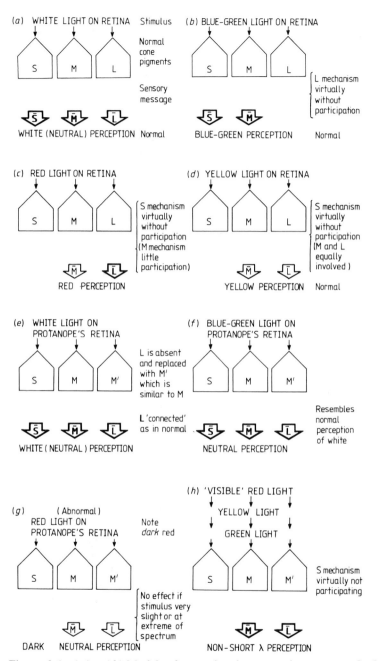

Figure 3.1 (*a*) – (*h*) Models of normal and protanopic responses. Such diagrams attempt to portray typical features such as pseudoisochromatic colours but are necessarily incomplete.

still exists but depends upon suitable initiation from absorption by the M' pigment. The situation in 3.1(f) resembles that in 3.1(e) and shows how the protanope's 'neutral point' arises. Figure 3.1(g) may be considered less satisfactory, since red lights occupy a wide waveband, and here the stimulus is supposed to be one which appears neutral (or non-existant if extremely long wavelength) rather than apparently isochromatic with yellow. Therefore in figure 3.1(h), a high intensity and relatively short wavelength 'red' is imagined as one of a series of pseudoisochromatic stimuli, all of which have the same retinal and perceptual effects or could be interpreted (according to their real or imagined intensities) as a percept associated with yellow *or* green, or some reds.

The principle of univariance, described (for example) by Mitchell and Rushton (1971a) relates the sensitivity of a colour mechanism to the bleaching of the appropriate photopigment; this bleaching is caused by the 'effective' quantum catch. Each mechanism registers its individual response to light, giving no indication of the spectral quality of such light, apart from an intensity variation which depends on the quanta absorbed. Two monochromatic lights shown alternately, one after the other, can be balanced in intensity if a single colour mechanism is present. Thus in figure 3.2(a) lights of wavelength 530 and 600 nm can be arranged to provide equal stimulation, so that points A and B are at the same level on the response curve of the long wavelength mechanism (R). The short wavelength mechanism is inoperative for wavelengths of 530 nm and above and protanopes and deuteranopes have only one effective mechanism in this region, G and R respectively. If a protanope views lights of 600 and 530 nm which are adjusted for equal brightness he sees no colour difference and if one of these lights is exchanged for the other he sees no difference; but the 600 nm light must have a higher luminance. This is clear in figure 3.2(b) where points C and D are at different levels. If the two lights are flickered, alternately, the luminances can be adjusted to minimise the flicker for a protanope but this adjustment will not suit a deuteranope. Having only the R mechanism operative the deuteranope (with approximately equal sensitivity to the two lights) will have minimum flicker for a different balance of luminances.

The OSCAR test ('objective' screening for colour anomalies and reductions) uses flickering diodes of variable luminance to apply the principle outlined above. Furthermore normal subjects detect flicker at both protanopic and deuteranopic settings, with the normal setting for minimum flicker being intermediate. Anomalous trichromats will make settings appropriate to their condition; deuteranomals will find minimum flicker on the deutan side of normal and progressively nearer to the deuteranopic position as they approximate more to this condition. For a description of the test see p. 345.

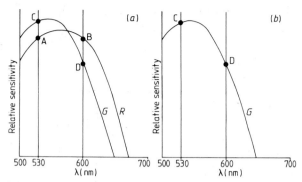

Figure 3.2 Photopigment and hence colour mechanism sensitivity. (*a*) Two 'long' wavelength responsive types. (*b*) A single type of response curve, assumed to belong to a protanope, sensitive to wavelengths longer than 500 nm.

3.3.2 Abnormal cone pigments—characteristics and *in vivo* measurements

The technique of fundus reflectometry has revealed the presence of two foveal photopigments in the medium to long wavelength range in normal eyes, and the absence of the long wavelength photopigment in protanopia and the medium wavelength photopigment in deuteranopia (see Rushton 1955, 1963a, 1965a). Studies of a psychophysical nature by Baker and Rushton (1965) and by Mitchell and Rushton (1971a,b) have provided additional evidence. Rushton proposed the terms erythrolabe (Greek 'red-seizing') for the pigment shown to have maximum absorption at approximately 560–70 nm, and chlorolabe (Greek 'green-seizing') for the pigment shown to have maximum absorption at approximately 540 nm. A comparison of the action and difference spectra indicated to Baker and Rushton the identical nature of erythrolabe in the fovea of the colour normal to that in the fovea of the dichromat. Previously Rushton had demonstrated similar characteristics between chlorolabe in the colour normal and protanope. The pigment cyanolabe (Greek 'blue-seizing'), assumed to control the short wavelength region of the visible spectrum, has eluded measurements to date. Cyanolabe is considered to be present in the eyes of colour normals, protanopes, deuteranopes and anomalous trichromats (with possible altered characteristics in tritanomalous trichromats). Such evidence comes from two sources, firstly the psychophysical determination of the blue cone mechanisms, demonstrated by Stiles (1959) (see §1.11) and secondly the presence of the short wavelength pigment measured by microspectrophotometry (see §1.10.4). However Alpern (1976) considers it too difficult to state categorically whether cyanolabe is present or absent in tritanopia. The limited energy in the short wavelength spectral region reflected from the fundus has made fundus reflectometric analysis of cyanolabe difficult. Furthermore the fovea has been shown to lack short wavelength cone photoreceptors (see Marc and Sperling 1977) so that measurements of cyanolabe must be sought in extra-foveal regions.

Individual variations in the density of the long and medium wavelength pigments in the fovea have been indicated by the psychophysical techniques of Rushton and Baker (1964) involving red and green colour matching. King-Smith (1973a,b) has made further analysis of the optical density of erythrolabe by two methods, following psychophysical estimates of the densities in dichromacy by Miller (1972).

Experimental details

The demonstration by Rushton (1963a,b) of chlorolabe alone controlling red–green vision in the protanope and that of erythrolabe alone controlling red–green vision in the deuteranope (Rushton 1955, 1965a,b) was performed by the following procedure. After a period of dark adaptation a series of six to eight wavelengths was introduced to the eye in turn with corresponding density measurements recorded indicating the proportion of light reflected from the fundus. A strong red light or blue–green light was used in turn to bleach the chlorolabe or erythrolabe respectively and measurements of the reflected energy repeated for the spectral range. The reflected energy was seen to be greater during the second set of measurements than the first, indicating the role of the bleaching source in the effective removal of pigment. Graham's (1965) textbook provides suitable details. This procedure was employed following the initial observation by Rushton (1955) that a red light used to bleach away pigment from a normal eye had no effect on one eye of a protanope, suggesting to him that the protanope lacked the red-catching pigment.

The absorption action spectrum of chlorolabe (being the relative quantum energy required to bleach a given amount) coincides in form to the foveal spectral sensitivity for the protanope. Similarly the absorption action spectrum of erythrolabe coincides with the foveal spectral sensitivity of the deuteranope. While the selective loss of photopigment is generally considered a plausible explanation for dichromacy, other views which lay emphasis on a neural, opponent fault, usually in conjunction with a photopigment change, have been proposed, see for example Hurvich and Jameson (1964). A neural fusion explanation for deuteranopia has been favoured by one school of thought (see §3.4.1). Thus Boynton (1979), for example, indicated his reservations concerning the statement by Rushton (1964) that no chlorolabe is present in the eyes of deuteranopes. Confirmation, using fundus reflectometry, of the absence of chlorolabe in fifteen dichromats was however, given by Alpern and Wake (1977) who reported considerable individual differences in the density kinetics regeneration and wavelength of maximum absorption among their sample. Colour matching measurements by deuteranopes undertaken by Alpern and Pugh (1977), with the purpose of evaluating their cone pigments, also indicated significant individual variations in erythrolabe. The suggestion that individual differences in pigment characteristics can be explained by differences in absorption of a prereceptor pigment, as suggested by Pokorny *et al* (1973), cannot be ruled out.

Psychophysical evidence for more than two kinds of cone photopigment types in at least some protanopes and deuteranopes has been recently presented by Frome *et al* (1982), who measured temporal and spatial features. It has been their suggestion that some dichromats might have the same three photopigments as anomalous trichromats. Boynton (1982) proposed that most of the long wavelength sensitive cones of protanopes might contain chlorolabe; at the same time he suggested that most protanopes have in addition some residual evidence of erythrolabe. Boynton *et al* (1959) used chromatic adaptation to give some support to the idea of red-mechanism loss in protanopia and fusion of 'red' and 'green' input in deuteranopes. Wright's chromatic conditioning method was used in deuteranopes and deuteranomals by Richards and Luria (1968), pointing to different possibilities as to the contributions by photopigments in such cases.

Stiles's method for π mechanism study (see §§1.11.3 and 1.24.3) was applied by Cavonius and Estevez (1978), who put the argument for π_5 having the characteristics of erythrolabe. Increment spectral sensitivity measurements by de Vries-de Mol *et al* (1978) showed how the mechanisms of dichromats and carriers can be differentiated. Test methods such as this are considered later on p. 360 and §7.11.1, having special use in acquired conditions (see, for example, Vola *et al* 1982). Figure 3.3 shows data by Hansen (1979), obtained by adapting to 'background' colours, in which normal, protan and deutan subjects reveal different photopigments and mechanisms.

3.3.3 Anomalous trichromatism (cause)

It is likely that an 'alteration' of one of the cone pigments is the cause of such conditions, particularly in the case of the protanomalous and deuteranomalous (see Alpern *et al* 1965). Anomalous trichromats would accept normal anomaloscope matches if their cone pigments were normal and since the time of Rayleigh (1881) it has been known that normal matches are not accepted. The possibility of a 'screening' effect was discounted by Alpern and Torii (1968a) and the V_λ performance of anomals is compatible with pigment alterations. The cone pigments in anomalous trichromats have been described by Pokorny *et al* (1973), after reconciliation of Schmidt's (1955b) data with the 'analytical anomaloscope' method used by Mitchell and Rushton (1971b). This instrument was used by Rushton *et al* (1973a) to measure the sensitivity curves of 'protanolabe' and 'deutanolabe'. The assumption that the anomalous pigment is the same in both protan and deutan anomalies was discussed by Hayhoe and MacLeod (1976). Pokorny and Smith (1979) reviewed other techniques by which such pigments have been investigated.

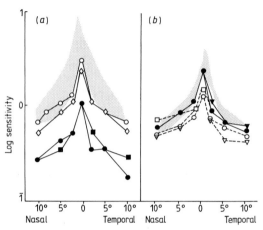

Figure 3.3   Perimetric response data. Protan and deutan subjects show different thresholds of response (protans indicated by filled symbols and deutans by open symbols). The shaded areas indicate the variation of seven normal subjects (mean value ± 1 SD). Static perimetry was performed with (*a*) a blue background using a 582 nm target and (*b*) in purple light using a 552 nm target. Courtesy Dr E Hansen, Oslo.

The 'alteration' nature of 'simple deuteranomaly' was established by de Vries (1946) using the increment threshold method.

Two staunch protagonists of 'opponency', Hurvich and Jameson (1964), applied the principle to an explanation of anomalous trichromatism. Pitt (1949) proposed reduced sensitivity of the red responsive mechanism as an explanation for protanomaly, Hurvich and Jameson indicated that this should cause 'unique yellow' (not greenish and not orange–yellow) to appear at a longer wavelength than is normal. On an opponent theory basis abnormal photopigments could accompany alterations or reductions of function of the red–green paired process; thus protanomaly should show unique yellow at a shorter wavelength than normal. Data were produced to show that unique yellow (between 575 and 589 nm for normals) was below 576 nm for protanomals and above 590 nm in deuteranomaly.

The following (simplified) account indicates how alteration from normal colour vision in anomalous trichromacy can explain the characteristic colour matching of these observers. Assuming that a normal trichromat has pigments G and R with absorption maxima and variations approximately as indicated in figure 3.4(a) these allow a 'Rayleigh equation' match with red, green and yellow lights. In this case monochromatic yellow, y, will stimulate the G and R mechanisms equally since it affects G and R pigments equally, in this diagram. A matching mixture of r and g stimulates the mechanisms equally since R absorbs r as much as G absorbs g and R absorbs g as much as G absorbs r. Thus G and R (mechanisms) excite a balanced sensation.

If a protanomalous trichromat has normal G pigment but a modified pigment PR as in figure 3.4(b) the match made by a normal will not be acceptable. In this case monochromatic yellow, y, will stimulate the G mechanism normally, but it is absorbed to an abnormally large extent by PR and the R mechanism is stimulated excessively; the yellow will tend to appear too 'red'. r is absorbed more by PR than by G. Since this is less than by the normal R, the R mechanism of the protanomal is affected less than the normal R mechanism. g is absorbed normally by G but more than normally by PR, so the G mechanism is influenced as much as normal while the R mechanism responds more to g. The diagram suggests that the combined r and g effects on the R mechanism are no more than in the normal observer. As a result the yellow appears too red and a match is obtained by the protanomalous adding red to the mixture.

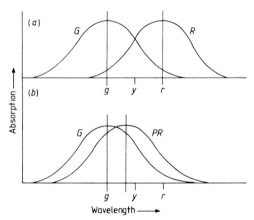

Figure 3.4 Photopigment sensitivities for 'long' wavelengths. (a) The normal pair of pigments which operate in this part of the spectrum. (b) The protanomalous situation, where PR is the altered form of R.

Protanomals match yellow with excessive red while, in a mixture of red and green, deuteranomals require excessive green. Thus for protanomals $R + g = Y$ and for deuteranomals $r + G = Y$. Colour confusions for anomalous trichromats tend to be similar to those of the appropriate dichromat; these confusions are enhanced by desaturation of colours. Many protanomalous observers have a selectively lowered sensitivity to red lights, hence they tend toward the protanope's relative luminous efficiency variations and to similar confusions of red (or blue–green) with neutral. Distinctions between tritanomaly and tetartanomaly were discussed by Hurvich (1972). The suggestion has been made of an association between rod receptors and blue cone receptors by Willmer (1949) but this has been discounted by Sperling (1980). The tritanomal studied by Schmidt (1970) showed good rod function. Incomplete tritanopia and its possible demonstration by absence of one of Stiles' three blue-sensitive mechanisms, was considered by Cole and Watkins (1967).

Post mortem studies of absorption properties of cones in the retinae of colour anomalous squirrel monkeys, presented by Jacobs (1981), have confirmed the alteration view. Behavioural measurements showed one monkey to be protan and later a single photopigment was identified, with peak absorption at 535 nm. In a 'deutan' animal two pigment types with maxima at 552 and 568 nm could be identified.

Psychophysical studies by Nagy *et al* (1981) involving colour matching among heterozygous female carriers of Daltonism suggest evidence of slightly abnormal cone pigment.

Clearly, among some monkey species at least, individual differences in colour vision can be largely explained by variations in cone pigment absorbance as shown by Jacobs *et al* (1982). The benefit of relating psychophysical measurements to microspectrophotometric analysis of the cone pigments in the same animals was indicated by these authors. In this study monkeys diagnosed as dichromats on behavioural tests were shown to lack one of the long wavelength pigments. Three deuteranomalous monkeys showed evidence of two pigment types in their retinae with spectral absorbance peaks close to those suggested by Pokorny *et al* (1979) as responsible for simple deuteranomalous vision among humans.

Pigments in anomalous trichromacy

Anomalous trichromats match spectral wavelengths with three reference stimuli, and unlike dichromats can distinguish between red and green and its mixtures. Such ability in matching colours implies the presence of three cone photopigments interacting together, one at least being abnormal. König and Dieterici (1893) recognised that a shifted spectral sensitivity function characterised anomalous trichromacy. The suggestion that deuteranomals have abnormal medium wavelength absorbing cone pigments, and protanomals abnormal long wavelength absorbing pigment, dates to von Kries (1897) who assumed that the two remaining pigment types were normal.

Evidence to be discussed in this section provides strong backing for this viewpoint. However, alternative approaches also justify consideration. Details of the exact characteristics of the anomalous pigment(s) in deuteranomaly and protanomaly will only be forthcoming in the light of additional experiments.

Virtually nothing is known of the proposed pigment, cyanolabe, absorbing strongly in the short wavelengths, or of the underlying mechanism(s) operating in tritanomaly. By inference from protanomaly and deuteranomaly a photopigment alteration would seem a likely explanation either in part, or in full, for this rare inherited colour vision defect.

Fundus reflectometry

Investigations by Rushton (1965c) using the technique of fundus reflectometry among anomalous trichromats were disappointing, in view of the success of this method with dichromats. Rushton (1970) thus concluded that severely deficient dichromats could not be distinguished from mild to moderate colour deficient anomalous trichromats using density measurements of the photopigments. The anomalous pigment(s) were considered either to be present in insufficient quantities to be detected by densitometry (despite sufficient presence to indicate colour matching differences) or to be so similar that the technique was not sufficiently sensitive to distinguish between the normal and anomalous photopigments subserving colour perception. Careful reflectometry/densitometry measurements by Alpern and Wake (1977) have revealed the presence of two cone photopigments in the red–green spectral range for simple deuteranomalous trichromats. Pigments of the extreme deuteranomalous trichromats bore similarities to those of deuteranopes; these authors further indicated marked individual differences in λ_{max}, density and regenerator/kinetics properties for erythrolabe in deuteranopes.

The significant individual differences in colour matching among anomalous trichromats (for example with the Rayleigh match of red plus green to match yellow in the anomaloscope) was even recognised by Rayleigh (1881) when he first identified that category of colour deficient–the anomalous trichromats. These could be readily explained by individual differences in the absorption characteristics of the visual pigments subserving red–green perception. Very poor discrimination in the red–green range could thus be explained by the close similarity of the abnormal pigment to that of the remaining normal pigment. Considerable variability in the medium to long wavelength sensitive pigment for colour normals has also been indicated (see Rushton and Baker (1964) for example) which complicates studies of anomalous pigments.

Theoretical computations

The characteristics of the photopigments have been studied extensively by means of the theoretical approach on account of the difficulty of obtaining reliable experimental data. Alpern and Torii (1968a,b) computed the spectral sensitivity features of the anomalous pigments for both protans and deutans. Smith and Pokorny related psychophysically determined colour mixture or matching data to the absorption spectra of the cone photopigments. Possible absorption spectra for anomalous trichromats were derived by Pokorny and Smith (1979) and, combining Rayleigh equation data with standard curves for the visual pigments, curves peaking at 541.6 nm for protanomals and 554.6 nm for deuteranomals were proposed (Pokorny *et al* 1973). Relative spectral sensitivity curves for human cone photopigments were proposed by Smith *et al* (1976). The proposed curves are in agreement with both colour-matching data of normal and colour defective observers and theoretical explanations of features of visual photopigments.

Using a colour-matching technique MacLeod and Hayhoe (1974) considered the possibility of a single anomalous pigment for deuteranomals and protanomals and provided an estimation of the absorption characteristics. Hypothetical sensitivity curves were proposed for the long wavelength absorbing photopigment (considered to be present in colour normals and deuteranomals), and for the medium wavelength absorbing pigment (considered to be present in colour normals and protanomals). They further proposed an intermediate anomalous pigment for both deuteranomals and protanomals. Pokorny *et al* (1973) however consider that red–green colour matching demands the presence of different anomalous pigments for protanomaly and deuteranomaly. Although unable to refute the single-pigment hypothesis these authors were unable to accept that the single-sensitivity function derived from the colour-matching functions of the protanomal and deuteranomal was the photopigment responsible for anomalous colour vision. A recalculation of the absorption spectra of anomalous pigments by Pokorny *et al* (1975) followed as did an evaluation of the single-pigment shift model by Pokorny and Smith (1977).

A different approach proposed by Ruddock and Naghshineh (1974) involved normal visual pigments in the cones of protanomals and deuteranomals but mixed together in each of the two receptor types. Variability in the λ_{max} of the normal visual pigments formed the basis for another viewpoint proposed by Alpern and Moeller (1977). Anomalous trichromats were considered to possess the normal visual pigment in both the medium and long wavelength sensitive receptors with some variability demonstrated. A range of both long and medium wavelength pigments was proposed even for normal observers. This conclusion followed from results obtained by three different psychophysical techniques used to measure the action spectra of the long wavelength sensitive cones among deuteranopes. All three methods indicated small, systematic differences in λ_{max}. Variability has been noted in microspectrophotometric measurements of visual pigments (see for example Bowmaker *et al* 1978, 1980). Although most investigators consider that deuteranomalous observers have normal erythrolabe and cyanolabe, the pigments controlling predominantly long wavelengths and short wavelengths, the concept of a single erythrolabe common to all deuteranopes and to the long wavelength sensitive cones of colour normals was rejected by Alpern and Pugh (1977) and Alpern and Wake (1977). The additional study by Alpern and Moeller (1977) suggested that the action spectrum of the medium, as well as the long wavelength sensitive cones of one deuteranomalous observer may each be identical with that of the long wavelength sensitive cones of different deuteranopes. Nevertheless all deuteranomals were not considered to have the same three pigments. These authors concluded that the same erythrolabe photopigment group controlled the medium and long wavelength sensitive cones of the deuteranomalous observer studied.

Using the exchange threshold technique, whereby the activity of one class of cones in the red–green range was inhibited, Rushton *et al* (1973b,c) provided details of the spectral sensitivity of the proposed abnormal pigments described as protanolabe (for the protanomalous observer) and deuteranolabe (for the deuteranomalous observer). The spectral absorption characteristics were only slightly different, but consistently so, with protanolabe λ_{max} at 550 nm and deuteranolabe λ_{max} at 555 nm. De Vries (1948) quoted an unpublished 1937 thesis of Schouten which proposed that the anomalous pigments of the deuteranomal and protanomal might be identical. This concept is refuted by Rushton *et al* (1973b,c). Piantanida and Sperling (1973a,b) hypothesised a change to the protein part of the pigment to account for differences to chlorolabe and erythrolabe in anomalous trichromats.

3.3.4 Opponency and Daltonism

An alternative approach to anomalous trichromacy compatible with the opponent theory of colour vision has been postulated by Jameson and Hurvich (1956b). These authors consider a shift in the absorption characteristics of one or more photopigments in both deuteranomaly and tritanomaly together with a reduction in the strength of the red–green opponent neural channel. They further suggest the likelihood of just one anomalous pigment for each of protanomaly and deuteranomaly. The absorption characteristics of the long wavelength pigment of the protanomalous could be similar to that of the normal 'green' pigment in the colour normal but slightly displaced towards the red spectral region. Their argument suggests that the absorption of the 'green' photopigment for the deuteranomalous is akin to the 'red' pigment for normal trichromats but shifted slightly towards the blue spectral region (see also Hurvich and Jameson 1974). However as Pokorny and Smith (1977) have indicated, a theory that proposes both photopigment and neural alteration among colour defectives poses the problem of how both factors might be inherited together.

Romeskie (1976) made the first direct measurements for anomalous trichromats and dichromats (see Romeskie 1978a and Romeskie and Yager 1978) with a cancellation technique, showing significant variations between observers. Explanations of protanopia and deuteranopia by the opponent colour model involves the absence of a photopigment class and of the red–green chromatic channel (see figure 3.5). Cancellation techniques such as those employed by Romeskie (1978a) thus required the use of only two instead of four stimuli. For both protanopes and deuteranopes the opponent response functions were similar to the blue–yellow function of colour normal observers. The functions for the protanope and the deuteranope were seen to be similar in shape on the shortwave side of the crossover point, close to the neutral point of each observer. On the longwave side, the function for the protanope is shown to have a considerably steeper longwave slope.

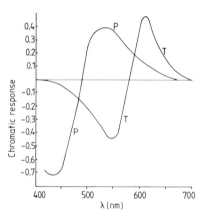

Figure 3.5 Responses of opponent systems in protanopia (P) and tritanopia (T). After Hurvich (1981).

The suggestion that evidence of opponent processing may be implicated in temporal chromaticity and modulation sensitivity led Jameson *et al* (1982) to examine this function in colour normals and abnormals. Marked individual differences were found among all groups between relative sensitivities to temporal luminance and chromaticity changes. The ability to detect temporal transient changes between stimuli of 540 and 640 nm was noted for one protanope who could not distinguish the wavelengths in terms of colour. A general reduction in sensitivity to all temporal frequencies (range 1.5 to 20 Hz) was noted for a protanopic observer compared with that of five observers with normal colour vision (Varner 1981).

Both protanomalous and deuteranomalous observers studied by Romeskie (1978a) showed a decrease in the height of the red–green function, relative to the blue–yellow colour opponent function, the magnitude of which correlated with the matching range of the Rayleigh equation. A progressive decrease in the height of the yellow function relative to that of the blue was also measured (see figure 3.6).

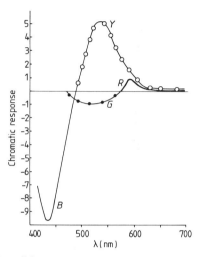

Figure 3.6 Possible protanomalous opponent mechanism sensitivities. After Romeskie (1976).

3.3.5   The cause of monochromatism (achromatopsia)

Many normal trichromats appear to use retinal information provided by the chromatic fringes of images as part of the stimulus to a change in accommodation (Fincham 1951). Using protanopes and deuteranopes Fincham (1953) showed that they respond no better to a red–green mixture than to monochromatic yellow, hence the chromatic element does not stimulate their accommodation, which suggests that they have a retinal disorder of colour discrimination. In the case of three 'cone' monochromats he found the accommodation reflex for heterochromatic light to be normal and concluded that in such monochromats the defect is above the chiasma; the site may therefore be cortical. Rod monochromats can be regarded as possessing a single set of normal cones with all others containing a rhodopsin type of photopigment. Section 3.5 should be consulted for more detail.

Heath (1956) reviewed retinal functions which might be attributed to either rods or cones in monochromats and the proposals that some of these observers have functional cones. Three 'achromats' with low visual acuity were examined by Heath, who used a modified optometer to study their accommodative responses. A myopic state (accommodative) was found and attributed to the low acuity.

3.4*   CONFUSION OF COLOURS

A striking feature of 'defective' colour perception is the abnormal colour names ascribed to many familiar objects. The historic accounts of the condition first appeared as a result of incorrect colour naming and colour confusions which led to differences of opinion between colour normals and colour defectives.

3.4.1*   Dichromats

Maxwell (1855a) first used his colour triangle (figure 3.7) to demonstrate the confusion colours by plotting lines linking the confusion colours; these radiated from a point (or were parallel, using different reference colours).

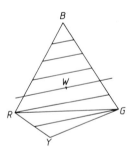

Figure 3.7   Maxwell's drawing of confusion lines superimposed on a colour triangle.

The missing mechanism of a dichromat was shown to be related to the point. Pitt (1944a) regarded such dichromatic confusion colour (isocolour) lines as meeting in a point representing a 'lost' fundamental sensation and suggested that, if parallel, they would be parallel to a line joining two 'fused' sensations. Helmholtz (1866) had earlier explained 'colour blindness' as a loss system, after Maxwell (1855a) had recognised the absence of a fundamental sensation on his colour triangle. Judd (1944, 1945) summarised the way in which experimental data from dichromats were used by König and Dieterici (1893). They derived the 'fundamental sensations' R',

G' and B' which were related by Judd to CIE stimuli as follows:

$$X = 0.2444R' - 0.058G' + 0.014B'$$
$$Y = 0.056R' + 0.150G' - 0.005B'$$
$$Z = 0.000R' + 0.000G' + |0.200B'.$$

From these figures an estimate of a copunctal point for a protanope was given as:

$$x_p = 0.81 \quad \text{and} \quad y_p = 0.19.$$

A series of dichromats studied by Pitt (1935) in Wright's London laboratory led to the confusion colour diagrams using the xy coordinate system of the CIE. These diagrams are shown in figures 3.8 and 3.9. Here the protanope's pseudoisochromatic lines radiate from the point $x_p = 0.747$, $y_p = 0.253$, while the corresponding deuteranopic point is for $x_d = 1.1$, $y_d = -0.1$ to a close approximation. The neutral part of the spectral locus occupied a pseudoisochromatic line with CIE standard illuminant B, the 'neutral' reference light used by Pitt.

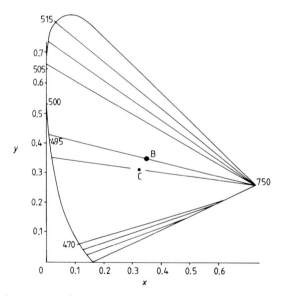

Figure 3.8 Protanopic confusion lines. The lines through standard illuminants B and C are shown. These lines limit one of the 17 isocolour zones described by Pitt. Six additional zones are shown, 3 above and 3 below.

Dichromatic wavelength discrimination data enabled Pitt to divide the 'dichromatic spectrum' into steps of just-noticeable-difference, which produced zones of isocolour, 17 for the protanope and 27 for the deuteranope.

Later Pitt (1961) developed an isocolour chart for tritanopia, in which the copunctal point was estimated to be very close to the short wavelength end of the spectral locus. Helmholtz (1866) recognised that colour mixing data and their geometric representation gave indication of the colours seen by the 'colour blind'.

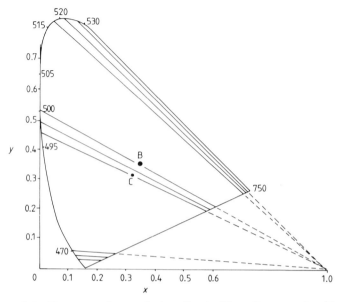

Figure 3.9 Deuteranopic confusion lines. The diagram should be compared with figure 3.8 noting that in this case Pitt identified 27 isocolour zones; again 6 are shown.

Tritanopia is a rare condition in its inherited form and some reports of 'cogenital tritanopia' may have been acquired conditions. Very slight uncertainty thus surrounds the exact location of the copunctal point (see figure 3.10). Wright (1952) and Thomson and Wright (1953) made important contributions, followed by Walraven (1974) who restated all three 'confusion centres'. Wright (1952) and Smith (1973) have detailed the colour confusions (related to naming) of tritans showing that blue and green are frequently confused, usually in the range of wavelengths 450 to 500 nm. Wright (1952) pointed out that 'yellow–blue blindness' is a misnomer since tritanopes require some red addition to blue for a yellow match.

The deuteranopic copunctal (or convergence) point was further established by Nimeroff (1970), who surveyed the data of several previous studies. Placing little reliance upon König's results, Nimeroff arrived at a mean value $x = 1.54$, $y = -0.54$ (for deuteranopia) and was able to compare this with the protanopic point (positioned as expected) in one remarkable subject who was protanopic in the right eye and deuteranopic in the left eye.

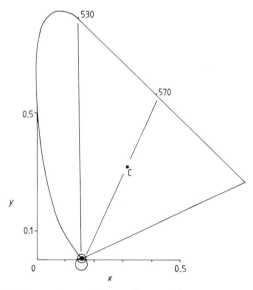

Figure 3.10 Tritanopic confusion lines. The convergence point is uncertain and is shown by the open circle touching the *x* axis. One well authenticated confusion line, 530–420 nm is drawn after Wright (1952) and Walraven's (1974) proposed point is shown as an encircled dot just above the *x* axis. Illuminant C is indicated so that the likely position of 'neutral' colours can be estimated.

There is need for caution in using these diagrams. Pitt's dichromatic data were averaged, and there could have been prereceptor pigment effects on the 'neutral locus', as indicated by Farnsworth (1961), who doubted the linearity of the isocolour lines. Pitt (1961) commented that the isocolour zones of a dichromat should be related to a white point and a colour triangle, both based on a trichromat with pigmentation similar to that of the particular dichromat; he also stressed the relatively wide dispersion of white points among deuteranopes. The 'classical' pseudoisochromatic lines are valuable, indicating likely confusions. Their uncertainty is reinforced by variations in the diagnosis suggested by many (if not all) pseudoisochromatic plates for some observers.

Tetartans are described as having two neutral points, a blue at approximately 470 nm and a yellow one at approximately 580 nm, with appropriate colour confusions. Judd (1943) suggested it was combined with 'impaired' red–green discrimination, although Willmer (1946) proposed that tetartanopes saw the spectrum only in terms of red and green. Walls and Mathews (1952) did not hesitate to consider the involvement of tetartanopia in some obscure cases; they also equated tetartanopia with absence of blueness receptors, including the small 'blue-blind' very central foveal region.

3.4.2*   Anomalous trichromats

A chromaticity diagram for the protanope can be drawn as a straight line, on which one part represents 'neutral'; this compares with the corresponding diagram for a normal trichromat which is triangular. Figures 3.11 and 3.12 show how intermediate forms of chromaticity diagram have been used by Farnsworth (1943) and Pitt (1949) for anomalous trichromatism. Such diagrams can be visualised as a tilting of the normal chart to an extent which is proportional to the degree of anomaly. The axis of tilt is at right angles to the 'neutral' axis in the projection of Farnsworth. Prereceptor pigment variations must be considered in both anomalous trichromatism and dichromatism as indicated by Pitt (1949) who discussed the difficulty of defining the boundaries between normals and dichromats, with respect to anomalous trichromatism.

Confusion lines for dichromats (figures 3.8 and 3.9) tend to be perpendicular to the major axes of normal MacAdam ellipses (figure 2.15).

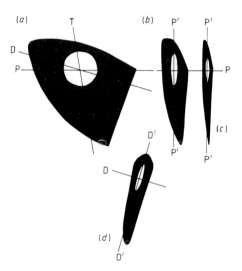

Figure 3.11 Anomalous trichromatism represented by a uniform chromaticity diagram (after Farnsworth 1943). (*a*) Shows a normal trichromat's diagram with a colour circle (white) on which the FM 100 Hue test colour samples can be found. Through the white point in this circle pass neutral isocolour lines for protanopes (P), deuteranopes (D) and tritanopes (T). The line P'–P' is at right angles to P and forms an axis about which an imaginary rotation of the diagram is possible to produce projections of successively greater degrees of protanomaly. Thus (*c*) shows greater protanomaly than (*b*) while in (*d*) a deuteranomalous diagram is shown, rotated about the axis D'–D' which is at right angles to D.

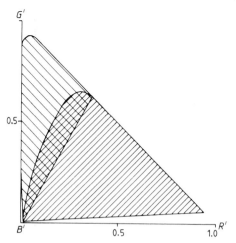

Figure 3.12 Chromaticity diagrams after Pitt (1949) based on Wright's R', G', B' axes. A protanope's diagram would consist of the line $B'-G'$ while a protanomalous diagram, PA, is intermediate between this and the diagram for a normal trichromat (NT).

Chapanis (1948) described the minor axes of these ellipses as being increased in anomalies, according to the degree of defect. According to this view the dichromatic confusion zones replace the ellipses.

In practice it is found that the confusions made by dichromats are followed to a greater or lesser degree by anomalous trichromats; in particular the just-noticeable step from white (which extends to the 'spectral neutral' for a dichromat) can be expected to be greater than normal. Hence the more desaturated forms of 'confusion colour test' such as HRR plates, or desaturated versions of D15, can detect the anomalous even if they do not confuse more fully saturated colours (see Chapter 7).

It is the line joining red and green on a typical colour triangle which contains the most striking confusion colours, for both protan and deutan subjects. Here we find the Rayleigh equation, used in the Nagel anomaloscope (see Chapter 7), and the total lack of wavelength discrimination show by both protanopes and deuteranopes. Anomalous trichromats show characteristic departures from normality for the matches involved. Nearby in the colour diagram is the locus of colours typical of a tomato during ripening, so readily confused by both protans and deutans.

The discovery by Lord Rayleigh (1881) of anomalous trichromatism led to the investigation of both protanomalous and deuteranomalous trichromats by means of their abnormal settings when yellow is matched by a mixture of red and green. Misnaming of colours, frequently desaturated colours, typical of the confusions of dichromats, is usual.

Von Kries (1911) indicated the poor wavelength discrimination frequently shown by the deuteranomalous between 535 and about 580 nm. He highlighted also the difficulty of colour recognition when the visual angle is reduced, for example when a signal light is presented at a greater distance. Since both spatial and temporal summation is critical, the viewing time must be greater than normal if an anomalous subject is to recognise a colour. Colour contrast is often accentuated in anomalous colour vision (see Chapter 7). The colour anomalous make errors of colour vision more readily when tired. The sum of these factors indicates clearly that it is hazardous for an anomalous trichromat to rely on recognising signal colours at a distance, in mist and when fatigued, when rapid decisions are essential for safety.

3.4.3 Abnormal colour perception

The sensations experienced by dichromats and anomalous trichromats when viewing coloured objects have been the cause of much speculation. Holmgren (1881) stressed the value of unilateral dichromatism in expressing the sensations in normal terms. He described the 'red-blind' as having yellow and blue colours in their spectrum, separated by a 'neutral belt' where blue–green is normally seen with 'yellow' applying to yellowish–green, yellow, orange and red. Similarly Edridge-Green (1920) gave the example of Dr Pole who was judged to be 'red-blind' by Maxwell, 'green-blind' by Holmgren and a 'simple dichromic' by Edridge-Green. Some older textbooks give coloured plates of the spectrum as seen by different individuals with defective colour vision.

Judd (1945) used the confusions and other responses of dichromats to derive functions called W_d, W_p and K to replace the X, Y and Z functions of colour normals. He used $W_d + K$ as a pair representing the vision of a deuteranope, while $W_p + K$ described a protanope, where W described the 'warmer' and K the 'colder' elements of the dichromat's vision. Later, Judd (1948) applied the data to the question 'what colours do colour blind people confuse?' and showed how many Munsell colours could be given 'deuteranopic' or 'protanopic' descriptions. In this way a nominal 10R 5/4 would have a renotation of 5Y 5.1/2.3 for a deuteranope with a 'deuteranopic reflectance' W_d ($= Y$) of 0.2028 and a deuteranopic chromaticity coordinate, w_d, of 0.5774. The same sample would have, for a protanope, W_p of 0.1749 and w_p 0.5477, with the protanopic renotation 5Y 4.7/1.7. The data indicate the lower response to this red stimulus on the part of the protanope. Thus $K = Z$, $W_p = 0.46X + 1.36Y + 0.1Z$ and $W_d = Y$ and the isocolours of the protanope are identical in K and W_p, while the isocolours of the deuteranope are identical in K and W_d. Extensive tables, examples of which are given below, show confusions.

Holmgren wool test confusions were described by Judd (1948) in such Munsell terms so that the red wool (IIb) was equivalent to the Munsell 5.5R

4.8/10 with protanopic renotation of Y 4.2/1.8 and deuteranopic renotation of Y 4.9/3.5. The Stilling and Farnsworth B-20 tests (see Chapter 7) were similarly considered.

	Munsell notation	Normal renotation†	$W_d = Y$	Deuteranopic renotation
(a)	RP 7/4	6.ORP 7.4/3.3	0.4885	N 7.4/
	N 7/	7.5PB 7.1/0.1	0.4433	5PB 7.1/0.2
	BG 7/2	3.5BG 7.3/2.1	0.4729	N 7.3/
			W_p	Protanopic renotation
(b)	N 7/—	7.5PB 7.1/0.1	0.4441	5PB 7.1/0.1
	BG 4/2	5.5BG 4.3/2.5	0.1475	N 4.4/

†Renotation refers to a revised and more equal spacing of the colours, produced in about 1943.

More recently Boynton and Scheibner (1967) and Scheibner and Boynton (1968) challenged the suggestion that the protanope does not experience red and green by their 'category assignment' method. In the later study it was established that a series of dichromats could distinguish between 'isochromatic' or confusion colours.

A lengthy denial of the spectrum of the protanope being confined to 'yellow and blue' was made by Walls and Mathews (1952).

3.4.4   Neutral or 'achromatic' points in the spectrum

It is frequently valuable to identify a dichromat as a protanope or deuter-anope, perhaps in order to study his family tree. Each is capable of iden-tifying a definite 'colourless' region of the spectrum as 'neutral', in hue and brightness, although previous colour difficulties may inhibit the admission that the name 'grey' is best. The spectral region (i.e. the wavelength, to the nearest nanometre) corresponds to that where wavelength discrimination is best. A neutral point estimation is thus a valuable feature of a wavelength discrimination test.

A different approach requires the subject to match a neutral standard (such as a filtered light) to one part of a spectrum. If a monochromator sup-plies one half of a field and the 'neutral' occupies the other, a rapid com-parison is made. The appearance of a neutral spectral zone by dichromats results from the position of neutral or white on the CIE chromaticity diagram relative to the confusion loci of dichromats (see §3.4.1). Thus protanopes confuse red with neutral and blue–green while deuteranopes confuse neutral with green and a non-spectral hue, a mixture of red and blue.

The first measurements of neutral points in dichromats were made by König (1884). The work of Pitt (1935) and Judd's (1944) contribution are essential reading for those concerned with precise differentiation. Walls and Heath (1956) stressed how a dichromat (and his classification) is proved only when a neutral point is established, emphasising to both geneticists and physiologists the need for this definitive test. Walls and Mathews (1952) provided an extensive discussion but in a context somewhat superseded by Walls and Heath in 1956, at which time these authors used disc mixing of Munsell papers (see table 3.2(a)).

Pitt (1935) found the neutral points for five protanopes to be 495.5 nm and for six deuteranopes 500.4 nm (white source 4800 K). Wright (1952) found the neutral point for five tritanopes to be 570 nm.

An extensive study of dichromatic neutral points by Massof and Guth (1976) involving both foveal and peripheral ($15°$ nasal) viewing, indicated values displaced towards shorter wavelengths for peripheral viewing (shift of 11 nm for deuteranopes and 4 nm for protanopes). Five hundred settings were made by each of three protanopes and three deuteranopes for each viewing situation. The results are presented in table 3.2(b). Protanopes thus locate 'neutral' in a blue–green part of the spectrum and deuteranopes in a region which is more green, some 5 nm longer in wavelength than the position for protanopes. The exact values depend on the apparatus used, in particular the colour temperature of the source. Pitt's values of 490 and 494 nm are typical but subject to variation of the order of ± 3 nm (see figure 3.13).

Table 3.2 Neutral points. See text for details.

(a)   Protanopes (39 observers)
       Illuminant C comparison 492.3 nm (490.9–494.6; with SD 0.70)
       Illuminant D comparison 490.3 nm (488.7–492.0; with SD 1.05)
    Deuteranopes (38 observers)
       Illuminant C 498.4 nm (495.9–501.7; with SD 1.21)
       Illuminant D 496.2 nm (493.7–498.8; with SD 1.25)
    (data from Walls and Heath 1956)

(b)   Protanopes (data from 3 observers based on 500 settings)
       Observer 1   Mean 485 nm   SD 4.9 nm
       Observer 2   Mean 493 nm   SD 4.0 nm
       Observer 3   Mean 492 nm   SD 7.4 nm
    Deuteranopes (data from 3 observers based on 500 settings)
       Observer 1   Mean 497 nm   SD 4.3 nm
       Observer 2   Mean 495 nm   SD 5.9 nm
       Observer 3   Mean 495 nm   SD 3.6 nm
    (data from Massof and Guth 1976)

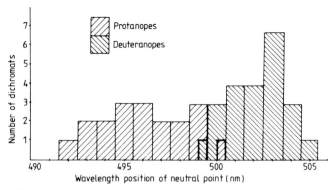

Figure 3.13   Neutral points chosen by dichromats, showing data for 19 protanopes and 24 deuteranopes (after Crone 1961).

Tritanopic dichromats identify a neutral point at about 580 nm, depending on the reference stimulus; if a 'least colour' position is located in successive parts of the spectrum, from a monochromator, a trichromat should choose a position between 570 and 598 nm. König (1884) studied six cases of acquired tritanopia and showed the neutral point to lie between 566 and 570 nm. The tritanopic observer examined by Gothlin (1943) gave a neutral point of 571 nm, that of Fischer *et al* (1951) at 570 nm for a white standard of 4800 K. Voke-Fletcher and Fletcher (1978) obtained a setting of 572 nm by this method with a less definite indication of achromatism around 402 nm; there is a tendency for the violet end of the spectrum to be colourless in the tritanope but tungsten sources give very little radiation in this part of the spectrum and the observation can fail to be made under photopic conditions.

Although neutral points do not appear to anomalous trichromats these subjects frequently confuse suitably desaturated 'blue–green', 'red' or 'purple' colours and neutrals of matching brightness, according to their type of defect.

3.4.5   Abnormal relative luminous efficiency

Foveal measurements under photopic conditions and those involving spectral regions from green to red are of greatest interest for a study of colour vision deficiencies. Measurements on the V_λ function of the red–green varieties of colour deficiency include the classical studies of Pitt (1935) on dichromatism and of Nelson (1938) on anomalous trichromatism. A century before, Seebeck (1837) noted two kinds of luminosity function among the colour blind.

Dichromats

The features for deutans and tritans are similar to those of normals.

Protans, however, show a marked loss in sensitivity in the long wavelength region, and as a result both protanopes and protanomalous observers see reds as unusually dark. This darkening or shortening of the long wavelength region for photopic vision was demonstrated by both Macé and Nicati (1879) and von Kries and Küster (1879), according to Graham (1965). The difficulties this may pose in the recognition of signal lights are discussed in Chapters 7, 11 and 12. Judd (1943) described a simple conversion for the luminosity values of protan defectives. While some investigators have noted a luminosity loss in some deuteranopes in the blue–green to green spectral regions (Hsia and Graham 1957) and in the blue spectral regions (Alpern and Torii 1968a) a luminosity gain for wavelengths above 520 nm was reported by Heath (1958, 1960) for some deuteranopes. A practical point raised by Chapanis (1944) was that dichromats tend to be quicker in carrying out V_λ measurements than normals; the more severe the colour defect the easier the subject seemed to find brightness equality judgments. Greater variations in the V_λ curve for deuteranopes than for deuteranomalous observers was noted by Collins (1959) when observations were restricted to cone function. The inter-subject variation among deuteranopes is thought to be accounted for by individual macular pigmentation variations (Alpern and Torii 1968b). The protanope shows a photopic peak about 545 nm, indicative of his 'missing' long wavelength system, while the deuteranope's peak is around 580 nm. Tritanopes are more normal but anomalous trichromats' curves occupy positions between the normal and the appropriate type of dichromatic V_λ curve. Figures 3.14 and 3.15 should be consulted.

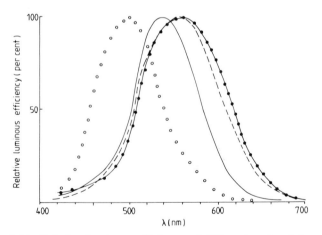

Figure 3.14 Relative luminous efficiency of different parts of the spectrum for: ———— protanope; •–•–• deuteranope; – – – normal photopic; ooooo normal scotopic.

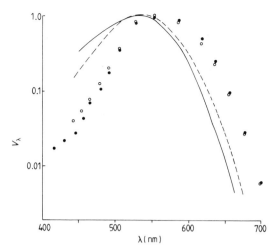

Figure 3.15 V_λ curves on a log scale for: ∘∘∘∘ normal; •••• tritanopic; −−−− protanomalous; ——— protanope.

The tritanopic V_λ function, studied by Wright (1952), showed only slight departures from the colour normal with minor luminosity losses in the short wavelength region below 500 nm, as interpreted by Farnsworth (1955b) and Fischer *et al* (1951). Wright's study of seven tritanopes showed 'little evidence of any significant lowering at short wavelengths' in the V_λ function. He noted too that the short wavelength mechanism is not usually thought to contribute greatly 'to the photometric value of light', so a reduction might perhaps not be expected. Some cone monochromats appear to have protanopic tendencies for the wavelengths longer than their peak of about 550 nm, with 'normal' characteristics for other spectral regions. Rod monochromats show 'scotopic' curves even for photopic intensities.

An interesting disturbance of the Purkinje shift, determined by Hough and Ruddock (1969) was apparent at different intensities.

Anomalous trichromats

The spectral luminous efficiency function for anomalous trichromats is similar to, though less pronounced than, that for dichromats. Early studies were made by Nelson (1938) on six deuteranomalous observers, and by McKeon and Wright (1940) on eleven protanomalous observers. The extreme anomalous trichromats show characteristics which are more akin to dichromatic vision, whereas many mild anomalous trichromats may show a normal V_λ curve. Alpern and Torii (1968) noted protanomalous observers to be less sensitive than protanopes in the blue−green region but more sensitive to reds. Verriest (1971), who used a flicker comparison, found few differences between the deutans (deuteranopes compared with deuteranomalous observers) and protans (protanopes compared with protanomalous observers). The protanomalous trichromat is similar to the protanope, with reduced sensitivity in the long wavelength region. Collins (1959) noted deuteranomalous luminosity to lie between normal and deuteranopic luminosities, although the results were not thought to be statistically significant. Kinnear (1974) believed the extent of the Rayleigh matching range on the anomaloscope to be a good indicator of V_λ variations from normal. A greater sensitivity in the blue–green region and less sensitivity to reds was noted by Alpern and Torii (1968b) and Kinnear (1974) for deuteranomalous observers compared with deuteranopes. Kinnear believed this observation is only significant at a limited number of wavelengths. Parsons (1924) noted that since all anomalous trichromats have essentially normal scotopic (V_λ') curves, their differences under photopic conditions should not be attributed to variations in crystalline lens transmission. Nevertheless the prereceptor media, including macular pigment, have some effect on photopic curves and should be considered in detailed assessments. The way in which V_λ performance aids detailed investigations of the colour anomalous is exemplified by Smith and Pokorny (1972, 1975) and by Pokorny and Smith (1981). The value of these methods dates back to Hecht and Shlaer's (1936) work and beyond.

Carriers

The important discovery by Schmidt (1934, 1955) that protan carriers frequently give a manifestation of their genetic make-up by a partial change in the V_λ curve (Schmidt's sign) provides a useful tool for the identification of such heterozygotes. Crone's (1959) data indicated the need for caution if a 'luminosity-quotient' is used in this context but he was able to identify some deuteroheterozygotes by this method. De Vries (1948) detected disturbances of the V_λ curve in carriers of a deutan defect in the shorter wavelengths of the spectrum, but Crone showed overlapping with normal results. The distortion of the normal V_λ curve which results from an adapting field of red light, shown by Ikeda *et al* (1974), provides a comparatively simple method of using a flicker photometer. This distortion was not evident in deutans but the technique enabled the identification of deuteroheterozygotes.

The V_λ curve in acquired defects

Foveal spectral sensitivity has been investigated in cases of acquired colour disturbance by a number of investigators, notably Francois and Verriest (1957a), Jaeger and Grützner (1963) and Marré (1973), although classification on this basis alone is difficult. The photopic spectral sensitivity represents a summation of the three basic colour vision mechanisms and where one colour mechanism is selectively impaired by disease the V_λ curve reflects the reduction in sensitivity accordingly. Marré (1973) noted that in most retinal diseases the V_λ curve indicates disturbances of all three colour channels. Although there is a general reduction in all spectral regions in optic nerve diseases, a distinction can be made between one or other of the two possible sites of the deficiency, because sensitivity in the long wavelength region is affected more in retinal than in optic nerve disease. In senile nuclear cataract Verriest (1970) noted a V_λ shift towards long wavelengths, while shifts towards short wavelengths were found in retro-ocular disease and strabismic amblyopia.

3.4.6 Wavelength discrimination

Dichromats

The wavelength discrimination for protanopes and deuteranopes was measured by Pitt (1935) and for tritanopes by Wright (1952). Typical curves are shown in figures 3.16. The protanope shows a minimum at 490 nm, where discrimination is maximum; this position in the spectrum corresponds to the neutral point for protanopes. Discrimination rapidly deteriorates for both shorter and longer wavelengths. The deuteranope's minimum is at 495 nm, which also approximates to the neutral point; discrimination is poor at other spectral regions, giving a characteristic U-shaped curve. The tritanope has two spectral regions where colour discrimination is best, near violet and near 570 nm as shown by Fischer *et al* (1951), Wright (1952) and Voke-Fletcher and Fletcher (1978) (see figure 3.17). These dips correspond to the tritanopic neutral points. Unusual curves for dichromats presented by Balaraman *et al* (1962) were similar in many respects to those found by Graham and Hsia (1958a,b); at wavelengths greater than 560 nm evidence of wavelength discrimination was sometimes found. The data could have been affected by 'saturation discrimination' or attributable to stray light. Hecht and Shlaer (1936) noted how dichromats may use saturation clues as a basis for wavelength discrimination judgments.

The wavelength discrimination peak in deuteranopia was shown by Walraven and Bouman (1966) to alter according to the intensity of the stimuli. The peak was at about 495 nm (for 10 trolands[†]) but nearer to 570 nm for 0.1 trolands, where sensitivity was lower. Trick *et al* (1976) confirmed a change with intensity.

[†]1 troland is the retinal illuminance when a surface of luminance one candel per square metre (1 cd m^{-2}) is viewed through a pupil one square millimetre in area.

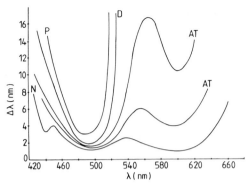

Figure 3.16 Wavelength discrimination data, assumed to be for 2° fields. N, normal; AT, anomalous trichromat; P, protanope; D, deuteranope.

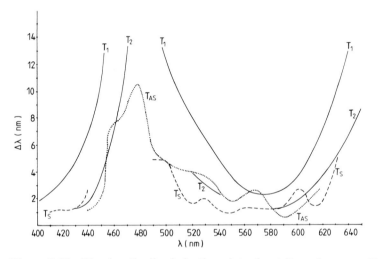

Figure 3.17 Wavelength discrimination data for tritan observers. T_1 represents the (bimodal) classical description of the performance expected of a tritanope. T_2 shows the subject F of Voke-Fletcher and Fletcher (1978). T_S is a tritanope's (INCD) performance (after Smith 1973). T_{AS} follows Schmidt's (1970) tritanomalous data.

Anomalous trichromats

Measurements of wavelength discrimination for 21 anomalous subjects were made by Engelking (1925), as reported by Jameson and Hurvich (1956b). The wavelength discrimination data for 6 deuteranomalous observers measured by Nelson (1938) was extended by measurements for 11 protanomalous subjects by McKeon and Wright (1940). There are large inter-subject variations but generally the curves lie between those of the normal and of the dichromat; the more severe the defect, the more similar the curve is to that of a dichromat. The region of best discrimination is usually between 490–500 nm. Frequently two minima are noted, one in this region and a second near 600 nm (see figure 3.18). Pitt (1949) related the absence of the 440 nm sensitivity dip in protanomaly to reduction of the secondary maximum of the 'red' response. The lack of one normal response system might be expected to shift the peak of discrimination at both 500 and 600 nm and such shifts were indicated by Motokawa and Isobe (1954). Hurvich (1972) predicted appropriate variations, as shown by McKeon and Wright (1940). In figure 3.18 the protan dip at about 580 nm (or the deutan at about 610 nm) is seen to be higher than the more constant 495 nm 'dip'.

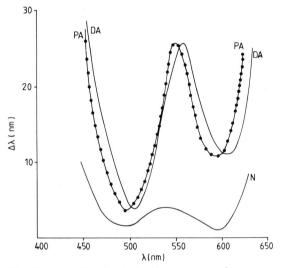

Figure 3.18 Wavelength discrimination curves for two anomalous trichromats one protanomalous (PA) and one deuteranomalous (DA), each of the same degree of anomaly. The curve N is for normals.

Although the severity of defect in anomalous trichromacy is evident from the wavelength discrimination curve, there appears to be little correlation between the wavelength discrimination curve of an anomalous trichromat and his Nagel anomaloscope setting. McKeon and Wright (1940) failed to find such a correlation. Schmit (1970) and Smith (1973) reported the hue discrimination curves of congenital tritanomals.

Acquired defects

Classification of acquired defects by means of wavelength discrimination is difficult since variable characteristics are found. Measurements were made by Grützner (1963) and more recent studies have been made by Marré (1973). In cases of optic nerve disease showing red–green disturbances, useful wavelength discrimination data can be obtained in two spectral locations, one near 480–500 nm and one near 590 nm, as shown by Grützner (1966). In optic nerve disease where the blue colour mechanism is involved as well as the red and green mechanisms severe impairment of wavelength discrimination is noted, with the 590 nm region being more affected than lower wavelengths. It is not possible to differentiate reliably between the blue and red–green acquired defects in retinal and optic nerve disorders on the basis of wavelength discrimination (see Marré 1973).

3.4.7*   Colour mixture—colour defectives

Dichromats

Using his colour box spectral stimuli, Maxwell (1860, 1890c) was the first to demonstrate dichromatic vision in a protanope who required only two matching stimuli to match spectral wavelengths. Individual variations were demonstrated. These measurements, together with those he made of colour normals, represented the first colour mixture data. With his colour top Maxwell further showed how the colour blind differed from normal and reported this in the text by Wilson (1855). The redundancy of one of the trichromatic stimuli for dichromats was shown further by König and Dieterici (1893) and Pitt (1935), for both protanopes and deuteranopes, and later by Wright (1952) for the tritanope. Wright (1946) described their matching as being with two controls only. Judd (1943, 1945) explained the need for a dichromat to involve mixtures of only two coloured lights to achieve his complete gamut of colour experience. Furthermore he noted that, just as certain lights of different spectral composition can appear to match in colour for normal subjects, as 'metamers', so a dichromat usually accepts these normal matches and has additional matches, or 'confusion colours' distinguishable by normal observers.

Pitt's measurements indicated that both deuteranopes and protanopes need the red 'primary' to be mixed with a blue test light for a match with the blue 'primary'. Deuteranopes require relatively little of the blue 'primary' to be mixed with the red 'primary' in a match for long wavelength test lights. Hecht and Shlaer's (1936) data indicated that the deuteranopes needed relatively more blue.

Scheibner (1973) asked a deuteranope and a protanope to view the colour matches of normal trichromats and anomalous trichromats, with similar results.

Using a tristimulus colorimeter employing interference filters Vogt (1973) measured spectral mixture curves for protanopes, deuteranopes, protanomalous and deuteranomalous observers.

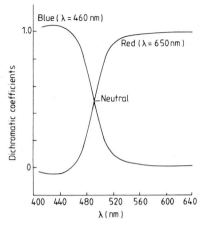

Figure 3.19 Colorimetric data typical of protanopes (after Pitt and Wright) where two stimuli (a red and a blue) were mixed to match in turn monochromatic test colours from parts of the spectrum. The use of red ensures that desaturation of the test colour is avoided between 460 and 650 nm. The neutral point, marked as found by Pitt (1935), corresponded to a relatively 'red' comparison 'white'.

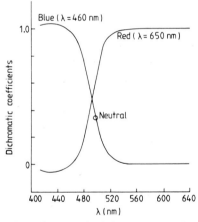

Figure 3.20 Data for a typical deuteranope after Pitt (1935).

Thus the terms 'trichromatism' and 'dichromatism' arose in the context of experimentation with colorimeters using three matching stimuli. Colour mixing, using wavelengths 458.7 and 570 nm, was reported by Hecht and Shlaer (1936) and formed the subject of comment by Fry (1944), including the disadvantage of altering the monochromatic test stimulus to match different mixtures. Hecht and Shlaer indicated how the colour mixture of the normal trichromat is poor in regions where wavelength discrimination is reduced. On this account only wavelengths between 480 and 520 nm give reliable results for dichromats. The variable results shown by a protanope are indicated in figure 3.21 indicating how variability is greatest in spectral regions where wavelength discrimination is poor. Corresponding results are noted in deuteranopes. The colour mixture data, taken with V_λ for the appropriate dichromat, can predict the spectral neutral point using a method described by Fry (1944) together with a method of differentiating between dichromats and anomalous trichromats; dichromats would readily match illuminant C with a mixture of the 458.7 and 570 nm lights. A much discussed account by Graham *et al* (1961) of a 'unilateral dichromat' included colour mixing curves resembling those described above. Most colour mixing today is with spectral stimuli, but Birch (1973) used a commercial subtractive colorimeter for matches of desaturated test objects, showing matching ranges. The relative imprecision of the method revealed normals with unusually large MacAdam ellipses; deuteranomalous observers gave the expected increases in range.

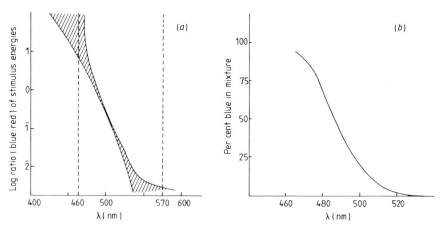

Figure 3.21 (*a*) Colour mixing by a protanope after Hecht and Shlaer (1937). The shaded area shows variability in the mixture. (*b*) Data in (*a*) expressed as a percentage of the blue stimulus (after Fry 1944).

Colorimetric data thus provide useful information about the characteristics of defective colour vision. The mixture curves for protanopes and deuteranopes are so similar that it is difficult to distinguish between them on such evidence alone. Nevertheless the ratios of the primaries for the long wavelength end of the spectrum do differ slightly for the two types of dichromats. Jaeger and Kroker (1952) measured the colour matching properties of dichromats using a Newton's top disc for central ($1\frac{1}{4}°$) small fields and large fields (22°). Their results indicated that some dichromats are dichromatic under both conditions while others show features of anomalous trichromacy for large fields.

Anomalous trichromats

The discovery by Rayleigh (1881) that some observers, although requiring all three tricolorimeter stimuli, made abnormal colour matches in the red–green spectral range led to the introduction of the term anomalous trichromatism. His observations were made using a simple colour apparatus involving the mixture of just red and green to match yellow, on which Nagel based his anomaloscope. The abnormally large amount of red required by the protanomalous observers, and the abnormally large amount of green required by deuteranomalous observers in a mixture of red and green to match yellow was recognised. These abnormal proportions for a match are unacceptable to colour normals and the 'normal match' is generally unacceptable to the anomalous observer. Nagel noted wide variability among both deuteranomalous and protanomalous observers for small-field colour mixing involving red and green (see von Kries 1897). Anomalous trichromats were examined by Nelson (1938) and McKeon and Wright (1940), disclosing considerable variations in trichromatic coefficients, e.g. frequent absence of negative red for protans and greater matching tolerances for yellow parts of the spectrum.

Relating trichromatic vision and colour mixture curves

Although colour mixture data give some insight into the sensitivity of the visual system to different wavelengths it is impossible to draw exact parallels between the physical stimuli and visual sensations. Visual response depends upon a great many factors such as adaptation effects: such information is not presented in a single set of colour mixture curves. Thus it is not possible to predict the sensitivities of the photoreceptors from such data alone. Thomson and Wright (1953) proposed a possible set of sensitivity curves of the retinal receptors based on colour mixture data, but emphasised that such curves 'are typical of mixture curves which could represent the receptor sensitivities'. Nevertheless they stressed that colour mixing data are determined by the spectral sensitivity curves of the retinal receptors and as such they do provide information about the absorption characteristics of the visual pigments. The trichromatic theory proposed by Young (1802a) received experimental backing from the colour mixture experiments of Maxwell and Helmholtz which served to set it on a strong foundation.

3.4.8 Abnormal saturation discrimination

Hue, saturation and luminosity interact in a colour and Hecht and Shlaer (1936) argued appropriately that a dichromat's wavelength discrimination is a matter of spectral saturation differences; this is particularly true since the dichromatic 'neutral point' is a central peak of 'desaturation'.

Anomalous trichromats have more difficulty with desaturated colours, which they are often liable to confuse with grey. Chapanis (1944) demonstrated a basis for the quantitative scaling of different anomalies using saturation discrimination. While his normal subjects easily noticed when two samples (not alike in brightness) were of the same hue, his Daltonic observers did not; they could easily match a colour to neutral by a slight brightness adjustment. When two greys of slightly different brightness were compared, Chapanis found that dichromats and some anomalous trichromats described the darker one as 'coloured', an effect which could be reversed by brightening the darker sample. In these experiments anomalous subjects were found clearly to differ from normal in saturation threshold. Protans gave data different from deutans and the amount of 'decrease in saturation' appeared to be related to the degree of anomaly, so this was seen as a possible quantitative scale for defects (see figure 3.22). Normal features of this threshold have been described in §1.17 and test methods are outlined in §7.10.4.

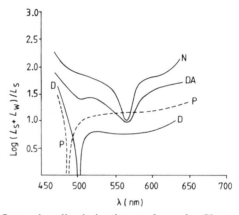

Figure 3.22 Saturation discrimination as shown by Chapanis (1944). N, normal subject; DA, deuteranomalous; D, deuteranope; P, protanope.

Both inherited and acquired defects of colour vision can be studied by saturation thresholds and while most attention has been paid to red–green defects, Cole *et al* (1965) described typical tritan characteristics. Abramov and Gordon (1977) found a tritan tendency in the apparent desaturation of eccentrically viewed coloured objects, more often associated with 'deutan' features of the peripheral retina. They showed that for stimuli of about 6.5° all colours appeared of high saturation. Red has more consistency in different locations, appearing saturated to normals. Frisen (1973) has maintained the superiority of coloured stimuli for perimetric study of early pathology.

Early measurements of just-perceptible-differences between white and spectral wavelengths were made by Jones and Lowry (1926), Martin *et al* (1933) and Wright and Pitt (1937).

A recent study by Kogure (1980) has utilised the Lovibond colour vision analyser (see §7.6.6) to measure saturation sensitivity among colour normals, particularly with respect to aging effects. A sample of 73 Daltonics examined similarly showed depressed saturation sensitivity corresponding to spectral regions 422–510 nm, and around 493 nm for protanopes with suppression in spectral regions 496–510 nm and at 510 nm for deuteranopes.

3.5* MONOCHROMATISM OR ACHROMATOPSIA

Two main types of complete defects are recognised—typical achromatopsia (rod monochromatism) and atypical achromatopsia (cone monochromatism). In the literature the term 'achromatopsia' has tended to be associated with American and Continental European investigators, while the term 'monochromatism' has been favoured by British clinicians and researchers.

The rarity of both types makes an accurate assessment of incidence extremely difficult; the figures given by Weale (1953c), Duke-Elder (1968) and Krill (1977) as 1 in 30 000 (0.003%) and 1 in 35 000 for typical achromatopsia and 1 in 10^6 and 1 in 10^8 for atypical achromatopsia, are usually taken.

3.5.1 Typical achromatopsia

Characteristics

Typical achromatopsia involves the gross malfunction/modification of the retinal cone mechanism, essentially the absence of functioning cone receptors. Vision is confined to an essentially scotopic mode mediated by receptors operating at low luminance levels. The exact status of the cone receptors is uncertain (see the next section) but the rod receptors operate normally giving a relative luminous efficiency curve of a scotopic form, with no Purkinje shift even at high luminance levels. As Smith and Pokorny (1980) indicate in their excellent review, the scotopic mechanism can therefore be studied in achromats at a level of illumination at which in normals it is masked by the photopic mechanism.

Other visual functions under the control of photopic vision, chiefly visual acuity, are impaired in typical achromatopsia. The alternative term 'rod monochromatism' arises on account of the dominant role of the rod receptor mechanism in these individuals.

Typical achromatopsia involving cone defects is characterised by the following.

Total loss of colour discrimination.

Low visual acuity—typically around 6/60 (20/200) although it can be better.

Eccentric fixation—although approximate foveal fixation may gradually become established with a decrease in nystagmus.

Electroretinographic anomalies—photopic component reduced or absent.

Flicker abnormalities—typically low fusion frequency in response to intermittent light stimuli.

Photophobia or light aversion.

Nystagmus—usually lateral/pendular which may diminish after the first two decades of life.

Family history of consanguinity.

Other associated characteristics which may be present include the following.

Visual field defects—typically a central scotoma because of impaired cone function.

Refractive errors chiefly myopia and squints, concomitant strabismus or exotropia, but this is questionable.

Fundus abnormalities.

Absence of photopic section of dark adaptation function.

Occasionally associated impairment of rod function.

Several characteristics need further clarification. Photophobia, a characteristic of typical achromatopsia which distinguishes it from the atypical variety, arises because low intensity receptors, which are unable to cope with high levels of illumination, predominate. All 19 cases (3 incomplete) studied by Sloan (1954) showed photophobia typified by frequent blinking and a tendency to close the lids. It is seldom truly painful however, and can be controlled adequately with tinted spectacles. As a consequence retinal illumination is reduced to the level appropriate for maximum visual acuity. Only four of the nineteen had normal fundi under close examination, although a casual glance with the ophthalmoscope and fluorescein angiography seldom reveals abnormalities. Slight disc changes, pallor and irregularities of the macula were noted in eight of Sloan's patients. Evidence of an absent foveal pit is suggested in some inherited cases of achromatopsia and macular aplasia—a decreased, absent or irregular foveal reflex is often characteristic of the condition.

Francois and Verriest (1961b) quoted 20% of cases as having optic disc changes—notably pallor. Auerbach (1974) found 'binocular amblyopia and severe refractive errors' in all 39 cases he examined. Krill (1977) noted that near vision may be better than distance vision possibly because convergence reduces the nystagmus.

Evidence for residual cone function

Although the characteristics of complete achromatopsia suggest a lack of cone function, psychophysical studies do indicate that a second receptor type operating at higher luminance ranges is present in many subjects. The nature of these 'day rods' is uncertain but it is generally believed that residual cone photoreceptors containing a visual pigment, similar to rhodopsin in its spectral attributes, may be involved. Hecht *et al* (1948) described a case of rod monochromatism showing a single-function dark adaptation curve but flicker and intensity discrimination studies both suggest evidence of a dual system. Recent investigators have also indicated the duplex function from visual acuity, flicker and increment threshold studies on achromats (Verriest and Uvijls 1977). Thus the spectral sensitivity of the so-called 'cone' component was of a scotopic nature. Earlier Wolfflin (1924) and Sloan and Newhall (1942) had shown a typically normal dual dark adaptation function, a characteristic which was confirmed in three siblings by Lewis and Mandelbaum (1943) and two cases by Walls and Heath (1954). Sloan (1954, 1958) found that 11 out of 14 rod monochromats showed similar evidence for receptors other than normal rods, particularly after high levels of pre-exposure. This leads to what Alpern *et al* (1960) call the 'enigma of total monochromacy'. Measurements on four subjects showed the distinct directional sensitivity effect (Stiles–Crawford effect I) typically associated with cones, yet displayed by receptors showing characteristic scotopic spectral sensitivity. Further evidence of a residual photopic function is present in the majority of 39 achromats studied through electroretinographic studies by Dodt *et al* (1967) and Auerbach and Merlin (1974) although non-conventional techniques are required to reveal it.

There are three possible explanations for these observations. Firstly that the 'photopic receptors' of the achromat lie inactive, as suggested by Sloan (1954). Secondly that the photopic receptors are normal cones in anatomical form but contain the rod pigment visual purple (rhodopsin) instead of the cone pigment (Hecht *et al* 1948, Alpern *et al* 1960). The difference in sensitivity at the the blue end of the spectrum between the scotopic curve and the 'photopic receptors' can be explained by the absorption of macular pigment—a fact that was known even to Hering (1891). This hypothesis cannot account for the duplex dark adaptation curve found, with careful examination, in some achromats.

Human cone pigment studies are not far enough advanced to provide a full understanding of total colour blindness. In the published discussion on the work of Alpern *et al* (1960) Rushton noted that a fault in the nervous organisation of the achromat must not be ruled out, neither must a fusion of blue and green cone pigments. He considered the 'high level photopic receptor' to be cone structures, perhaps appearing to have the strange combination whereby a pigment with the kinetics of a cone pigment but displaying the spectrum of a rod pigment is present. Since Conner and MacLeod (1976) have been able to measure a duplex flicker function from the rod receptors of colour normals it is possible that the duplex function in achromatopsia may be mediated by rod receptors in some way.

The final explanation, by Walls and Heath (1954), suggested that the second receptor class was 'blue cones' but these receptors did not contribute to brightness.

Histology of complete achromatopsia

Anatomical evidence is sparse and the findings of four studies to date are in some respects contradictory, particularly observations on peripheral 'cones'. Reports agree on the presence of rods of normal appearance. Foveal cones are typically shorter, fatter and distorted more than in normal trichromats but variations in their density are wide, including two reports that they are very sparse or even absent (Harrison *et al* 1960, Glickstein and Heath 1975). Larsen (1921), the first to publish histological findings from a complete achromat, noted normal peripheral cones but abnormal foveal receptors. The combination of a psychophysical study followed by histological analysis from the same eye is most valuable; Falls *et al* (1965) tested one eye psychophysically and the other histologically. Foveal cones were abnormal but of normal density with no evidence of a foveola. The proportion of extrafoveal cones was markedly lower than normal.

In almost every case there is the difficulty of changes in the region, either *post mortem* or degenerative processes. This situation revives interest in functional aspects, including variations during life. Anatomy certainly points the way to the strictest control of visual fixation during testing. More studies would be highly advantageous.

Electrophysiological studies

Because various electrophysiological functions of the visual system reveal photopic and scotopic contributions, the ERG (electroretinogram) and the VECP (visual evoked cortical potential or response) are valuable clinical tools for the examination of achromats. The ERG essentially reflects neural activity at an early level of processing in the visual system. The VECP indicates much 'higher' activity, predominantly photopic in nature, which can be used to establish the duality of visual function at scoptopic and photopic levels through a measure of the latent period following light stimulation. A complete lack of photopic function has been shown for the ERG in complete achromats firstly by Vukovich (1952) and more recently by Francois *et al* (1963) but variations do occur.

Smith and Pokorny (1980) indicate how the high-radiance scotopic ERG can vary with the different types of achromatopsias—complete and incomplete.

3.5.2 Atypical achromatopsia

Achromatopsia with normal visual acuity is exceedingly rare; it was considered to have an incidence of around 1 in 10^8 by Pitt (1944b) or 1 in 10^6 or 10^7 by Weale (1953c) and only 15 cases have been reported in the opinion of Francois *et al* (1955). These individuals have 'normal' cones, which accounts for their good visual acuity, show no nystagmus or photophobia and have normal dark adaptation and ERG functions, but have a gross (but not always complete) loss of colour perception. The retinae of these individuals appear normal by ophthalmoscopy. In early literature they are referred to as cone monochromats. The fault of the colour defect is considered to be at a higher level than the retina. It is as if normally processed information in the retina is relayed to higher centres of the neurone system where it does not register in colour (see Willmer 1946). Fincham (1951) confirmed this cortical site by noting an accommodation response to mixed colours in the blurred retinal image which showed 'retinal' but not 'cortical' colour differentiation. The presence of normal retinal cone photopigments in the cone monochromat has been shown objectively by analysing the light reflected from the fundus (fundus reflectometry) and by selective adaptation techniques combined to measure subjectively the spectral sensitivity of the individual colour vision mechanisms (Stiles mechanisms). Weale (1959) was able to show the presence of the red and green pigments by the first method and Gibson (1962), using the second, showed the presence of three colour mechanisms. However, some confusion arises, because Alpern (1974) could trace only the red cone pigment in his subject and the presence of a blue retinal mechanism by psychophysical means, which led him to conclude that a retinal defect was combined with a post-retinal cause.

Two sub-types are recognised based on the photopic spectral sensitivity characteristics. One type shows a protanopic V_λ curve with a reduced sensitivity to reds, the other a normal V_λ curve. Colour matching indicates monochromatism, for all spectral wavelengths can be adequately matched by one varying control. Colour discrimination tests such as the FM 100 Hue test indicate significant errors (200–300 total error score) with a distinct tritan axis as noted by Krill and Schneiderman (1966). Red–green screening tests can indicate either deutan or protan confusions. As for other achromats, a large field of view is recommended for examination which can frequently yield residual colour perception.

3.5.3 Colour vision examination of typical achromatopsia

Colour vision test performance shows gross abnormality with frequent anarchic responses, and there is a characteristic oddity about the way in which persons who have never seen colours use colour names. One of Sloan's patients, for example, used the term 'hunter green'. Although there is by definition, a significant and usually complete absence of colour discrimination, Sloan (1954) reports that many achromats can name colours correctly; she advocated simple matching tests as being more valuable than colour naming.

The distinction between the inherited and acquired forms of total colour blindness cannot be made solely on the basis of clinical colour vision tests; other clinical symptoms, and in particular careful recording of history, must assist in the diagnosis. Sloan noted that in cases of inherited total colour blindness different colour names may be given to a series of greys; this rarely occurs with a person who once possessed normal colour perception.

Complete forms

Pseudoisochromatic plates can give a misleading diagnosis and classification as red–green anomalous. Some achromats can pick out the figures on account of brightness clues, but Sloan (1958) noted that the HRR plates differentiate reliably between achromatopsia and the red–green defects.

The Farnsworth D.15 test has proved a valuable tool for diagnosis. Despite wide variations among subjects and in repeated performance of the same observer, a typical 'scotopic' axis is frequently noted lying randomly between the deutan and tritan axes, to give what Sloan (1954) called 'an order of decreasing scotopic reflectance'.

Sloan designed a simple achromatopsia test comprising seven sets of Munsell papers. Each reveals a small spot of colour (or neutral) usually of Chroma 5 on a series of 17 neutral backgrounds which range from N1 to N9 (in 0.5 steps). Thus R5/12 tends to appear as bright as N4 (or less) and G5/8 appears as bright as N5.5. The task required is the matching of a colour to some level of neutral grey varying from black to white. While this is a most difficult and confusing task for the colour normal or red–green deficient, the achromat finds this perfectly acceptable.

The FM 100 Hue test typically reveals anarchic scores in excess of 500 for the total error score with no distinct axis and large errors for individual caps. Sloan's (1954) experience with the Nagel anomaloscope suggests that the typical achromat can obtain a perfect match between yellow and green and between yellow and red by adjusting the relative luminance of the two colours—in effect the results typical of dichromats. A distinguishing factor, however, was the luminance of the yellow required to match red, which was lower than that used by the protanope; the luminance of yellow to match green was found to be typically higher than for the red–green defective. Complete achromats have a loss of brightness appreciation under photopic conditions for reds. Smith and Pokorny (1980) found a variety of brightness matches for complete achromats—greens appear brighter than red. They give detailed characteristics for colour matching for both complete and incomplete types of achromatopsia. Typically, complete achromats can match all spectral colours with one 'primary' (hence the term monochromat) by the adjustment of the radiance of one control as shown by Alpern *et al* (1960). These authors state that incomplete types require two reference stimuli, i.e. are essentially dichromatic at moderate photopic levels. The nature of the second visual pigment is discussed in the section dealing with incomplete achromatopsia.

The low brightness of red lights must be borne in mind when attempts are made to relieve the photophobia by prescribing dark lenses. Thus Weder (1975) showed the advantage of using medium density purple tints which enable brightness contrasts of different colours to be maintained.

Incomplete achromatopsia

The incomplete achromat requires special consideration because these defects are perhaps the most difficult to diagnose among all colour vision anomalies. Because other visual functions are often near normal the role of a careful clinical colour vision assessment is stressed by Smith and Pokorny (1980). Krill (1977) noted that small central scotomas are sometimes found in the visual field of incomplete achromats so that small test objects, such as the $2°$ field typically used in psychophysical examination, may cause problems for this group. Smith and Pokorny (1980) advocated an $8°$ field for the examination of achromatopsia. Different groupings are recognised for the incomplete forms based on their genetic transmission pattern. One such is described in detail below.

X-chromosomal recessive incomplete achromatopsia. Blackwell and Blackwell (1957, 1961) first described such cases as blue cone monochromacy. Alpern *et al* (1965a) preferred the term π_1 monochromacy since active blue cones were demonstrable by the presence of the π_1 cone mechanism of Stiles. The mode of transmission is described on p. 197 and in §3.7.3.

Retinal function of the blue cone monochromat appears to be under the control of normal rod photoreceptors at low luminances and normal blue cones at high luminances, as shown by their spectral sensitivity function which peaks at 440 nm under photopic conditions. The long wavelength (red) and medium wavelength (green) cone receptors seem to be lacking or malfunctioning, but evidence for normal short wavelength (blue) receptors is suggested because the visual acuity of these receptors is similar to that of blue cones in the normal fovea, as suggested by Green (1972). A close correlation between the absolute visual sensitivity of the blue cone monochromat and the normal π_1 mechanism is found (see Alpern *et al* 1965a).

Visual acuity is poor, around 6/18–6/60, but variable between family members, and never as low as complete typical achromats (rod monochromats) because of the blue cone activity. Other typical visual malfunctions, such as photophobia, minor fundus changes on close examination, and an extinguished photopic ERG, are reported. Francois *et al* (1966a) include also concomitant myopia as a sign; there is some disagreement concerning the presence of nystagmus. Blackwell and Blackwell (1961) claimed it to be absent but Pokorny and Smith (1979a) included pendular nystagmus among symptoms. Results on clinical colour vision tests among this group are described by Francois *et al* (1966a) and by Pokorny and Smith (1979a). Hue discrimination as assessed by the FM 100 Hue test indicates some residual yellow and blue–green perception so that a deutan axes of confusion is typical on this and the Farnsworth Panel D.15 test with errors of 400+ on the former. A typical red–green defect is indicated on PIC (pseudoisochromatic) tests. Two colours are used for colour mixing at low photopic levels, indicating dichromatism, but at low levels subjects become monochromatic showing complete achromatopsia. Similarly at high photopic levels the short wavelength sensitive photopigment becomes dominant and vision is again monochromatic. This group therefore has the unusual characteristic of a 'reverse Purkinje shift' from 500 nm at low levels to 440 nm at high levels of illumination, as was first shown by Blackwell and Blackwell (1961). A loss of brightness sensitivity at the red end of the spectrum is characteristic, in the opinion of Pokorny and Smith (1979a), with a spectral neutral point at 476 nm. The Rayleigh–Nagel anomaloscope match is similar to that of the complete achromat. A number of investigators indicate that female carriers of the X-chromosome linked achromatopsia often show slight colour vision losses—specifically a deutan defect on the 100 Hue test and a 'deutan' Nagel anomaloscope match (see, for example, Spivey *et al* 1964).

Autosomal recessive incomplete achromatopsia. Smith *et al* (1978b, 1979) indicated two distinct classes of this group of achromats. They may show slight retinal changes as are typical of complete achromats, such as a reduced photopic ERG in association with photophobia and pendular nystagmus, with visual acuity varying from 6/18 to 6/60. The division is made on the basis of their photopic spectral sensitivity response, V_λ; this can show a protan bias with reduced sensitivity to reds and a peak at around 540 nm, or a deutan bias which is normally less severe giving a photopic peak at 570 nm. This indicates some residual colour discrimination in both groups at photopic levels, hence their classification as incomplete types. The protan group show a protan Nagel match, setting pure red as a dark hue, and colour discrimination tests such as the FM 100 Hue test and Panel D.15 give a protan or scotopic axis, indicating a severe loss of colour perception. Red–green screening plates are normally failed. The deutan group usually indicate a scotopic axis on matching tests. Nevertheless both groups can name some colours correctly. The deutan group function as trichromats at a colour matching task involving spectral wavelengths and the protan group act as dichromats, requiring two controls to match all spectral hues. It is believed that in both groups the blue mechanism must be operating.

Considerable difficulties arise in diagnosing the incomplete types of achromatopsia. Confusion with acquired types must be avoided since a possible identification of the site or cause of an acquired cone degeneration condition may lead to treatment and to partial or complete recovery. Careful compilation of the history of colour perception in the patient is of particular importance. There are many possibilities for acquired defects compounded with inherited ones to give rise to achromatopsia, particularly in the elderly. Furthermore Pinckers (1972b) noted that in some cases of inherited achromatopsia progressive changes to photoreceptors may be evident. A combination of rod and cone malfunction suggests the likelihood of an acquired achromatopsia.

A simple test using coloured arrows as PIC objects was described by Berson *et al* (1983) as a means for distinguishing between X chromosomal blue cone monochromatism and autosomal recessive rod monochromatism. Several cases were identified with this method, as an alternative to the approaches suggested by Pokorny and Smith (1979).

3.6 DISTRIBUTION OF INHERITED DEFECTS

3.6.1* Incidence

Perhaps the earliest study of the frequency of defective colour vision was made by Dalton (1798) who examined 50 observers and found 3 to have unusual colour perception; he was unable to locate a female example. Seebeck (1837) located 5 examples among 41 students and it was he who diagnosed different categories of anomaly, recognising not only variations in colour confusions between anomalous observers but also variations in the severity of the anomaly. It was Wilson (1855) a Professor of Technology at the University of Edinburgh who compiled the first extensive survey of defective colour vision. Using crude matching and naming tests he noted 5.1% of the male population to have anomalous colour vision.

Racial differences in the incidence of colour vision anomalies were reported originally by Clements (1930) and have been confirmed by the great many surveys which have been carried out in various parts of the world. It is impossible to summarise all. In general the incidence is high in urban, industrialised areas, and is low in isolated rural communities and nomadic populations ranging from 8% in Europe, North America and Australia to 2% among Eskimos, desert nomads and tribal peoples, as shown by Vernon and Straker (1943), Skeller (1954), Mann and Turner (1956), Dutta (1966) and Voke and Voke (1980). One of the most complete reviews of geographical variations in red–green defects presented in tabular form is given by Iinuma and Handa (1976). These authors quote the incidence for males as $8.08 \pm 0.26\%$ (mean and standard deviation) in the Caucasian races, $4.90 \pm 0.18\%$ in the Asiatic races and $3.20 \pm 0.40\%$ in other racial groups. The differences are statistically significant for males, but not for females, who show the corresponding incidence of $0.74 \pm 0.11\%$, $0.64 \pm 0.08\%$ and $0.69 \pm 0.07\%$ for the groupings indicated. The ratio of the type of defect is fairly constant throughout the world in a $1:3$ protan to deutan ratio. However Adam (1980) found a $1:1$ ratio among Malay males living in South Africa, with 2% of the group examined showing protan defects and 2% deutan defects. For males of European origin (Francois et al 1957a) deutan defects occurred approximately three times more frequently than protan defects. Adam (1980) was unable to find any evidence of protan defects among Zulu and Nama tribesmen.

Calculations of the male/female incidence ratio can be made on the basis that the female incidence should be the square of the male incidence. Using the percentage frequency for males, which is usually taken to be 0.08 (8%) the square of this is 0.0064 (0.64%) which is approximately what is found in Northern European populations. Discrepancies have to be accounted for in different ways (see Franceschetti and Klein 1957 and p. 205).

3.6.2 Influences

It is well established that inbreeding raises the incidence of genetically-carried abnormalities. For example a typical urban Japanese community studied by Ichikawa and Majima (1974a) showed an incidence of Daltonism of 5.05% males and 0.27% females (sample 3000) whereas the incidence was slightly higher (6.83% males, 0.8% females) among residents of a small, relatively isolated Japanese island where inbreeding was noted in 2.5% of families examined. The same authors noted a male incidence of Daltonism of 14.47% ($\pm 4.03\%$) among tribesmen of a remote island south of Taiwan where the prevalence of inbreeding was 41.2%.

Considerable variations have been shown among peoples with different ethnic origins and those pursuing a different life-style in the same region. Thus Clements (1930) found an incidence of only 1.9% colour anomalous among American Indians, but 3.7% among Negro Americans and 8.2% among Caucasian Americans. Studies among hunting tribal and non-tribal peoples confirm this finding. Dronamraju and Meera Khan (1963) reported an incidence of 2.5% among tribal peoples of Andra Pradesh, India, but 6.5% among non-tribal people in the same area. Fiat (1967) found 2.24% of hunting tribesmen in North Thailand to have colour vision defects, but more than twice as many rice-growers in the same region (5.17%). Two investigations made in Israel by Kalmus *et al* (1961) and Adam *et al* (1967) show the wide ethnic variations giving an overall incidence of between 3% and 10%. Adam (1980) found an incidence variation of between 1 and 3% among populations in South Africa which were of either Negro, Malay or Khoikhoi origin (total sample 5000 males). Voke and Voke (1980) reported a very low incidence for desert dwellers of the Central Arabian Peninsula (2.3 ± 1.3%) compared with those from the Eastern regions where industrialisation is taking place (4.7 ± 0.67%), total sample 1000 males.

Population studies using large numbers have been made by a number of authors early this century including Waaler (1927), von Planta (1928), Schmidt (1936) and Nelson (1938). More recently extensive surveys have been carried out by Koliopoulous *et al* (1976) among 30 000 Greek males, and by Kuypers and Evens (1970) among 10 000 Belgian males. A selection of these studies is summarised in table 3.3.

3.6.3 Possible explanations

The wide geographical variation in the incidence of defective colour vision is somewhat of a mystery but can be explained, at least in part, by biological factors such as gene flow, selection mechanisms and the rise of mutant genes, together with features such as the migration and mixture of races. Post (1962) considered there to be a correlation between the frequency of colour vision defects and the distance in time of any population from the hunting stage. Because of the need for good colour perception in hunting and gathering, a natural selection mechanism against the defect among such groups may operate. It is his hypothesis that when agriculture became organised, to some degree around 3000 to 5000 years ago, the selection against defective colour vision suddenly relaxed to zero. During hunting and gathering eras defective colour vision would cause an obvious handicap and it would be preferable for the incidence to be low. To some extent this is borne out today—pastoral agricultural communities such as are found in Europe, North America and Australasia are settled peoples who operate on a highly organised basis. They have less need for perfect colour discrimination than the tribal forest dwellers and desert nomads, who show significantly lower incidence of faulty colour perception.

By a gradual process of unchecked mutation the characteristic of defective colour vision has emerged to the high level found today. Kalmus (1965) suggested that since Neolithic times the frequency of the protan gene has increased from 0.005 to 0.02 (0.5% to 2%) and the deutan gene from

Table 3.3 Incidence of red–green colour vision defects—a selection of studies (mostly among males).

Author	Study area	Sample size	Protanopes (%)	Deuteranopes (%)	Protan-omalous (%)	Deuteran-omalous (%)	Total (%)
Waaler (1927)	Europe	9049	0.88	1.03	1.04	5.06	8.01
von Planta (1928)	Europe	2000	1.60	1.50	0.60	4.25	7.95
Schmidt (1936)	Europe	6863	1.09	1.97	0.68	4.01	7.75
Nelson (1938)	UK	1338	1.27	1.20	1.27	5.08	8.82
Kuypers and Evens (1970)	Belgium	10 000	–	–	–	–	7.07
Koliopoulous et al (1975)	Greece	30 000 (males) 8754 (females)	1.00 0.01	1.14 0.02	1.20 0.03	4.61 0.35	7.95 0.42
De-Vries -de Mol (1977)	Netherlands	1586	1.4	0.9	1.4	4.0	7.7

(Noted incidence varies considerably with test method.)

0.016 to 0.06 (1.6% to 6%) though there is little real scientific or mathematical backing to these projections. The same author has gone as far as to say that 'the sex-linked forms of defective colour vision provide the only clear example of a single inherited defect, the frequency of which has significantly increased with the advent of civilisation'. Li (1955) points out that the most favourable conditions for biological change are those in which a large population is subdivided into numerous isolated groups with migration occurring between them. Since the desert dwellers of the Arabian Peninsula fall into this category, it will be interesting to see if the incidence increases as a greater proportion of the population take up a settled way of life with greater industrialisation in the future.

3.6.4 Global distribution

Population studies of the frequency of Daltonism made in different regions of the world have led to the construction of a tentative map of global 'colour vision isophenies' by Cruz-Coke (1970) but such an effort must, at this stage, be crude (see figure 3.23). Surveys have generally been conducted with the Ishihara plates, frequently used in isolation, and the different criteria adopted for pass/fail levels and differences in results on account of variations in editions (printing) will undoubtedly have led to individual bias. Nevertheless, surveys carried out by different investigators among similar

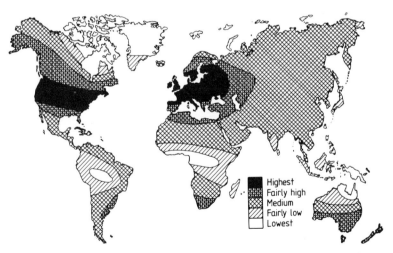

Figure 3.23 Distribution of defective colour vision (after Cruz-Coke 1970). Cruz-Coke's plotting of isophenies was based on surveys using Ishihara tests for 'red–green' defects by various authors. These approximate boundaries are related to frequencies (in males) of 8% at the highest level decreasing to 6%, 4% and 2%.

population groups have yielded, in general, similar percentages (see, for example, Pickford 1951 and Cameron 1967). Table 3.4 lists some of the studies that have been made to date in a variety of countries. In some instances several population samples have been studied within one country by different investigators; for brevity only one or two within each country are included, and the list is therefore far from comprehensive.

The best review of African variations is given by Roberts (1967) from the laboratory of human genetics at Newcastle University. His own study of two populations, the Abua and Ogoni tribes, showed 2.0% with anomalous colour vision, though the sample sizes were less than 400 in total. In earlier studies in northern and southwest Nigeria a similar low incidence was noted, but again the sample sizes were always small. In a study of 500 Ugandans in Kampala defects were found in 1.8% of the group and a survey in Rwanda of 100 Bahutu tribesmen and 1000 Batutsi showed a 2.6% incidence in both groups. The African incidence is comparable with Australian aborigines and Papuans, both of whom are heavily pigmented (Mann and Turner 1956).

Turning eastwards, numerous studies have been reported among groups in India (e.g. Dutta 1966). The majority of the Indian surveys involve extremely small samples. A recent study of just 450 tribespeople in Andra Pradesh noted marked variations in the percentages of colour deficiency. Out of 270 males 1.4% were colour defective and among 180 females only one (0.5%) was Daltonic. These authors, however, quoted figures for these small samples to four places of decimals! They present a review of over 19 studies among Indians, but sample numbers are rarely above 200, and frequently involve groups of fewer than 100 observers, so few conclusions can be drawn. When all the literature reported on India is collated, over 10 000 males can be assessed, giving an overall 3.6% showing abnormalities. Among 4000 females the integrated incidence amounts to 0.2%. This indicates a higher proportion of colour deficiency than in the African continent.

A study by Bhasin (1967) involving 500 females and 500 males representing one of the most ancient ethnic groups in Nepal, the Newars, showed a 4.23% incidence among males and 0.19% among females.

In China variations from between 4.0% to 6.9% are reported among males and between 0.4% and 1.7% for females. A survey among 7300 males shows a 5.22% incidence (Chung et al 1958).

Defective colour vision has clearly emerged as a characteristic 'easy' to study in population groups. This has unfortunately led to some abuse, particularly among those who are not well versed in its features, or who have little experience in testing procedures. A perusal of the literature indicates a significant number of rather slipshod surveys involving small samples. In many cases the sex of the group is not recorded and few conclusions can be drawn; every investigator should realise the much lower

Table 3.4 Global incidence of colour vision defects.

Population	Investigator	Incidence (%) Male	Female
Norway	Waaler (1927)	8.01	0.44
Switzerland	Von Planta (1928)	7.95	0.43
France	Kherumian and Pickford (1959)	8.95	0.50
Greece	Koliopoulos et al (1971)	7.95	0.42
Belgium	Francois et al (1957)	8.37	—
Great Britain	Vernon and Straker (1943)	7.25	—
Netherlands	Crone (1968)	7.95	0.45
Germany	Schmidt (1936)	7.75	0.36
Scotland	Pickford (1947)	7.80	0.65
USA	Thuline (1964)	6.18	0.45
Australia†	Mann et al (1956)	7.35	0.61
Japan			
Tokyo	Sato (1937)	3.93	0.61
Nagoya	Majima (1961, 1969)	5.85	0.50
Sado	Nagasima (1949)	4.41	0.39
China			
Peking	Chang (1932)	6.87	1.68
Kweiyang	Fang and Liu (1942)	5.58	1.50
Taiwan	Chang (1968)	5.34	0.23
Korea	Yung et al (1967)	4.24	0.21
Philippines	Nolasco et al (1949)	4.28	0.20
Turkey	Okte (1959)	5.22	1.10
India			
South		3.7	—
East	Dutta (1966)	3.48	—
West		3.75	—
Nepal-Newars†	Bhasin (1967)	4.23	0.19
Eskimo†	Skeller (1954)	3.89	0.51
'Pure' Mexico (tribal)†	Garth (1933)	2.28	0.61
Colombia	Mueller and Weis (1979)	2.53	0.13
American Indians†	Clements (1930)	1.9	—
American Indians			
Navajo tribes†	Garth (1933)	1.12	0.56
American Negros†	Clements (1930)	3.7	—
Iran	Plattner (1959)	4.5	—
Iraq†	Adam et al	6.1	—
Saudi Arabia	Voke and Voke (1980)	4.7	—

†Sample size less than 100

incidence of Daltonism among females. Although the Ishihara test has been employed in almost every case reported since its emergence in the first quarter of the century, the testing criteria adopted undoubtedly vary and this makes comparisons particularly difficult. The population frequencies of defective colour vision in Western countries have been largely derived from studies involving recruits for employment purposes, volunteers for the Armed Services, or schoolboys (e.g. Vernon and Straker 1943). The former groups show some self-selection so are not truly random samples, and the data collection in these studies has usually been made for reasons other than for the purpose of obtaining reliable population estimates. Within isolated tribal regions frequent inbreeding is liable to upset population statistics.

The most valid complaint which can be levelled against many published studies concerns the smallness of sample size; Clements (1930) voiced this criticism. Sample size is one of the most important factors in the evaluation and planning of population studies. Samples of several hundred are necessary to make one confident of the accuracy of small percentages. The significance of a difference in percentage between samples from different ethnic groups also depends on the size of the sample as well as the magnitude of the percentage. Using the figures presented by Kalmus (1965) from *Biometrical Tables for Statisticians*, a sample size of 1000 is necessary to ensure that the incidence lies between 2.7% and 6% at the 0.99 confidence level for an expected incidence of 4%. Using a smaller sample size of 400 the figure would span 2.3% to 7.0% at the 0.99 level. Post (1962) in his review of studies made among the races to that date, cited the variation in incidence worldwide as being between 0% for Fiji islanders to 10.5% among Czechoslovakian males, but the sample numbers in these cases were only 200 and 650, respectively. Clements (1930) compiled the first table indicating the racial variation in the incidence of colour defects, drawing from the data of Rivers (1901a,b, 1905) and Collins (1925). Sample numbers quoted varied from 11 up to over 32 000. Holmgren surveyed the latter sample, presumably in his native Scandinavia, with his wool test, for the incidence of anomalies quoted as 3.16% is consistent with such a test. The Eskimo percentage is one of the lowest, being 0.8% but the sample tested was only 125. Since the publication of these early figures Skeller (1954) has confirmed a very low incidence among Eskimos. Swedish army surgeons reported an incidence of 6.3% among Lapp males and 0.9% among females, which Clements regards as accurate.

There is a need to avoid inbred groups for a realistic evaluation of incidence, and Kalmus (1965) also points out the need for an unbiased sample, avoiding therefore groups such as recruits for the armed services, jobs in transportation etc.

Even within a small country such as Great Britain marked regional variations are reported (Vernon and Straker 1943). The most marked differences are for east Scotland (5%) and the southwest of England (9.2%). Cruz-

Coke (1970) suggested that in the short distance between London and Southampton the three isophenies of 6, 7 and 8% cross. Invaders of Britain during the 5th and 6th centuries remained in the eastern areas only and their western line of advance coincides almost exactly with the 6% colour vision isophene line of today. The Ancient Britons, however, who occupied Wales and the Southwest and the western Scottish mainland and Isles, seem to have given rise to a higher overall incidence in colour deficiency. These observations are interesting but a good deal of research would be necessary to substantiate the argument.

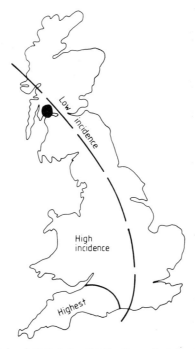

Figure 3.24 Mainland Britain's distribution of defective colour vision (after Cruz-Coke 1970). The isophenies are based on data by Vernon and Straker (1943) gathered from male recruits. The Scottish west coast area indicated was reported by Haughey and Haughey (1976) and Cobb (1980) to have unusually high incidences of defective colour vision among school children.

Since population differences can be attributed largely to migration and racial mixtures it seems likely that the early settlers of Europe took the defective gene to other regions and selective mechanisms have controlled and further refined the incidence of Daltonism in different world regions.

3.7* INHERITANCE OF ABNORMAL COLOUR VISION

Defective colour vision was one of the first genetic characteristics to be identified in man, and geneticists agree that this condition is an important means of studying human variation among the X-chromosome linked congenital abnormalities. Observations on the way in which defective colour vision is transmitted to several members of families were made as far back as 1777 by Huddart, but it was 100 years later before Horner (1876) outlined the hereditary pattern of deuteranopia with non-deuteranopic mothers as transmitters and a further 50 years before Morgan (1910a,b) gave a full scientific description of the mode of transmission; Wilson (1911) suspected the involvement of the X chromosome in transmission of red–green defects.

As genetics became better understood it became clear that there are modes by which each of the types of colour vision are perpetuated on account of the slight variations in inheritance of photopigment chemistry. Some writers use 'congenital' as a preferred term despite its association with the event of birth and anomalies originating then. This text will use 'inherited' as distinct from the 'acquired' features of colour vision covered in Chapter 4.

The genetics of colour vision is complex, as Kalmus (1965) and others have shown, but most texts on medical genetics, such as that by Fraser Roberts (1970) give prominence to 'colour blindness', thus it is frequently taken as an example. Associations are found between colour vision anomalies and other inherited conditions and colour anomalous genes can be used as 'markers' (see McKusick 1962). Although it is strictly the prerogative of geneticists and those caring medically for patients to enquire fully into abnormal cell characteristics, the investigation of Daltonism is frequently enhanced by a simple appreciation of genetic characteristics and special attention is deserved in cases where special conditions may appear or in distinctive geographical localities. Furthermore, modes of inheritance can be useful in assisting in the choice of colour vision tests.

3.7.1* X chromosomal recessive inheritance

Females possess a pair of X chromosomes in the composition of their cells, males having one X and one Y, as the basis of sexual differentiation. The genetic units forming the X chromosome, seen microscopically as a chain, carry at distinctive positions the materials which dictate certain bodily characteristics; genes for protan and deutan defects are located on the short arm of the X chromosome some distance apart (McKusick 1962) with the mode of transmission being described as the sex-linked recessive form. The feature is expressed or manifested only when it is present on both X chromosomes of a woman (homozygous) or on the single X chromosome of a man (hemizygous). When a single X chromosome of the female is affected women can 'carry' the defect to their sons without showing outward sign of the abnormality themselves (theoretically). This is known as the heterozygous condition. Their one affected X chromosome was transmitted to them from their Daltonic fathers or their 'carrier' (heterozygous) mothers. A son exhibits his maternal grandfather's condition or his maternal grandmother's expressed or carried condition, carried through his mother's X chromosomes since he obtained the Y element of each cell from his father. On this basis most red–green inherited deficiencies are explained with reasonable confidence, as in the examples given in figure 3.25 which indicate the possibilities which are translated into a population largely by chance. Theoretically only 50% of the sons of carrier mothers display the defect. There is a fifty–fifty chance that the daughters of a carrier mother and colour defective father will have the defect with the remaining daughters being carriers; this explains the low incidence (approximately 0.4%) of colour deficiency among women. All offspring of two parents showing the characteristic are themselves Daltonic (figure 3.25(f)).

Protan and deutan defects are due to genes in separate loci of the X chromosome (Waaler 1927, Pickford 1951).

*Dominance pattern for severity of defect

Mild defects (e.g. anomalous trichromacy) are dominant to the severe forms (e.g. dichromacy). A female carrier of genes for both deuteranomalous trichromacy (DA) and deuteranopia (D), for instance one on each X chromosome, will produce half her sons with DA and half with D. Accepting that an intermediate condition, extreme anomalous trichromacy, exists, the genetic transmission is unstable and the extreme anomalous state alternates with the simple anomaly.

Compound of genes

A woman who is a 'carrier' has an abnormal gene at the appropriate locus of one X chromosome and a normal gene at the same locus on the other X chromosome. The expression of the normal gene usually supercedes, or dominates, expression of the abnormal one (the normal gene is said to be 'dominant').

Interesting possibilities arise where a female carries both protan and deutan genes. Kalmus (1965) estimates the incidence of such women in a Western-type population (where the incidence of Daltonism is high relative to other world regions) is about 1 in 250. Such compound heterozygotes could be expected to show some outward anomaly but many appear to be normal. Thus Franceschetti and Klein (1957) noted that daughters of parents who each showed a different dichromatic form of defect (e.g. one parent was a deuteranope and one a protanope) had normal colour vision,

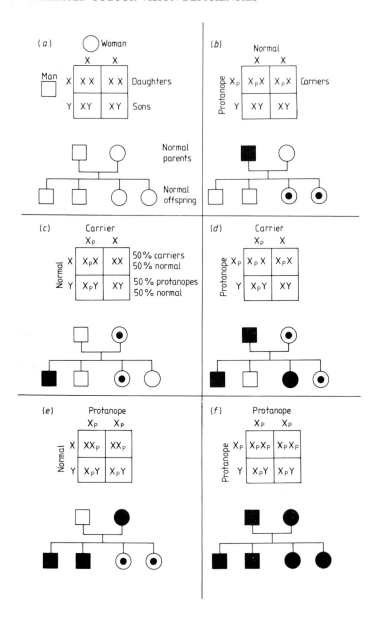

Figure 3.25 Different genetic combinations of male and female chromosomes, showing possible offspring. X, normal X chromosome; Y, normal Y chromosome; X_P, chromosome involving protanopia.

(*a*) A normal male (XY) has a normal genotype and phenotype (appearance and functional state). With a normal (XX) female normal offspring are produced.

while being compound heterozygotes genetically. Kalmus (1965) estimated that in a Western-type population 1 in 17 of the colour defective brothers of a male with a P defect would be of a D type and 1 in 50 of the brothers of a male with a D defect would be of a P type. An interesting pedigree was reported by Vanderdonck and Verriest (1960) where a protanomalous woman (with a deuteranopic father, a normal mother and a normal husband) produced two deuteranopic, one protanomalous and two normal sons, and one protanomalous and one normal daughter. Jaeger (1972) considers that by crossing-over (switching of genetic material between chromosomes during cell division) some of her X chromosomes changed into a normal chromosome, and one affected with D and PA, explaining the normal sons. The 'cancelling' effect of the two genes, (one for deutan defect and one for protan defect) as indicated in the case of compound heterozygotes is seen here also in the case of the normal offspring, but not in the mother herself. A schematic example is shown in figure 3.26 of a woman with a compound heterozygote composition, such as protanopia at the protanopia locus of one X chromosome and deuteranopia at the deuteranopia locus of the other X chromosome. Such situations suppose two loci, one for each condition, and following observations of families where women produced sons with different types of Daltonism, there has been considerable support for this idea. Waaler (1967) reviewed his 'two-locus hypothesis' of 1927, with some reservations, but for the present purpose the hypothesis, as supported by Walls and Mathews (1952), Kalmus (1965), Pickford (1965) and Jaeger (1972) may be accepted.

An order of precedence applies to the genes for the different types of colour vision; just as the normal gene is dominant over the abnormal and the lesser defect of anomalous trichromatism is dominant over dichromatism, for the two red–green types, deutan defects are dominant over protan. The 'dominance hierarchy' has been considered in terms of the variation of photopigments, and of the shapes of their absorption spectra. Piantanida

Figure 3.25. *continued*

(b) A protanopic male (X_PY) and a normal woman will produce normal sons but heterozygous 'carrier' daughters (X_PX).

(c) A normal father and carrier mother would be expected to produce 50% normal daughters, 50% carrier daughters, 50% normal sons and 50% protanopic sons.

(d) A carrier mother and a protanopic father could produce a protanopic daughter. Other daughters could be carriers and 50% of the sons would be expected to be protanopic. X_PX_P females are described as allelic or homozygous for the X_P gene.

(e) The expected offspring of a protanopic mother and normal father.

(f) The expected offspring of a protanopic mother and protanopic father.

(1974) used psychophysical data to explain different series of genetic forms on this basis. A diagram suggested by Waaler is modified in figure 3.27 to suggest a model of this dominance sequence.

A non-allelic heterozygote for both protanopia and deuteranopia, with a normal male, could produce children as shown in figure 3.28.

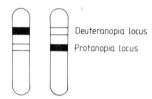

Figure 3.26   Two schematic X chromosomes showing two loci, one pro-tanopic, one deuteranopic. Defective loci are shaded. The example is of a woman with both defects, in whom the normal genes may ensure she has a normal phenotype. She can be regarded as doubly heterozygous.

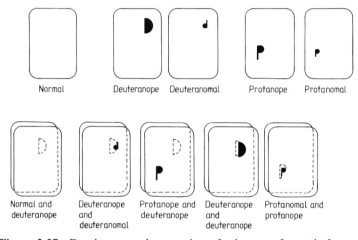

Figure 3.27   Dominance and expression of mixtures of genetic features. The model, modified from Waaler, uses tickets placed in pairs. At the top individual tickets are seen, punched through with coded holes. A pair held up to the light reveals how the phenotype may be expected to appear; thus a P and D compound appears normal.

There have been notable studies of compound defects, such as that by Jaeger (1951) in which, despite unexpected tritan defects in two sisters they manifested different types of red–green defect. Figure 3.29 shows how protanopia and deuteranomaly appeared together and were found in the last generation. Observations by Walls and Mathews (1952) were the basis of doubt expressed by them (p. 147) as to the genuineness of the tritanomaly mentioned above. Assuming two loci, the more varieties of red–green defect that are involved, the more the possible combinations. Since it is reasonable to consider protanopia, protanomaly, extreme protanomaly, deuteratopia, deuteranomaly and extreme deuteranomaly Kalmus (1965) suggested 117 possible female compounds. Mixtures of genes suggest the phenomena of 'crossing-over' of parts of chromosomes during splitting and reforming and the difficulty of clearly differentiating some Daltonic individuals into clear protan or deutan categories. A pair of brothers reported by Walls and Mathews (1952) presumably had compound hemizygote features since one was a protanope and the other was a deuteranope; each had added unusual features, indicative of the opposite condition but (the RDP Maxwell spot observations under the conditions then used can be discounted) appeared to have a protanopic defect at one locus on his X chromosomes with a deuteranopic defect at a separate locus.

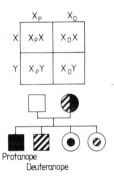

Figure 3.28 A woman combining protanopia and deuteranopia might produce sons of either type. Her daughters will carry either feature.

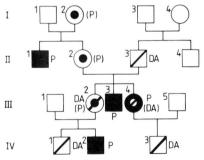

Figure 3.29 Protanopia and deuteranomaly compounds, part of a pedigree of Jaeger (1951) modified after Walls and Mathews (1952). The sisters III₂ and III₄ were revealed as, respectively (atypically) slightly deuteranomalous and protanopic. Each, with IV₂, showed a tritan deficiency as well.

Possibly the 'least manifesting' condition might be explained by mutation to a lesser form and these authors added 'there is many a slip twixt the gene and the phene ...'. The Vanderdonck and Verriest (1960) pedigree (figure 3.30) often quoted in this connection, included five male siblings with two normals who were considered to be 'recombinants'. In this series the precaution had been taken to ensure that the apparent sex was confirmed by staining of cell nuclei. Assuming that the mother of the seven children was affected on the D locus of one X chromosome and the PA locus of the other (and atypically had a PA phenotype) it is simple to explain her D sons and her PA sons. The normal sons and daughter could be the result of crossing-over of the protan and deutan loci. Jaeger and Laver (1976) and Went and de Vries-de Mol (1976) examined families with compound defects, protan and deutan.

A final example, by Pickford (1962) (figure 3.31) showing the son II₃ who combined PA and EDA features, presents him as a 'double hemizygote'; also the possibility of three different defects among brothers. This demonstrates the need for great care in arriving at conclusions.

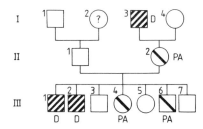

Figure 3.30 Protan and deutan combination in the pedigree of Vanderdonck and Verriest (1960). The woman II₂ is described as doubly heterozygous.

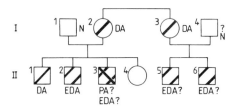

Figure 3.31 Pickford's (1962) pedigree involving three sons with different defects.

Heterozygous carriers

Daughters of Daltonic males and normal females, being heterozygotes, are usually expected to display normal trichromatism because of the dominance of their normal genes; this is frequently the case and often all carriers in a family are reported as normal. Waaler (1927) reported mild traces of Daltonism in carrier females and in 1967 was conscious of the possibility that 'uncertainty in reading Ishihara charts by girls might indicate heterozygote'. He recognised that even those with colour defective brothers could not be diagnosed as carriers by this 'partial manifestation' and sought to adopt a method reported by Rubin (1961), the apparent bimodality of distribution of the 'green' part of the spectrum. Waaler's (1967) results were promising although his apparatus (a Model II Schmidt and Haensch anomaloscope) was difficult to adapt.

Schmidt (1934, 1955a) noted a decreased sensitivity to reds by some protan carriers and a slight shifting of the photopic spectral sensitivity curve V_λ typical of that found in protan defectives. Crone (1959) computed a luminosity ratio of a green wavelength to a red as a quantitative index of the decreased sensitivity. The quotient in protanopia is between 40 and 50; in protanomaly between 30 and 40 and in carriers between 10 and 25; normal observers have a value of below 10. Walls and Mathews (1952) made early observations. Pickford (1959) detected mildly anomalous matches made on the anomaloscope by 71 heterozygotes, but was careful to stress that such a carrier cannot be identified with certainty by this method.

Careful examination with spectral apparatus is likely to show most, if not all, carriers to make slight departures from normality. Krill and Beutler (1964) and Krill and Schneiderman (1964) demonstrated elevated red thresholds in protan carriers, and conventional tests, used with care, were shown to be indicative of carriers. A great many investigators of Daltonism have noticed the failure of very sensitive pseudoisochromatic plates by manifesting heterozygotes. In the authors' experience the dual figure read as 74 by the colour normal and seen as 21 by the colour defective in the Ishihara series is frequently read as 21 by carriers. A definite protan axis bias to a pattern of low overall hue discrimination was noted by Krill and Schneiderman (1964) among 13 protan carriers.

Piantanida (1974) used a replacement model to explain how cone pigments could be produced in abnormal forms, according to the type and severity of the genetic defect; the threshold for red parts of the spectrum is raised (Schmidt's sign) by less erythrolabe appearing, with a quantum catch of greater than normal quantities of light of shorter wavelengths. Figure 3.25 (b, d, e) indicated how a straightforward involvement of a carrier can be detected in a family.

Jaeger's (1972) experience shows that a heterozygote woman (her son acting as the propositor) may display a slight colour vision defect which is actually acquired. There must always be doubt until reasonable satisfaction is reached that the condition has not changed. Ikeda et al (1974) used adaptation to red light in order to separate protoheterozygotes, deuteroheterozygotes and normals. Ichikawa and Majima (1974b) could distinguish about 80% of carriers from normals, using a portable flicker photometer and these authors believed that a carrier can have unstable colour sense.

The spectral location of 'unique' green has been investigated by Richards (1967) with bimodal results similar to those of Rubin (1961). Richards' criterion for 'normal colour vision' was, apparently, adequate performance on Ishihara plates but even this test may have been applied to males alone; questions of heterozygous composition may arise with respect to his female subjects. De Nazaré Trinadade Marques (1977) was not able to demonstrate manifestation in any of the carriers in her study; this is a reminder of the difficulties sometimes met.

A strong case was made for 'incontrovertible evidence from multiple tests' by Hill (1980) in a study of a female normal on some tests but deutan on others. A comparison of normal women with carriers clearly shows the greater frequency of small colour defects, which Pickford (1965) found to correspond to relatives' defects, as would be expected. Pickford (1959) considered the 'double heterozygote' (e.g. protan and deutan genes on different loci) where a woman's sons could be Daltonic but differently so. Kherumian and Pickford (1959) pointed out the likely variations of the different 'types' of women in different populations and difficulties encountered since testing approaches are different. On the basis of 7.8% of defective men in a population Pickford (1959) calculated that there would be frequencies of 0.041 586% protan 'allelic compound defective heterozygotes', 0.158 778% deutan 'allelic compound defective heterozygotes' and 0.341 192% of defective women. He demonstrated exactly this proportion (correct to one woman) in a suitable area.

The double defective allelic compounds, such as deuteranomaly combined with extreme deuteranomaly, may not always follow the expected manifestation of the less recessive (more dominant) gene as in Pickford's (1959) case of a protanomaly and protanopia compound, where the woman appeared as a protanope instead of the more likely protanomal. Franceschetti and Klein (1957) reported a protanopic father and deuteranopic mother with three sons as deuteranopic and one daughter as normal; the girl therefore demonstrated the normality of a mixture of the two genes, each on a separate X chromosome (figure 3.32). Figure 3.25(d) should be compared with figure 3.33 as a substantiation of expectations. A few rare cases of true Daltonism in heterozygous females have been reported; Waardenburg (1963b) gives a review of such cases. In a population sample the number of carriers should be approximately twice as numerous as the incidence of Daltonism.

Patterns of inheritance influence the numbers and the proportions of men and women falling into different categories of Daltonism in a population (see Judd 1943, Platt 1943, Waardenburg et al 1961, Waaler 1977 and most texts on medical genetics). With a male incidence of 8%, we expect the square of this, 0.64%, to be the incidence of female defects. McKusick

(1962) modified this 'one-locus' hypothesis expectation of 0.64% to 0.4% using 'two-locus' assumptions, giving 0.36% female deutans and 0.04% female protans, figures fairly well borne out in practice. Manifesting heterozygotes probably inflate the reported female Daltonic number slightly. Walls (1959a) drew attention to the apparently manifesting carrier 'females' with an actual XY composition and testicular feminisation. Other ostensible women, XO hemizygotes, have gonadal dysgenesis. Colour vision tests therefore have some use for geneticists predicting X chromosomal anomalies and which parent features come from. The complexity of these matters and their sensitivity make it necessary for the non-geneticist to avoid asking too many questions.

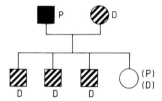

Figure 3.32 Normal daughter of protanope and deuteranope with deuteranopia in the sons. After Franceschetti and Klein (1957).

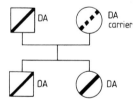

Figure 3.33 Pickford's (1964) pedigree of a carrier for deuteranomaly manifesting a 'red–green weakness' married to a deuteranomal with two deuteranomalous children.

Lyon (1962) drew conclusions (from the mosaic behaviour of the colours of mice) which have a bearing upon colour vision; the retina is composed of a mosaic of different receptors, and the X chromosomes governing their photochemistry may be influenced by inactivity of varied types. Possible differences between the colour vision of a woman's two eyes have been mentioned in this connection, as by Kalmus (1965) and Pickford (1965), while differences between manifesting heterozygotes were suggested by Pickford (1967a); for instance, different women could have different amounts of X chromosomal inactivation which could make their colour vision different. Krill and Beutler (1964) discussed mosaicism of red receptors in Schmidt's sign and Piantanida (1974) reviewed the possible varieties of retinal grain in relation to acuity and other functions in dichromats.

Intersex states

The Y chromosome has particular connection with sex determination but the role of the X sex chromosome is of interest particularly in those 'intersex' states where colour vision is under consideration. A single chromosome may be present (Turner's syndrome individuals have an XO composition) or there may be an excess of X elements as in Klinefelter's syndrome (an XXY composition). The sterile Turner's syndrome patient is usually described as having one X chromosome 'active' and shows somatic peculiarities such as a short body; female Turner's patients will be as likely to manifest Daltonism as if they were male. Such people have underdeveloped secondary sex features, or female gonadal dysgenisis.

*Association with other conditions

Certain genes close to each other on the X chromosome produce close linkage of different abnormalities and the marker attributes of Daltonism have already been mentioned. Kalmus (1965) and others suggested that on the short arm of the X chromosome protan and deutan loci are fairly close to each other. One gene probably lies near (or between) them, that for G6PD deficiency (a blood condition, sometimes associated with colour vision defects and found in some Mediterranean dwellers), known to be closely linked with them; McKusick (1962) and Kalmus (1962), for example, showed the association of Daltonism with this abnormality. Again, the possibility of an acquired associated condition must be considered and when Cruz-Coke (1970) publicised an association with cirrhosis of the liver connected with alcoholism it was clearly difficult to separate cause and effect. Presumably the recognition of alcoholism as a pathological condition, inherited in some X-linked way, fits the reported tritan defects among alcoholics, with females carrying the two conditions. Yet the autosomal nature of tritan colour vision abnormality and the prevalence of acquired tritan difficulties of perception argue otherwise.

3.7.2 Blue inherited defects

Relatively few investigations have been possible on account of the rarity of both tritanopia and tritanomaly in inherited forms. The exact mode of transmission of tritanopia and tritanomaly is therefore uncertain, and there is always a suspicion of an acquired defect. Cole *et al* (1965) considered that there are no separate genes for tritanomaly and tritanopia.

Several reviews of the literature have been made, such as that by Henry *et al* (1964), and tritanopia is likely to be inherited by autosomal dominant means, often manifesting in an incomplete way, according to Kalmus (1955, 1965), whereas tritanomaly is sex-linked. Walls and Mathews (1952) regarded the dichromatic state, tritanopia, as being likely to be sex-linked but Crone (1955) supported Kalmus's view by identifying autosomal tritan features in a family tree; this actually involved monochromatism.

Tritanopia

Congenital tritanopia is considered to be transmitted in an autosomal dominant form which may be incomplete (Kalmus 1955). However a case found by Fischer *et al* (1951) appeared to be isolated, since both parents and three sisters had normal colour vision, as did a case described by Voke-Fletcher and Fletcher (1978) where both parents were normal. Autosomal dominant infantile optic atrophy produces a blue defect and the need to exclude this condition in suspected cases of congenital tritanopia was stressed by Krill *et al* (1971). Voke-Fletcher and Fletcher were careful to rule out this condition in their observer.

Variable expression is seen in a single family, clouded, naturally, by differences in the light transmission of the ocular media according to the different ages of the individuals. Thus Cole *et al* (1965) showed three distinct dichromats in an extensive pedigree where another three appeared to be tritan but not dichromatic. The use of colorimetric apparatus proved valuable in this investigation and there was a suggestion that 'incomplete tritanopia' can be confused with, or is, tritanomaly (see also Smith and Pokorny 1973 and Alpern 1976).

Sixteen tritans, mostly males, were included in several families seen by Neuhann *et al* (1976) which confirmed the inheritance as autosomally dominant with incomplete penetrance. Dominantly inherited juvenile atrophy of the optic nerve was ruled out.

Autosomal dominant inheritance of a tritan defect could explain two cases of van de Merendonk and Went (1980) both showing deutan defects as well; the families concerned revealed unusual difficulties in reading pseudoisochromatic plates.

In the light of these resports we may expect sons or daughters to appear as tritanopes with or without similarly tritan parents of either sex.

The criteria proposed by Krill *et al* (1970) for tritanopia are chiefly of use in avoiding acquired tritan subjects; these criteria invoked all those in the family and normality of fundi, central fields and visual acuity. Cole and Watkins (1976), by the method of increment thresholds, showed blue mechanisms to be affected.

Tritanomaly

Tritanomaly was first described by Engelking (1925) and has been described as 'a more or less recessive sex-linked trait' (Engelking 1925, Oloff 1935, Meitner 1941 and Jaeger 1955 as quoted by Kalmus 1965). However a surprisingly high incidence of affected women has been found in families investigated by Jaeger (1955). The opinion of Cole *et al* (1965) that tritanomaly may not exist as a 'separate entity' must be set against many other opinions that it is caused by a sex-linked, separate, gene. Crone (1956) supported the 'autosomal dominant' view of Kalmus.

3.7.3 Monochromatism or achromatopsia

Mixtures of colour vision defects probably produce many of the 'monochromats' which are described, although there are special features in cone monochromats. A complete absence of colour vision is very rare, especially as an inherited condition. Smith and Pokorny (1980) have reviewed classes of achromatopsia.

Complete forms

Complete achromatopsia shows an autosomal recessive inheritance pattern (see Gothlin 1941, Francois *et al* 1955 and Waardenburg 1963b). Occurrences are most frequent among siblings of consanguineous parents. Duke-Elder (1964) quoted consanguinity present in 20% of cases and Sorsby (1970) suggested 30% of instances. Five of the sixteen complete achromats described by Sloan (1954) had parents who had intermarried; in two separate families this involved parents who were second cousins. Twelve knew of siblings or other relatives with markedly defective colour vision. The parents of a brother and sister showing typical achromatopsia, studied by Voke (1978a), were first cousins; a further son of the marriage was a Naval officer reputed to have perfect vision.

Incomplete forms

The incomplete types show a variety of inheritance patterns. Family pedigrees frequently show coexisting red–green defects. Thus Francois *et al* (1955) state that female carriers for cone monochromatism are usually compound heterozygotes for both protan and deutan anomalies. Blackwell and Blackwell (1957) described a monochromat of specific type with an X-chromosomal recessive mode of inheritance (see also Spivey 1965 and Spivey *et al* 1964). A recent statement of the autosomal recessive inheritance of incomplete achromatopsia was made by Pickford *et al* (1980), heterozygotes being revealed by several standard tests as having definite colour vision defects.

3.8 UNILATERAL DEFECTS

Studies of individuals experiencing abnormal colour vision in one eye only are valuable in indicating, to some extent, the colour sensations experienced by persons with anomalous colour vision, since such observers can report differences in their experiences. Variations between the perception of colours in each eye is especially important in cases where acquired defects exist. It is for this reason that the testing of each eye separately should be encouraged (where time permits) for suspected inherited defects. All acquired defects should be examined monocularly.

Judd (1948) estimated that 39 publications on unilateral defects had appeared by 1948, involving 36 individuals; he provided an excellent analysis of the cases. De Vries-de Mol and Went (1971) reported older cases, such as the investigations of a girl by Becker (1879) and a male by von Hippel (1880). Their analysis of Judd's reported cases suggests that eight had inherited red–green varieties, and that the others were of pathological or traumatic origin.

A study of the inherited unilateral defects is valuable. Genetically transmitted unilateral true dichromats are extremely rare. Judd (1948) noted only one case in the literature of a unilateral protanope. The unilateral deuteranopic woman reported by Graham and Hsia (1958b) may have involved some acquired disturbance. Sloan and Wollach (1948) described one of the few true unilateral cases. Although the observer examined by de Vries-de Mol and Went (1971) was considered to be either deuteranopic or extreme deuteranomalous in one eye, with a 'normal' other eye on the anomaloscope, he failed the Ishihara plates with this eye. Studies of other members of the family showing inherited colour vision defects provide a more complete picture.

The observer studied by MacLeod and Lennie (1974) is probably an inherited example; in this observer one eye was deuteranomalous and one deuteranopic.

4 Acquired defects

4.1* INTRODUCTION

The inherited departures from normal colour vision are generally predictable in their characteristics, usually affecting all parts of the visual field in both eyes. Attention in this book has so far been directed to the characteristics of such inherited colour vision defects.

There exist, in addition, a wide group of colour vision disturbances which are acquired during life, predominantly the result of ocular or general disease, the consequence of exposure to a chemical, toxin or medication, or resulting from physical injury to the head. The incidence of such disturbances is uncertain (see Lyle 1974), although the estimate by Smith (1972a) that 'at least 5% of the population have an acquired defect as severe as the 8% with a congenital defect' is a valuable guideline. The general aging process in itself brings with it subtle insidious changes to vision including colour perception; these are discussed in §1.13.4 and must be considered to be an acquired change of a physiological nature.

Alterations to colour vision which arise throughout life from a cause or causes other than the normal physiological processes have received far less general attention than the inherited defects. Although physicians were aware about two centuries ago of changes to colour perception resulting from disease, accounts were confined, in general, to isolated cases up until Kollner's 'classical' description in 1912, and renewed interest was shown thereafter by German ophthalmologists. The application of the Farnsworth–Munsell tests to acquired defects of colour vision in the 1950s and 1960s led to renewed interest in the causes, manifestations and testing procedures for this separate group of colour vision disturbances, chiefly through the influence of Belgian ophthalmologists; a new era of investigation thus began.

During the period leading up to this renewed interest and insight the colour changes accompanying ocular disease were prominent, although the toxic effects of tobacco and certain chemicals had received some attention in the last century. Lyle (1974) cites the example, given by Helmholtz in his *Treatise on Physiological Optics*, of de Martini, who in 1858 noted the effects of santonin (a naphthalene compound) on colour vision. The greater awareness of, and concern for, industrial health in recent times has highlighted the need for control over exposure to industrial toxins. Furthermore the development of synthetic drugs and the greater availability of medication has alerted pharmaceutical manufacturers and doctors to the many side effects of drugs, among which effects on colour vision are prominent. It is now realised that colour vision changes provide a valuable means of monitoring the progress of a disease, or the toxic effects of a chemical substance, whether exposure is deliberate for therapeutic purposes or unintentional in the case of an industrial hazard. The effectiveness of treatment can occasionally be assessed by the continued monitoring of recovery of an acquired colour vision disturbance. Renewed interest has thus been concentrated around the use of colour vision as a diagnostic tool although there have been differing estimates of the efficiency of colour vision tests as the 'earliest' index of malfunction.

Disease in this context has a wide meaning, frequently involving some inherited conditions. It embraces, dystrophies (many of which are associated with metabolic disturbances), varied consequences of trauma or of tumours and degenerative conditions (age-induced change is one example). Many other pathological conditions, involving damage to the nervous system, can be included. Very detailed surveys and specific studies are available in the literature. Excellent surveys are provided by Lyle (1974) and by Pokorny *et al* (1979). Voluminous information has resulted from the work of Verriest and collaborators in Belgium and Lakowski in Canada. The present text does not attempt to duplicate such comprehensive accounts; it presents selected aspects covering the main varieties of conditions with restricted references to valuable literature. Some such conditions have a basis in heredity and a few can be called 'congenital', hence the separation by some writers of 'acquired' defects from 'congenital' defects. The latter have an hereditary basis. Although the approach to the investigation of a suspected acquired colour vision disturbance will closely parallel that taken for an inherited defect in the early stages, subtleties exist which demand very careful adaptations to the normal methods of examination.

4.1.1* Characteristics

Acquired disturbances of colour vision are a highly varied group of defects with frequent departures from established patterns. They can progress, for example, from normal trichromatism to anomalous trichromatism on to a dichromatic stage and even to monochromatism where most colour vision is lost, or they may be relatively stable. On recovery or withdrawal of the cause, colour vision may typically revert to normality through these phases if the colour loss had been considerable. A variety of related visual disturbances may accompany the colour vision change, principally visual acuity (VA) and field losses. Verriest (1963) noted as a general rule that a colour disturbance is usually evident as soon as central VA falls below 0.6. Lyle (1974) commented that reduced VA may not always be present, and no correlation is apparent between the VA loss and the type or degree of colour vision loss. Listed below are some of the important characteristics which will assist in the judgment of whether a presented colour vision disturbance has an inherited or acquired origin, for in many instances these features contrast to the stable, more predictable manifestations of inherited colour anomalies.

(*a*) Differences in colour perception between eyes. Colour loss may be confined to one eye and/or localised in one part of the visual field.

(*b*) Colour loss may be accompanied by deficiencies in other visual areas, notably reduced visual acuity, visual field defects, impaired dark adaptation, brightness perception, spatial contrast sensitivity or flicker sensitivity, ERG changes, nystagmus.

(*c*) Disturbances of blue–green–yellow vision are as common, or more common than, red–green vision in acquired forms.

(*d*) Females are affected in the same proportion as men.

(*e*) The elderly are particularly susceptible on account of the cumulative effects of toxins and increased incidence of ocular and general pathology.

(*f*) Severity of the defect is variable according to the progression of disease or degree of exposure to the drug or chemical precursor.

(*g*) Transient chromatopsia (appearance of colour to white surfaces/objects) may be present.

(*h*) Colours can often be named correctly by patients with an acquired defect on the basis of their memory for colours prior to the defect.

(*i*) Confusion in diagnosis, particularly classification of the defect when clinical tests are applied, frequently on account of an 'anarchic response'.

(*j*) Hue discrimination is typically impaired.

(*k*) Relative spectral luminous efficiency is abnormal.

(*l*) Colour perception is frequently improved for the patient with an acquired defect when the size, luminance, exposure time or saturation of the test colour is increased. The appearance of induced colour contrast effects is frequently less marked than for persons with inherited colour vision defects or normal colour vision.

(*m*) A neutral zone may be identified in the spectrum but this may not necessarily correspond to the neutral zone of inherited dichromats.

(*n*) An acquired defect may be superimposed on an inherited defect.

(*o*) The severity of an acquired defect depends on whether the cause is active or inactive.

(*p*) Some acquired defects may imitate inherited defects; thus very careful examination is required.

(*q*) Any 'unusual' colour vision disturbance or report of a change in colour perception whether of long-standing origin or of sudden onset suggests an acquired anomaly.

4.1.2 Testing of acquired defects

One of the greatest difficulties in the investigation of acquired defects has been the widespread application of tests which were designed for the patient with an inherited colour vision disturbance. The reading of plate charts, for example, frequently takes an uncharacteristic pattern when compared with the responses given by the inherited Daltonic; in addition figures seen by the colour normal may not be read. Ohta (1972) designed special charts for acquired defectives but in general chart tests alone are an inadequate form of assessment. Pinckers (1980a) recommended the Tokyo Medical College (TMC) test in a battery of tests for acquired defects, its tritan plates being particularly sensitive. Undoubtedly the use of the FM 100 Hue, D.15,28 Hue test and other Munsell-based tests are of greatest value for the detection, diagnosis and monitoring of acquired defects. A striking tribute to the 100 Hue test has been made over a decade in the suggestion that for acquired defects it is 'the most important test' (Francois and Verriest 1961a, Dubois-Poulsen 1972). A varied flexible approach, especially for the anarchic defects, is important. The variable nature of these disturbances indicates the value of frequent retesting. Monocular examination is essential.

The use of norms for different tests and conditions, as discussed by Lakowski and Kinnear (1974), is an important factor. Their examples, drawn from diabetic and ocular hypertensive conditions, show the frequent superimposition of 'acquired' and 'congenital' disorders of colour vision. Frisen and Kalm (1981) recommended the rapid Sahlgren test (see Chapter 7) for acquired defects.

Although monocular examination is usually favoured for patients with suspected or confirmed acquired defects it is often worthwhile to permit an initial binocular run when the colour vision test involves an unfamiliar or difficult task. The FM 100 Hue test is such a test; Verriest *et al* (1982) noted how fewer errors were made among subjects of all ages viewing the 100 Hue test binocularly than those restricted to monocular vision. Half the colour normal sample underwent binocular testing followed by monocular testing, while the other half carried the test out monocularly before a binocular repeat. The greatest number of errors were made on monocular tests among subjects without previous experience with the task.

4.2   THE NOMENCLATURE OF ACQUIRED DEFECTS

Several accounts of the development of 'classes' and explanatory descriptions of these deficiencies have been given, for example by Ball (1972), Grützner (1972), Verriest (1963, 1964, 1974b) and Schmidt (1973). The considerable diversity of approach leads to difficulties both in communication and in understanding the literature. The pragmatic definitions of 'types' by Verriest and his colleagues are frequently used and following the example of Pinckers *et al* (1979) the present text will adopt these in principle.

Dubois-Poulsen (1972) used a different approach which is worthy of note, dividing the field into five categories.

(*a*) Chromatopsias with 'positive' characteristics.

(*b*) Dyschromatopsias resembling 'congenital defects', being reduction systems.

(*c*) Achromatopsias showing mixed features.

(*d*) Reductions, resulting in the isolation of either a remaining scotopic or photopic mechanism.

(*e*) Colour agnosias, presumably resulting from cortical brain defects.

Francois and Verriest (1957a,b) differentiated between 'alteration' and 'absorption' causes of colour deficiency. Alteration types involve a changed photopigment, and cases such as protanomaly are characterised by a failure to accept all colour mixing matches made by normal trichromats. Absorption defects are more usually acquired, involving a filtering action before the retinal receptors are reached by light, and can be imitated by holding, for example, a yellow filter in front of the eye. The other popular term 'reduction' applies to a loss (possibly a 'collapse' or 'fusion' when inherited conditions are considered) or impairment of a retinal mechanism; these are often seen to progress from trichromatic vision, through dichromatic stages, to possible monochromatism and may be attributable to a defect of nervous conduction. A neutral region (or two) can be expected to be identified in the spectrum with such 'dichromatism'. The typical reduction defect is characterised by agreement with colour matches made by normals, as in protanopia.

Verriest (1964) used the term 'scotopisation' for the usurping by rods of activities in practical (photopic) situations, where the scotopic V'_λ curve supplants the V_λ curve with a shift towards short wavelengths; the condition may be associated with eccentric fixation. Another term used by Verriest was 'mesopisation', for a tritan-type defect resembling the physiological deterioration of normal colour vision in the lower ranges of illuminance where colour vision becomes difficult. Such classification tended to be based on results using Farnsworth's tests.

It must be stressed that cases seldom fall into precise 'types' with features closely resembling the non-acquired forms of Daltonism. Mixtures or 'anarchic' features are frequent. While acquired conditions progess and worsen, many improve showing reversal of the defects, with treatment or other 'cure' being at least potentially complete.

4.2.1   The Verriest types

Type I. This red–green affection involves scotopisation, accompanied by loss of cone function and lowered visual acuity and can progress to a form of monochromatism. Verriest (1974b) showed how 'central retinal degenerations' could be involved in such cases.

Type II. Primarily associated with Kollner's description of conduction affections with red–green (and possibly mixed) defects, this type of condition tends to maintain a normal V_λ curve, although it may progress towards dichromatism and even to monochromatism. Visual acuity loss may be expected but there can be considerable recovery of the optic nerve function.

Type III. A tritan-type defect, usually described as blue–yellow, caused by any of a wide variety of disorders, including the changes in the 'normal' crystalline lens. Mesopisation and (on anomaloscope readings) protanomaly may be found.

More mixed defects with indefinite distinctions of colour confusions may be added to the main three types; there is some merit therefore, in presenting this as a 'fourth type' (see Francois and Verriest 1957a,b).

Some idea of the distribution of the varieties of acquired defects is useful. Such a study was given by Pinckers (1972a) in analysing the performance of 314 patients with a test battery. Eighty-two eyes were shown to have blue–yellow distubances, with 'retinal' and 'media' causes; 57 eyes showed protan changes, typically with receptors being involved, while 38 eyes revealed deutan defects which were less retinal and more conductive in origin.

A decade ago Marré (1973) noted the great similarity between the classifications of acquired defects proposed by Jaeger and Grützner, on one hand, and Verriest, on another. She interpreted Kollner's 1912 classification in detail along the following lines.

(*a*) Red–green defects (usually associated with the optic path, starting with the ganglion cells and extending to the cortex) tend to be 'progressive'; they involve all colours, but red and green suffer most.

(*b*) Blue–yellow defects (usually retinal in origin) tend to safeguard the perception of red and green longer than that of blue and yellow. They must be combined with (*a*) in order to progress to total loss of colour perception.

4.2.2   Other forms of defect

Achromatopsia. This is sometimes a synonym for monochromatism.

Chromatopsia. Postive colour introduced into a normally white scene is termed 'chromatopsia'. The different forms have been described in detail by Dubois-Poulsen (1972). These range from the 'yellow vision' induced by santonin to the 'cyanopsia' found in the aphakic. Possible reddening of the visual field by blood in the eye has been suggested but not well substantiated.

Dyschromatopsia. This term is widely used as an alternative to 'disturbance of colour perception' etc, but more frequently in connection with 'acquired' than with 'congenital' forms.

4.3   EFFECTS OF THE OCULAR MEDIA

Light is lost as it passes through the refractive elements of the eye by absorption and scattering; attenuation is wavelength selective in both cases. It is important in the older eye on account of the yellowing of the crystalline lens and the possible increase in macular pigmentation with age since, along with natural pupil constriction with age, less light finally reaches the retina. In §1.13 the physiological aspects which relate to 'acquired' defects are covered. Figures 1.40 and 1.41 should be considered in this context.

4.4   COLOUR VISION DEFECTS IN DISEASE

Many systemic and ocular disorders have been shown to induce colour vision change. A survey of 92 ocular diseases involving over 1000 affected eyes, presented by Francois and Verriest (1968) discussed the causes, manifestations and variety of methods available to examine these abnormalities. Readers should consult the account of Verriest (1974b). The text by Pokorny and Smith (1979) is a valuable source book, written primarily from an ophthalmological viewpoint. An exhaustive survey in this text is not possible. Readers should consult table 4.1 for a list of representative conditions. Testing methods for the investigation of acquired defects are covered in Chapter 7 while Chapter 8 includes some typical profiles of patients with acquired defects. Consultation of these sources will indicate profound variety among patients. In general it is most suitable to interpret patient records through an application of general principles. Attempting to 'match' the clinical records of one patient against another is not recommended. In some cases acquired colour vision defects might result from a dysfunction of the interaction between cone mechanisms, as suggested by Zrenner (1982).

4.5   COLOUR DEFECTS CAUSED BY INJURY

Partial or complete loss of the colour sense can occur through cranial trauma without other visual functions being affected. A blow to either the

Table 4.1 A selection of disorders related to the eye which affect colour vision. Note that while this table provides an overview of the possibilities it lacks detail and should be used as a guide only.

Condition	Variety of colour vision defect	Possible bias	Probable Verriest type no	V_λ	Neutral region
Amblyopia (other eye may be involved)	None, mixed or $R-G$			Possible red loss	
Cataract	$B-Y$	Tritan	III		
Central path lesion	$B-Y$	Tritan	III		
Central retinal vessel disturbance or other vascular retinal condition	$B-Y$	Tritan	III		
Centroserous retinopathy	$B-Y$	Tritan and protanomal	III	?	
Chorioretinitis	$B-Y$	Tritan	III		
Diabetic retinopathy	$B-Y$	Tritan	III		
Glaucoma	$B-Y$ possibly mixed	Tritan	III		
Hysteria	$R-G$ or mixed				
Juvenile optic atrophy (dominant)	$B-Y$	Tritan but vague	III		$B-G$
Leber's optic atrophy	$R-G$	Deutan	II		$BG-R$
Macular degeneration					
Senile	$B-Y$	Tritan	III		
Juvenile without pigment changes	$R-G$ possibly $B-Y$ rarely	Protan	I or quasi-III	Scotopic shift	
Juvenile with pigment disorder	$B-Y$	Tritan	III		
Macular dystrophy	$R-G$	Protan	I	Scotopic shift	
Optic nerve trauma or compression	$R-G$ possibly mixed	Deutan Tritan	I or II possibly III		$Y-G$
Retinal detachment	$B-Y$	Tritan	III		
Retrobulbar neuritis	$R-G$	Deutan	II	Normal	
Retinitis pigmentosa	$B-Y$	Tritan	III	?	
Stargardt's disease	$R-G$		I	Slight scotopic	

front of the head or the occipital lobe at the back can give rise to colour vision changes; these may be transient or permanent. One of the earliest reports of 'colour blindness' caused by a blow was by Fontan (1883), cited by Jennings (1896), and various war injuries have given rise to such disturbances. Extreme sensory losses, including total loss of colour vision, were experienced by a colonel who had been struck on the left temple by a musket-ball, according to Posada-Armiga (1885). Although the sensory faculties of smell and taste returned, it was two years before colours could again be recognised; firstly reds and lastly greens. Other examples in the literature during the last century include a case quoted by Jennings of a physician who was thrown from his horse, resulting in severe concussion and long-term 'cerebral excitement'. On recovery he was unable to resume his hobby of sketching because his colour vision had permanently become both 'weakened and perverted'. A severe red–green defect was indicated by Wilson who examined this patient. A further case described by Jennings (1905) involved a soldier wounded on the forehead from an explosion. The following day vision appeared to be tinted blue and visual fields were poor. On the second day yellow was mistaken for red and light yellow for blue. By the third day all colours seemed to the patient as blue and on the fifth day all but very light colours returned to his perception normally, followed by complete recovery.

More recently a patient showing loss of colour perception for blues, reds and greens resulting from a blow to the back of the head was studied by Young *et al* (1980). An anarchic arrangement was shown by this patient on the FM 100 Hue test with an error score of approximately 250. With the Ishihara test correct identification of the demonstration plate only was possible.

Perimetry using coloured targets can be valuable for investigating such patients. A patient examined by Schneider (1968) showed blue and red field visual field constrictions corresponding to the side where the blow to the head had occurred. In the same patient the Ishihara and HRR plates were failed when testing was carried out at the normal viewing distance (0.75 m) although the test was passed at 2 m; tritan confusions were made on the D.15 test. Careful visual field investigation in a patient examined by Hansen (1972) revealed a peripheral scotoma in one eye for blue and red targets the day following slight concussion. Normal fields were evident for white targets.

Cerebral vascular accidents can also change colour vision; Abney (1895) reported a unilateral case. Furthermore cerebral lesions can induce complete colour loss, as seen in the case of 1899 reported by Mackay and Dunlop (cited by Green and Lessell 1977).

When the integrity of any part of the visual pathway is threatened, by whatever cause, colour vision may be impaired, and the presence of multiple perceptual abnormalities can make colour vision examination difficult (see Critchley 1965). Vascular lesions in the visual pathway have been shown by Vola *et al* (1972) to be associated with red–green and 'tetartan' defects when such patients are examined with the Farnsworth–Munsell 100 Hue, the D.15 test and the HRR pseudoisochromatic plates. Other examples of dyschromatopsia in patients with neoplasms or vascular lesions are discussed by Green and Lessell (1977). In many patients early random disturbances of colour vision were noted.

Patients for whom sight is restored after several years of blindness and those recovering from strokes are frequently unable to recognise, use or name colours in an ordered manner. Colours seen in everyday scenes are described as degraded, desaturated or muddled or even extend beyond the expected confines of their projected images. A century ago Wilbrand (see Critchley 1965) enunciated features of agnosias involving colours, including errors of naming. In some patients naming was identical for different stimuli, while in others correct colour names were ascribed to objects associated with the colour, such as red post boxes, but incorrect names were given to an identical colour out of context. Colour agnosia is often associated with disturbances of form perception. Partial lobectomy of the occipital cortex can give rise to blue–green colour defects with varied presentation, as shown by Dubois-Poulsen *et al* (1952).

4.6 GENERAL RETINAL CONDITIONS

Conventionally, pigment epithelium and cone receptor disorders have long been associated with tritan defects. While 'outer' neurones frequently induce protanomalous defects, disorders of the 'inner' neurones frequently give rise to deutan acquired defects. Tritan defects are often classically associated with outer layers of the retina and red–green defects with inner layers. Such findings are supported by recent investigations, for example, by Grützner (1972), while Marré (1973) has shown early loss of blue sensations in retinal disease followed by red–green disturbances.

General lesions to the retina on account of therapeutic photocoagulation or vascular changes will bring about colour vision change. Thus colour vision is frequently disturbed for red, blue and green in artherosclerosis, the degree of disturbance increasing with the progression of the disease (Trusov 1972).

4.6.1 Retinal degenerations

A great variety of conditions can be included in this category. Tapeto-retinal degenerations can produce tritan defects though significant macular involvement will often superimpose protan disturbances. A family described by Taylor (1977) is interesting since an acquired tritan defect was superimposed on an inherited colour vision defect.

Dystrophy of cone receptors in the retina can give rise to greatly modified receptors. A retinal dystrophy with delayed symptoms initially affecting cones, but later involving rods, will cause lowered visual acuity and colour deficiency. Frequently results on clinical colour vision tests are anarchic, such as in the case of the patient examined by Norden *et al* (1978); dark adaptation studies in this patient demonstrated clearly the cone disturbance.

The degenerative rod disease *retinitis pigmentosa* is one of the most common of all inherited eye disorders. It leads eventually to blindness because rods outnumber the cone receptors by far, and very often cone involvement is typical in the later stages. The cone dystrophies are more complex; they too are normally inherited, often resulting from consanguineous marriages, although, unlike the rod diseases, cone dystrophies are usually apparent from an early age, and they are seen less often in a complete stage. Achromats normally have great difficulty in fixating objects because their foveal cone vision is impaired. Nevertheless some vision is always possible, and in general the handicap resulting from cone disease is less marked than that associated with rod malfunction. Progressive cone dysfunctions do occur, though rarely, usually in an inherited form. There is a progressive loss of visual acuity but no nystagmus, with onset usually in childhood. Colour vision can deteriorate quite rapidly and an abnormal ERG is found as the disease progresses.

An unusual case of permanent cone dystrophy of sudden onset, following therapy for systemic infection after Caesarean section was reported by Siegel and Smith (1961). Although visual acuity was reduced and photophobia was reported, the peripheral cones appeared to be predominantly affected. Constriction of the peripheral colour fields was evident with milder losses shown by foveal wavelength discrimination. Clinical colour vision tests indicated poor hue discrimination on the 100 Hue test although on both the D.15 test and HRR test normal responses were seen. Figures could be seen on the Ishihara test by this patient and ERG changes were evident.

A fungal infection prevalent in the USA produces histoplasmosis effects in the central retina with macular effects and scar tissue disseminated around the fovea; in such cases colour vision can be expected to be disturbed (see Schlaegel 1977).

4.6.2   Testing for retinal degenerations or intoxications

Although PIC plates can be useful as the initial indication that colour vision is disturbed, the Farnsworth-Munsell tests, such as the 100 Hue test, and more especially the desaturated D.15 test, give more valuable results. Grützner (1972) has advocated using spectral sources among this group since hue discrimination measurements are valuable. In many instances saturation discrimination measurements will reveal disturbances in the short wavelength region.

4.6.3 Macular affections

Since normal trichromatic function is chiefly associated with the foveal region it is inevitable that disturbances of the retina around the 'fixation area' are often manifested by colour vision defects. While vision and visual acuity are often affected simultaneously, the more fragile colour sense may suffer selectively, earlier than form sense. The use of special non-coloured tests, such as contrast sensitivity with gratings can be valuable. Since some central retinal disorders affect one eye only the unaffected eye can serve as a valuable comparison for the assessment of acquired loss of colour vision; retrobulbar neuritis and centro-serous retinopathy are examples.

In many cases crude colour vision tests will fail to reveal any abnormality in either eye, while more refined testing methods will indicate clear differences between the two eyes; such was the experience of Scheibner and Thranberend (1974) examining patients with neuritis. Colour vision tests can often prove more sensitive than visual field examination, indicating colour losses which precede visual field defects. Such value for colour vision testing was indicated by Williams (1976) and Williams and Leaver (1980) in a comparison of 100 Hue test errors to the extent of slight detachment in the macular region. The use of a broad range of tests for *'la choroidite séreuse centrale'*, including form sense, Farnsworth's colour tests, perimetry and electrophysiology has been reviewed by Haché *et al* (1972).

(PAGE)	SUBJECT'S CHOICE OF MATCH	NORMAL	PROTAN	DEUTAN	TRITAN
1	TB	B ⬡	R	L	(T)
2	TR	R ◊	B	L	(T)
3	TL	(T) ⬡	B	R	L
4	LR	R ◊	T	B	(L)
5	LB	(L) ◊	R	T	B
6	TR	R ◊	L	B	(T)
7	RL	L ◊	T	B	(R)
8	TL	L ◊	R	B	(T)
9	RB	B ⬡	L	T	(R)
10	LR	(L) ◊	T	R	B

(PAGE)	SUBJECT'S CHOICE OF MATCH	NORMAL	PROTAN	DEUTAN	TRITAN
1	T	B ⬡	R	L	(T)
2	T	R ◊	B	L	(T)
3	L	T ⬡	B	R	(L)
4	R	(R) ◊	T	B	L
5	L	(L) ◊	R	T	B
6	R	(R) ◊	L	B	T
7	L	(L) ◊	T	B	R
8	L	(L) ◊	R	B	T
9	R	B ⬡	L	T	(R)
10	B	L ◊	T	R	(B)

Figure 4.1 Results of the CUT (1st edition) for two patients with centro-serous retinopathy, each showing tritan defects on a medium sensitivity clinical test in the right eye only. The observer with the greater number of tritan errors tended to mix these with normal responses; in this record the more marked preference was ringed and the second preference was underlined. Macbeth illumination was used in both cases.

The so-called 'macular color defect' described and investigated by Pokorny *et al* (1980) was accompanied by near normal visual acuity with significant disturbances of colour perception. A variety of protanomaly was suggested by examination with the Rayleigh standard equation and with the 'extended' version. A comparison between the performance of the two eyes was undertaken by Pokorny *et al* (1980) among their patients using the Moreland anomaloscope, and variations in matching ratios in the affected eyes were interpreted in terms of an effectively reduced range of spectral absorption of photopigments.

For patients over the age of 50 macular degeneration of the 'senile' variety can have serious consequences. Blue—yellow defects were noted in such patients more often than red—green confusions, in a series described by Francois and Verriest (1961a). By contrast the juvenile presentation of macular degeneration typically involves red—green confusions. Although tests such as the FM 100 Hue and D.15 standard and desaturated versions are of great value in investigating these patients, Stilling's PIC plates can reveal early senile changes (Heinsius 1972). Disorders of the young macula (described as Stargardt's disease) are varied and frequently associated with red—green defects, although tritan confusions may occasionally be revealed. Ariel (1979) has indicated that anarchic colour vision changes are not unusual.

Lens and macular pigments

The yellow pigment, present particularly in the aged crystalline lens and macula, frequently complicate the clear identity of tritan defects associated with maculopathy. Moreland *et al* (1978) attempted to minimise this problem by the careful choice of stimuli for their anomaloscope. They emphasised the importance of central fixation particularly in such patients who would prefer to use extra-foveal regions for colour judgments.

Pigment and membrane related opacities in the anterior segment of the eye, possibly associated with retinal developmental anomalies, were considered by Sarwar (1961); he noted that fewer errors on the Ishihara test were made by some children when polarised filters reduced irregular reflections.

Sloan (1942) described a patient with macular pigment 'changes' who was able to pass the Ishihara test but failed on some Stilling plates and other tests when tritan confusions were present.

Methods of examination

The widely used Friedmann visual field analyser has been adapted by Friedmann (1969) to measure macular thresholds with and without red filters. By means of a suitable comparison of macular thresholds for normal observers and patients in the same age group it is possible to monitor the progress or deterioration of a condition in most patients. In some cases colour perception and visual acuity can remain normal.

Chisholm (1969) evaluated the FM 100 Hue test in acquired defects, and gave added support to its clinical use for the differentiation of varieties of acquired dyschromatopsia. The trend to improved scores on retesting was noted and examples were given to show how retrobulbar neuritis (in the affected eye) presents a different profile from that of untreated tobacco amblyopia. The increase of score variance at high levels for initial testing appeared to be significant only when scores in excess of 600 were encountered. The detection of macular changes with the D.15 test was described by Collin (1966). Several reports by Bowman (1978, 1980a,b) have concentrated on senile macular degeneration (SMD), with respect to colour vision changes and their relation to illuminance and to visual acuity. The FM 100 Hue test was shown to be useful in monitoring changes in patients with senile macular degeneration showing less severe reductions of visual acuity. In general, the normal variation of total error scores (between 10 and 1000 lux) is accentuated in these patients with tritan defects, being indicated most easily at low levels. In patients with low visual acuity Bowman recommended the D.15 test as reliable. 'Slight' SMD patients typically made errors (often tritan) on the D.15 test at 1.0 lux while 'definite' degenerations produced errors at 10 or even 100 lux.

The use of small apertures in a lantern test was shown by Giles (1950a) to be capable of detecting certain central colour scotomata. He stressed the importance of monocular testing where a unilateral macular condition may exist. Such cases may be cystic or toxic in origin, or the result of radiant energy damage, including exposure to solar eclipses and lasers.

The coloured scotoma plates designed by Ohta (1972) allow hue and 'luminosity' perception to be investigated in detail. Another Japanese test, resembling the FM 100 Hue test, was used by Hukami (1959), demonstrating that *chorioretinitis centralis serosa* produced distinct loss of discrimination resembling tritan defects.

A variety of test methods described by Jaeger (1956) included the anomaloscope, the colorimeter and a simple pigment 'luminosity test'; of particular interest was the appearance of a neutral point (yellow) in the spectrum in macular oedema.

Colour naming is another approach, though not always reliable for the acquired deficiencies. Lanthony (1980) investigated this approach among 21 patients and noted disturbances of the normal colours and the location of neutral appearances.

4.7 AMBLYOPIA

The classification of amblyopia is complex and variable (see Duke-Elder 1949 and Burian and von Noorden 1974). The types are numerous depending on the cause but the majority result from disuse of an eye or inappropriate or inadequate stimulation as a consequence of an infantile or childhood squint, cataract, congenital lid defect (drooping upper lid), or a large difference between two eyes in power. Amblyopia can be defined as low visual acuity without any apparent optical cause. Roth (1966a, 1968) has provided an excellent review of typical colour vision characteristics among such patients.

Typically, colour vision disturbances in squint amblyopia are akin to the colour losses associated with eccentric retinal stimulation among colour normals rather than classical acquired disturbances (see Francois and Verriest 1967, Roth 1968, Marré and Marré 1978). The reduced sensitivy to long wavelength regions in the amblyopic eye, as shown by a depressed V_λ curve in this region, was originally demonstrated by Wald and Burian (1944). In many cases colour vision can be relatively normal, as suggested by examination results using plate tests, Farnsworth-Munsell tests and the Nagel anomaloscope by Roth. Even when foveal fixation was maintained and poor visual acuity resulted, there can be an apparently normal colour sense. With eccentric fixation the expected decrement of colour vision found normally was shown also in amblyopia. In those cases where an inherited colour vision defect was present, the dominant eye and amblyopic eye appeared to have equal 'dyschromatopsia'.

Marré and Marré (1982a,b) have noted that for foveolar fixation (within 20') a red–green acquired defect is not reported since any associated acquired colour deficiency attacks the blue mechanism. Eccentric fixation by a diseased eye may involve red–green defects but blue is still most disturbed. These authors discussed the V_λ shift (towards typical scotopic values) which sometimes manifests 'rod intrusion' in amblyopia combined with eccentric fixation. Such a shift when optic nerve disease is present results from 'eccentrisation'; in a retinal affection displaying the same displacement of fixation 'scotopisation' is more typically seen, although foveolar fixation may remain in retinal disease. A flicker photometer with a $12°$ (photopic level) surround was used for this investigation. Eccentrisation of the usual small amount can reduce visual acuity to approximately 50% while a $4°$ shift lowers acuity to 10% of foveal values. Retinal disease associated with scotopisation can leave 80% levels of visual acuity intact.

Using increment threshold techniques Marré and Marré (1978) demonstrated a decreased presence of the red and green colour vision mechanisms, with enhanced sensitivity of blues, in amblyopic eyes showing extra foveal fixation modes. Hue discrimination may be only slightly impaired in such patients, and sensitivity is often generally better than would be expected, considering the retinal region involved for fixation. Marré and Marré found that the saturation discrimination of amblyopic eyes was not significantly disturbed in amblyopes. Israel and Verriest (1972), using increment thresholds of a limited series of coloured targets on a white background, have also noted distorted colour perception of amblyopic eyes.

4.8 OPTIC NERVE AFFECTIONS

A variety of conditions are involved in optic nerve disease which may produce red–green and tritan defects. Dominant (autosomal) hereditary optic atrophy was studied by Jaeger (1956) in an extensive series of cases. In such patients acquired 'tritan' characteristics can be demonstrated with low visual acuity for distance and minor visual field disturbances. Using the ophthalmoscope, pallor of the temporal aspect of the optic disc is often seen. The colour vision defect was described by Schmidt (1973) as bilateral and equal. Bilateral effects on colour were also shown in patients with retrobulbar neuritis, examined by Griffin and Wray (1978) using the FM 100 Hue test, even when VER data appeared normal in one eye.

Francois et al (1961) made a valuable comparison of different hereditary affections of the optic nerve, showing the great value of colour vision investigation. Tritan defects accompanied their cases of dominant infantile optic atrophy while mixed defects were found in other conditions. In cases of 'dominant inherited juvenile optic atrophy' reported by Smith (1972a) acquired tritan disturbances were seen. The FM 100 Hue test was considered highly valuable for diagnosis and the presence of tritan dyschromatopsia was regarded as 'critical' in the diagnosis. Smith showed how peripheral fields for white can be expected to be normal while colour fields can be 'inverted'. Völker-Dieben et al (1974) confirmed the variability of types of achromatopsia in dominant inherited juvenile optic atrophy.

According to Kollner's dictates, diseases affecting the ganglion cells and the optic nerve fibres are usually associated with red–green defects. A notable exception is a dominant atrophy of the optic nerve, as suggested by Krill et al (1970) and discussed by Schmidt (1973). Such patients often show good visual acuity (certainly for reading) with a typical early tritan defect and other distinguishing features such as minimal pallor of the temporal aspect of the optic disc. Krill and his colleagues even questioned whether 'congenital tritans' were not hereditary dominant optic atrophy subjects while Schmidt defended certain of her own diagnoses.

Typical early onset (infantile) autosomal dominant atrophy was judged by Scott (1941) to be associated with colour vision loss, often of a tritan nature. Nevertheless some cases reveal protan or deutan defects as shown by Kok van Alphen (1960) and Jaeger et al (1972).

Zanen (1972) used a device for determining spectral thresholds for the fovea with which disturbance of the photochromatic interval in optic nerve conditions, such as retrobulbar neuritis, could be determined relatively easily.

The value of neutral points (or bands) in the spectrum has been shown by Grützner (1972) in connection with varied optic nerve conditions. Using a reference light of 3200 K he found changes in the spectral bands as conditions altered. In Leber's optic atrophy, or in *tabes dorsalis*, the neutral area may extend to 700 nm, or to 400 nm in the infantile dominant atrophy, where the 'achromatic' region may be split. Hue discrimination disturbances are likely to accompany these features.

Colour vision test results were used by Chisholm and Kearns (1981) in the management of pituitary tumours. They divided a series of patients into two groups. The first group, with a moderate loss of visual acuity, showed typical red–green defects with the FM 100 Hue test but had no central scotomata; these patients recovered with treatment in about a month and their colour vision improved considerably. Group two had suffered severe loss of visual acuity and had central scotomata as well as anarchic FM 100 Hue results; even so, after treatment they recovered well although this improvement was slow.

4.9   RAISED INTRAOCULAR PRESSURE

Excessive pressure in the eye and glaucoma in its many forms threaten retinal function. Early detection by means of a variety of techniques is essential for a favourable prognosis. Subjective tests, for example perimetry, are used in conjunction with direct physical measurements of intraocular pressure. In recent years it has become clear that colour vision disturbances, which can be monitored quickly and easily, constitute an early warning. Nevertheless a combination of approaches, including ophthalmoscopy, is always required.

In general, blue–yellow perception fails early, but Austin (1974), Foulds *et al* (1974) and others have shown how optic nerve fibre damage might have indicated a typical red–green disturbance; this indeed *can* occur, as shown first by Engelking (1925) and Verrey (1926) and more recently by Maione *et al* (1976) using sensitive increment threshold techniques. However, 'chronic simple' glaucoma is more frequently associated with tritan disturbances. A survey by Verriest (1964) of patients with established chronic simple open angle glaucoma suggested that approximately 20% had normal colour vison, 72% had blue–yellow defects and only 6% showed red–green disturbances. It is noteworthy, however, that some glaucoma patients can display reduced colour discrimination of a mixed variety. Even normal subjects exposed to artificially raised intraocular pressure were shown by Foulds *et al* (1974) to develop long-lasting tritan defects.

An excellent summary of the literature has been given by Francois and Verriest (1959). In this early study of the relation between acquired dyschromatopsias and glaucoma the incidence of monocular tritan defects was striking, both in cases of open and closed angle glaucoma.

Errors in colour discrimination can be reduced when greatly increased illuminance is possible (such as with dilated pupils). Conversely a patient on miotic treatment may develop a pseudo or enhanced tritan defect. Lakowski and Oliver (1974) have provided an excellent account of the effects of pupil area. Ourgaud et al (1972) demonstrated the importance of dilated versus undilated pupils in glaucomatous eyes with regard to colour discrimination. The latter study showed a preferential reduction of blue–yellow deficiencies and of some red–green defects as the retinal illuminance was markedly raised. No 'achromatic' deficiencies altered, even at 2200 lux.

The 'ocular hypertensive', with little to show except suspiciously raised intraocular pressure but nevertheless presenting a risk of developing glaucoma, presents a problem which may sometimes be resolved by an assessment of retinal damage by colour vision testing. Lakowski (1972), for example, showed how anomaloscopic losses, supported by FM 100 Hue data, can be a valuable index of colour loss. Lakowski and Drance (1979) noted that some 20% of ocular hypertensives showed colour losses equivalent to the glaucomatous groups and of these some subsequently developed chronic simple glaucoma. Thus these authors consider that potential glaucoma patients might well be identified using colour vision as one indicator.

Kalmus et al (1974) used the D and H Color-Rule on glaucoma and ocular hypertensive patients, finding many enlargements of 'range' of acceptance, chiefly tritan in nature. Nevertheless, negative results and the tediousness of the test gave rise to caution.

The doubt as to whether field defects are preceded by colour deficiency was expressed by Poinoosawmy et al (1980) who showed that FM 100 Hue errors accompanied glaucoma with field defects, the differences in error scores between a patient's eye being significant. Monocular field defects accompanied such intraocular features as colour and cup/disc ratio. However, a number of investigators have demonstrated the presence of acquired colour vision defects before visual field losses (Fishman et al 1974) and a simple relationship between the degree of colour vision defect and the amount of field loss. The clear utility of a colour vision change as an early diagnostic tool for glaucoma and ocular hypertensives has emerged in recent studies by Adams et al (1981) who, using sensitive increment threshold techniques, have shown sensitivity losses to both the colour and brightness pathways in glaucoma patients showing good visual acuity and minimal field changes. It is suggested by these authors that both tonic and phasic ganglion cell fibres may be affected by the disease. The short wavelength mechanism is confirmed as being at greatest risk in glaucoma.

As Chisholm (1979) has pointed out the detection of acquired defects of colour vision in chronic simple glaucoma is clearly 'test dependent'. While the FM 100 Hue test shows a prominent tritan defect in over 30% of eyes, raised error scores without prominent axis are reported in a further 30%. Metzstein (1982) showed that box 4 of the FM 100 Hue test gives higher error scores than box 1. Francois and Verriest (1957b) were perhaps the earliest investigators to show the value of PIC tests and the Farnsworth D.15 in detecting glaucoma. In one of the most recent studies using the D.15 test Adams and Rodic (1981) have seen how minor modifications to the scoring of the Farnsworth D.15 panel test can separate out glaucoma and glaucoma-suspect patients from age-matched normals with a high degree of reliability. Using a pass/fail criterion of failure based on one single place-ment error (minimum) these authors found that for the normal D.15 test 53% and 32% of glaucoma and glaucoma-suspect patients, respectively, failed compared with 1.3% of age-matched colour normals. By use of a desaturated D.15 test giving greater sensitivity and thus difficulty in achiev-ing a pass, 78% of glaucoma and 58% of glaucoma suspects failed com-pared with two colour normal failures.

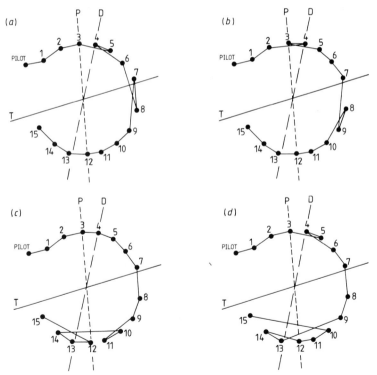

Figure 4.2 D.15 'failure' criteria used by Adams *et al* (1981). (*a*) and (*b*) show more than one single-place error and (*c*) and (*d*) errors greater than single place.

Clearly the development of more detailed measurement techniques and clinical tests of colour vision (in particular the increment threshold approach) is of vital importance for the early detection of disease caused by raised intraocular pressure. Very careful examination is required, however, since both small pupil-size and normal aging changes can mimic the colour vision changes reported in glaucoma (Grützner and Schleicher 1972, Lakowski and Oliver 1974, Marmion 1978).

The value of a desaturated D.15 test in glaucoma suspects was described by Adams *et al* (1981). While they discovered that some 11% of normals failed the (Value 5, Chroma 2) test, 78% of 19 glaucoma patients failed and 58% of glaucoma suspects failed; the last group showed about half the rate of failure on the standard (5/4) D.15. In this study failure was taken as 'more than one single-place error or any error greater than single-place' (see figure 4.2).

4.10 DIABETES MELLITUS

Retinal functions are damaged by diabetes in the first instance by a nutritional deficit which affects the neurones and later by vascular disturbances associated with microaneurysms. Blood and exudative masses, which may displace tissue, later damage sight and the coagulation of blood vessels used as a treatment can itself affect colour and form vision. Lakowski *et al* (1972) reviewed the history of colour vision losses associated with diabetes and gave an analysis of different test methods in many cases; in general, blue losses are most severe. They stressed the fluctuating nature of some colour vision difficulties, in the event of variation of blood sugar level. It was seen to be difficult to predict the state of vision or retinal disturbance from either normal or abnormal colour vision but a useful approach was found to be a combination of weighted data (from different colour tests) and duration of condition; by such means there was reasonable prediction of minor retinopathy likely in diabetics in the under 30 and over 60 age groups.

Colour vision tests of the PIC, anomaloscope and Farnsworth varieties have a valuable place in the detection, care and continuous assessment of diabetics. A widespread reliance upon the FM 100 Hue test has to be regarded in the light of the reasonable criticisms of Smith (1972a,b), relating to the hesitance of some patients doing this time-consuming test, which must be conducted monocularly for greatest value.

Various groups of 100 Hue colours were studied with diabetics by Barca and Vaccari (1977) showing how many patients had predominantly blue—green discrimination difficulty. Verriest (1964) noted that a red deviation on the red—green anomaloscope is typical of many diabetics. Of the 500 diabetics surveyed by Kinnear *et al* (1972) the majority had reasonable visual acuity. When compared with non-diabetics, using a range of ages between 20 and 65, these diabetics clearly gave 'considerably poorer' results with 100 Hue errors, those diabetics with visible fundus oculi changes having poorest discrimination. Ishihara and anomaloscope results suggested similar tendencies; tritan deficiencies were revealed as being most marked. Data on changes detected after a few years, with predictions which arise, leave little room for doubt that the deterioration of colour vision in diabetics may resemble senile changes, but is more rapid.

A definite relationship between severity of colour vision loss (as monitored by the FM 100 Hue test) and severity of retinopathy, particularly with respect to retinal vascular non-perfusion, has been noted by Condit *et al* (1982). Those with a higher FM 100 test score tended to have a greater number of microaneurysms, soft exudates and dilated capillaries as evidenced from fundus photographs. A very strong statistical correlation between FM 100 Hue test scores and significant capillary drop-out shown by fluorescein angiograms was found. Blue cone sensitivity dysfunction is even evident in juvenile diabetics with good visual acuity (Adams and Zisman 1980). In some adults the blue loss can closely parallel exacerbations and remissions of diabetic control, suggesting that this particular dysfunction may be related more to diabetic control than to retinopathic changes.

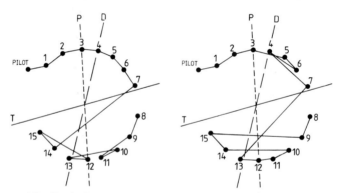

Figure 4.3   Typical D.15 plots for a diabetic. This observer made 7 errors with HRR plates in the right eye (right) and 9 in the left eye (left).

Very recent research using the sensitive technique of increment thresholds (white background) has been able to highlight losses in the blue mechanism of diabetics before visual field changes or visual acuity become impaired (Adams *et al* 1981). Selective colour vision loss was reported very early in the course of the disease, suggesting that it far precedes retinopathic abnormalities. Furthermore these authors consider that the colour losses thus measured probably precede early changes detectable by the FM 100 Hue test, interpreting their findings in terms of a selective damage to the tonic retinal ganglion cell fibres. Blood sugar levels which vary can produce day to day fluctuations of colour vision. The macular involvement at an early stage of the condition was considered to be caused by a selective impairment of tonic ganglion cell fibres. A 3.4 log troland white background was used, with a 1° test flash which was at 1 Hz for chromatic and 25 Hz for achromatic sensitivity measurements.

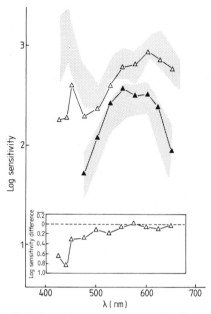

Figure 4.4 Six diabetic subjects compared with nine normals by Adams *et al* (1981) all subjects being under 40 years. Chromatic sensitivity shown in open triangles, achromatic sensitivity in closed triangles with the shaded zone indicating the normal mean, plus or minus 1 SD. The lower inset gives the mean spectral sensitivity difference between normals and diabetics. Courtesy Dr A Adams.

Kinnear (1965) considered the implications of lowered colour discrimination in young diabetics with respect to their employment. The risks and disadvantages were set against postive factors for employing diabetics for tasks involving colour. Among the guidance given was the need for testing for both acquired and congenital colour defects with coloured light stimuli as well as surface colours.

4.10.1 Domestic urine tests

Taylor (1972) drew attention to the hazard for a diabetic with significantly defective colour vision who uses coloured indicators for self-testing urine sugar. Thompson *et al* (1979) investigated a range of comparison colours in a reagent tablet system where there are several possible confusions. The orange end of a range (the 2% level) may progress to a red–brown in higher concentrations. Smith *et al* (1982b) confirmed the likely difficulties of diabetics and all indications are that the matter must be taken seriously. Colour vision testing, regularly repeated, is essential for diabetics. If they have to test their urine and lack the necessary colour discrimination they must be assisted, or use methods which are not dependent on colour.

4.10.2   Oxidase-peroxidase test strips

The 'BM-Test-Glycemie, 20-800' method of testing for glucose in blood is a semiquantitative application of small test strips which are compared to coloured samples. These samples are blues, blue–greens, greens and yellows and may give difficulty where a tritan-type colour deficiency exists. Thus a colour approximating to 5B 5/6 is paired with 7.5G 5/2 in the middle of the range, which extends from about 2.5Y 8.5/4 and 7.5BG 8/4 to 2.5B 4/4 and 10BG 2/4. Figure 4.5 shows such colours.

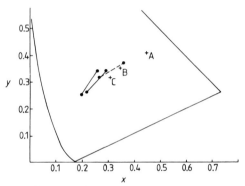

Figure 4.5   Approximate positions of pairs of colours in the 'BM-Test-Glycemie, 20-800' reactions. A, B and C are positions of standard illuminants.

An alternative view was expressed by Graham *et al* (1980), who concluded from clinical trials that routine colour vision tests are not, in fact, necessary for diabetics using this method. They used 48 diabetics who estimated their own blood glucose with the strips. Neither diabetic retinopathy nor the state of colour vision appeared to affect the procedure, even though some of the subjects had 'severe' FM 100 Hue test defects amounting to scores of 250 or more.

4.11   FIELD MEASUREMENTS IN ACQUIRED DEFECTS

In addition to detailed examination of central colour vision using clinical tests, perimetric examination for both central and peripheral locations is valuable in acquired defects of both central and retinal origin.

Some thirty years ago current opinion favoured the study of isopters by the manipulation of size, luminance and contrast with white rather than coloured stimuli (Dubois-Poulsen 1952); by 1981, however, this distinguished clinician had become convinced of the utility of suitable coloured objects.

Filters in a standard device such as the Tübingen perimeter permit the determination of disturbed red (or other) isopters, as shown by Sloan and Feiock (1972); other contributors to the same volume added further evidence, including the use of field studies in conditions such as trauma, optic neuritis and macular affections, as well as age effects.

Chapter 7 covers in detail testing methods employing colour thresholds at different retinal positions and readers are referred to this. Perimetric studies to date have concentrated heavily on the acquired defects. The charts of Amsler (1953) are valuable to supplement such data. An inherited cone and rod dystrophy case, reported by Norden et al (1978) included useful comparisons of Amsler grid findings in two eyes, with marked colour vision defects.

Kollner (1912), one of the first to describe pathological changes in colour vision in any detail, noted the constriction of the blue isopters in diseases affecting the retina and choroid, and the impairment of red–green perception associated with optic nerve disease. This led to Kollner's law which, although known to have many exceptions, is an approximate description of acquired colour vision defects.

Coloured stimuli have been used for the detailed plotting of changes in central field defects. Walsh and Sloan (1936) showed how blue and red stimuli could be applied, including reported 'greenness' of the blue stimulus in some cases.

The application of the Goldmann perimeter with red, green and blue stimuli superimposed on a neutral ground is well illustrated in a series of cases of Oguchi's disease reported by Francois et al (1953).

A simple cylindrical type of perimeter was used by Francois et al (1963a) in a study of normals, and acquired defects of central colour vision; this allowed a comparison of central and 'peripheral' colour discrimination, substituting D.15 or PIC objects for more conventional coloured stimuli.

One revival of perimetry (using red, blue and green filters with the Goldmann instrument) by Francois et al (1964) provided data in the horizontal meridian with I, II and III size stimuli. The field out to some 45° was shown, for instance with the central peak featured by other colours being inverted by the blue plots. Macular degenerations were shown to lower central sensitivity to red, also to blue in juvenile cases.

Verriest and Uvijls (1977a,b) found that in juvenile macular degeneration and progressive cone dystrophies the spectral sensitivity curve measured in affected retinal areas is depressed in sensitivity in the long wavelengths. In central serous retinopathy the short and medium wavelengths are affected. Some pigmentary retinopathies show short wavelength depression but many optic nerve diseases give rise to an enhanced sensitivity of blue regions.

Important studies by Hansen in the mid 1970s showed depressed sensitivity of the red and green mechanisms in chloroquine retinopathy with little impairment to the blue system (Hansen 1974b). In two cases of cone dystrophy the green mechanism was selectively faulty (Hansen 1974c). In the early stages of *retinitis pigmentosa*, associated with rod impairment, a marked reduction in sensitivity to the blue mechanism is typical, with frequent absence altogether indicating significant disease; the red and green systems are affected to a lesser extent (Hansen 1979). Hansen also found a central red and green cone depression is found in many amblyopes. In cases of diabetic retinopathy, cataract, optic neuritis, myopic retinopathy and retinal detachment the typical pattern observed was for blue impairment, with the degree of involvement depending on the severity of the disease. It appears that π_2 is most sensitive at the beginning, giving rise to a distorted function when measured which eventually tends towards a reduction overall. In retinal detachments the red mechanism π_5 shows a characteristic shortening of the long wavelengths and the green mechanism is abnormal in cases of cataract. Tobacco optic neuritis was typified by a lack of measurable blue function and also reductions in the sensitivity of the red and green mechanisms (Hansen 1979). Overall, the green π_4 mechanism is the least vulnerable in acquired disease.

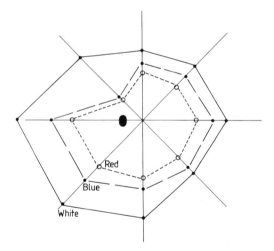

Figure 4.6 Early defect of visual field of left eye, a typical result of a tumour of the pituitary body. Despite the small number of plots, the isopters for different colours are disturbed in the upper temporal area, a feature called 'disproportion'.

Involvement of the colour mechanisms beyond the limits of the field indicated by a scotoma was demonstrated by Maione *et al* (1978). The same authors had shown a reduction of the π_5 mechanism to be general in retinal disease, with a depression of the π_4 mechanism in optic nerve pathology.

Vola *et al* (1980), using the Tübingen perimeter, confirmed that blue impairment is the typical characteristic of many ocular diseases, with the degree of involvement being related to the severity of the disease. In their opinion the blue π_2 mechanism is most sensitive at the beginning of the disease; the π_5 (red) mechanism is also frequently involved. Greve *et al* (1974), who initially modified the Tübingen perimeter for two-colour measurements, measured a depressed blue function in cases of macular oedema.

Arias (1981) showed the value of detecting raised thresholds for short wavelengths in both central and peripheral field locations in acquired tritan defects using the Goldmann perimeter. Arias adapted the Goldmann perimeter to measure thresholds in both normal and colour defective observers. Using different locations for presenting the test stimulus in the field of view, he indicated the relative contribution of the different cone mechanisms at the low background luminosity level of $10 \ \mathrm{cd} \, \mathrm{m}^{-2}$. Weale (1951a,b) noted a higher peripheral sensitivity of the normal eye for short wavelengths which was confirmed by Arias for both protans and deutans. Protan observers showed higher central and peripheral thresholds for long wavelengths (615–650 nm). Deutans displayed high sensitivity to stimuli between 443 and 547 nm and between 615 and 660 nm for central viewing although normal results were seen at $15°$ nasally.

4.12   THE EFFECTS OF CHEMICALS AND RELATED SUBSTANCES

A wide variety of chemicals and drugs affect colour vision indirectly, usually as a consequence of damage to the retina and/or optic nerve. Santonin, quinine and tobacco were among the earliest examples, and today there is a growing awareness of the side effects of many prescribed substances. Optic nerve damage is typical too as a consequence of exposure to a great many toxins among them thallium, carbon tetrachloride, carbon disulphide or bisulfide, methyl and ethyl alcohol and lead compounds. The general statement made by Lyle (1974) that almost any drug, toxin or disease causing amblyopia, central scotomata, optic neuritis or atrophy is likely to affect colour vision, is a valuable guideline. Inflammation and atrophy of the optic nerve is a frequent consequence of long term medication of a great many drugs, notably the anti-inflammatory agents used to treat rheumatoid arthritis, cardiac agents, antimalarial and antituberculosis drugs, and to a lesser extent some antidepressants such as the hydrazines and analgesics. Evidence that ethambutol affects the colour-opponent neurones leading to colour vision losses is presented by Zrenner and Krüger (1981); in the patients studied the three cone receptor mechanisms were still operable.

In many cases the colour vision disturbances resulting from intoxication involve both eyes. The separation of cause and effect is often difficult since it is often impossible to differentiate the colour vision effects resulting from a specific medication from those resulting from a disease for which the medication was prescribed. Careful monitoring is required of the possible harmful effects of toxic substances to which individuals may be exposed at work, or by accident or deliberately as medication. The following tables (tables 4.2, 4.3, 4.4 and 4.5) give a broad outline of associated colour vision changes before a detailed approach follows.

Table 4.2 A selection of substances causing red–green colour disturbances. Partly after Fraunfelder (1976).

Alcohol	Methotrimeprazine	Quinine
Aspirin	Nialamide	Rifampin
Chloroquine	Opium	Streptomycin
Chloropromazine	Paramethadione	Tobacco
Ethambutol	Pargyline	Thiethylperazine
Ethopropazine	Perazine	Thiopropazate
Fluphenazine	Pericyazine	Thioridazine
Hydroxychloroquine	Perphenazine	Tranylcypromine
Isocarboxazid	Phenelzine	Trifluoperazine
Isoniazid	Piperacetazine	Triflupromazine
Mesoridazine	Promazine	Trimeprazine
Methdilazine	Promethazine	Trimethadione

Table 4.3 A selection of drugs often causing significant blue–yellow colour disturbances.

Use	Drug
Cardiac glycoside and heart stimulant	Acetyldigitoxin
	Digitalis
	Digitoxin
	Gitalin
	Lanadoside C
	Ouabaine
Antimalarial	Amodiaquine
Anti-inflammatory	Chloroquine
Anti-arthritic	Hydroxychloroquine
	Indomethacin
	Novaquine
	Plaquenyl
Antituberculosis	Ethambutol
	Isoniazid
Ovulation inhibitors	Anovlar
	Lyndiol
	Ovulen
Antibacterial	Erythromycin
	Streptomycin
Anticonvulsant	Trimethadione
Antiepileptic	Paramethadione

Table 4.4 A selection of drugs often causing significant red–green colour disturbances.

Use	Drug
Monoamine oxidase inhibitor	Nardelzine
	Pheniprazine
Antidepressants	Hydrazines
Antituberculosis	Ethambutol
	Myambutol
Cardiac glycoside and heart stimulant	Digitalis
	Digitoxin
Sulfonamides ⎫ Antibacterial ⎬	Streptomycin
Anti-inflammatory ⎫ Antimalarial ⎬	Chloroquine
Antidiabetic	Chlorpropamide
	Tolbutamide
Antipyretic	Ibuprofen
	Phenylbutazone
	Salicylates

Table 4.5 A selection of toxins often causing red–green colour disturbances.

Toxin and use	Likely colour defect
Carbon disulphide (rubber, rayon, explosives, tanning industries)	Red–green
Carbon monoxide	Red
Carbon tetrachloride (fire extinguishers, dry cleaning agent)	Green
Chlorodinitrobenzene	Green
Dinitrobenzene (explosives)	Red–green
Dinitrotoluene (dry cleaning fluid)	Red–green
Ethylene glycol (antifreeze)	Red–green
Lead and lead compounds	Green
Manganese	Red
Methyl alcohol (methylated spirit)	Red–green
Methylbromide (fire extinguishers and fumigant)	Green
Thallium (rodent poison)	Red–green

4.12.1 Alcohols

Methyl alcohol has profound effects on visual function, associated with neurological damage. Colour perception losses, monitored by colour perimetry, show an initial tendency to losses of green then red vision, followed by yellow and blue. Optic neuritis is typical of severe poisoning with ethyl alcohol and irreversible optic atrophy accompanies severe methyl alcohol exposure. A reversible central scotoma may be present in mild cases of intoxication. Jaeger (1955) noted a blue–yellow defect with a neutral spectral region around 568 nm and a tendency towards optic atrophy. Changes in a variety of visual functions, including colour perception, following ingestion of ethyl alcohol were reported by Walsh and Hoyt (1969). Such agents are examples, along with ethambutol, where colour vision can be affected more profoundly than form vision (Sakuma 1973). A case of disturbed colour vision on account of ingestion of a half pint of antifreeze (a mixture of ethyl alchohol and glycerol) was reported by Ahmed (1971); bilateral optic atrophy developed and two weeks after ingestion the patient was unable to see blue, red or green and visual acuity was around 2/60. The colour defect appeared to be a permanent feature.

Exposure for over five years to alcohol hydrolyzate, furfurol and hydrochloric acid fumes can lead to reduced sensitivity to reds and greens with a narrowing of isopters for white, red and green. Ryabushkina (1970) considered such colour vision changes to be important early signs of chronic poisoning.

Alcoholism and liver disease

Although differing opinions have been reported as to the type of colour vision disturbance which results from alcoholism, a marked change in many cases seems clear; Verriest (1974b) and Cruz-Coke (1970) should be consulted for review. Saraux *et al* (1966) report a characteristic red–green variety, while Thuline (1967) and Ugarte *et al* (1970) consider a 'blue–yellow' disturbance to be more characteristic. Mixed defects are reported, for example by Sassoon *et al* (1970). There is some evidence that the colour vision defects may be semi-permanent (Swinson 1972). The same author noted that the frequency of abnormal colour vision defects found in alcoholics depended significantly on the test method used.

A possible genetic link between cirrhosis of the liver and colour vision defects was first suggested by Cruz-Coke (1964). In a review paper Cruz-Coke (1972) suggests the possibility that alcoholic cirrhosis or alcoholism tends to interfere with retinal pigment formation. Carta *et al* (1967) noted blue–yellow defects among many cirrhotic patients (sample 16) as did Dittrich and Neubauer (1967).

4.12.2 Effect of vitamins

Vitamins are vital to health and vitamin A in particular is essential for the maintenance of the integrity of the retinal photoreceptors. Chronic liver disease can affect colour vision, since most vitamin A retained is stored in the liver. Abnormalities of the visual field are found in vitamin A deficiency, indicating both rod and cone dysfunction. An early investigation of colour vision loss associated with night blindness (a symptom of vitamin A deficiency) was made by Stephenson (1898), who described the contraction of the visual fields for red and to a lesser extent for green. The study by Sloan (1947) provides a valuable review. Changes to the visual thresholds for both cone and rod vision were noted by Wald and Steven (1939).

Indian children with signs of conjunctival xerosis, on account of vitamin A deficiency, examined by Reddy and Vijayalaxmi (1977) showed normal colour vision scores on PIC plates although 18 of the group of 28 had signs of night blindness and all showed biochemical signs of the deficiency.

Other authors have noted colour vision changes and Chisholm and Pinckers (1979) should be consulted for a review. Although symptoms may be slow to develop, a blue–yellow loss has been shown as a typical consequence of reduced vitamin A intake (Cruz-Coke 1972 and Bronte-Stewart and Foulds 1972) and clinical tests such as PIC plates (HRR) and the FM 100 Hue have been used effectively to monitor such changes. Colour vision disturbances can be used as a valuable indicator of deficiency before other visual symptoms occur (Bronte-Stewart and Foulds 1972). The same authors were able to demonstrate the reversible nature of the colour and visual losses, using vitamin A medication. The FM 100 Hue total error score was reduced by approximately a quarter following therapy for six weeks.

Studies by Chisholm (1972) with patients with Addisonian pernicious anaemia indicated that either vitamin B_{11} or B_{12} was responsible for a permanent blue–yellow loss.

4.12.3   Lead

Lead and its compounds, extremely toxic substances to the body, have been studied for both their neural and peripheral effects for three and a half centuries. Baghdassarian (1968) considered only 1–2% of cases of severe exposure to show ocular effects although there are complications when they do arise. Colour vision is a likely visual function to be affected since optic neuritis is a common consequence. Although Sax (1975) suggested that such poisoning is not uncommon the withdrawal of lead compounds from many paints has clearly reduced the risks in recent years. Plunkett (1976) cites visual disturbances involving both the optic nerve and retina. The specific colour losses involved in cases studied have been varied, possibly depending on the extent of toxicity to the retina and optic nerve. Schmidt (1973) suggests that optic nerve involvement would tend to lead to red–green confusion while Lyle (1974) noted blue–green disturbances or losses of unspecified colours. Typical cyanopsia, chloropsia and xanthopsia as subsidiary effects were cited by Dubois-Poulsen (1972). Cobb and Shaw (1980) emphasised the controversy on the subject when they found poor colour discrimination in several workers associated with lead which they did *not* attribute to poisoning. The patient studied by Baghdassarian, a middle-aged paint shop worker, achieved normal vision following medication. Grant (1974) notes how symptoms may take some time to develop.

4.12.4 Chromatopsias

The ingestion of various substances has been associated with 'chromatopsia'—the perception of colours on a white area in part of the visual field or its entirety. Such perceptions, which are not well understood, may have retinal or central causes. They can be likened to a pseudo-colour experience and are always transient. The Greek derivation is used to describe the main colour appearance—cyanopsia, chloropsia, erythropsia, xanthopsia, etc. Dubois-Poulsen (1972) gives such phenomena prominent coverage. Frequently the appropriate colour perception may precede the onset of a distinct acquired colour vision defect, thus erythropsia and chloropsia prior to a red–green defect, or cyanopsia prior to a blue–yellow disturbance (Lyle 1974). Table 4.6 indicates some typical substances which evoke such perceptions.

4.12.5 Quinine and derivatives

Quinine and derivatives, such as chloroquine and indomethacin, have a profound effect on ocular tissue, in particular the cornea, the retina and the optic nerve. Quinine was the favoured antimalarial agent before safer synthetic derivatives such as Novaquine and Plaquenil became available. It has also proved valuable in treatment of night leg cramps and been used for attempted abortions. Chloroquine and associated drugs (e.g. hydroxychloroquine under the proprietry names Elestol, Nivaquine, Avalen, Avloclor) have found wide application in the treatment of collagen and rheumatoid arthritis, being anti-inflammatory agents, and as antiparasitic agents for amoebiasis. Acute intoxication from quinine has resulted in total achromatopsia, although either a selective blue or red–green disturbance can be typical. Small central scotomata may expand so that peripheral vision is acutely affected.

Grützner (1969a,b) was able to detect colour changes using PIC plates and a leaning towards protanomaly on the anomaloscope ($R-G$ match) in some patients, though he stressed that (though useful for a diagnosis of toxicity) colour vision disturbances were not necessarily an early symptom. Hue discrimination thresholds are often raised in the medium and short wavelength region, indicating blue–green deficits (Grützner 1972). A normal V_λ curve is frequently found, as by Zanen (1971), which led to a conclusion by this author that the site of pathology is the retina. The varied nature of the colour loss was confirmed by Francois *et al* (1972) who successfully used both Ishihara and HRR plates to indicate red–green losses and the D.15 test to show either blue or mixed disturbances. Colour disturbances are frequently at least partially reversible.

Table 4.6 Experience of chromatopsias associated with chemical or drug ingestion.

Colour experience	Substance
Cyanopsia (blue perception)	Digitalis
	Ethyl alcohol
	Amyl alcohol
Erythropsia (red perception)	Eosine
	Atropine
	Ergotamine
	Quinine
	Creosote
	Nicotine
Xanthopsia (yellow perception)	Santonin
	Lead
	Arsenic
	Carbon monoxide
	Nicotine
	Sulfonamides
	Methyl salicylate
	Thiabendazole
	Thiazide diuretics
	Chromic acid
	Amyl alcohol
	Chromic acid
	Arsenicals
	Chloroquine
	Digitalis
Chloropsia (green perception)	Lead
	Carbon sulphite
	Griseofulvin (anti-fungal agent)
	Quinine
Ianthinopsia	Nicotine

Large doses of chloroquine taken for prolonged periods have been associated with profound visual consequences through deposition of the drug in the ocular media, principally the cornea. Night blindness, photophobia and a general reduction in visual acuity are among such consequences. Serious retinopathy results from the destruction of both rod and cone receptors, with optic atrophy. At an early stage these changes may be reversible (Carr *et al* 1966). Long term monitoring is therefore important. The retinal damage caused by hydroxychloroquine is usually less severe than that by chloroquine.

The relation between colour vision disturbance and dose was stressed by both Carr *et al* (1966) and Carr *et al* (1968) but medium daily doses of approximately 200 mg were sufficient to raise retinal thresholds. Thresholds found were significantly impaired foveally and peripherally, even though the early ophthalmoscopic signs tend to be confined to the macula. Lakowski (1972) demonstrated how minimum doses in young patients showing good visual acuity were sufficient to induce impaired colour vision of both red–green and blue–yellow varieties. Lagerlöf (1980) found clinical colour tests to be valuable in showing such disturbances in 79% of his patients on chloroquine. The superiority of measuring thresholds over the better established clinical tests of colour vision was shown by Friedmann (1969).

Indomethacin (also used widely for rheumatoid arthritis) has a similar effect to chloroquine. Decreased retinal sensitivity, primarily blue–yellow defects as well as other visual deficits (ERG, dark adaptation changes and field losses) could be found with daily doses of 150 mg taken over fifteen months by fifty females (Burns 1968). Approximately oné year without medication was required for normal colour vision to return.

4.12.6   Antihelmintics

Antihelmintics such as quinacrine and mepacrine (used widely to treat tapeworm infestations and amoebiasis, as well as malaria) can have visual effects which are similar to chloroquine. Since 1959, an association of retinopathy with chloroquine has been accepted, and over 200 cases have been reported. Most cases occur when the daily dose is in excess of 500 mg. It can occur also at lower levels, but then the incidence is low. The onset may be insidious and therefore not noticed by the patient. It affects peripheral vision first and central vision may remain good. An early sign is an abnormal finely granular appearance of the retinal pigment layer. Later this is followed by a clump of pigment in the perimacular area, giving rise to the 'doughnut' or 'bulls eye' appearance. Rods and then cones are destroyed. The narrowing of arteries is thought to be a secondary effect. Field defects may appear before pigmentation is noted. Electrodiagnostic techniques (ERG and EOG) are valuable to indicate early retinal changes.

4.12.7   Antituberculosis drugs

Antituberculosis drugs such as myambutol, based on ethambutol, ethionamide and isoniazide frequently give rise to optic neuritis when taken in daily doses in excess of 25 mg kg^{-1} (of body weight) for prolonged periods. In turn colour vision is affected, although most changes are reversible. A central loss with scotomata for green, and sometimes red, may be involved, or there may be a periaxial involvement affecting peripheral fibres without loss in visual acuity. The cause of the effect is unknown, although it may be due to a demyelination of the optic nerve fibres. In combination with alcohol ingestion the toxic effects can be especially severe.

Chisholm and Pinckers (1979) surveyed the literature, covering almost 3000 patients on ethambutol therapy; rather few developed optic and retrobulbar neuritis. Fewer than 10% of patients of Koliopoulos and Palimeris (1972) (sample size 138) showed colour vision losses, these being equally distributed between red–green, 'blue–yellow' and anarchic varieties; no other ocular side effects were noted, indicating how colour vision may be a sensitive indicator of pathological change. Indeed some believe colour vision disturbances can be the first indicator of optic neuritis and recommend a three-monthly colour vision examination of patients undergoing therapy. While colour changes can involve all parts of the visual field, Verin et al (1971) emphasised that induced colour perception disturbances can persist even after normal visual acuity has been restored.

4.12.8 Cardiovascular agents

Cardiac glycosides, principally digitalis and the associated drugs digitoxin or digoxin (one proprietary UK name is Ouabaine) which are widely used to increase cardiac output, or to correct arrhythmias, have been known for some time to affect colour vision at therapeutic levels. The complaint of chromatopsias, frequently xanthopsia, is common, objects appearing tinged with yellow, blue, green or red; blue halos can appear around lights. Alken et al (1980) found a colour vision disturbance in 80% of toxified patients.

The prolonged action of digoxin and digitoxin gives rise to the most pronounced colour perception disturbances, although even these are generally reversible. Ouabaine, a shorter acting agent, gives less significant colour vision disturbances. A close correlation between the extent of colour vision disturbance and serum digoxin concentration is reported by Alken et al (1980).

Optic or retrobulbar neuritis is the underlying cause for colour vision disturbance, with prominent macular cone damage; central scotomata are thus typical. On evidence that the photochromatic interval is normal Zanen (1971) proposed that the site of disturbance is retinal. While several investigators noted a red–green disturbance with PIC tests and the FM 100 Hue test in cases of severe intoxication (Cozijnsen and Pinckers 1969, Verriest 1974b), others report a blue-like defect (Grützner 1969a,b, 1970, Babel and Stangos 1972). Towbin et al (1967) found a significantly decreased sensitivity to green using the Nagel anomaloscope, although Alken et al (1980) stress other colour vision changes also. A specific disturbance of the Stiles π_5 (red) mechanism, was observed by Gibson et al (1965). Their patient showed an inherited deuteranomalous defect, with a superimposed protan and tritan involvement, as a consequence of the drug toxicity. The photopic relative luminous efficiency (V_λ) showed a protan bias with decreased sensitivity to reds. Hue discrimination gradually improved following withdrawal of the drug, with only the inherited deutan defect remaining after seventeen months.

The need to investigate and specify carefully the dosage and specific drugs involved in studies concerning acquired colour vision losses has been stressed by Alken (1982). He confirmed protan-like defects (of an uncharacteristic type on the Nagel anomaloscope) in patients taking single 1 mg doses of digoxin, which had earlier been reported by Kittel (1957) and Cozijnsen and Pinckers (1969) in digitoxin-intoxicated patients. Repeated daily doses of 0.375 to 0.75 mg for two to three weeks revealed green disturbances in his sample confirming the general findings of Towbin et al (1967).

4.12.9 Other drugs

Vasoconstrictors/hypertensive agents, antibiotics, anticonvulsants, analgesics, derivates of ergot, such as ergotomine and ergonovine (proprietary UK names Methergin, Fermergin and Lingraine) used for the treatment of vascular headaches such as migraine, and uterine stimulants for abortion, are reported to give objects a red tinge (Grant 1974).

Monoamine oxidase inhibitors such as iponiazid, isocarboxazid, phenelzine, pargyline (proprietary name Eutonyl) used to treat hypertension, can give red–green anomalies and optic neuritis (Gillespie et al 1959). Antidepressants used for the treatment of endogenous depression (e.g. hydrazines) can give rise to reversible red–green defects on account of optic neuritis, according to Grant (1974).

Pinckers (1975) notes that antipyretics can produce red–green defects as a result of optic neuritis. The psychiatric drug Melleril (thioridazine) can give tritan defects, as shown by Grützner (1969a,b). Ethchlorovynol, a hypnotic non-barbiturate, can give reversible colour vision changes.

Antimicrobial/antibiotic drugs (e.g. streptomycin, the sulfonamides, tetracycline, oxytetracyline, oleandomycin and penicillin) have also been associated with minor colour vision changes as indicated on the 100 Hue test (Laroche and Laroche 1970). A reversible colour perception change is reported as a side effect of trimethadione, an anticonvulsant agent (Sloan and Gilger 1947).

Disturbances in colour perception of the red–green variety have been noted with significant prolonged medication of aspirin (acetylsalicylic acid), a mild analgesic (Grant 1974). Objects can appear tinged with yellow.

4.12.10   Tobacco amblyopia

This condition, although relatively uncommon in the UK at the present time, is here treated separately from retinal and optic nerve defects. It is chiefly associated with a disturbance of the papillomacular bundle of fibres. It is a classical toxic amblyopia, aggravated by the associated ingestion of alcohol. Abney (1891) made an extensive study of several patients; one young man, a smoker of half an ounce shag daily together with four pints of beer, reported poor vision for two months but showed no optic disc changes. A central and centrocaecal scotoma (between the fixation point and blind spot) was typically noted. Red and green lights were both mis-named white, and a shift in the V_λ curve towards shorter wavelengths show-ing reduced sensitivity to reds was noted.

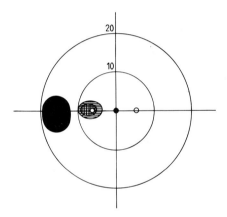

Figure 4.7   Left central scotomata, described as 'centro-caecal', typical of tobacco amblyopia. The oval limits of the larger scotoma could be located with a 10/1000 red stimulus, the smaller with a 5/1000 red or even with a 1/1000 white . Two small red objects are placed about 6° each side of fixation, showing that they differ in apparent colour.

This progressive condition is usually binocular and insidious in onset. Symptoms include a general dimness of vision, often more pronounced in one eye, and frequently night vision is superior to day vision, indicating cone involvement. Ophthalmoscopic signs are often not marked, but may include a dark red appearance of the optic disc in the early stages and a slight paleness at the temporal region of the disc in advanced cases. Pro-vided total abstinence from tobacco is maintained for several months normal vision is restored, although visual acuity often returns to normal before colour vision (Chisholm *et al* 1970). The site of action is unclear, though the retina is likely to be involved (Schepens 1946); it is possibly the optic nerve (Foulds *et al* 1970). The possibility that it may be a cyanide-induced optic neuropathy must not be ruled out, since it responds to treatment by hydroxocobalamin. Vitamin B_{12} may also be critical since its successful use in therapy has been indicated.

Small coloured targets, particularly red stimuli have been used in classical perimetry to detect tobacco amblyopia (Holth 1928). Modern investigations using clinical colour vision tests reveal a red–green disturbance which may start as a protanomalous-type manifestation advancing towards protanopia (Bhargava 1973). A deutan bias was reported by Chisholm *et al* (1970), although these and other authors acknowledge an anarchic colour disturbance on the FM 100 Hue test as being quite typical when visual acuity is greatly disturbed (see Francois and Verriest 1961a). A correlation between the FM 100 Hue error score and the decline in visual acuity, age and the duration of symptoms was found by Chisholm *et al* (1970) using a large number of cases. In allied conditions such as poisoning by combined tobacco and alcohol, or syphilis, Sloan (1942) was able to show PIC plate errors. Anomaloscope data might be expected to be useful and Bhargava (1973) showed 'extreme protanomaly' in some subjects where FM 100 Hue results were affected, although the binocular conditions may have modified the defects.

4.12.11*   Oral contraceptives

Steroid ovulation inhibitors, now taken regularly by an estimated 55 million women worldwide, have been associated for some years with a variety of ocular side effects including, in a few isolated cases, thromboembolic disturbances, retrobulbar neuritis and optic neuropathy, with temporary loss of vision in some parts of the visual field. Evidence that clinical colour vision tests could detect colour vision changes, in both blue and red perception, among women taking the Pill, was first presented by Neubauer (1973). Ishihara plates, the Nagel anomaloscope and the D.15 test were found to be valuable. By monitoring colour discrimination by means of the FM 100 Hue test in women who had used the Pill for varying lengths of time, Marré *et al* (1974) reported that the duration was critical, an aspect which Lakowski and Morton (1977) were able to confirm. Blue disturbances were typical, though red–green defects were not significantly different from the control group. Five years of use seemed a critical period, with colour vision changes being detected more frequently thereafter. The value of the D.15 test was also confirmed. Lagerlöf (1980) examined 74 women on the Pill (148 eyes). Of these only 15% had normal colour vision, although a further 28% showed only minor changes. Over half the eyes examined had a definite blue–yellow disturbance with five further eyes showing mixed changes. Lakowski and Morton (1978) noted higher error scores on the FM 100 Hue test for young diabetics (without retinopathy) taking oral contraceptives, compared with age-matched non-diabetic women using the medication. Predominantly blue, but some red–green, disturbances were detected by clinical tests in Pill users. Diabetics taking the Pill were characterised by a greater incidence of blue–green defects and an enlarged red–green colour matching range, typical of the colour vision changes in diabetics with vascular complications. Diabetics on the Pill were shown to be approximately five times more likely to develop extreme colour vision changes than those not using an oral contraceptive.

Although the colour vision question clearly needs further study with larger patient numbers, the limited evidence to date has been sufficient to convince Chisholm and Pinckers (1979) that, bearing in mind the other ocular complications of the Pill, damage to the optic nerve is a strong possibility. Lakowski and Morton (1977) add a further suggestion; in view of the known effect of ovulation inhibitor therapy on carbohydrate metabolism, impairment of the function of the retinal neurones might be involved. The suggestion that vitamin deficiency can be associated with users of ovulation inhibitors might also have a bearing on the colour vision element, since colour vision losses are typically associated with hypovitaminosis.

4.13 MISCELLANEOUS DISTURBANCES

Lyle (1974) quotes a number of miscellaneous examples of acquired colour deficiency arising in conjunction with the production of loud sounds, and the stabilising of the retinal image; these are transient and usually inconsequential. Other more physiologically-based inducements to colour vision disturbances, often of a transient but re-occurring nature, include exposure to chlorine in swimming pools and oedematous effects to the corneal epithelium from glaucoma.

5 An approach to testing

This chapter deals with some ways to optimise the testing for colour vision defects. Details of the methods, charts and instruments involved are considered in Chapter 7 and much of the present chapter is relatively elementary. While it may be obvious to those experienced in such procedures, many points are likely to help those entering the field and are for consideration by personnel being instructed in the techniques of screening or testing; such personnel are not concerned with the relative merits and more theoretical aspects of tests, which will be described in Chapter 7.

The 'subject' or 'patient' may submit to the tests for any of a variety of reasons. In a school, in the armed forces or at a factory, the test may be a routine or part of a wide assessment of visual and other features. In an optometric or ophthalmological context the patient will be actively seeking evaluation advice and possible remedy.

The term 'patient' will be used here as being most suitable, except where a distinctly scientific or laboratory context makes a term such as 'subject' or 'observer' appropriate.

5.1* THE ROOM

If an ordinary office, part of a medical care area or an optometric office is involved, it may be useful to make slight modifications such as the use of a separate table and special illumination. A test area devoted to colour vision examination is ideal and the proposals for its ambience necessarily must be modified in each case. Desirable features are listed below

(*a*) ample size, e.g. at least 3.5 m long;
(*b*) quietness and reasonable privacy;
(*c*) adequate illumination plus wall sockets;
(*d*) comfortable chairs (preferably adjustable height stools);

(*e*) desk and/or table, matt medium grey surface;
(*f*) matt grey or white walls, plus small pictures to give some contrast;
(*g*) almost complete blackout, if required.

Figure 5.1 shows a suitable plan.

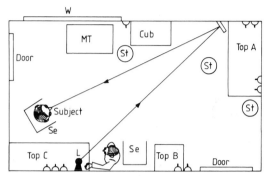

Figure 5.1 Plan of test room/office. L, wall-mounted lantern; W, window — blackout with ventilation; Se, seats with backs and armrests; St, stools, varied of adjustable heights; Tops, built-in worktops; A, for daylight lamp, PIC and other 'loose' tests; B, for separate instruments such as colour vision analyser; C, for notes, anomaloscope etc; MT, movable table for accessory equipment; Cub, store cupboard for apparatus, record sheets etc; Light sockets are placed 20 cm above table and bench tops and pictures are on the walls near the door.

5.2 APPOINTMENT

Unless the investigation follows on from another interview or encounter the appointment should be made with the purpose and likely duration of the tests in mind. Simple screening may need five minutes, a fairly careful comparison of three short tests could occupy twenty minutes and an hour would be needed for extensive testing and discussion.

An appropriate selection of tests must be at hand when the patient arrives. Additional apparatus known to be required may be kept elsewhere, in which case it should be booked for the time necessary.

5.3 RECORDS

Notes are made, legibly, as the tests progress and preliminary details such as name and address and occupational data must be entered carefully with the date of testing. Many patients will appreciate a view of what is being, or what has been, recorded, especially if they are nervous about the consequences. Copies are usually by carbon or photocopy and the record form should be designed accordingly. Examples of records and record forms can be seen in Chapter 8.

It is important to know the facts about ownership of the record. In some firms, local authorities or schools, rules and legal features prohibit the removal or unauthorised use of records. The record might be considered to be the property of the person doing (or paying for) the tests but this will depend on the circumstances. Hospitals sometimes regard the tests as part of a patient's confidential documents; a consultant in charge of the 'case' may have to give permission if the data are to be quoted or used elsewhere.

Frequently a written report is based on the record, when the two are filed together. In this situation notes should indicate the gist of any information or notice given to the subject.

Coloured pencils or ink can be used to good advantage when separate results are obtained for individual eyes. Thus red denotes 'right eye only', blue shows 'left eye' and either black or green is used for binocular records. A note of the code should appear clearly on the record.

5.4.   RECEPTION OF THE PATIENT

No reference to 'colour blindness' must be made but 'colour tests', 'colour vision' or merely 'tests' could be mentioned in conversation. An appointment card, for example, could be presented which indicates

You should attend at .... for screening/examination/ discussion of your colour vision on ....

Any fees likely to be charged must be made clear.

A friend or relative is often seen together with the patient. This is usually an advantage to all concerned; it provides a witness for obvious mistakes made by the patient if colours are named and it puts him more at ease. A parent often fills in family background and helps with the expression or description of difficulties, real or feared.

Privacy, quietness and reasonable freedom from interruption are required. Seats should be comfortable and as informal as possible but office-type seating at a desk or table is suitable when the tests begin. A high stool and a low chair can be ready as alternatives, with extra loose cushions for small children. Children need the consideration of special toys, pictures, etc which can often become involved in the discussion of colour difficulties.

Illumination, under ready control by switches and blinds, must meet all the requirements without glare, e.g. light shining into the patient's eyes from a window or lamp, or indirectly reflected from a polished table top.

5.5* TESTING

Different situations and needs will dictate variety in tests used but the following features should always be involved if possible. Alternative approaches should be available. Familiar ground is often covered if an Ishihara book is used as a preliminary. Discussion can then flow from the errors revealed.

Occupation. School or hobby activity and future career plans are noted as real and important features of life, initially avoiding direct reference to colour vision. Photography, for example, could be noted as a hobby and following this the question 'just black and white?' may produce a useful response.

The obvious *gender* of the patient should have been noted and an appropriate title is to be used at intervals. It is essential, even in the case of a small child, not to question the declared sex because colour vision and associated genetic considerations can intrude into complicated areas, often of great sensitivity.

Details of the personal *history* of the patient may have a bearing on colour vision and a careful selection has to be made. For instance, present or past treatment for conditions which influence colour perception is noted. The parts of the world where the person has lived can be elicited and reference to spectacles or contact lenses needed is usually helpful. Family tree details and possible inheritance data can be sought. A ready made 'genetic tree' is conveniently printed on the record form.

The *person(s) conducting* the tests should be identifiable from the record, since later a small difference of technique or a personal method may be involved. The date should be carefully recorded. The route by which the patient or subject was 'referred' or required to attend should be clearly established.

Assuming that the context will dictate the initial approach, it is now useful to ask if the person has any *difficulty* 'with colours'. Are they good or bad in recognising colours? The question may be phrased in terms of worst colours to 'tell apart' or the best conditions which are preferred. Rapidly, it is necessary to elicit the chief colour confusions, to assess the person's own estimate of the situation and to list special difficulties which affect work safety or dress choice. The defence mechanism of many younger patients often makes them shy to admit to confusions but they may be prompted by a more acceptable (indirect) approach. A sympathetic but non-surprised reaction is best, with a non-commital attitude at this stage.

Parents usually enter the discussion at such a point. Often the subject of a three-cornered exchange will appreciate a protective, kindly, reaction if the parent is apt to be destructively critical.

When did the patient first notice colour vision problems? At what age and in what way? How has the situation changed? Such questions must be adapted to the individual and to the progression of the encounter; gentle domination of the situation is essential if the person tends to speak too much. By probing it can be established whether, for instance, a red cricket ball is lost easily in green grass, if shopping and choice of 'matching' items of dress are difficult and how the normal working day is affected. Positive ways in which the individual overcomes or minimises his 'deficiency' should be extracted and noted. Does he use special lights or filters or does he find it helpful to blink rapidly? Is there a difference in his performance when tired, or does he find that rest periods or changing his tasks are helpful?

5.6 EXPLANATIONS

At suitable intervals, or at the end, some comment is necessary. But it may be contraindicated in a few special conditions, when polite avoidance may be needed. The patient and/or parents should be given a short, simple (firmly positive but not dogmatic) account of the condition, in the form of an opinion. The word 'opinion' is usually distinctly emphasised. It is useful to seek questions and to encourage the patient to bring up, later if need be, any doubts or fears. The nature and extent of this episode in the testing procedure naturally depends on the role and professional status of the 'examiner' or 'practitioner' and care must be taken not to assume undue authority or to poach on the preserves of other professionals. Nevertheless, kindly and guarded explanations are often reassuring even if more is conveyed by their friendly, positive and limited nature than by their real content. Misunderstanding must be anticipated and minimised by clear and thoughtful statements. This will be discussed further in §5.8.

5.7 RETESTING

If fatigue or other duties halt the testing before it is finished do not hestitate to make another appointment, with proposals that 'it will be better to come back fresh' or 'a second attempt often gives a better result'. This may give an opportunity for the patient to bring spectacles he had forgotten, to volunteer more considered information or to come back accompanied by a 'friend'. The examiner can reconsider whether extra tests should be brought into play, which tests should be repeated and whether some palliative aid or filter should be prepared in advance. Ideally, all that is likely to be useful should be available at the initial visit.

In rare cases, or in patients with drug treatment, poisoning or other variable conditions, a few months interval may have to elapse between appointments.

5.8 ASSESSMENT OF THE PATIENT'S COLOUR VISION

Often it is necessary to describe the type and the extent of a person's colour vision 'deficiency' and to explain this simply. A judgment must be made, which some practitioners will do only after using several tests, although some may have confidence in one or two approaches which they consider well tried.

A decision as to 'protan' or 'deutan' is seldom difficult since most tests give indications and the effective brightness of red light is low for protans, as can be judged with an anomaloscope, by a comparison between two lights or with the dark red filter in a lantern. Yet some cases remain doubtful.

Dichromatism can be detected by the extensive confusions made, as on an anomaloscope, or by the neutral point in a spectrum; selective wavelength discrimination can be used. Anomalous trichromats, however, sometimes present difficulties when an estimate of their safety and vocational liability is needed and there is a wish to grade them as 'mild', 'medium' or 'severe'. Plate tests are usually difficult to interpret for this purpose and the anomaloscope 'range' takes time. It is sometimes useful to note that a fail on a sensitive test, 'contradicted' by a pass on an easier test such as D.15 or CUT roughly suggests a good range of occupational potential. Yet protanomalous trichromats can be expected to be more 'dangerous' with signals than their deutan counterparts.

Error scores for the FM 100 Hue test are some use and as a guide a score over 100 represents low performance; protan or deutan 'peaks' may appear even with scores as low as 50, usually with 'mild' anomals, while 'marked' protans or deutans are expected to score over 100, as are dichromats. A person with less than 25 has exceptionally good colour discrimination.

Patients usually need an estimate of the potential difficulties, which prospective employers may be interested in. Employers tend to err on the safe side and to reject Daltonics. One disadvantage of giving such opinions is that it can lead to involvement if a person who fails to secure a post asks for support. Oral communications are used with the subject but sometimes these are necessarily guarded and limited. Maximum information is both kind and efficient, subject to the constraints of medical needs, of employers' requirements and other discretion.

5.8.1 Written guidance

It is very helpful, though time-consuming, to give the patient and/or those otherwise involved some simple explanation of the type of colour vision found in addition to suitable advice. To avoid missing aspects, for convenience and to enable the comments to be digested at leisure, a written account is most suitable. One Australian procedure, described by Alexander

(1976), combines an estimate of the 'colour recognition' of the patient in terms 'normal', 'slightly defective', 'defective' or 'seriously defective'. Hue discrimination is, in addition, graded from 'above average' to 'poor', and sensitivity to red light is estimated as 'normal', 'below normal' or 'defective'. At the end of the report a box is selected for one of several alternative 'recommendations' e.g. 'suitable only for tasks involving simple recognition of colour'.

It is frequently suitable to select and to recommend simple reading material which can give more information. Accounts of colour vision such as those by Voke (1980a, 1982) have been used in this way. Purpose-written sheets are easily produced but they should be revised at intervals; a specimen sample is indicated below.

5.8.2 A specimen text

Varieties of colour vision and their significance

Tests show that about one in every twelve men cannot see colours perfectly normally. Among women the figure is much less (about one in two hundred) but mothers can transmit the condition to their sons. A few people have acquired defects of colour vision, mostly on account of disease, and many of us find that colour vision tends to deteriorate in later life.

The inherited condition, called Daltonism, usually gives rise to confusion between colours such as red, green, brown and yellow. Pale colours or the red lights in distant signals often give trouble. Technical names are used, as follows.

(*a*) 'Protan' subjects have difficulty with reds, but made distinctive errors with other colours.

(*b*) 'Deutan' subjects confuse green or purple with grey.

Both (*a*) and (*b*) can appear in different grades, so one person may have a very slight anomaly while another may have a severe disability. There is a rare 'tritan' condition, involving confusion of yellow and other colours which are more blue or grey. It is seldom that anyone has complete 'colour blindness'.

Pigments in the cone cells of the retina in the eye determine the type of colour vision which we inherit and colour vision tests can show up the effects. Tests differ in sensitivity and are used for various purposes. The railways and Merchant Navy emphasise signal lights by using coloured lantern lights and those with particular occupations in mind must discover what special tests have to be passed. Some jobs demand outstanding ability with colours for reasons of safety or to ensure products of high quality.

Inherited colour vision defects cannot be cured, but in some cases the use of coloured filters is helpful for short periods; for instance, they can assist in the use of colour codes in maps or board games. Expert advice and guidance is really necessary.

Further details about the occupational importance of colour vision can be read in the publications listed below, which most libraries can obtain.

Voke J 1978 Colour vision defects—occupational significance and testing requirements *J. Soc. Occup. Med.* **28** 51

Voke J 1979 Colour blindness and the electrical industry *Electr. Rev.* **204** 38

Voke J 1979 Colour bars of an occupational kind *Occup. Saf. Health* **9** 10–12

Voke J 1980 *Colour Vision Testing in Specific Industries and Professions* (London: Keeler Instruments) (Monograph)

Fletcher R and Voke J 1985 *Defective Colour Vision—fundamentals, diagnosis and management* (Bristol: Adam Hilger)

5.8.3 A specimen report for the patient

Simplified accounts of the findings, dated and signed, may follow the example given below, which can be given to the patient. A copy should be filed with his record.

There is something to learn from the fact that one patient possesses two different 'official statements' from the same institution as to the state of his colour vision; evidently the first time he 'passed' an Ishihara test, having learned what to say, and was issued with a 'normal' certificate. At a later date, after a dispute elsewhere, he was examined with more care and certified as significantly Daltonic.

REPORT OF COLOUR VISION TESTS

Name: Mr/Mrs/Ms   Age: Date of test:
Address:

1   The following tests have been selected for use and have been applied under suitable conditions. It is not usual for all such tests to be given to each person.

Test (*delete* those not given)	Result or score	Comment, if any
Ishihara (edition)		
CUT (edition)		
100 Hue		
D.15		

Continued

Test (*delete* those not given)	Result or score	Comment, if any
Anomaloscope		
Naming of wires, etc		
Lantern (type)		
Others (specify)		
a		
b		
c		
d		

2   Opinion as to variety of colour vision
(Delete as appropriate) Normal

Protan	Slight	Definite	Marked
Deutan	Slight	Definite	Marked
Tritan	Slight	Definite	Marked
Mixed			

3   It is expected that this person might have the following difficulties recognising colours:
    a   None which can be predicted from the tests
    b   Very few, possibly when coloured objects are pale or small
    c   Significant (except when colours are bright) and possibly dangerous
    d   Many, including those which may be dangerous

4   Red or similar lights are likely to be recognised (Delete as appropriate)
(a) normally   (b) not well   (c) very badly

5   The following methods may assist the person for some aspects of colour recognition, and to some extent (Delete as appropriate)
    a   None appears to be helpful
    b   Filters, such as
    c   Extra light, particularly 'warm' light such as a tungsten lamp
    d   Using . . . .

6   A further examination of the colour sense is recommended
    a   If indicated at the next 'eye-examination'
    b   In   months
    c   In   years
(Note: colour vision may alter in time, for several reasons)

7   Occupational implications. As far as can be predicted on the basis of the tests applied, he or she is likely to be (Delete as appropriate)
    a   Capable of making critical colour judgments, particularly if conditions are well controlled
    b   Capable of most tasks needing normal colour vision, given adequate illumination and time

c   Capable of relatively easy coloured tasks

d   Unsuitable for jobs involving colour decisions

8   With particular reference to work involving ....
which this person does/wishes to do, the following comments can be
made ....

9   Additional notes:

10   *Note:* Those with minimal variations from 'normal colour vision'
should not be excluded unfairly from occupations which they can really perform yet those likely to make errors over signals and quality
judgments are obviously well advised to accept their limitations.

Signed .... Date ....

5.8.4 Other reports

A written report may be required by employers or colleagues. Usually a
report in duplicate is sent, keeping a third copy on file. In this way a family
doctor, employer or careers adviser can pass on a ready-made copy to
suitable people without the bother of photocopying. Wherever possible,
show the patient the report or read out a synopsis.

The contents of the report should include the following.

(*a*) Identification of the parties involved and date(s).

(*b*) Brief statement of reason for test.

(*c*) Summary of tests used, basic conditions, results and any individual
interpretations applied.

(*d*) Definite but guarded clues to type and extent of Daltonism etc,
including possible indications of acquired defects.

(*e*) Comments, usually given as an 'opinion' and within the ethical and
legal constraints of the person writing.

(*f*) Charts and other data, all codes clearly explained.

Thus, while it is inadvisable to state 'this person's colour vision is normal'
one could write 'today my examination of Mr Smith disclosed no errors on
the tests listed below, therefore in my opinion his colour vision is good'.

Instead of 'he is colour blind' or 'he is a deuteranope', a careful 'my
examination today revealed errors shown in the attached charts, suggesting
that he is likely to be dichromatic, probably a deuteranope' can be used.
Where a certificate calls for a 'pass' on one or more tests, the opinion that
he has 'passed' might well be qualified by 'since today he made no errors'.
Excessive caution is tedious for all, but most of those who deal with this
sort of patient sometimes come to wish that they had been more cir-
cumspect. Some recipients do not appreciate wordy and 'technical' descriptions. The following suggests how a report can be worded.

'Today, 25 December 1999, I have applied the colour vision tests listed below to Mr A Patient and the results are summarised below. He is interested in electronic work and although he has a small departure from normal (slight protanomaly) I believe that he is likely to perform well under most circumstances. His success in his hobby of building 'hi-fi' devices reinforces my opinion. The tests given were as follows:

10th edition Ishihara, under daylight, 10 errors with first 24 plates.
TCU test 1st edition, under daylight, 2 protan errors.
Giles lantern, without dark adaptation, few mistakes, then only on small sizes.
Nagel anomaloscope, mid-point 49, range 12.
Twelve typical used resistors correctly named in daylight.

Signed John Smith (qualifications and address)

5.9 REFERRAL

Medically qualified and paramedical personnel such as optometrists (ophthalmic opticians) should refer patients for specialised attention when necessary, e.g. where pathological conditions are suspected on account of anarchic or changed colour vision defects. A suitable report or telephone communication is made to the ophthalmologist, hospital or neurologist. In the UK it is normal procedure under the NHS to refer initially to the family medical practitioner, except in emergency. Careful instruction should be given to the patient, avoiding alarm but impressing the wisdom of seeking 'another opinion' without delay.

5.10 RELATION TO LOW VISUAL ACUITY

Vision (e.g. letter chart performance, without any required spectacles) and visual acuity (letter chart viewed *with* suitable spectacles for the distance) may be affected by many factors. Ocular media changes and pupil size are typical influences. These are described in many texts, including Fletcher (1961) and Duke-Elder (1970).

It is important, as emphasised by Giles (1960), to give the patient the best chance of seeing colour tests *clearly* so correct spectacles for the test distance must be used.

A colour vision defect may appear to be present because visual acuity (or vision) is insufficient for the coloured elements to be distinguished. Blur experienced by the myope may cause failure on a lantern test; even 'night myopia' in a person usually emmetropic can disturb recognition of tiny coloured lights. Low visual acuity caused by cataract, corneal disturbance etc can give a low performance on colour tests. A combination of colour impairment and poor visual acuity is easily produced by retinal or optic nerve conditions and this type of involvement must be considered. The influence of tinted or coloured spectacles or contact lenses, although obvious, must not be forgotten.

6 Illumination for colour vision tests

Daylight has been the traditional and 'natural' source of light under which colour tests have been viewed, but daylight varies considerably. Colour normal and anomalous observers find that the appearance of colour vision tests depends on several factors, as indicated in Chapter 7. Three influences of note here are:

(*a*) the relative energy in different parts of the spectrum,
(*b*) the illuminance, whether high or low,
(*c*) the directions of the light paths and of vision, with respect to the plane of the test.

Other factors, less involved with the light source, are viewing distance and exposure time, but these may have some relation to the mounting of lamps and allied mechanical considerations. Where filters are used the ideal position is to place them between the source and the test, otherwise minor 'goniophotometric' characteristics of pigments intrude and there are subtle perceptual influences; such aspects can usually be ignored in practice and the difficulties of heating or fading can be avoided if filters are placed some distance from lamps.

The quantity of illuminance is important, although provided it is adequate for the task this aspect is not so critical as the colour characteristics of the source.

6.1 DAYLIGHT

Solar radiation is the obvious choice when low cost light is required but there are significant disadvantages. Daylight is variable in content and in amount with important regional differences.

A complete survey of daylight and its spectrum has been made by Henderson (1970), including the imitation of daylight by artificial means. Cloud cover, the season and a host of other factors influence the composition of daylight. Judd *et al* (1964) identified five 'typical daylight phases' each with a related colour temperature (4800, 5500, 6500, 7500 and 10 000 K). These were compared to spectral distributions and colours in tropical regions by Sastri and Das (1968). With the sun there is a 'warmer' daylight (figure 6.1). North-sky daylight is a relatively blue phase of indirect sunlight plus light from the sky, and is commonly described in terms of the mean colour temperature of 6500 K roughly equivalent to the CIE standard illuminant C. BS 950:Part 1:1967 is drawn from the Judd *et al* (1964) data with the inclusion of ultraviolet radiation, as tabulated in this British Standard, equivalent to standard illuminant D65.

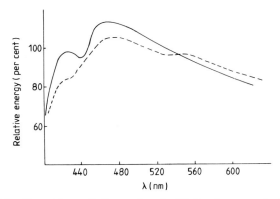

Figure 6.1   Sky without (full curve) and with (broken curve) visible sun, showing energy distribution.

In practical colour vision testing it is tempting to rely on daylight, and provided a good north-sky is available and testing is confined to the middle of the day with around 300 lux available, no difficulties are likely to be encountered. Experience however shows how often daylight spoils standardised test conditions. Accordingly it is more sound to use substitutes for daylight, under control.

6.2   TERMS AND NOMENCLATURE

Several technical terms are associated with descriptions of illuminants and the following notes will clarify their later use.

6.2.1 Colour temperature

The visual quality of a source, matched with a 'black body radiator' at a particular temperature and measured in terms of that temperature in kelvins (K) is sometimes expressed in this way. Two sources may have the same colour temperature while differing in the energy they emit in different parts of the spectrum. At lower colour temperatures the source appears more red, or warmer, than higher on the scale.

6.2.2 Mired value (micro-reciprocal degree values)

A source with a certain colour temperature has a mired value equal to $10^6/T$ where T is the colour temperature measured in kelvins. The term 'reciprocal megakelvins' has been applied and the unit is used for the power of a filter to alter colour temperature; that is the range through which the filter changes the mired value of transmitted light. At the same time, the filter absorbs some light.

> 2000 K represents mired 500
> 2850 K represents mired 350
> 5000 K represents mired 200
> 6000 K represents mired 167
> 6500 K represents mired 154
> 7000 K represents mired 143.

Figure 6.2 illustrates the position.

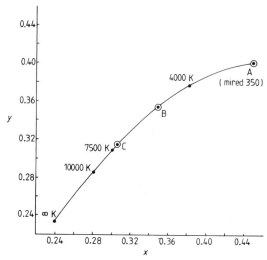

Figure 6.2 Correlated colour temperature of main points on a black body locus in relation to three standard sources, A, B and C.

6.2.3 Filters

Since a typical tungsten source (approximately standard illuminant A) has a colour temperature of 2850 K, blue filters are often used to alter the distribution of energy across the spectrum. A common example is when illuminant A is to be changed to illuminant C, requiring a colour temperature of 6700 K; this represents a change of mired value from 350 to 149 and a filter with a mired of 201. Yellow filters increase the mired value (figure 6.3).

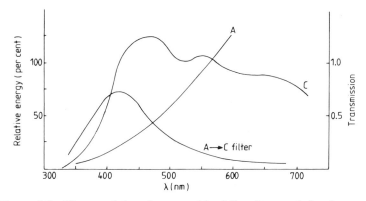

Figure 6.3 Characteristics of sources A and C and transmission features of filter used for conversion (after Davis *et al* 1953).

6.3 STANDARD SOURCES

Three sources, coded A, B and C were defined in 1931 by the CIE using A (with a colour temperature of 2848 K) as the basis (see Davis *et al* 1953). By 1966 a new source, D65, was specified, and Wyszecki (1970) has given relative spectral irradiance distributions for several related sources. Several of the standard illuminants are listed below.

Illuminant A corresponds to light from tungsten filament lamp. Colour temperature 2854 K (mired 350). 40 W lamps tend to be as low as 2800 K; 100 W up to 2900 K. CIE $x = 0.4476;/y = 0.4075$.

Illuminant B corresponds to yellower phases of daylight. Colour temperature 4870 K (mired 205).

Illuminant C corresponds to bluish daylight. Colour temperature 6700 K (mired 149). CIE $x = 0.3135; y = 0.3236$.

Illuminant D55 corresponds to natural sun plus sky radiation. Colour temperature 5500 K (mired 182).

Illuminant D65 corresponds to daylight most frequently encountered in the UK. Colour temperature 6500 K (mired 154).

Illuminant B is now little used, and figure 6.4 compares the spectral distribution of energy for the three more important sources. In Chapter 7 it will be shown how the choice of illuminant affects colour vision tests, by variations in quality and quantity.

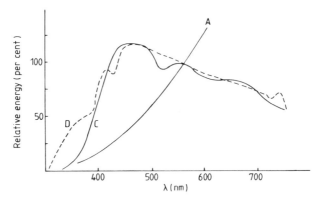

Figure 6.4   Distribution of spectral energy for sources A, C and D65.

6.4   PRACTICAL FILTERS

The 1931 CIE definitions for the three light sources A, B and C, used a gas-filled lamp at 2848 K as A and liquid filters contained in glass vessels to modify this A source to B and C, according to the filters used. Davis *et al* (1953) provided details, bringing the data up to date. Figure 6.3 shows the conversion A to C but by 1957 the CIE committee concerned was considering new definitions using more convenient glass filters. Judd (1961a) described filters then available and their departures from perfect performance. In the USA Corning glass filters have been popular and a Macbeth lamp incorporating an easel has gained wide use; it is a tungsten lamp filtered by a blue glass hemisphere. Comparisons can be seen in *The Science of Color* (1953) and in Judd and Wyszecki (1975).

A British glass with suitable properties is the Chance Pilkington OB8 glass which has approximately 50 mired per millimetre; thus to convey a source 2850 K (mired 350) to 6500 K (mired 154) would require about 4 mm thickness. Where a high level of illuminance is required, a heat filter, or a suitable heat-selective reflector, should protect the blue glass filter. The 100 W tungsten bulb used in the Macbeth easel lamp provides a magnitude of illuminance of 300–400 lux. Fletcher (1972) used a 300 W standard slide projector (mired 294) altered to a mired value of 152 (6600 K) by 2 mm of OB8 glass. A wooden mount with an adjustable mirror to direct the light makes a convenient piece of office furniture, including a store for tests, as seen in figure 6.5. With this device an illuminance level of 400–600 lux is obtained.

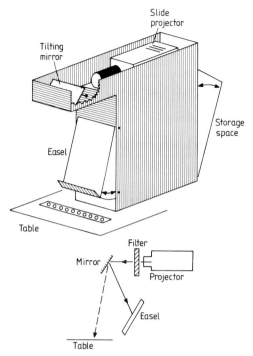

Figure 6.5 A projection lantern illumination device. This uses a standard slide projector held in a plywood mounting. Light via the tilting mirror provides even illumination on the table or on the easel. The easel can be retracted and a storage space is covered by a rear flap held by a magnetic catch. A glass filter is fitted, but can be removed.

Various plastics filters have proved to be satisfactory in practice, notably those produced for theatrical lighting. They vary in material, in heat resistance and in thickness. When used in front of the eye rigidity may be added, with protection, by sandwiching the filters between glass.

A study of figures 6.3, 6.4 and 6.6 and the assurance that practical trials demonstrate the feasibility of these filters for colour vision tests, show their practical nature, as attested by Gayle (1978) and others. Thus, Pokorny *et al* (1978) treated Wratten 78B and 80B filters for glass mounting and used these with a 200 W bulb and Higgins *et al* (1978) showed the practicability of a Wratten 78AA filter. Our own preference is for a combination of three thicknesses of LEE filters; two, each of LEE 203 (mired 35 and 35 = 70), plus one only of LEE 201 (mired = 137) giving a total mired value of 207. With the reasonable assumption that the position of the filter is of little importance (figure 6.7) it is feasible to provide hand-held or clip-on filter holders as shown in figure 6.8. The monocular form recommended prompts the testing of one eye at a time, which is desirable. There are reversible types of spectacle frame which can be adapted if it is desired to provide sides which fit over the ears to leave the subject's hands free; it is sometimes an advantage to occupy the subject's hands.

Data tables from manufacturers such as Kodak give valuable reference graphs, tables and suggestions. Davis and Gibson (1931) gave notes on stability and other features of gelatin and liquid-cell filters, and a short treatment of the control of spectral energy by filters is given in Crawford *et al* (1968).

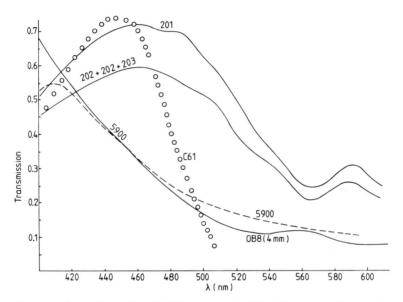

Figure 6.6 A glass filter '5900' compared with OB8 4 mm thick, with other filters, and with one combination of filters (202 + 202 + 203).

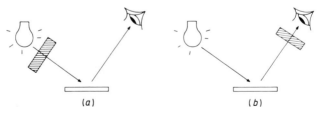

Figure 6.7 It is possible to equate, for most practical purposes, the two situations (*a*) and (*b*), despite small differences in the goniophotometric features when the filter is placed in the two positions.

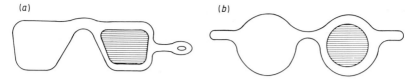

Figure 6.8 Occluder and filter holders for blue glass or plastics filters, in which the opaque side can be placed in front of each eye in turn. Design (*b*) permits the spectacle to be placed behind the observer's spectacles.

6.5 FLUORESCENT AND OTHER LAMPS

In Parts 1 and 2 of BS 950:1967 consideration is given to fluorescent tubes for colour matching and graphic arts purposes. We will see in Chapter 7 that such tubes have had some acceptance for colour vision tests, despite their notoriously non-uniform spectral energy distribution; up to about 2000 running hours some tubes conforming to BS 950:Part 1 can have a correlated colour temperature close to 6500 K and may have a spectral distribution approximating to standard illuminant D65. The main question is whether peaks such as 'mercury lines' are adequately damped by phosphors so that a sufficiently continuous spectrum is provided. Clark (1973) found it acceptable to use the Philips de luxe colour matching Daylight 55 fluorescent tube.

Recent comparisons of different lamps by Richards *et al* (1971) and Lange *et al* (1980) showed some to be inadequate for critical evaluation of colour vision; protans would benefit from more, and be handicapped by less, red light. In these series 'Vita-Lite' was closest in performance to the Macbeth filtered lamp for protans and deutans, and cool white was poor.

Using a 15 W 'daylight' fluorescent tube Schmidt and Fleck (1952) applied some 6% excess volts to achieve '6500 K' and up to 600 lux. Combined with a reflector, and an automatic rotator for PIC plates, this gave satisfactory results in most cases.

Frisen and Hedin (1972) drew attention to the relatively narrow shape of tests such as the FM 100 Hue test and the advantage of a tubular lamp, and designed a box with a sloping lectern and a shade for the tube; they recommended an annual change of tube. Another approach by Walraven and Leebeek (1962a) made use of a circular fluorescent tube (Philips 34) in a desk-lamp plus a blue plastics filter, which provided 500 lux.

Several 'viewing cabinets' are produced commercially using fluorescent tubes as sources and with slight modification they could be used for colour vision tests, although they are designed for viewing coloured prints. Some add filament lamps and tubular sources of ultraviolet for special purposes and it is important not to use most so-called 'daylight' lamps available commercially. The best safeguard in the UK is to avoid tubes not conforming to BS 950 Part 1: 1967 and not approved by the DHSS for hospital clinical purposes.

Xenon arc light has been used by Thomas (undated communication). A close approximation to BS 950:Part 1: 1967 was obtained by filtering the very powerful light with Chance-Pilkington HA3 and SC2 filters, plus an aluminised mirror. Arcs are likely to be used only in experimental situations on account of the heat, breakage danger and cost.

High pressure sodium lamps have appeared with a broader spectrum than the monochromatic yellow (low pressure) type. As Serra and Mascia (1978) discovered, unpredictable results such as artificial 'tritan' defects appear when the high pressure sodium lamp is used for the FM 100 Hue test.

6.6   ILLUMINATION LEVEL

When the level of illumination is reduced markedly to mesopic values colour vision deteriorates and tritan-like defects are typically indicated. Nevertheless wide illuminance levels can be tolerated for clinical use by both normal and anomalous observers. Hill *et al* (1978a) who examined the effect of changing the illuminance level from 600 to 200 lux with 10 colour vision tests used by 20 Daltonics, found no significant differences in performance although more errors were made with the HRR and American optical plates under low illuminances and poorer discrimination is found on the FM 100 Hue test. In addition the confusion axis of a defective is more difficult to discern at low illuminance levels. These authors suggest 400 ± 100 lux as being optimum for clinical evaluation.

The deterioration of colour discrimination with decreasing illuminance level is significantly larger in the colour anomalous (Aarnisalo 1980). The effect of this factor in testing is considered further in Chapter 7.

6.7   MEASUREMENT OF ILLUMINANCE

There is a type of small portable meter often based on the selenium cell which measures illuminance and another variety, also used for photographic work, which combines a cadmium sulphide cell with a battery and a meter. The application of such photographic meters to determine levels of illuminance has been explained in practical terms by Long and Woo (1980). Strictly, care should be taken that the meter used has a spectral filter or that its own characteristics are suitable for the spectral distribution of the light measured; there is usually a small enough error for this factor to be ignored.

It is very common practice to apply a 'colour rendering index' to lamps in the way laid down by the CIE in 1965 and 1974, giving the degrees of change of colour appearance of objects when there is a change from the lamp concerned to a 'standard' lamp which is used as a reference point. Exact reproduction of the standard gives 100%. An official report by the

Medical Research Council (1965) pointed out that only fluorescent lamps corresponding fairly closely to the spectral distribution of energy of a 'full radiator' give satisfactory colour rendering for critical clinical decisions. Two tubes available in the UK at present pass the DHSS specification for clinical areas in hospitals; Thorn Kolor-Rite and Philips Trucolor 38. Thorn's Artificial Daylight (4365/580/Aug 1775) complies with BS 950:Part 1 but when we compare its power distribution we find it to have greater variation in the blue part of the spectrum (and around wavelength 555 nm) than Kolor-Rite and therefore it is doubtful which would be better for colour vision tests. The tube characteristics of two of these lamps are shown in figures 6.9 and 6.10.

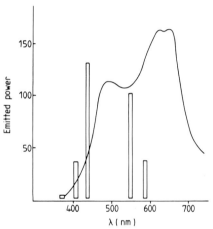

Figure 6.9   Tube characteristics of Philips Trucolour 38.

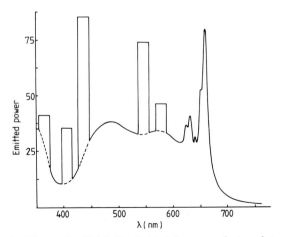

Figure 6.10   Thorn Artificial Daylight tube manufactured to BS 950 Part 1.

6.8 SUPPLIERS

L Hubble Ltd, Roecliffe Road, Woodhouse Eaves, Leicester, make 'Verivide' colour matching cabinets, to BS 950 in various sizes.

Chance-Pilkington OB'OB8 blue glass can be obtained from Chance-Pilkington Ltd, Glascoed Road, Clwyd LL17 0LL.

Multi-Light Ltd, a division of Instrumental Colour System at Kennet Side, Kennetside Park Industrial Estate, Newbury, Berks RG14 5TE, supply a range of lighting cabinets with a variety of light sources to BS 950.

Precision Optical Instruments (Fulham) Ltd of 425–433, Stratford Road, Shirley, Solihull, West Midlands B90 4AE, supply a range of coloured filters.

Lee Filters Ltd, Central Way, Walworth Estate, Andover, Hants SP10 5AN, manufacture a full range of colour control and colour effect polyester lighting filters for film, TV and theatre industries and also manufacture a full range of resin and polyester camera filters.

Rank Strand Ltd, PO Box 51, Great West Road, Brentford, Middlesex TW8 9HR, supply Cinemoid self-extinguishing cellulose acetate filters to BS 3944 and Chromoid high-temperature, flame-retardant polycarbonate filters to BS 3944. Reference books are available.

7 Detection and diagnosis

7.1* INTRODUCTION

Anomalies of colour perception are often easy to detect by simple, subjective screening tests taking a few minutes and administered by an
inexperienced person. Diagnosing and classifying a colour vision defect is
a more lengthy task which needs experience. McLaren (1966) estimated that
nearly 200 methods have been devised over the years, but today only about
20 are commonly encountered. An 'ideal' colour vision test, suitable for all
purposes, and providing an unequivocal diagnosis is probably an impossibility; so the use of two or three independent means of assessing colour
function is preferable, though not always practicable.

Very detailed analysis of colour vision function requires accurate
colorimetry involving monochromatic colours. The determination of
fundamental characteristics is confined to vision laboratories, usually in
universities.

Colour is a subjective phenomenon which requires a psychological
interpretation of a physical stimulus after detailed physiological coding; it
is appropriate that an assessment based on the subjective responses of an
observer should be given, particularly in the clinical context, but later in this
chapter 'objective' approaches will be considered.

7.2 EARLY TESTS

Early colour vision tests involved the naming of coloured ribbons, wools,
beads, cloth, paper or glass. Huddart (1777) examined the Harris brothers
with coloured ribbons—probably the first colour vision test. Dalton (1798)
also employed this method, choosing twenty specimens. It would seem
likely that such materials were chosen because of their ready availability and

the known problems that arose in choosing suitably coloured fabrics for attire by those with colour defects. Sherman (1981) notes that colour naming was the only testing method used until the beginning of the 1830s, ending with Colquhoun (1829), although of course colour naming tests are still used today.

Herschel (1845) reported on his use of spectral colours to examine a Daltonic observer in 1827, and although this provided some attempt at standardisation, colour naming was still required. An extensive summary of the observations which resulted from this investigation is presented by Sherman (1981). It appears that the possibility of a variety of colour vision disturbances both in type and severity had not been suggested until the publication of Seebeck (1837).

Seebeck required his subjects to sort into order 300 coloured papers using sunlight. He apparently used the ordering or misplacing to classify those with anomalous colour perception (Sherman 1981). Spectral colours and those produced by interference were also used by Seebeck to confirm the difficulty and make a rough measurement of the colours where sensitivity was poorest and best—a crude estimation of wavelength discrimination. He determined their colour limits or gamuts using a spectrum and compared such a measure with the results of observers with normal colour vision. Such a test provided him with the knowledge that some observers were essentially blind to reds.

It was Seebeck who first carried out a systematic assessment of colour vision defects. He noted too that the qualitative type of defect is maintained in one family by showing similar colour vision characteristics in a grandfather and grandson. According to Sherman, Seebeck was aware that colour discrimination depends on luminance as well as hue perception.

Seebeck, using his 300 papers for ordering in a sequence, concluded that there are at least two classes of defective colour vision.

(a) Poor overall colour perception, mostly for red but also for green. Confusion of red and green with grey. Confusion of blue with grey. Best perception of yellow.

(b) Similar characteristics to those in class (a) but more marked red and blue confusions. 'Only a weak perception of the least refracted rays' (e.g. red, therefore protanopes).

He was also aware of mild cases of colour deficiency, near-normal cases, and concluded that Daltonic observers could fall in a range between near-normal and significantly deficient. As Sherman points out, he thus anticipated the work of Rayleigh (1881) in the discovery of dichromacy and anomalous trichromacy.

Tests were designed towards the end of the last century, specifically to eliminate dangers from faulty colour recognition in transport. Holmgren, is usually credited with the first test using skeins of wool around 1876–7 following a major rail accident in Sweden at that time. He emphasised the need to test not only railwaymen but also pilots, lighthousekeepers, sailors and school-children. Some resistance was offered by the railway officials, but when Holmgren demonstrated the poor colour matching ability of 13 out of 266 employees (i.e. 4.8%) these were all dismissed and the King of Sweden issued orders that every railwayman should take the test (Holmgren 1877).

Topley (1959) recorded that a primitive form of test using coloured glasses and cards was introduced by the Merchant Navy in the British Isles, also in 1877. Like Holmgren's the test lacked standardisation. Despite the introduction of an oil-lamp flame illuminant in 1885 and the improvement of the glasses and cards, this method was superseded in 1892 by Holmgren's wool test on the advice of a committee of the Royal Society. Holmgren's wool test caused much dispute over the case of a Mr Trattles at the turn of the century (see Chaper 11). This was an example of the difficulty in placing an individual on the proper side of the safety line and also the need for standard viewing conditions and a universal procedure. Intense activity in the design of new tests occurred during the Second World War and since then frequent innovations have appeared. This chapter gives a very selective account of important tests and guides the reader to an informed choice for individual needs.

7.3 TESTING FOR A PURPOSE

Normally advice must be given if a defect is found. Good advice is based upon correct diagnosis of the degree and, to some extent, on the type of Daltonism. Present practice relies too much upon screening tests, in particular the pseudoisochromatic plates, which are often used in isolation. These tests rarely provide a diagnosis of the severity of defect, although this is often of far greater importance in the occupational context than the type of defect (Farnsworth 1957, Cole 1964, Voke 1976a, 1978c).

The patient requires an indication of any appreciable colour vision defect which will affect daily life and occupation; this may be important for the teenager planning a future career. Diagnosis of type and severity of a defect and advice based on interpretation of results for the individual is seldom easy, but enquiries into past confusions and difficulties with colours may assist; experience plays a major factor.

7.4 COMMON VARIATION IN DIAGNOSIS

A definition of normal colour vision is difficult and although defects are usually classified into convenient groups (see Chapter 3) a whole spectrum of anomalies is met in practice. Many do not fall unambiguously into any one class. This difficulty is marked in relation to severity of defects, since there is no agreed scale defining the type or severity of Daltonism and the clinical tests available frequently vary slightly in their diagnosis. Tests are designed on different principles and each type of test measures a different ability. Murray (1943) considers that no individual should be diagnosed as colour anomalous on a single test; three different tests are requisite. The example of Mr Trattles and the many present day cases where the diagnosis is unclear, confirm this view.

Tests can be grouped in different ways. Few involve direct colour naming, a difficult and confusing task for the Daltonic. Most require recognition of a coloured symbol constructed to be confused with the background, or the estimation of small differences in hue in a selection of coloured samples. Successful tests have been based often on the confusion loci of dichromats (Chapter 3). Colour reproduction for plate tests is difficult and there are frequently marked differences between editions; certain individual plates are more accurately reproduced than others. Brightness differences must be minimal to avoid discrimination on this basis.

7.5*   PSEUDOISOCHROMATIC TESTS (PIC)

Being simple, quick and the cheapest tests available these are the most commonly encountered in industry, in the professions and in examination of school-children (Voke 1976a). The confusion 'false colour' charts consist of patterns of coloured dots often varying in size (usually 2–5 mm) in which a symbol has to be recognised against a background. By a suitable choice of colours which lie along confusion loci the figure will be invisible to the Daltonic; most are read by colour normals. Often complex and ingenious in design they are frequently far from ideally constructed since the original forms were produced before Pitt's (1944a) data on colour confusions. The principle of the PIC chart is shown in figure 7.1. Wide variation between editions makes standardisation difficult; few guidelines are given for scoring and what should constitute a pass and failure. Great caution is needed in interpretation. Few allow diagnosis based on the type or degree of defect, so use is restricted to screening. Such tests can reveal a high incidence of colour deficiency, therefore classification is highly uncertain and some margin must be allowed for errors and variations. The number of plates failed seldom bears any relation to the degree of defect, since their sensitivity often ensures that a very mild Daltonic will fail most plates. Lakowski's (1969a,b) opinion echoes the view of most authorities, that 'diagnosis of the extent of defect based on the misreading of pseudoisochromatic charts alone is always difficult, never certain, and at best only probable'. Their value in predicting occupational suitability has been compared by Cowan (1942) to that of a visual acuity chart containing only 6/6 letters.

PIC tests, particularly, have been the object of many comparisons, notably Sloan and Wollach (1948), Wright (1957a), Belcher *et al* (1958), Verriest (1968) and Paulson (1973). Such studies become dated as new editions and new tests appear, but often repay study. Lakowski (1965b, 1966) made valuable critical evaluations dealing with the individual stimuli in detail. Hansen (1963) evaluated different tests using an anomaloscope as a reference point and Taylor (1977) compared several PIC tests for 'screening reliability'. As a result the reader's choice can be a small number of particularly useful plates; Sloan (1945), Sloan and Habel (1956) and Green and Sloan (1945) have provided useful guides and it is appropriate to introduce a suggested 'set', although individual tests will be described later (see table on p. 276 and later text).

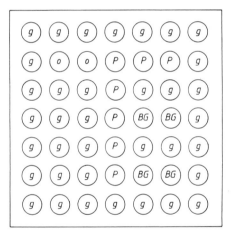

Figure 7.1 The principle of the PIC chart. A mosaic of coloured spots is represented, assumed to be displayed on a black background. Most are an olive-green, *g*. A vertical line of bluish purple spots, *p*, forms the upright of a letter T; this can be read as T by normal observers since four blue–green spots merge with the *g* spots and two orange, *o*, and two more *p* spots give sufficient contrast with *g*. Protan and deutan subjects are liable to confuse *o* with *g* but they see the bluish elements in *p* and in *BG* sufficiently well to trace a letter E.

Since some PIC plates appear 'blank' to Daltonics it is best to intersperse them with plates providing 'positive' symbols at judicious intervals, following Crawford's (1955) remark that resentment stems from a subject being unable to read all of a series of plates.

Origin	Plate number	Normal	Red–green defect	Blue–yellow defect
Dvorine	1	48	48	48 (Introduction)
Dvorine	9	74	Nil	Not applicable
Dvorine	2	67	Nil	Not applicable
Dvorine	15 †	39	Nil	Not applicable
Dvorine	13 †	46	Nil	Not applicable
Farnsworth	F2	2 squares	Green square	Blue square
Ishihara	6	5	2	Not applicable
Ishihara	8	15	17	Not applicable
Ishihara	5	57	35	Not applicable
Ishihara	14 †	5	Nil	Not appliable
Ishihara	15 †	77	Nil	Not appliable
CUT (1st edn)	1, 2, 5, 6 †, 7 †		Results for normals, R–G and B–Y subjects shown on record sheets.	CUT 2nd edn 1, 2, 3

† Indicates plates omitted if a set of 10 is used.

PIC plates vary in design and it is sometimes possible for normal subjects to detect an ambiguity. Defective observers are apt to improve their performance by moving their heads to take advantage of the different reflecting properties of parts of the pictures, or by hesitation and thought. In some cases an altered distance of viewing helps. Thus plates must be presented at the recommended distance, often beyond the reach of patients' fingers (which frequently soil or 'polish' the surface) and for a limited period, usually 3 seconds. Illumination must be controlled (see Chapter 6), the light should fall on the plate from the subject's side and the gaze should be at right angles to the surface of the plate. Table 7.1 shows a selection of distances and times recommended.

7.5.1*   Ishihara test

First designed by the Japanese, Ishihara, in 1917 the test has undergone numerous reprintings in both the British Isles and Japan and is widely used in British industry and schools (Voke 1976a). The test is remarkably efficient as a screening test for red and green defects, but does not test the blue anomalies. It uses one or two numbers, designed on the confusion principle. Most editions contain four plates constructed so that normals see no number but the Daltonic sees one. Some tracing paths are included for use among illiterates and young children. Now 'diagnostic' plates are included to separate deutans from protans. The numbers are displayed, one a pure red, the other a purple-red, both on a neutral background; strong protans fail to see the red number and deutans miss the purple figure. The diagnostic plates are not wholly satisfactory and the test is considered to be valid overall only in a screening capacity (Pickford 1949, Hardy *et al* 1954, Crone 1961, Cole 1963). Wright (1957b) noted that some protanopes show an insensitivity to blue, possible from lens or macular pigmentation, frequently missing both numbers. Where the protan/deutan differentiation plates give uncertain results a majority of protan (or deutan) responses may be sought or the patient is asked which of the two figures on each plate is 'less distinct'. Missing the purple figure indicates a deutan observer, while low visibility of the red figure against the neutral background indicates protan.

Table 7.1 A summary of PIC recommendations.

Test	Viewing distance (cm)	Viewing time
Nagel (*c.* 1934) (circular dot display)	75–150	—
Stilling (1939) (Hertel)	60–100	—
Rabkin (1939)	—	5 s or more
AO Co. (1940)	60–90	Prompt response needed
AO Co. (1965)	75	2 s
Boström and Kugelberg	30–50	15 s max
Boström (1935)	50–75	15 s max
Ishihara (1951)	Under 150	—
Ishihara (1978)	75	3 s
Dvorine	75	5 s
CUT (Fletcher 1975)	35	3 s
Guys (Gardiner 1972)	60	Slowly
TMC (Umazume 1957)	45	2 s (qualitative plates, no limit)
Ichikawa 1978	75	3 s
Velhagen 1974	80 (50 if over 50 yrs)	15 s
HRR	75	2–3 s
Farnsworth and Kimble (1939)		5 s or more
Schmidt and Fleck 1952		3 s with neutral interleaves

Numerous editions make standardisation difficult; the 10th (1951) edition is considered to be one of the best and the 1978 edition is satisfactory. Until the 15th (1961) edition no guidance was given on the pass/fail criterion, leaving the possibility of multiple interpretations. A general rule that 'colour normals' are permitted up to four errors is reasonable. No quantitative classification is possible. Frequent misinterpretation arises with the special figures which appear as one numeral to the colour defective and a different one to the colour normal. The test is available in 24 plate and 36 plate editions, but instructions for different editions vary. Cole (1963) has explained the limitations of the test.

7.5.2   The Hardy, Rand and Rittler (HRR) test

This was described by the designers (1954) and probably arose from dissatisfaction with the tests available in the USA during wartime; it provided red–green and blue–yellow tests, while attempting to demonstrate 'tetartan' deficiency. The first (published) edition of 1955 was superseded in 1957 by a second edition, apparently a rearrangement of identical plates from a single original printing. It has now been discontinued but novel features were the use of cross, triangle and circle symbols, always on a neutral background of dots of different sizes and intensity, as well as a variation in the saturation (purity of colour) of the symbols. Opinions of the test varied, although few have disputed its use for acquired defects, and since secondhand copies exist a brief account is indicated. Walls (1959b) made trenchant criticisms—that the test was liable to misdiagnosis between protan and deutan and as to extent. He thought the six blue–yellow plates might better have been avoided. Vos *et al* (1972), however, tended to defend the HRR test in a study of 200 defective subjects also using the TMC test. They found plates 1, 2 and 5 to be useful for screening; 14 others were preferred for protan–deutan sorting, the 'degree' of defect being judged according to total errors. Some faults over incorrect protan–deutan distinctions were recognised.

Belcher *et al* (1958) considered the four protan and deutan screening plates to be of low efficiency; they are of very low saturation and difficult even for some colour normals to detect. Crone (1961) and Farnsworth (1957a) found the test to be reliable in distinguishing protans from deutans and to indicate the severity of defect when compared with the anomaloscope. Dreyer (1969) similarly noted poor diagnostic ability.

Hardy *et al* do not report having given their test to the rare tritan observers. Walls (1959b) found that three tritanopes and two tritanomalous observers gave strong tritan scores, although Cole (1964) reported that three out of nine inherited tritan observers were diagnosed as normal by the test when other methods, including a colorimeter, had classified them as having definite blue–yellow defects. Sloan and Habel (1956) found that the test sometimes misclassified normal subjects. An imaginative observer is apt to 'manufacture' a circle in any of the quadrants of the plates.

Pinckers (1980a) found the HRR plates to be more reliable than the TMC or Ishihara for normal diagnosis. 75% of colour normals between the ages of 10 and 59 years made no errors on the TMC plates whereas 95% read the HRR perfectly without error. Up to five errors are considered by Pinckers to be within the normal limits for reading the Ishihara plates. His findings with the 1970 edition indicated that fewer than 50% of normals perform without any misreading. Pinckers notes the particular difficulty with the Ishihara test of those under ten years.

7.5.3 Dvorine test

This was produced in the USA in 1944 for school use and the second, much revised, edition appeared in 1953, after criticism of the original plates (Sloan 1945). Dvorine (1963) showed how 14 plates with numbers and 7 with tracks enabled protan and deutan differentiation to be attempted. Peters (1954) reported a diagnostic study, as did Belcher *et al* (1958). Sloan and Habel (1956) gave a satisfactory evaluation. Crawford (1955) found it an effective screening test. Lakowski (1966) made spectrophotometric measurements and pointed out the reason for Dvorine plates to be 'failed' in a 'tritan' manner by many subjects over 40 years of age, although they purport to be for red–green defects.

Dvorine included two discs presenting single and somewhat desaturated colour spots for colour naming purposes, berated somewhat by Murray (1945) and Crawford (1955). There is good reason for including the more 'effective' plates in the selection to be used for screening, described above.

7.5.4 Tokyo Medical College (TMC) plates (Umazume 1957)

Hand printed colours, arranged in a rectangular matrix, give this test an appearance of separate coloured dots about 4 mm in diameter. The 13 plates include 5 for screening red–green defectives and 2 for tritans. Qualitative plates are designed to distinguish between protans and deutans, but probably present similar difficulties to the Ishihara plates. A series of quantitative plates follow, based on the quantitative plates of the HRR test, to distinguish between mild (grade I), medium (grade II) and severe (grade III) defects. All are of a similar orange hue but differ in saturation; the background is a constant yellow–green. Umazume and Matsuo (1962) found the overall diagnosis to be highly correlated with the anomaloscope, giving an accuracy of 91% for protans, and 97.4% for deutans. Vos *et al* (1972) pointed out some of the difficulties in using the test.

7.5.5 Stilling plates

These early pseudoisochromatic charts were produced in 1877. They are currently available in a modified form in the new American Optical 13375 series of plates. The original form included tracing paths for children. Trendelenburg and Meitner (1941) evaluated these plates. Lakowski (1965b) noted that the 12th edition of this series used colours similar to the Ishihara plates. Later editions have been varied, e.g. by Hertel (1939), Engelbrecht (1953) and Velhagen (1974).

The tritan plates have been assessed on the basis of colorimetric data by Lakowski (1976) to be 'poor', and their usefulness in detecting acquired defects therefore seems limited.

7.5.6   B–K tabulae pseudoisochromaticae

In 1913 Göthlin designed some PIC plates which were not published, on which Boström based his own test in 1934. Subsequently Kugelberg cooperated with Boström and B–K plates appeared in 1944. A second edition incorporated the modification (Kugelberg 1972) of a white background. Hedin (1974) analysed the performance of the test and noted that some older subjects normal on the anomaloscope appeared to be anomalous with the B–K test. Three plates have no figure at all, for malingerers, and fifteen have single numerals. There are two 'labyrinths'. The test is confined to red–green deficiencies and in Sweden it is used as an official technique for evaluation. Two or more mistakes comprise a failure and one error raises serious doubt. Each plate is shown twice. Frey (1958) found that 19% of normals were classed as anomalous reinforcing the views of others who suggested that the plates were too sensitive.

7.5.7*   Farnsworth's 'F2' PIC plate

Farnsworth (1955a) designed a successful and now famous plate for tritanopia. A limited number were given away by the New London Submarine Base and a useful copy was described, then printed, by Kalmus (1955, 1965). Taylor (1970) expanded its use in protan and deutan cases and in 1975 (Taylor 1975b) explained how to make such copies with Munsell papers. Lakowski (1966) plotted the colours used and elaborated on its use. It is a very useful screening test; variations in results, as usual, are probably attributable to different methods of use.

A background of purple is the setting for two intertwined squares, one green, one blue (see figure 7.2). Normals see both squares, almost irrespective of age, the green being more pronounced. Tritanopes (and some acquired tritans) do not see the green square under 'daylight'; possibly the green square is faintly visible. When using the plate for protans and deutans, they often find the blue square difficult to see. Protans are more likely to be revealed.

The F2 plate was shown by Ohtani et al (1974) to be 100% effective for the detection of protans, whereas deutans of different types often saw both squares. These authors gave a small-size colour reproduction of the plate.

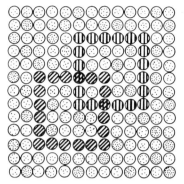

Figure 7.2 A representation of Farnsworth's F2 plate. Most dots are purple. One square is green, one is blue (formed in each case by contrasting dots). Normal subjects see both, but abnormal observers only one, of the squares.

Two studies (Legras and Coscas 1972, Pinckers 1972c) have shown the Farnsworth tritan plate to be unreliable for detecting blue–yellow defects. These authors stressed its value in screening for *red–green* defects, both inherited and acquired.

7.5.8   Other PIC tests and general comments

Many PIC tests have been devised and forgotten. Vierling (1935) produced a variation of Nagel's 12 test cards in which 22 or 23 small dots were printed on a white background in a circle. An undated translation in English by Breuer indicated how the patient had to identify cards on which all the colours were the same and those where colour differences were seen. Thus green and orange confusions, blue and yellow etc were elicited.

Velhagen's test, which has seen 26 editions, is considered by Neubauer and Harrer (1976) to be reliable for detecting acquired tritan defects (24th edn).

A colorimetric analysis by Lakowski (1978b) of the 26th edition suggests that only a few plates are able to provide qualitative information, although plate 20 is well designed to diagnose tritanopia.

Velhagen's 26th edition appeared unsatisfactory to Pinckers (1980a) in connection with tritan testing, since many normals gave false positives. Ichikawa *et al* (1978) published fifteen plates using modern digital display figures. Ten plates are used for screening, of which present experience shows numbers 7 and 13 to be very satisfactory; these have greenish numbers on brown backgrounds. There are no tritan plates but five attempt a protan–deutan distinction. Many slightly anomalous subjects easily read all parts of the plates.

New pseudoisochromatic plates for acquired defects have recently been developed by Ichikawa *et al* (1983) with nine plates for screening blue–yellow defects, five plates for red–green defects and two for scotopic vision. The plates were shown to provide a successful diagnosis for 22 of 36 central serous retinopathy patients with blue defects giving almost identical classification to the HRR and D.15 tests.

Several tests have been devised with animal or human figures instead of numbers. Gardiner (1972) produced a series of plates with loose plastics letters which children can match with the charts. Unfortunately the colours used are poor and confusion arises from the conflict of different letters seen simultaneously. The idea of separate cut-out figures is worthy of emulation. Plates 5 and 6 are the least reliable although they usually reveal dichromats and are unlikely to suggest that a normal is Daltonic. The fourth plate is most likely of all to detect defectives but 50% of normals give 'false positives'. Plates 1 and 3 often given false positives with normals.

Berson *et al* (1983) showed how Munsell papers can be used as a backing to outlines of arrows to form a simple PIC test which can be managed by some young children. This test was developed for monochromats, as described in Chapter 3 (p. 187).

It is necessary to have a flexible approach when using PIC charts, although a strict protocol may be essential if the administration is delegated. Experience greatly improves the interpretation of responses and the apparent simplicity of such tests prompts users to expect more information than can actually be obtained (most PIC tests are for screening purposes). Editions may vary in quality and this makes standardisation difficult. Finally, it is important to note that faded and soiled charts become unreliable.

Importance of illumination

Care must be taken when illuminating PIC and all colour vision tests lacking an inbuilt illuminant. Kalmus (1965) considered this to be one of the most important sources of error with colour vision tests. He noted that the colour defective may be helped to read PIC charts correctly by oblique light or by an illuminant such as a tungsten bulb by which blues and blue–greens are suppressed and the red end of the spectrum is enhanced. A daylight source of colour temperature around 6500 K is essential. Waaler (1927) first noticed that deuteranopes could read Stilling's plates in tungsten illumination and Hardy *et al* (1954) confirmed this with both deuteranomalous and deuteranopic observers. Since natural daylight is variable, a 'daylight' source is preferable. Variations in natural daylight may be sufficient to account for differences in error scores of deuteranomalous subjects. Chapters 6 and 9 elaborate on this matter. Kalmus (1971) showed the interdependence of intensity and composition of the illumination of

Ishihara (9th edn) charts in relation to exposure times of 1, 5 and many seconds.

Comparison of tests

Throughout the literature there are examples of comparisons between different tests such as PIC plates by different inventors, and each new test is subjected to this ordeal. Criteria against which the new tests are measured have to be established and a common method is first to classify each subject by a battery of tests. Paulson (1973), for instance, has shown how a slightly anomalous trichromat can pass some tests and fail others. Different editions of an individual test vary in quality; some War Department PIC plates produced between 1939 and 1950 are not good.

The obvious variety of modes of presentation of tests introduces a complication; for instance a subject can assist himself to 'pass' a PIC plate by tilting the plane of the test relative to the light and introducing helpful reflections.

Using photometric and colorimetric data for three PIC tests (AO Co (AOPP), HRR and the 1st edition of the Standard (SPP) plates by Igaku-Shoin) Chioran and Sheedy (1983) evaluated their design with respect to two illuminants. Dichromatic confusion loci and luminance contrast criteria were used, showing the difficulties of meeting all criteria. A remarkable conclusion was that the AOPP and SPP should perform better under tungsten A illumination than the specified Macbeth illuminant C.

Among a sample of 1000 cases of acquired dyschromatopsia Lagerlöf (1980) found pseudoisochromatic charts to detect 10–30% of all patients (10–15% for the Ishihara test, 25–30% for the Boström–Kugelberg test). Farnsworth's tritan plates showed poor sensitivity since the blue square was missed in advanced cases of dyschromatopsia.

Reproduction of PIC plates

Care is needed in using other than rigorously standardised plates. There is a temptation to copy plates by colour printing or colour transparency and the risks are great. Projection is not the answer to mass screening since visual angles and colour rendering in a slide and projector combination are fraught with difficulty. Paulson (1973) has shown how the colour slide used in the vision screener called the Ortho-Rater (No 71-21-21) performs. The slide has 9 coloured dots and the official scoring method 'fails' only 12% of Daltonics; a stringent method still passed 72% of defectives. Somewhat better detection was possible with Ortho-Rater slide 71-21-50 comprising 6 PIC digits.

Taylor (1975b) and Fletcher (1978b) have given instructions for construction of PIC tests in a simple manner from standardised coloured papers.

7.6   SORTING OR MATCHING TESTS

In these tests small colour shapes are arranged in order to make a colour sequence; an obvious advantage is the use among illiterates.

7.6.1   Bead, cloth, wool and paper tests

The use of beads, wool or cloth samples was the first means for examining the human colour sense, now superseded by methods using carefully standardised colours. The Holmgren (1877) wool test is the best known. The subject is presented with four skeins, a grey, a red, a green and a purple. From many other wools of varying hues and brightness one or two must be found which match.

Daae's arrangement (1877, 1878) of 70 coloured wool patches, in rows, was described in detail by Hansen (1978); such a test had an 'efficiency' comparable to the D.15. Roberts' (1884) 'Berlin wool' samples resembled Daae's and Jennings (1896) describes Donders (1880) as being among those using wool tests. Thompson (1894) and Oliver (1893) carefully explained their variations. One of the chief antagonists of 'wools', over a long period, was Edridge-Green. His objection in 1891 and the controversy with the Board of Trade, described in 1933, pitted the lantern against other tests. Certainly bad standardisation and other disadvantages leave only a limited scope, today, for such *ad hoc* trade tests. Abney (1895) stressed the value of the subject 'doing something' without words with Holmgren's test and he used filters to modify the viewing conditions; conversely, Edridge-Green (1891) was of the opinion 'that a person cannot be efficiently tested' if the names given to colours be disregarded. He devised an extensive 'classification' test including threads and fabrics and squares of coloured glass; also a 'pocket test' comprising 112 single wool threads, 13 twisted silks etc and a piece of green velvet. Orange, next violet and then six other colours had to be matched from samples small enough to detect central scotomata.

The heightened simultaneous (and successive) contrast effects demonstrated by the colour defective have prompted the design of special tests.

Jennings (1896) mentioned several simultaneous colour tests including 'tissue paper' methods.

Chapanis (1949) reviewed extensive literature on simultaneous contrast effects and conducted observations by normal and abnormal subjects with a lantern showing two lights. Contradictory opinions have been prevalent and data such as Chapanis produced are helpful; when shown a red and yellow light simultaneously, anomalous trichromats tend to choose 'green' as the name for the yellow light. Normals show this tendency as well. Similarly, yellow is called 'red' in the presence of green. A white light also is likely to be named as 'green' when seen with red, and 'red' when shown alongside (or following) green. This apparently enhanced colour contrast is not always found and it has been interpreted by Chapanis as the consequence of decreased 'differential sensitivity', giving the anomalous effectively a larger range of colours from which to choose an appropriate name. The phenomenon can be demonstrated in practice with an anomaloscope or a lantern. Figure 7.3 reinforces this important point.

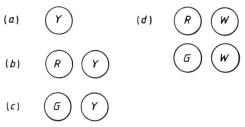

Figure 7.3 Simultaneous colour contrast. Anomalous trichromats are most likely to perform as shown. A similar tendency is possible with successive colours. (*a*) Yellow alone — gives a normal appearance. (*b*) Yellow with (or after) red. The yellow tends to appear 'greener'. (*c*) The same yellow with (or after) green acquires a 'reddish' appearance. (*d*) Similarly white is more likely to be called green when seen with red or red when seen with green.

Hansen (1963) tested heightened colour contrast by Cohn's 'tissue paper' covering technique, by Stilling's contrast plates and with a lantern subtending about 11′ of arc. Velhagen (1974) used modifications of Stilling's plates for contrast testing. Hansen (1976a, b) extended the tissue paper approach to show the excellence of the method as a supplementary test.

Additional comment on simultaneous contrast has been given by Pickford (1951). Edridge-Green wrote in 1891 that such tests 'have very little value'.

7.6.2*   The value of trade tests

Where a clinical test is also a close simulation of a specific job or task in which the patient is involved daily, there is some justification for making a decision purely on the basis of the performance at this one test; the examination of a pilot or train driver by a lantern test is an obvious example, as is the identification of colour-coded resistors and wires by an electrician. Potential ability is then clearly established, and such an approach might also be suitably recommended for careers guidance, although it must be remembered that employers may require a pass on any test they choose for examination. Results on trade tests must be treated with caution however and Taylor (1971) suggested valuable provisos for the use of trade tests; that the conditions must be those encountered in the actual situation, e.g. dirty cables and poor illumination, and the results must not be regarded as of a general application but only to the trade situation. This is sound advice.

7.6.3*   Dichotomous D.15 test ('Panel D.15')

Farnsworth (1943) devised this as a simplification of his 100 Hue test and used it to enable a transformer company to avoid employment of men with 'severe' colour vision detects in colour coding jobs. It involves a colour circle made from a blue 'pilot' or reference Munsell paper (B 5/6) and 15 other coloured (5/4) numbered discs of effective diameter about 12 mm. Each is mounted in a plastics dropper bottle cap, the colour being visible through the aperture but relatively safe from soiling. The specifications are as follows:

   Pilot, 10B 5/6; 1, 5B 5/4; 2, 10BG 5/4; 3, 5BG 5/4; 4, 10G 5/4; 5, 5G 5/4; 6, 10GY 5/4; 7, 5GY 5/4; 8, 5Y 5/4; 9, 10YR 5/4; 10, 2.5YR 5/4; 11, 7.5R 5/4; 12, 2.5R 5/4; 13, 5RP 5/4; 14, 10P 5/4; 15, 5P 5/4.

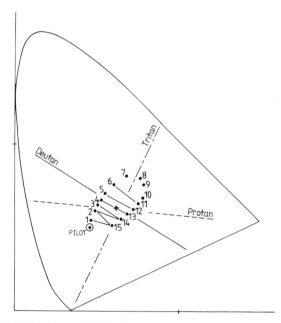

Figure 7.4   Positions of D.15 colours. The pilot colour is followed by 1, which is confused with 15 by a deutan. Neutral N5 is marked by +. Major confusion axes are shown.

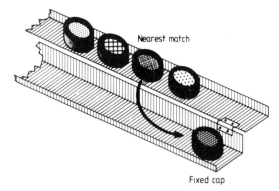

Figure 7.5 D.15 caps in tray with lid open. The fixed pilot cap and the 'nearest match' for the colour of the pilot, which is to be moved next to the pilot, are indicated.

The authors (following Sloan) add an extra cap, an N5, which is not a standard procedure. Figure 7.4 indicates how the positions of the colours permit confusions to be made across the colour circle. A cross shows the N5 location. The caps 1–15 are randomly arranged on the tray formed by the lid of their box and the pilot is placed at one end of the box base (figure 7.5). The subject chooses the colour appearing to match the pilot colour most nearly and places it next to the pilot; he completes the sequence, always matching the one he placed last. He is not pressed for time and may change the order. If the N5 is used it is now shown saying 'put this colour next to any you think is nearly the same colour'. The lid is closed, the box is inverted and on opening it the numbered bases of the caps are seen; they are recorded and a diagram is completed (noting the position(s) found for the N5, on one or both sides of the circle). A normal trichromat correctly makes a sequence 1 to 15. Deutans typically use an order which may proceed 1, 15, 2, 3, 14, 13, 4, 12, etc, or 1, 15, 2, 14, 3, 13, 4, 5, 12, 6, etc; that is using confusions nearly parallel to the 'deutan axis'. Deutans tend to match N5 to 4, 14, 13 or 3. Protans and tritans make characteristic errors along their own axes. Figure 7.6 shows the approximate axes for the conventional types of disorder, also a 'scoptic vision' axis described by Verriest *et al* (1963) and the nominal position of N5 for normal observers. It is usual to obtain less than the ideal sort of pattern suggested by figure 7.7 for a protanope or severely protanomalous subject; therefore it should be no surprise that figure 7.8 is the actual record of a protanope and that figure 7.9 is one for a protanomalous subject who made 15 errors with 24 Ishihara plates. Figure 7.10 shows a subject with tritanopia, the two eyes giving almost identical results. Cole (1964) found the test to be unreliable with inherited tritans but in acquired cases it is useful.

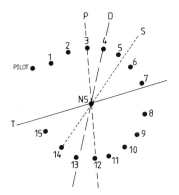

Figure 7.6   D.15 record sheet, outlining a colour circle, with main confusion axes.

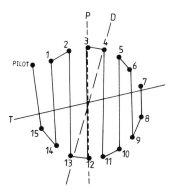

Figure 7.7   Stylised confusions of a protanope.

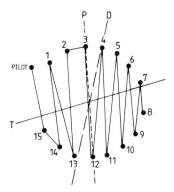

Figure 7.8   Actual confusions of a protanope.

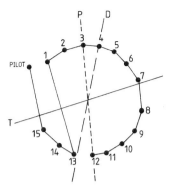

Figure 7.9   Confusions by a slightly protanomalous observer.

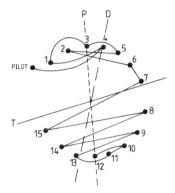

Figure 7.10   Confusions (right eye only) by a tritan observer.

It is possible to modify the test, and using Munsell papers with Value 8 and Chroma 2 makes it more difficult. Figure 7.11 displays the 8/2 results of a slightly protanomalous subject who obtained normal performance on the standard 5/4 D.15 test; he made 22 errors on 24 Ishihara plates, had served in the wireless branch of the RAF and was good at wire sorting. He had an anomaloscope range of 5, mid-point 53. A lower chroma set of D.15 colours was described by Lanthony and Dubois-Poulsen (1973).

A selection of saturation (Chromas 2, 4, 6 and 8) with neutrals, in sets similar to D.15 tests, was introduced by Lanthony (1975) with a view to evaluating acquired defects.

Dichromats can be distinguished by a variation of the D.15 test, the H.16, used in the USA naval service. This test uses colours of relatively high saturation.

Recent reports, such as given in papers at the IRGCVD conference in Geneva, 1983, show a definite acceptance of the great value of 'low saturation' D.15-type tests.

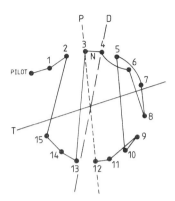

Figure 7.11 Subject PS giving unclear confusions with a Value 8, Chroma 2 D.15 test. The standard 5/4 D.15 test gave normal results. N = position chosen for N8.

The Sahlgren test described by Frisen and Kalm (1981) uses twelve paper discs of diameter 18 mm, five being green–blue, five being blue–purple and two being neutral. Different saturations are included so that when the subject identifies 'grey' discs it is possible to estimate saturation discrimination. Data reported made comparisons with Ishihara, D.15 and Lanthony tests and suggested a quick screening device for acquired defects which concentrates on bluish colours.

Achromats arrange the colours in an order of decreasing 'scotopic luminosity' such as 1, 2, 3, 4, 5, 14, 15, 12, 7, 13, 11, 10, 9 and 8, as shown by Sloan (1954).

The D.15 test 'fails' about 5% of a typical male population, the criterion of 'fail' being two or more major crossing confusions. Normal subjects sometimes interlace their choices along minor parts of the circle, such as choosing an order 1, 2, 3, 4, 6, 5, 7 and the linking of 7 and 15 sometimes takes place in the presence of correct responses between 1 and 7 and between 15 and 8. Farnsworth (1947), Sloan and Wollach (1948), Sloan and Altman (1951), Crone (1955) and Linksz (1966) have shown examples of the use of the test. Bowman (1973) emphasised the simplicity and reliability of the test, suggesting its use in conjunction with a PIC test. He gave data and tables, including colour difference values for the cap intervals. Verriest and Bozzoni (1965) found diagnosis separating deutans from protans to be clear in 91% of subjects and the extent of defect as judged by the number of crossings was about 75% correct. Sassoon and Wise (1970) used it successfully with children as young as three although the concept of a colour match leading to a progressive order may be difficult. Francois and Verriest (1961a) and many others have found it to be suitable for acquired defects. Helve and Krause (1972) used it on elderly people and found that 8 out of 209 could not perform the test, but that 16 failed having completed;

some cases of macular degeneration failed, 20 cases of glaucoma passed and low visual acuity made the test difficult.

Roth's (1966a, b) modification involves 28 Munsell colours as a colour circle.

The D.15 is often called the dichotomous test since it was designed to separate significantly Daltonic people from the rest. One difficulty is the fact that since a cap is committed to a place it is difficult to use it as a confusion colour for others which come later.

A method called the total colour difference score (TCDS) was adopted by Bowman *et al* (1983) in which the D.15-type tests were quantified by means of calculated colour differences between the coloured elements. Thus the effects of age on performance can be evaluated, since the TCDS increases as errors by a subject increase. Since the colour steps are likely to differ in individual tests, such as desaturated versions, a colour confusion index (CCI) was introduced. This CCI compares the TCDS for a subject with the 'perfect' one; it can be used for comparing tests.

7.6.4 The Farnsworth–Munsell 100 Hue test

This is a widely used hue discrimination task in which four separate quadrants of a colour circle are presented one at a time. The format and box construction resemble the arrangement of the D.15 test, using Munsell papers (Value 5, Chroma 5) in plastics caps; unlike the D.15, the 100 Hue test prevents confusions across diameters or large chords of the colour circle since each quadrant is shown alone. The 100 colours described by Nickerson and Granville (1940) were reduced to 85 by Farnsworth (1943) in order to give a better series of just-noticeable hue differences for young subjects. Lakowski (1966) found the colour differences to be between 0.6 and 5.7 NBS units.

Each of the four boxes has two reference colours fixed, one at each end of the tray and 21 loose (numbered) caps are placed in random order in the lid. The subject rearranges the loose caps, between the reference caps, to form a colour series. As shown in figure 7.12 cap number 1 has the Munsell specification 5R 5/5 and the contents are as follows:

Box 1	85–21	Pink, through orange, to yellow
Box 2	22–42	Yellow to blue–green
Box 3	43–63	Blue–green to blue–purple
Box 4	64–84	Blue to reddish purples to pink.

Farnsworth (1957b) commenced with Box 1; Taylor (1974) recommended Box 2, to assist red–green Daltonics to grasp the test quickly, as they have few difficulties in this region. As in the D.15 test the colours are numbered on the reverse side and a score of accuracy can be derived from a subject's order of arrangement. A perfect sequence with consecutive numbers is achieved only by normals with unusually critical colour perception; the number of errors gives an indication of the severity of any defect, while poor performance in two opposite parts of the colour circle suggests a type of Daltonism, by an orientation *at right angles* to the usual confusions *across* the colour circle.

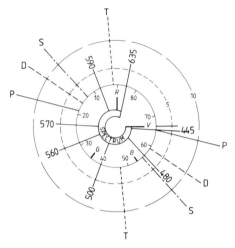

Figure 7.12   The FM 100 Hue test diagram. Concentric circles denote the scores for successive colours. Dominant wavelengths are given in nm.

Administration

The test is illuminated by 'daylight' with an illuminance between 200 and 1000 lux on a table and within easy reach of the subject. Instructions can be 'Please find colours which match these two end colours; transfer colours from the top tray to form a colour series between these two ends. Take care not to touch the actual colours—this will spoil them. You have about two minutes for each tray but accuracy is more important than speed'. The form of words must be adapted to suit the individual.

Scoring

Each colour receives a 'score' found by adding the difference between its own number and the numbers of the two colours placed on either side of it. Colours in the correct sequence each have a score of 2. As an example the subject arranges nine caps as 1, 2, 3, 5, 4, 8, 7, 6, 9. The scores are as listed below.

Score for cap 2 = (2 − 1) + (3 − 2) = 2 (normal)   partial error score = 0
Score for cap 3 = (3 − 2) + (5 − 3) = 3 (error of 1) partial error score = 1
Score for cap 5 = (5 − 3) + (5 − 4) = 3 (error of 1) partial error score = 1
Score for cap 4 = (5 − 4) + (8 − 4) = 5 (error of 3) partial error score = 3
Score for cap 8 = (8 − 4) + (8 − 7) = 5 (error of 3) partial error score = 3
Score for cap 7 = (8 − 7) + (7 − 6) = 2 (normal)   partial error score = 0
Score for cap 6 = (7 − 6) + (9 − 6) = 4 (error of 2) partial error score = 2

These seven caps can be plotted on a polar diagram or on a rectangular graph as shown in figure 7.13. A subject usually alters his score, and profile, on retesting, so despite the time involved it is advisable to average two attempts (see figure 7.14).

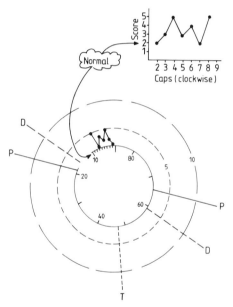

Figure 7.13 Part of an FM 100 Hue test score, showing the polar diagram plots for colours 2 to 8 inclusive, and plots for these colours on an ordinary graph. The score 2 is 'normal' for each colour.

A 'total error score' is sometimes used. This is the sum of the individual 'partial error scores', in each of which the normal score of 2 is *not* counted. Such total error scores assist when acquired defects are monitored. Lakowski (1968) showed how errors increase after age 20 and summarised norms in relation to age and conditions. The elderly frequently show tritan errors and Verriest (1963) and Kinnear (1970) pointed out that a tritan defect can appear in young observers. Krill and Schneiderman (1964) noted a mean error score of 51 for fifty colour normals with caps between 35 and

55 giving most difficulty. They also plotted error scores linearly, still show-ing errors in different regions, and thus recorded significant loss of discrimination in carriers of Daltonism. McLaren (1966) showed how train-ing over eleven months could reduce the normal error score from 68 to 12.

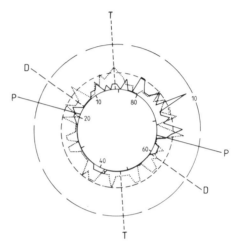

Figure 7.14 Polar diagram of a very slightly deuteranomalous observer, using both eyes at once. Two successive attempts differ from each other.

Two studies have indicated superior performance on the FM 100 Hue test by females compared with males, both among chidren (Verriest *et al* 1981a, b) and young adults (Verriest *et al* 1962), although a further study by Ver-riest *et al* (1982) failed to show any sex differences in performance.

The way in which familiarity with the task can influence performance at a difficult clinical test such as the FM 100 Hue was shown by Verriest *et al* (1982). The greatest number of errors was made among subjects with no previous experience of the task.

The establishment of mean error scores for normals as a norm has been important for a realistic assessment of colour function using the FM 100 Hue test. Pinckers (1980a) quotes a typical mean error score for normals up to age 30 years as 37.5, for normals between 30 and 50 years 67.1 and for normals aged over 50 years as 94.4. These are higher than equivalent norms given by Verriest *et al* (1962, 1982) and Aspinall (1974a).

The repetition of the FM 100 Hue test on eight subjects (wearing clear contact lenses) in a study by Harris and Cabrera (1976) emphasise the variability of the scores. The mean error score became significantly lower over fourteen repetitive trials extending over two days. Chisholm (1969) showed the influence of the level of score on variance in retesting.

Kinnear (1970) proposed a simplified use of the scoring sheet to give slightly altered but acceptable graphs and showed how skewness of certain scores can be treated. Aspinall (1974a, b) proposed that an individual box producing an error score over 200 might represent a random arrangement, e.g. by a child.

The test gives useful information in many cases and it has been adopted as a criterion of potential industrial colour discrimination in some instances. A real disadvantage is the time involved, particularly in working out scores, so attempts at automation have been made. Parra (1972) developed a Fortran program to study defects and a commerical electronic procedure has been described by Taylor and Donaldson (1976) and Taylor (1977). Digital computation of 'resistor coded' caps controls a rapid polar graph plotter. The 'Chromops' device gives a linear printout in about three minutes (figure 7.15).

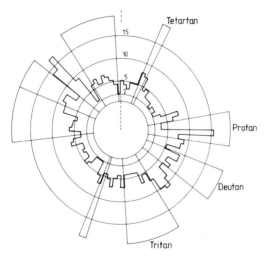

Figure 7.15   Chromops device plot for an observer who has an error score of 305.

The distinct tritan-like tendencies which become evident in the normal eye with advancing age have been fully studied by Verriest *et al* (1962, 1982). The 1982 study, a performance of the FM 100 Hue test on 109 males and 123 females between 10 and 80 years of age with at least 6/6 (20/20) acuity, no systemic or ocular disease and normal colour vision as assessed by a perfect reading of the Ishihara plates, confirmed the 1962 data from 480 normals ranging in age from 10 to 64 years. Young adults (20-29 years) show the least number of errors with error scores increasing in both younger and older age groups.

Interpretation

Dichromats give distinctive profiles of poor colour discrimination. Boxes 1 and 3, for example, permit tritanopes to confuse caps along confusion loci, and Box 1 shows the poor discrimination by protanopes and deuteranopes of the red, orange and yellow caps. Thus 'confusion' or 'poor discrimination' spikes appear, as follows:

> protanopes at cap 18 or 19 and at 66
> deuteranopes at cap 16 and at 59
> tritanopes at cap 1 or 2 and at 46 or 47.

A 'scotopic' axis appears in some retinal disturbances.

Cole (1964) considered the test to be most reliable as a qualitative diagnosis for tritan defects and to 'fail' most dichromats, and many deuteranomals, if a pass score is under 100. 'Normals' with a score over 100 but no defined profile must be considered to have poor discrimination but in some cases it may be poor application or lack of interest. Although protanomalous and deuteranomalous trichromats may display correctly orientated 'spikes' it is more likely to be the 'extreme' anomalous who do more than show large error scores (Lakowski 1968). Very poor performance with no discernable polar pattern (a 'bomburst' star) is given the apt description of 'anarchic' (figure 7.28).

Taylor (1974) proposed a redistribution of the colours thus: Box 1, 7–27; Box 2, 28–48; Box 3, 49–69; Box 4, 70–6. This makes the protan and deutan spikes fall clearly within two boxes with tritan errors in a third box. Figures 7.16 to 7.28 show examples of typical profiles.

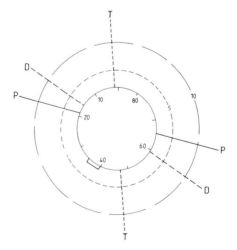

Figure 7.16 Observer with almost perfect FM 100 Hue record, slight errors appearing between 35 and 40.

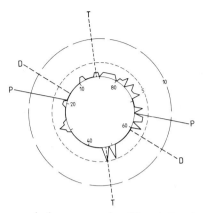

Figure 7.17   A normal observer scoring 42 and showing well distributed 'acceptable' mistakes.

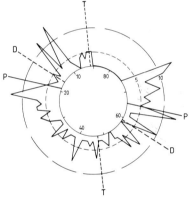

Figure 7.18   Record of a protanope, scoring 332. The major axes or wings lie nearer to the P directions than to the D axes.

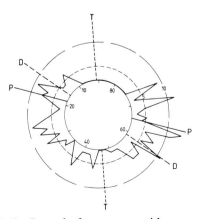

Figure 7.19   Record of protanope with score of 189.

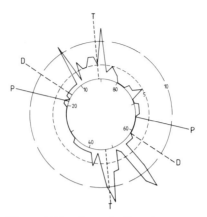

Figure 7.20   Record of a tritanope.

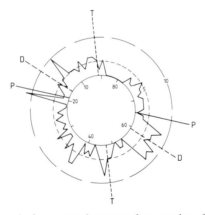

Figure 7.21   Record of a woman known to be a carrier of deuteranomaly.

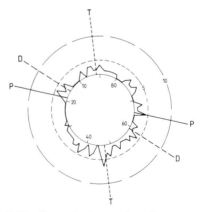

Figure 7.22   Record of a carrier of protanomaly.

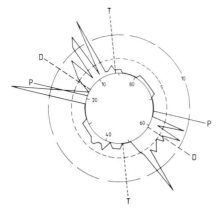

Figure 7.23 Deuteranomal with score of 290.

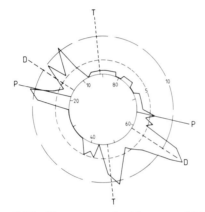

Figure 7.24 Deuteranomal with score of 209.

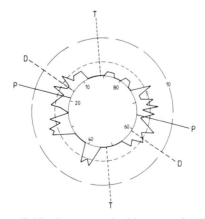

Figure 7.25 Protanomal with score of 107.

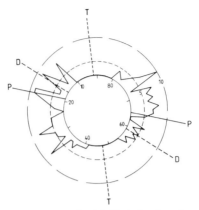

Figure 7.26 Protanomal with ill-defined regions of error.

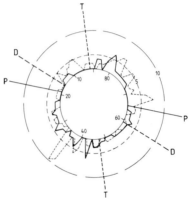

Figure 7.27 Tritanomalous record (after Young 1974) showing the results with the naked eye (full curve) and through 1.30 neutral density filter (broken line).

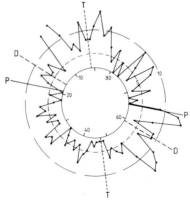

Figure 7.28 Anarchic plot which may be gross carelessness or may be caused by an acquired defect.

Other features of the FM 100 Hue test

No two colour vision tests measure exactly the same thing and Lakowski (1969a) found unique cognitive and variance features with the FM 100 Hue test; his spectrophotometric data of 1966 show how it differs from the Colour Aptitude Test (CAT). The CAT (Dimmick 1956) is a set of plastics objects with which subjects can manipulate saturation differences in small areas of colour space. Both tests are of use to indicate proficiency with colours.

Clarke (1968) drew attention to the way in which two FM 100 Hue sets can be compared for 'split field' hue discrimination and an adaptation by Burnham and Clark (1955) forms a test of colour 'memory'. Another use, by Boyce and Simons (1977), was to show how different lamps and illuminances influence 'performance'. The worst performance was by older subjects but they improved by using 1200 lux rather than 400 lux.

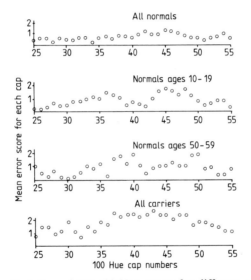

Figure 7.29   Variation of FM 100 Hue scores for different cap numbers, after Krill and Schneiderman (1964), showing their mean score for each cap at different ages for normals and for carriers. Note that most errors for normals are within the group of caps 35–55.

Surveying the literature which examined the effect of illuminance levels on FM 100 hue test scores for normals and for defective observers, Bowman and Cole (1980) noted the need for a study involving a larger range of levels; they gave the test to ten subjects who were all over 60 but who had good ocular media and normal colour vision. Five younger controls were used, and the four illuminance levels involved were 1, 10, 100 and 1000 lux. The results agreed well with earlier reports, including the 'tritan tendencies' of the older observers below 100 lux. Bowman and Cole concluded that 100 lux is a suitable level for all ages because this maintains a good sensitivity to loss of colour discrimination and it is not necessary to illuminate for the absolute maximum of normal performance.

Krill and Schneiderman (1964) who favoured the non-polar graph plotting of FM 100 Hue test data, indicated their finding that errors for normals tend to be greatest in the region of caps 35 to 55 and that the width of the area shown in their graphs (figure 7.29) was proportional to the age of the subjects. They demonstrated the sensitivity of the test by showing how 'carriers' of Daltonism make more errors, and have a greater width, in this area.

Pinckers (1971) noted how 16 components of the FM 100 Hue test correspond closely to the caps which form the D.15 test, as shown below.

100 Hue number represents	D.15 number
58	Pilot
56	1
50	2
46	3
41	4
36	5
31	6
27	7
19	8
12	9
6	10
2	11
84	12
77	13
73	14
70	15

He proposed that by using the appropriate FM 100 Hue numbers, joined across the central circle of the FM 100 Hue score sheet, a composite record can be made for a subject; this gives confusion axes for the D.15 (inner) record perpendicular to the axis representing the spikes formed by the (outer) FM 100 Hue results.

7.6.5* The City University test (CUT)

Ten plates each provide a display of five coloured spots (figure 7.30) which gives a test approximately the same difficulty as the D.15 test, since some of the colours are the same as in the D.15. The subject chooses the spot most closely matching the colour of the central spot, using 'top', 'bottom', 'left' and 'right' for location. He does not touch the plates. One colour is a natural choice for a normal subject, another is more likely to be chosen by a person who is 'significantly' protan, a third is a good 'deutan match' and the fourth is designed for a tritan confusion. Sometimes more than one may be chosen, perhaps a deutan as well as the normal spot. Results entered on a record sheet suggest both type and severity of any defect. The first edition, described by Fletcher (1972, 1975, 1978b) showed some variations between plates. The second edition (Fletcher 1980a,b) uses the best six original plates plus four slightly more difficult plates in which size and saturation are modified (figure 7.31).

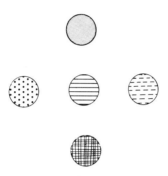

Figure 7.30   Display of City University Test plate showing a central coloured spot surrounded by four spots of different colours.

Lanthony (1977) found 'good indications' as to type and severity of hereditary and acquired defects with the the test, as compared with D.15 and HRR tests, such as complete agreement with D.15 and about 96% agreement with HRR as to hereditary 'type'. A dichromatic score on D.15 corresponded to 5 or more errors on CUT. Acquired defects gave a 50% exact agreement with HRR (almost complete agreement with 'tritan' scores) and most of the discrepancies indicated partial accord. A CUT score of 5 or more always corresponded to a very altered D.15 trace.

Independent studies by Hill *et al* (1978b), Ohta *et al* (1978a) and Verriest and Caluwaerts (1978) have indicated other findings. Ronchi *et al* (1978) applied the CUT to 98 subjects, detecting the four known Daltonics and finding that 20 of the 94 normal individuals gave 'tritan' indications; plates 8, 7 and 3 appeared to have greatest sensitivity to age-induced tritan tendencies and this was confirmed with appropriate FM 100 Hue caps.

The test is used at 35 cm under 'daylight', allowing 3 seconds per page. In the case of a child or handicapped adult it is sometimes necessary to ask first 'what is the colour of this spot in the middle?' then 'where is there another colour which looks a little bit like it?'. In some cases eliciting a colour 'name' for each spot gives a clue to performance but the normal, quick method of use is preferred.

Maggiore (date uncertain) designed a display of five (2.5 cm) holes in a mechanical device. A confusion colour of suitable intensity was turned into position and matches were sought by variations of the colour in other apertures.

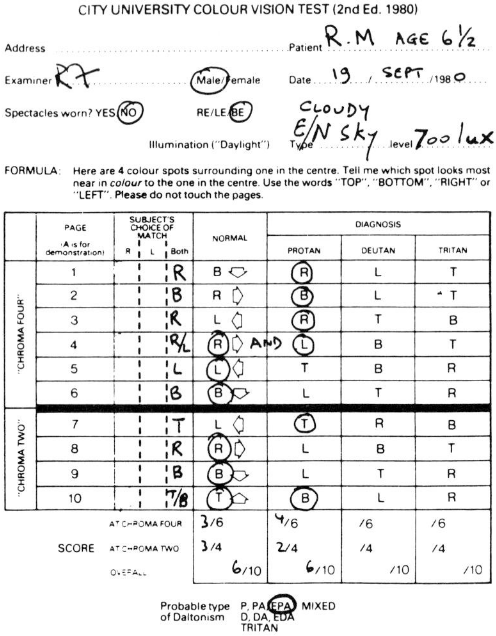

CITY UNIVERSITY COLOUR VISION TEST (2nd Ed. 1980)

Address .... Patient R.M AGE 6½

Examiner R⁴ .... (Male/Female)   Date 19 / SEPT /198 0

Spectacles worn? YES (NO)   RE/LE (BE)   CLOUDY
                                                   E/N sky
Illumination ("Daylight")   Type   level 700 lux

FORMULA: Here are 4 colour spots surrounding one in the centre. Tell me which spot looks most
            near in *colour* to the one in the centre. Use the words "TOP", "BOTTOM", "RIGHT" or
            "LEFT". **Please do not touch the pages.**

PAGE	SUBJECT'S CHOICE OF MATCH			NORMAL	DIAGNOSIS		
A is for demonstration)	R	L	Both		PROTAN	DEUTAN	TRITAN
1			R	B ▽	(R)	L	T
2			B	R ◊	(B)	L	• T
3			R	L ◊	(R)	T	B
4			R/L	(R)◊ AND (L)	(L)	B	T
5			L	(L)◊	T	B	R
6			B	(B)▽	L	T	R
7			T	L ◊	(T)	R	B
8			R	(R)◊	L	B	T
9			B	(B)▷	L	T	R
10			T/B	(T)▷	(B)	L	R

(left side labels: "CHROMA FOUR" for rows 1–6, "CHROMA TWO" for rows 7–10)

SCORE	AT CHROMA FOUR	3/6	4/6	/6	/6
	AT CHROMA TWO	3/4	2/4	/4	/4
	OVERALL	6/10	6/10	/10	/10

Probable type   P. PA (EPA) MIXED
of Daltonism   D. DA. EDA
                TRITAN

Figure 7.31   CUT (2nd edn 1980) record. The patient, age $6\frac{1}{2}$ years, gives six normal and six protan responses.

While dealing with the CUT, which 'passes' some of the subjects who 'fail' a more stringent test such as Ishihara's, it is suitable to emphasise again how the degree of Daltonism can be estimated. No errors on CUT but distinct errors on Ishihara suggest a slight anomaly likely to be 'safe' in many occupations. Two or three CUT errors (almost invariably accompanied by many Ishihara errors) show a significant defect, usually of a definite variety. Five or more CUT errors suggest the likelihood of dichromatism or its equivalent acquired condition.

7.6.6   The Lovibond colour vision analyser (CVA)

Dain (1971, 1974) described a portable instrument with several attractive features, including built-in illumination. Coloured Tintometer glasses form a circular array of 27 filters surrounding a neutral which can be altered in brightness. The subject rotates the colours to select those which match the central neutral. Since the saturation of the display can be altered it gives both qualitative and quantitative data for colour deficiency and assesses the colour matching ability of normals, usually taking between three and ten minutes. Ensell (1978) gave data comparing Ishihara and CVA performance with 127 subjects including Daltonics. The test is good for evaluating mild anomalies. Figures 7.32 and 7.33 illustrate the device and the colours presented.

Figure 7.32 The Lovibond colour vision analyser (CVA) (courtesy of the Tintometer Ltd, Salisbury). An observer gazes through the aperture as if reading at a normal distance. There is a control near the observer's right hand and the controls seen in the photograph are operated by the person conducting the tests.

First the subject has to find a 'match' for the neutral (illuminant C) in the centre, with saturation set at 20. Next, saturation is increased slowly from zero until 'colours appear' and then attention is directed to 'neutral matching' colours; the subject may alter the central spot's brightness but he finds his 'threshold', the saturation at which the match breaks down. The two colours giving the highest threshold saturation indicate the confusion axis (e.g. protan), and by a simple calculation a quantitative 'range' is obtained. The range is greatest for dichromats, approximately zero for normals and intermediate for anomalous trichromats and some acquired defects. A protanope gave the following results:

Chosen filters	13	1	14
Saturation thresholds	80	80	80
Neutral set at	0.7	0.7	0.7
Range	$(80 - 10) + (80 - 10) = 140$		

A recent study by Ohta *et al* (1978b) gave data on lowering of some saturation thresholds in normals over 40 years of age and showed good detection of Daltonism. There was some discrepancy between cases shown as protan by the anomaloscope and deutan by the CVA.

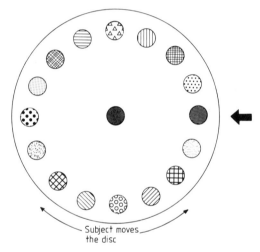

Figure 7.33 Coloured discs seen by the observer using the CVA. A central (neutral) colour has been matched by the small disc which has been moved to the level of the arrow.

7.7*   THE ANOMALOSCOPE

In 1881 Lord Rayleigh's colour mixing experiments, using a device dating from 1877, led him to identify deuteranomaly for the first time. Nagel's 'anomaloscope' developed about 1907 was based on Rayleigh's equation

$$Red + Green = Yellow.$$

A colorimeter device such as is shown in figure 7.34 can be used to make a red + green mixture to match a yellow, assuming that the latter can be adjusted to a suitable brightness match. The normal match is rejected by deuteranomalous trichromats, who require more green, while protanomalous subjects need more red. Dichromats accept any red + green mixture (including the normal setting) but protanopes are differentiated from deuteranopes by the fact that protanopes have to set the yellow field at a low intensity before it appears as dim as pure red.

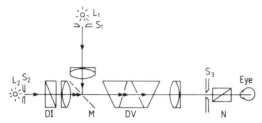

Figure 7.34 Rayleigh's anomaloscope. Lamp L_1 sends a yellow image of slit S_1 (via a direct vision prism, DV) onto slit S_3 and thence to the eye. Lamp L_2 sends red and green images of S_2 onto S_3 via a double image prism DI and DV; these two images are polarised at right angles to each other and the Nicol prism N is rotated about the instrument's axis to achieve different mixtures of red and green. Mirror M produces the split field of view.

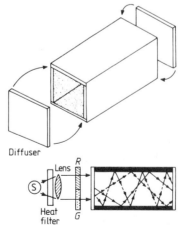

Figure 7.35 Diffuse light mixture by mirror box. Four inward facing rectangular mirrors of the proportions shown totally mix incident light. Ground glass or opal plastics covers each end. One diffusing screen receives light and the other emits it uniformly mixed. If red and green filters admit light a uniform yellow mixture emerges. A source S may be placed close or may be imaged by a lens. A heat filter can protect coloured filters.

The anomaloscope is assumed to use the Rayleigh equation, thus detecting red–green defects only, unless otherwise specified; some instruments can employ blue + green to match cyan, for tritans, as explained later.

Most writers rate the anomaloscope as the best means of classifying Daltonism. Properly used, with great care, it is indeed the 'final arbiter' but the remaining part of this section shows how much caution is required and how errors persist. Informed opinions should be heeded, such as that of Walls and Heath (1956) that the device is seldom so used as to make the distinction between 'extreme anomalous' trichromat and true dichromats. It is costly, although a relatively simple anomaloscope (figure 7.35 and figure 7.36) can be made to work adequately. Such mirror 'mixing devices' have been used extensively, for instance by Houston (1932), as a 2.5 cm square mirror corridor 25 cm long or as a wedge of glass 12 cm long and 12 mm thick throughout, tapering from 25 mm to 6 mm in width, with each end ground matt.

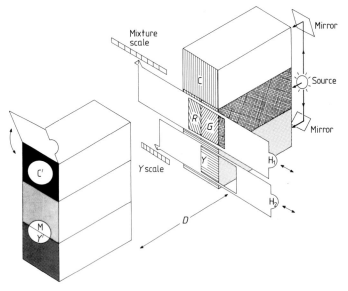

Figure 7.36 An anomaloscope system. Six individual mirror boxes (see figure 7.35) receive light from a single source, such as a car bulb. Distance D has been exaggerated for clarity. Mirrors divert light to upper and lower channels. C, a neutral adaptation light, is transmitted to C', to be exposed by a shutter. Y is a yellow filter transmitted to Y', the latter being altered in intensity by horizontal movement of handle H_2 and the sliding aperture. H_1 alters the mixture of red and green reaching M, the upper half of the test field. For initial experiment Wratten filters, perhaps in combination, can be used; e.g. 29 or 70 as red, 58 or 74 as green and 9 or 90 as yellow. The source will modify requirements. The height may be 12 cm, overall. If viewing is at 50 cm the diameter of the test field may be about 15 mm.

Every single instrument must be calibrated, using many 'normal' subjects, ascertaining their normality with several other tests. Provided the anomaloscope is not held to be infallible, it greatly assists the detailed study of a Daltonic subject.

In a context of controlled adaptation to a neutral light, the subject accepts or rejects a series of settings presented to him. The 'range' and the mean or 'mid-point' of the settings which appear to match the yellow field are considered. An outline of the instrument will now be given, followed by its use, after which some finer points must be considered.

7.7.1   Anomaloscopes

Spectral lights used by Rayleigh were the sodium yellow (589.3 nm) and 535 nm green mixed with 670 nm red; the mixture was slightly desaturated, therefore 546 nm was adopted in 1950 by some authors. The widely used Schmidt and Haensch Model I Nagel instrument combines 670.8 nm with 546 nm to match 589.3 nm. Figure 7.37 shows how the lights can be combined, and those wishing to construct a simple anomaloscope combining Rayleigh's and Nagel's principles may usefully follow Forshaw (1954) who modified a spectroscope.

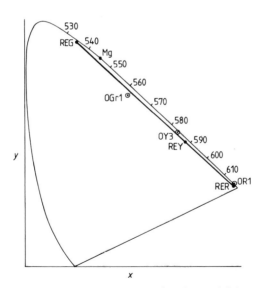

Figure 7.37   Different anomaloscope stimuli. The Rayleigh spectral lights are REG, RER, and REY. Mg = mercury green, adopted in 1950. OGr1, OY3 and OR1 represent Chance glasses suitably illuminated. The figure is slightly distorted around 530 nm to emphasise the advantage of 546 nm over 535 nm.

The mechanical features of the slit changes in the Nagel anomaloscope were discussed by Pokorny *et al* (1975) since asymmetrical movements involved tend to produce shifts of the 'bandpass' related to each stimulus. Transmittances, related to wavelength and slit widths, were seen to vary according to the type of Rayleigh equation mid-point being used. The data were used to calculate anomalous pigment sensitivities, and were featured in the discussion of a possible pigment common to protans and deutans (see Hayhoe and MacLeod 1976).

Non-spectral stimuli are 'adequate', as shown by Sloan (1950) and Willis and Farnsworth (1962). Sloan's comparison of Nagel's results with those from dichroic filters showed the advisability of high purity filters not overlapping in transmission if red—green ratios are to give good differentiation between normal and anomalous subjects. Comparable saturation of the mixture and of the yellow is best. Willis and Farnsworth gave details of different devices, each with an arbitrary scale dictated by mechanics, and their attempt at relating the scales hardly achieved its aim; however a useful colorimetric transformation was proposed by Lakowski and Aspinall (1972). The difficulties experienced by anomalous trichromats with desaturated confusion colours lend some weight to the use of non-spectral lights.

Other filter anomaloscopes, described by Shaxby (1944), Sloan (1950), Crawford (1951), Pickford and Lakowski (1960) and Lakowski and Tansley (1973), have low luminances, usually on account of poor design. Thus the Pickford—Nicolson device employs glass filters and presents varied field sizes. Pickford's (1976b) data with this device are extensive, but no pre-exposure is used and some mechanical aspects have been changed. Where the natural pupil is limited by the exit pupil of an instrument the retinal illuminance is usually affected and some subjects may have difficulty in keeping their eye in the right place.

Successful filters include several sets. Ilford filters 250, with 807 and 812 were used in about 1950. Pickford's Chance glasses (OR1 + OGr1) matched a combination of OY3 and ON32; his tritan set was (OGr1 + OB10) with (OB2 + ON32). Shaxby (1944) used Wratten (74 + 29) with 73 and in about 1940 Hecht and Shlaer used Wratten 27 + 33 mixed with 15 + 61, to match 73. Provided reasonable light levels reach the retina (the Nagel instrument typically gives 120 trolands, and about half this retinal illuminance is really needed) there is some latitude of choice. By constructing an anomaloscope with Chance—Pilkington glasses RG2, OGr1 and OG1 one of the authors (Fletcher 1971) gave acceptable results despite the need for slight desaturation of the yellow field. However, glass filters can alter their transmission when heated, necessitating a 'warm-up time'.

Interference filters give good transmission as shown by Moreland and Young (1974) and Roth *et al* (1978); in the latter design the 'Besançon' anomaloscope uses interchangeable interference filters, fibre optic presentation of the split field and controlled exposure time, as well as alternative field sizes, making it a versatile and useful device (figure 7.38) (see also Ohta *et al* 1980).

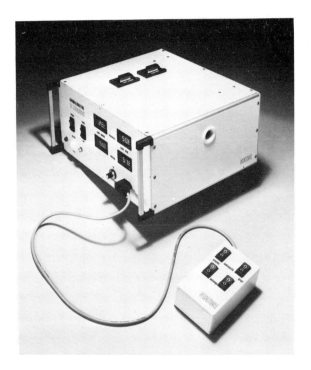

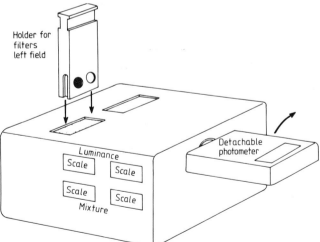

Figure 7.38 Roth's Besançon anomaloscope. View of instrument and control box with test field aperture and schematic view showing photometer in place.

Light emitting diodes now have some potential as anomaloscope sources, as described by Saunders (1976) and Dain *et al* (1980a), however the peak transmissions may present difficulties. The green should be close to 542 nm, with the red having λ_{max} at least 650 nm.

Up-to-date reviews of such 'metameric' matches as used in anomaloscopes were given in 1983 by Pokorny and Roth and will appear in the transactions of the Geneva conference of the IRGCVD †.

7.7.2* The anomaloscope range

One of the most scathing comments about the use of this device was by Walls (1959b) who suggested that neither Willis and Farnsworth, nor Hardy, Rand and Rittler had understood how to use the Nagel anomaloscope to detect extreme anomals. Walls pointed out the instability of the extreme anomal and how on a few 'settings' he may appear to be a dichromat by accepting the whole red–green scale. Normals, presented with a suitable number of settings, accept only about six of these (on a Nagel scale) which represents their 'range'. Two subjects MMM and CJB are now described, who were shown 18 and 12 mixtures, respectively, in the orders explained; the results were obtained with the Nagel anomaloscope, where setting 73 is pure red and 1 is pure green. There is also a scale for yellow.

The results show how SFM (subject's first match), an average of three matches made by the subject, may or may not be a reliable guide to his acceptance of settings subsequently shown at the examiner's choice.

Subject MMM (normal) SFM = 45, Y = 11 (to match SFM of 45), right eye only.

38	39	40	41	42	43	44	45	46	47	(Part of red–green scale)
12×	2×	3✓	13✓	9✓	4✓	6×	1✓	11×	5×	
14×	7✓					10✓	8×	15×		
						16✓	17×			
						18×				

Result: range = 5; mid-point = 42; AQ = 1.12 (see later text). Yellow brightness 21 matches pure G; Y 17 matches pure R. Note: order of each setting and whether 'yes' or 'no' in response to 'do they match?'

Subject CJB (normal) SFM = 45, Y = 12, right eye only.

41	42	43	44	45	46	47
3×	1✓	5✓	9✓	2✓	4?	6×
7×	11✓			8✓	10×	
				13✓	12×	

Result: range = 4; mid-point = 43.5; AQ = 1.05. Yellow brightness 31 matches pure G; Y 36 matches pure R.

†These transactions were published in late 1984 by Junk Publishers, The Hague.

Dichromats accept any setting from red to green as a colour match for yellow, as shown in the two subjects DB and RJF whose data follow. Whereas deuteranopes make almost normal settings of yellow to match the brightness of pure red, protanopes set the yellow to match the lower brightness they obtain from red.

Both protanopes and deuteranopes accept matches made by colour normals; furthermore deuteranopes accept colour matches made by deuteranomalous observers and protanopes find matches made by protanomalous observers acceptable. Such observations date back to König and Dieterici (1893).

Subject DB (protanope) SFM = 31, $Y = 21$, right eye only.

0	10	20	30	40	50	60	70	73
6✓	9✓	5✓	3✓	2✓	1✓	7✓	4✓	8✓
11✓								10✓

Result: range complete. Yellow brightness 34 matches pure *G*; *Y*1 matches pure *R*.

Subject RJF (protanope) SFM = 55 (mean of 62, 55, 18), right eye only. Matches complete range. Yellow brightness 27 matches pure *G*; *Y*2 matches pure *R*; *Y*6 matches '60' on scale.

Anomalous trichromats of the more usual variety accept nearly normal or slightly increased ranges which are displaced. Deuteranomalous subjects require an excess of green, protanomalous need more red and their characteristics can be studied in figures 7.39 and 7.40 and in the following data.

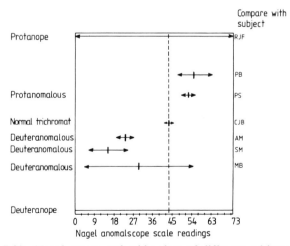

Figure 7.39 Nagel ranges and mid-points of different subjects, selected for comparison. 0 represents pure green and 73 represents pure red on the usual scale. The details of each observer's series of settings are given in the accompanying text.

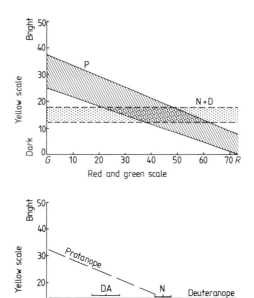

Figure 7.40   Anomaloscope data graphs for typical modern Nagel-type scales. The upper graph displays the area within which a protanope's complete range is likely to lie and another area which should embrace a deuteranope's settings; the normal settings can be expected to appear in the middle of the common region. The lower graphs illustrate typical ranges plotted with their most likely brightness settings.

Subject MC (deuteranomalous) SFM = 30, *Y* = 20, right eye only.

	17	18	19	20	21	22	23	24	25	26	27–32	33	34	35
(not showing	✗	✓	✓	✓	✓	✓	✓	✓	✓	✓	✓ ✓	✓	✓	✗
order of	✗	✓	✓									✓	✗	✗
settings)	✗	✗										✓	✗	✓
		✓											✗	✗

Result: range = 15; mid-point = 25.5; AQ = 1.72. Yellow brightness 17 matches pure *G*; *Y* 26 matches pure *R*.

Subject AM (deuteranomalous) SFM = 26, *Y* = 10, right eye only.

	20	21	22	23	24	25	26	27	28	29	30
	✗	✓	✓	✓	✓	✓	✓	✓	✗	✗	✗
	✗	✗	✓					✓	✗	✗	
	✗	✓						✓			
		✓									

Result: range = 7; mid-point = 24; AQ = 1.88. Yellow brightness 28 matches pure G; $Y55$ matches pure R.

Subject SM (deuteranomalous) SFM = 9, left eye only.

1	2	3	4	5	7	9	10	14	19	25	26	27	30	40	60
12×	5✓	11✓	9×	3×	15×	13✓	4✓	8✓	7✓	6✓	18×	14×	10×	1×	2×
				17×											
				16✓					18×						

Result: range = 18; mid-point = 16. Yellow brightness 21 matches G; $Y25$ matches R. Note order of settings and inconsistency. Subject tended to reject settings as 'matches' immediately after short neutral adaptation then, after 1 second viewing, to accept them.

Subject MB (deuteranomalous) SFM = 15, right eye only.

5	15	35	40	41	45	49	55	60
✓	✓	✓	✓	✓	✓	✓	✓	×
	✓							×

Result: range = 50 estimated; mid-point = 30; AQ = 1.53. A rapid assessment using only 11 settings.

Subject MDR (age 11) (deuteranomalous) SFM = 22, right eye only.

2	4	5	7	10	17	19	20	24	31	35	37	40	45	60
✓	✓	×	✓	✓	✓	×	✓	×	✓	✓	✓	×	×	×
	✓										✓			×

Result: range = 36; mid-point = 18; AQ = 2.5.

Subject PB (protanomalous) SFM = 50, right eye only.

45	46	47	48	51	63	64	65	66
✓	×	×	✓	✓	✓	✓	×	×

Result: range = 16; mid-point = 56; AQ = 0.8.

Subject PS (slight protanomalous) SFM = 49, right eye only.

35	40	41	45	46	47	49	50	51	52	53	54	55	56	57	58	60	65
×	×	×	×	×	×	×	×	×	✓	✓	✓	✓	×	✓	×	×	×
	×	×				×		✓				✓		×			
						×											

Result: range = 5; mid-point = 53; AQ = 0.85.

The usual definition of an 'extreme anomalous trichomat' is that his range of settings for red–green mixtures which match yellow reaches one end of the scale. This assumes he has equated brightness satisfactorily at each point. One variation, by Pickford and Lakowski (1960), suggested that the EPA (extreme protanomalous) and EDA (extreme deuteranomalous) have ranges 'several times' the normal range, with ranges usually including the normal mid-point. Schmidt (1955b) refined the definition to describe a subject whose range reaches one or both ends of the scale when unadapted to neutral pre-exposure (see §7.7.4) and whose range reaches only one end of the scale with neutral adaptation. Walls (1959b) described the condition as one where neutral 'tuning' gives a temporary smaller range.

Clearly, therefore, the range and its mid-point (see below) are related to each other and to pre-exposure. Cole (1964) warned that range may not be an accurate index of difficulties in practical life, such as errors with traffic signals.

7.7.3   The mid-point

This is the mean or central setting within the subject's range. No mid-point exists when a range reaches either end of the scale and this must be related to the possibility that an extreme anomal may accept either red or green as a match for yellow, as emphasised by Walls and Mathews (1952) who placed reliance on neutral points for separating the extreme anomalies from dichromatism. Walls and Heath (1956) reinforced the difficulty of using the anomaloscope in this connection.

A mid-point is an attractive way of separating deuteranomalous from protanomalous subjects and it is sometimes expressed in terms of an 'anomalquotient' proposed by von Kries in 1899 and Trendelenburg in 1929, which relates the subject's mid-point to the normal's mid-point on a certain instrument. Beyerlin's slide rule simplifies the calculation

anomaly-quotient (anomalquotient) = AQ = subject's ratio green/red
divided by standard ratio green/red.

Thus a normal subject has AQ = 1, a protanomalous AQ is less than unity and a deuteranomalous one is more.

This quotient is applied to protan and deutan subjects but Cole *et al* (1965) and Henry *et al* (1964) described a deuteranomalous anomalquotient in one tritan subject and protanomalous data in another using the Nagel anomaloscope.

Pickford (1951, 1967a, b) used the figure of ±3 standard deviations, from the mean of (say 100) randomly selected subjects' Rayleigh equations, to define 'normal trichromatism'. Those with mid-points between + 3 and + 2 or − 3 and − 2 SD from the mean, he described as 'deviant' and anomalous trichromats' mid-points according to this are more than 3 SD from the mean. A seldom-used term 'colour weak' was applied to those with fairly large matching ranges. Schmidt (1955b) used the description 'colour asthenope' for those with an enlarged U (unadapted) range and 'colour amblyopia' where both U and N (neutral) ranges were enlarged.

A combination of 'yellow scale' and 'mixture scale' readings is easily made with the graph shown in figure 7.40. Its use displays the data, which should fall within the areas appropriate to each type of subject. This graph prompts the use of brightness settings at the expense of little extra trouble.

7.7.4 Pre-exposure

A neutral adaptation to a $20°$ field has been used since it was introduced by Trendelenburg but many anomaloscopes do not provide this. Schmidt (1955b, and in personal communications in 1973) and Jones (1970) have shown that a colour temperature of 5500 K is suitable. In the Nagel anomaloscope a 60 W lamp provides a reddish light giving a retinal illuminance many times greater than the test field; this produces after-images and complaints that it is too bright. An Ilford 810 filter plus Neutral Density 0.5 gives 5600 K and 80 cd m^{-2}, which works well. A suitable thickness of Chance's OB8 glass can be used in addition to the instrument's diffusing glass.

Preliminary N adaptation of two or three minutes is advisable and final decisions on settings of the instruments should be made in 0.5 seconds viewing time, immediately after 10 seconds more N adaptation. Hansen (1963) used his anomaloscope with subjects in the U state, with 5 settings, and readers must now be aware that the literature is complicated by differences in practice. N adaptation is seldom provided, possibly because of ignorance or neglect, but it is only in the EDA and EPA subjects that serious difficulties arise.

The following criteria were used by Steen and Lewis (1972) with a Nagel scale 0–73; 'normal' where mid-points were 38–44 (range up to 9), range of 10 counted as mild defect, 11–25 as moderate and over 25 severe. They compared the range after 30 seconds neutral adaptation between matches with the 'chromatic adaptation' range, calling anomalies 'extreme' where the chromatic was larger than the neutral range.

Shifts of the setting of the anomaloscope by normals, after 'neutral' adaptation, were compared by Verriest and Popescu (1974) with adaptation effects when the *test* field was viewed for long periods; most observers drifted to a green setting to a greater extent than their 'red drift'.

7.7.5 Procedure in practice

Most adults use anomaloscopes reliably but some find the judgments impossible. Opinions differ as to how children perform; often under the age of 10 they show difficulty. It is best to offer a 'forced choice', after the subject has been given a little practice, asking for a 'yes' or 'no' response to 'now the brightness is equal in both halves of the field, is it also a colour match?'

A good technique, based on Schmidt's (1955b) guidelines, follows this procedure. First ensure that appropriate (but non-coloured) spectacles are used.

(*a*) Show a 'normal match', then pure green, then pure red, asking the subject always to alter the yellow for 'equal brightness'.

(*b*) Refer to the 'top and bottom' or 'right and left' parts of the field but do not use colour names.

(*c*) Ensure that the subject is comfortable, can see properly and that he understands the instructions. Give three minutes looking at the neutral field in a relaxed way.

(*d*) Let the subject control both sides of the field to make three matches. Average these to obtain the subject's first match (SFM) for guidance but do not use them for the range.

(*e*) Obtain ten readings with no further pre-exposure, as U settings. Use these for the U range.

(*f*) Repeat the settings in 'e' allowing 0.5 seconds for viewing after 10 seconds N adaptation; these, if not also those in 'e', must be set by the examiner, although where there is doubt the subject can first alter the yellow control for equal brightness. Use the results for the N range.

(*g*) If neutral adaptation prevents any setting from being acceptable seek positions which appear 'too red' and 'too green', respectively for a range and its mid-point.

7.7.6 Difficulties and variations

Willis and Farnsworth (1952) considered 'neither range nor mid-point... alone ... to be diagnostic'. They successfully combined the range with the anomalquotient and a 'practical' means to classify the degree of defect, which did not use the anomaloscope, showed reasonable agreement with their 'combined' data. Protans and deutans were usually well separated but some exceptions were noted and ascribed to 'a weakness of anomaloscopes'.

The wide range of individual differences in matching range among anomalous trichromats was first noted by Nagel, as quoted by von Kries (1924) and has been verified on many occasions since. All those who use the anomaloscope on large samples of colour deficient observers are aware of this fact. Variations in the action spectra of the visual pigments involved tends to be the usual explanation with the most severe colour discrimination confusion arising in observers whose abnormal cone mechanism displays an action spectrum of similar characteristics to the normal mechanism involved in the red–green match. Abnormal neural colour-coding mechanisms have been suggested to play a part, for example by Hurvich and

Jameson (1974). A third proposition, based on the observation that large field colour-matching involving the Rayleigh equation, without rod activity contributing, does not give rise to great individual variations, is suggested by Nagy (1982). He attributed the differences to the relative number of abnormal cones present in each retina.

Among 103 Daltonic subjects Sloan (1950) found four with normal Nagel anomaloscope responses. In one (unilateral) case reported in detail by de Vries-de Mol and Went (1971), and confirmed by other well equipped investigators, substantial errors were made on several tests but anomaloscope readings were normal. Crone (1968) showed subjects with normal anomaloscope data, yet making Ishihara errors. Kinney (1967a) identified as a 'deuteranope' a person with an incomplete Nagel range, accounted for by saturation differences appearing in the spectrum beyond 550 nm.

Uncorrected refractive errors, or excessive accommodation when looking through an eye-piece, tend to bias the data. Using an instrument without an eye-piece in darkness may result in a young person accommodating up to three dioptres. Weinke (1960) showed how myopia produced 'deutranomalous' and hyperopia produced 'protanomalous' tendencies.

In one study Verriest (1968) based his subjects' classification on the anomaloscope noting, however, that 2% of his 'deuteranomals' gave a normal response on the instrument.

Normal trichromats make 'deuteranomalous' settings when the field is reduced from the usual $2°$ subtense to $15'$, as shown in figure 7.41. Horner and Purslow (1947) used filtered light with large condensing lenses so that field size could be varied. Sloan (1950) discovered subjects (probably extreme anomals) who behaved like dichromats when the Nagel field she was using was reduced from $3°$ to $1°$.

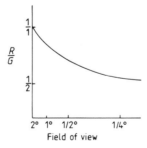

Figure 7.41 Effect of field size upon anomaloscope settings made by normal trichromats (after Horner and Purslow 1942).

Observers with normal colour vision were used by Pokorny et al (1980) to obtain green to red ratios for the Rayleigh equation for fields ranging from 0.5 to $8°$. Twelve subjects all used relatively more red as the field size increased, there being a marked variation as the size was changed from 0.5

to 2°. This study was connected with mixtures of red (670 nm) and green (545 nm) used to match a series of 'extended Rayleigh matching stimuli' being respectively 570, 580, 589 (the usual one), 596 (sometimes 600) and 610 nm.

Different instruments produce different data and a given anomaloscope can alter in vital characteristics, perhaps because the electrical supply varies or the lamp ages or is replaced. Some of the historic aspects of these changes were covered by Hallden (1974) who determined how precision of measurement improves if a lamp uses 75% of its rated voltage. In this context he showed difficulties in setting limits between normals and anomalous subjects.

Up to the age of 60 there is little influence on the Rayleigh equation through ocular changes, apart from pathological conditions, as shown by Boles-Carenini (1954) and Young (1974). The equation can be influenced by selectively absorbing filters in front of the retinal cones, including pigments in the ocular media. Wyszecki and Stiles (1967) found that filters can alter performance, but in young and healthy eyes pigment effects are negligible. Weale (1963) discussed the effects of prereceptoral filters, including a slight tendency for very old eyes to need more green in the mixture. Minor effects of filters such as tinted contact lenses were evaluated by Moreland (1974), who also described how asymmetry of slit mechanisms (particularly in Nagel Model 1 instruments) gives measurable variation from expected red–green ratios, but hardly influences normal use. Moreland (1972a) has also discussed the choice of stimuli to avoid effects of ocular pigmentation on anomaloscope data, particularly in tritanomaly. In 1978 the same author showed the lenticular causes of age effects with blue + green mixtures.

Seasonal alterations can be expected since Richter (1948, 1950) gave details of normals who add more red to their Rayleigh equation in the middle of the year, with a corresponding diminution of the yellow matching field. Figure 7.42 shows the tendency, which has been confirmed by others. Schmidt (1955b) suggested how an 'average' subject can be used to eliminate the effect in calculations. Remarkably, Schuster (1890) had noted that observers who used 'Lord Rayleigh's colour box' over a period of some months, gave marked variations of the same type.

The difficulties mentioned above must be kept in perspective and do not descredit a useful tool. To conclude this section some finer points are now considered.

Normals find that R/G mixtures, either very red or very green, appear brighter than mixtures more balanced to a yellow colour; the Helmholtz–Kohlrausch effect is operating. Therefore, when a 'brightness' match is made by adjusting the intensity of the yellow half of an anomaloscope field to match a series of mixtures, there is a tendency to choose progressively higher yellow settings as the mixture tends towards either red or green. This happens when the usual criterion of 'conspicuity' is used, where the two half-fields are equated for vividness. If the alternative criterion of 'equal luminance' is used, perhaps as the result of training, Schmidt (1974) showed how there is a different performance. The conspicuity method is preferable. The phenomenon is a low order one but it may account for variations experienced, for instance, in the careful examination of heterozygotes.

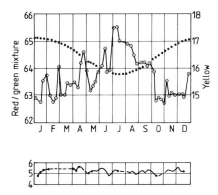

Figure 7.42 Variation of anomaloscope data according to season (after Ritcher 1950). Top graph: joined open circles represent red–green mixtures by normals; dotted curve gives normals' yellow settings. Lower graph: anomalquotient of one deuteranomalous observer.

The perfect Rayleigh match has been described by Varner *et al* (1977) in terms of equal quantum catches from each half of the test field, by each of the observer's cone photopigments. The use of counterphase sinusoidal flickering red and green (in excess of critical frequency) was shown by these authors to present cases in which a 'three-dimensional Rayleigh match space' is explored; this depends on the relationship between modulation amplitude, frequency and wavelength.

7.7.7 Blue–green equations

Anomaloscopes are not confined to the Rayleigh equation and the Schmidt and Haensch Model II, a rare device, uses alternative stimuli as described by Waaler (1969). Filter instruments have been used by Pickford and Lakowski (1960), Lakowski (1971) and Moreland and Kerr (1978). In the last anomaloscope blue and green stimuli are mixed to match cyan plus yellow and results varied according to the subject's density spectra of macular pigments. Engelking and Trendelenberg originally used the equation 470 nm + 517 nm = 490 nm but the mid-point matching made with this by normals is liable to be affected by variations in prereceptor pigments. Blue lights tend to be reduced by the crystalline lens and macular pigment.

Those who wish to investigate tritan defects, including acquired conditions may wish to employ blue–green equations. A modified tritan-omaloscope using 470 nm + 510 nm matched to 490 nm + white was used by Neuhann *et al* (1976) with confidence.

7.7.8   Substitutes for the anomaloscope

In 1974 a device was produced by Rodenstock of Munich, providing six transilluminated split fields for monocular or binocular colour vision testing. The question is 'are the two halves the same colour, or not?' It appears that most of the colours embraced the yellow to red gamut. Guilino and Wieczorek (1976) compared this with an anomaloscope. This 'Farbentestscheibe' (F173) was evaluated by Burggraf *et al* (1981) on a large number of protan and deutan subjects whose features were determined by a test battery. This report showed a high rate of success, especially in separating protanopia from deuteranopia and gave xy coordinates for the test colours used.

A simple slide rule device known as a 'colour rule', normally used to show metamerism, has possibilities for colour vision testing. Kalmus (1972a) described how this portable 'alternative to an anomaloscope' can be used with some success, although it has limitations. Matches alter in a manner to be expected when age and the yellowing of the crystalline lens are considered and tritan defects appear to be picked up. The average match, combined with 'tolerance area', provides very useful information.

The origin and design of the D and H Colour Rule are described by Kaiser and Hemmendinger (1980) and earlier Biersdorf (1977) gave data on its use as a screening test for colour vision. At the present there is little doubt that the Rule is a convenient device, somewhat uncertain with regard to deuteranomaly.

7.8*   LANTERN TESTS

A neglected but very valuable test of colour vision, the lantern requires the subject to name his colour perception. Errors display the dangers of Daltonism, not least to the onlooker, and provide an obvious fairness when standardised imitations of practical signals are used. Subtleties intrude in the form of variations of order, the presentation of pairs of lights and by false clues from brightness.

In 1877 mercantile marine candidates met colour vision tests for the first time, since Board of Trade examiners were issued with cards and glasses in six colours, for tests. A kerosene lamp with a large aperture was described in the 1912 report of a Departmental Committee, also nine coloured glasses to be named in front of this aperture both with and and without ground glass. Thus the Board of Trade's lantern eclipsed the wool test in 1912, after the earlier lamp had given way to the wools in 1894.

Non-British lanterns existed before 1900, notably Donders' model, described by Jennings (1896), using a candle behind a disc of apertures glazed red, green, blue and white, plus a slide with holes from 1 to 20 mm used at a distance of 5 metres. Jennings (1905) knew of the introduction of neutral filters with coloured lights and reported early lanterns of North America such as Williams' 1903 design and Thompson's variation of Donders', including a double aperture. Abney (1895) had his double-lantern method but as chief protagonist Edridge-Green (1891, 1933) stands supreme. His early system has persisted in an instrument still produced, originally using 13 slides which combined colours, neutrals and diffusers (figure 7.43). The contemporary reports of committees and his own forthright accounts bear witness to much controversy after which the Board of Trade adopted the method.

Figure 7.43 Edridge-Green's lantern (courtesy of Rayner Ltd, Hove). There is a single aperture and different controls for the filter discs.

In order to test Norwegian railway workers in a practical manner which appeared fair to the men involved, Schiøtz (1925) produced his own simple lantern (figure 7.44). This owed something to a Danish device by G Norrie (cited by Schiøtz) and to Edridge-Green's lantern. The effective aperture of 3 mm was modified by two pivoting arcs of filters which included red, green, orange and blue.

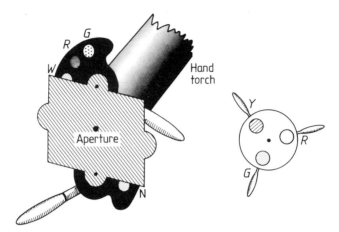

Figure 7.44 Schiøtz's lantern, from a sketch in Oslo by R Fletcher.

Military and transport users have proved the value of lantern tests but even the simplest design is capable of good results, with careful application. The test essentially identifies subjects whose colour perception, even if not completely 'normal', is acceptable since it conforms to a certain level of performance. Of course, variations from strict protocol can spoil the results but with correct administration and maintenance the test is in a special class, with the following benefits

(*a*) A 'trade test', practical and realistic, which separates those capable of reliable performance.

(*b*) An indicator of certain forms of Daltonism: protans are indicated by red light failures, anomalous trichromats show contrast effects and deutans typically confuse white with certain greens.

(*c*) A convincing demonstration of faulty naming to the normal observer, who may be a friend of the subject.

Lanterns imitate real signals in colour, size and intensity. Double lights represent port and starboard with mixed brightness clues. They deserve a place in a complete battery of tests and recent data by Paulson (1973) show how the Farnsworth lantern (Fa-Lant) salvaged some 30% of Daltonics even though they had *failed* PIC plates, showing they are as accurate as normals for service needs. Solandt and Best (1943) had reported similar success with a Canadian Navy lantern and Laxar (1967) found that none who fail D.15 pass Fa-Lant, while none who fail Fa-Lant pass stringent PIC plates. The Fa-Lant uses 18 pairs of lights and one error (or less) passes the subject although a person's performance on screening gives no direct guide to his type of defect; Farnsworth (1957) showed that three degrees of defect are easily disclosed by three tests, as in the table below.

Type of colour vision	Tests		
	Very sensitive PIC	Lantern	D.15
Normal	Pass	Pass	Pass
Mild or safe Daltonic	Fail	Pass	Pass
Medium Daltonic	Fail	Fail	Pass
Severe Daltonic	Fail	Fail	Fail

Wright's (1957b) opinion was that lanterns are important indicators of a defect's presence and the use of her Color Threshold Test (CTT) lantern enabled Sloan (1944) to show how a quantitative test can be applied.

An extensive survey of lanterns and an evelution giving special attention to the Farnsworth model was published by Cole and Vingrys (1982) in which variations between lanterns were stressed. These authors gave support to the application of a lantern to select acceptable deuteranomals.

Separation of degrees of defect by lantern has been partially successful in Japan. In 83 deutans (38 anomalous) Hukami (1967) had a third of the 'mild' cases failing more than 4 pairs of lantern colours, while among 14 subjects who failed less than 3 pairs all but one were 'mild'. A similar point was used by Majima (1972a,b) who used Ichikawa's lantern with results sometimes at variance with the consensus of three PIC tests; in 164 subjects, all Daltonic, 'mild' ones made up to 3 errors on the lantern and those likely to be moderate or strong made 4 or more errors.

7.8.1 Using a lantern

The subject's adaptation often influences results. Giles (1954) maintained that tests are more severe if given immediately room lights are put out and for several lanterns protocol includes 15 minutes dark adaptation; it is often omitted if an immediate 'pass' can be achieved without. Larger apertures seldom present difficulty and these are usually shown in sequence to acclimatise to the colour gamut involved. For Fa-Lant, prompting includes the statement: 'Only three colours appear, red, green, white; they look like distant signal lights; name them as soon as you see them', indicating how Daltonics often miscall colours when in haste.

Misnaming of red or green is a serious error, especially at large or medium apertures. Some normals cannot see the darkest smallest reds, particularly when a mirror is used to achieve the standard distance, but protanopes and many protanomalous observers often fail to identify a dark (or long wavelength) 'red'. Where yellow is used (and orange signals occur in real situations) it should not be miscalled as 'green' or 'red'.

Deutans were more often correct than protans in naming green lights, according to Neubert (1947) and Chapanis (1948), while the latter found that deutans made more red–white confusions and fewer green–white errors than protans; protans tended more frequently to confuse green with white or yellow. It must be borne in mind that many of the confusions involved simultaneous viewing of paired lights.

Anomalous trichromats frequently hesitate or vary in performance; when they are forced to give quick decisions in succession they tend to make errors. A yellow (or the somewhat warm white often used if yellow is omitted) may be named 'green' when shown after a red, or 'red' following a green. Similar 'exaggeration of colour contrast' also reveals the anomalous when suitable pairs of lights are shown and it is difficult to establish precise rules.

7.8.2 Examples

(*a*) *Subject PJW* (a very slight deuteranomal who made 10 errors on 24 Ishihara plates, did well on FM 100 Hue and had a Nagel mid-point 24 with range 7.
Giles–Archer lantern, small (1 mm) aperture gives:
Standard red (*SR*) = 'red'
Light green (*LG*) = 'yellow'; 'green' at different times
Yellow (*Y*) = 'yellow'
Dark red (*DR*) = 'red'
Standard green (*SG*) = 'red'; 'green'; 'red'; 'green'
Standard yellow (*SY*) = 'yellow' at times but
$\qquad\qquad\qquad\qquad$ = 'green' when following rapidly after *SR*
$\qquad\qquad\qquad\qquad$ = 'red' after *SG*

(*b*) *Subject NB* (heavily deuteranomalous)
Giles–Archer lantern, 1 mm aperture gives:
DR = 'red'; *Y* = 'yellow'; *SR* = 'red'
White (*W*) = 'white'; *SG* = 'green', but later 'white'
Blue–green (*BG*) = 'blue'

(*c*) *Subject DW* (heavily deuteranomalous) Nagel mid-point 36 with range 58; 9/10 deutan on CUT 1st edition
Giles–Archer lantern, 3 mm aperture gives:
SR = 'red', later 'green'; *W* = 'yellow'
DR = 'red'; *Y* = 'yellow'; *SG* = 'green'
SY = 'red orange' when following *SG*
$\quad$ = 'green' when following *SR*
LG = 'green'; *BG* = 'green'

(*d*) *Subject DRS* (PIC plates suggest slight protanomalous or possibly deuteranomalous)

Giles–Archer lantern, 5 mm aperture gives:
LG = 'red'; DR = 'nothing'; SG = 'green' or 'red'
SY = 'green' or, immediately after SG, 'red'

(*e*) *Subject FB* (deuteranomalous)
Pairs of 0.5 mm lights seen as follows:
Green + green = 'white' + 'white'
DR + green = 'red' + 'white'
White + DR = 'red' + 'white'
White + green = 'white' + 'white'
DR + red = 'green' + 'red'
Green + DR = 'white' + 'green'

(*f*) *Subject SM* (slight deuteranomalous)
22/24 Ishihara errors. Nagel mid-point 16, range 18. CUT 1/10 deutan.
Giles–Archer lantern, 5 mm aperture:
DR = 'red'; Y = 'yellow'; SR = 'red'; SY = 'green'; SG = 'green'
With 3 mm aperture:
DR = 'green or yellow'; SG = 'difficult to say'
With 1 mm aperture:
DR = 'indefinite'; SR = 'yellow', then 'red', then 'yellow'
SG = 'nothing there'
SY = 'yellow' but 'green' following SR

(*g*) *Subject DC* (protanope)
Giles–Archer lantern, 5 mm aperture:
SG = 'red' later 'green'; LG = 'yellow'; W = 'white'
DR = 'nothing' later 'red'; SR = 'red'
SY = 'red'; Y = 'yellow' later 'red'; BG = 'green'

(*h*) *Subject CGA* (protanope)
Giles–Archer lantern, 5 mm aperture:
LG = 'yellow'; W = 'green' repeatedly; DR = 'nothing' often, but sometimes 'red'; Y = 'yellow'; SR = 'green' or 'red'; SY = 'indefinite' or 'yellow'
'yellow'
SG = 'white' or 'green'; BG = 'green'

7.8.3 Colours and apertures

Enough has been said to prompt interest in extra detail without going into
all possible designs. Most lanterns concentrate on white, red and green,
using different intensities and various dominant wavelengths but usually
with regard to signals standards such as British Standard BS 1376:1974 on
colours of signal lights. Naming is often confined to a choice between

'white, red or green' but in some lamps the warm light prompts the name 'yellow' when a clear aperture is shown. Thus Martin (1939) used two different reds, one green and one white in the RN lantern, a modification of the original Board of Trade (BoT) design. The latter is an important landmark and deserves attention. Topley (1959) gave a full description in which he showed the severity of the test; 8% of 2777 men who passed Ishihara plates were failed by the BoT lantern and form vision clearly affected the results, as is usual with lanterns, since one quarter of the lantern failures had less than 6/6 visual acuity with both eyes.

The BoT design with an oil lamp was used from 1912 to the late 1970s. It showed 12 colours through a 'large' (0.2 inch, 5 mm) aperture at 20 feet, with pairs of 0.5 mm lights, 25 mm apart, being shown after 15 minutes dark adaptation. A total of 36 pairs was used, comprising colours of equal 'luminosity', four red, four green and four white. Significant errors are: red called 'green', not seen, or called 'white'; green called 'red' or 'white'; white called 'green' or 'red'.

The specification called for imitation of ships' oil navigation lights one and a half miles away on a dark clear night, a difficult combination of form and colour vision.

The extension of colours beyond the BoT has few advantages when merely screening with the lantern using an official criterion. It has been seen how yellow is used with anomalous trichromats, for instance in the Giles–Archer lantern. The Edridge-Green lantern test incorporates yellow, blue and purple, plus six apertures (1 mm to 12.7 mm); the ground, ribbed or neutral glasses give great range of presentation and the fact that three separate discs of seven colours (including two reds and two greens) can be superimposed provides almost perplexing combinations; some are necessarily 'checked by a normal person' in case of doubt.

Single apertures were found by Neubert (1947) to be less effective for detecting 'unsafe' subjects than double or even triple lights. Among 30 subjects naming single lights correctly 21 made errors with two. Neubert's opinion was that a person failing to name correctly two 5 mm lights (25 cm apart) at six metres was not 'safe'.

A deuteranope who failed Fa-Lant demonstrated, in data by Kinney (1967a), distinct 'induced' colours, notably red coloration of illuminant A of 1° field size, when surrounded by a 5° annulus comprising, in succession, a range of green (and blue) lights. Differences in saturation appearing in the spectrum beyond 550 nm were said to account for his incomplete Nagel range.

The RCN lantern (one or more errors) was the criterion of colour deficiency for Chapanis (1948) in comparing tests. Large 5 mm (0.2 inch) apertures gave only slightly better results than apertures one tenth the size (see figure 7.45).

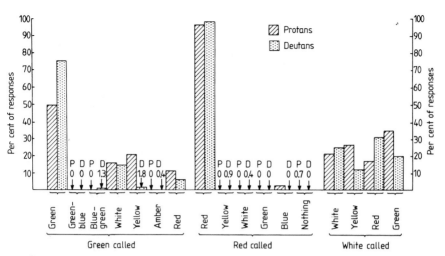

Figure 7.45 Colour names, after Chapanis (1948). Individual lights, each 0.02 inches in diameter, were named when shown in pairs with the RCN lantern.

During the Second World War, US Air Force needs for a reliable screening test, with a supplementary quantitative grading of defects, brought into being the Color-Threshold Tester (CTT) of Sloan (1944). This provided two small blue lights to control fixation, two working apertures, which subtended just under 4′ at one of two distances, and the following filters: two reds, two yellows, two greens, one blue and one white and eight ND (neutral density) filters zero to 2.1 in 0.3 steps. At low brightness blue may be confused with green by normals but most normals correctly identified all lights with the darkest setting, at which all eight colours were equally difficult. The method of penalising errors at high rather than low intensities is a noteworthy feature of the scoring, because many errors tended to appear at high values; notwithstanding this Sloan and Habel (1956) established conflicting evidence related to retinal illuminance and degrees of defect.

Table 7.2 summarises information about selected colour vision lanterns.

7.8.4   The Holmes–Wright model

The need for a modern version of the BoT lantern, to be used also by the Services, resulted in a new instrument employing a random sequence of nine pairs of colours being developed in the 1970s (see figure 7.46). Two red filters are used, with one blue–green, a yellow–green, also a white which is produced by a tungsten halogen lamp, just inside the boundaries recommended by BS 1376:1974. A yellow was considered for inclusion but eliminated. All colours have the same photopic luminance at each level; a

Table 7.2 Selected colour vision lanterns—a summary.

Lantern	Light source	Colours and intensities	Single or pair	Conditions
Edridge-Green (1891 and later)	Non-stabilised electric	$R(1)$, $R(2)$, clear, Y, G, signal Gr, blue, purple 5 neutrals + diffusers	Single. Sizes 0.04 to 0.5 inch	20′
Board of Trade (1912)	Oil lamp	4 whites, 4 reds, 4 greens each with neutral	Pair—1 inch apart 0.02 or 0.2 inch. Horizontal arrangement	15 minutes dark adaptation can be used
Martin (1939)	Stabilised electric 40 W	$R(1)$, $R(2)$, G, neutral	Pair—0.02 or 0.2 inch. Horizontal or vertical	
Giles–Archer (1934, 1954)	Non-stabilised 15 W bulb	Current model: DR, Y, SR, SY, SG, BG, LG, clear	Single—1, 3 and 5 mm. 1 mm with ND	
Canadian (1943)	Headlight bulb (aged) 32 Cp	White, G, R in 9 combinations	Horizontal pair 1 inch apart, 0.02 or 0.2 inch	Use at 20 feet. Preferably after 10 minutes dark adaptation
CTT (Sloan 1944)	Non-stabilised replaced after 6 months use	$R(1)$, $R(2)$, $Y(1)$, $Y(2)$, $G(1)$, $G(2)$, blue, white 8 intensities by ND. Range 2.1 log units in 0.3 steps	Single ⅛, ¼, 1 inch. 2 guide lights	Gradual transition to dark room. Smallest 3′40″. 5 second response
Farnsworth (1945)	Bulb	$R(1)$, $R(2)$, $G(1)$, $G(2)$, white	Vertical pairs	2 second response
Holmes Wright (c. 1975)	Stabilised bulb	$R(1)$, $R(2)$, $G(1)$, $G(2)$, white	Vertical or horizontal pair. 1.6 mm apertures	5 second response 20′

choice of three intensities is provided, the lowest giving near-threshold point brilliance at 6 m. The deep red, close to British Rail signal red, is readily misnamed by protanopes. The two greens, nearly equal in saturation, represent the different dominant wavelengths at the practical limits of the BS chromaticity boundaries. Viewing is in a lightened room (using high brightness) or in total darkness (with low brightness), following demonstration of three pairs at the very high demonstration level. A working distance of 6 m is used, including a mirror reflection if necessary and naming of each pair of lights is required within 5 seconds. These pairs may be vertical or horizontal, according to the model and there is no variation of the 1.6 mm aperture for all lights. The Services and Civil Aviation have adopted standards of colour perception using the lantern and, with horizontal lights, it is used by the Merchant Navy (see Voke 1980b)

Figure 7.46 The Holmes–Wright lantern (courtesy of MS Precision Ltd) showing the lamp removed from its normal position and the cylinder of filter holders.

The Holmes–Wright lantern is portable and has simple controls; it includes a voltage stabiliser, illuminated indicators and a long-life reading light, a great advantage compared to some previous lanterns, yet retaining representative 'field conditions' similar to the 1912 BoT model; it is not a diagnostic test. A description of their lantern was given by Holmes and Wright (1982), with comparisons of the chromaticity of colours used in this and in other lanterns. This includes a justification of the size and intensity choices.

7.8.5*   The Giles–Archer lantern (colour perception unit)

Giles (1934) designed a very simple lantern, a modification of Donders', with a mains 15 W bulb. This has been used extensively in the UK but is still not widely known. Originally six colours could be shown, at two intensities, using a 0.5 mm single aperture; a 5 mm demonstration aperture was also fitted. The latter 'aviation' model has now replaced the first model, using gelatine filters as follows: clear, dark red, yellow, standard red, standard yellow, standard green, blue green (a signal green) and light green, arranged in that order. The colours are close to recent civil aviation signals and in addition to the 5 mm hole there are 3 mm and 1 mm apertures; these smaller apertures were intended to carry either a diffuser or a neutral filter, but care should be taken to ensure that they are present (figure 7.47). Giles recognised that simultaneous contrast effects with SY might be ignored if there was no misnaming of Y, or other significant error. The convenience of the device is attractive despite its obvious limitations and the examples of subjects given above indicate the practical sensitivity which can be achieved. DR with a 1 mm size is often invisible to normals at 6 m and they may call the clear aperture 'yellow'. A record form is shown opposite.

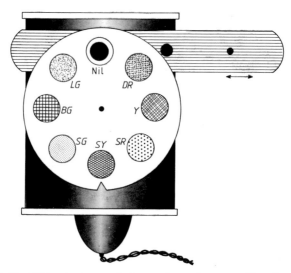

Figure 7.47 Giles–Archer aviation colour lantern. The lamp-house cylinder contains a 15 W diffused bulb and has a slide with three apertures. The disc of filters presents seven colours and a clear aperture.

G.A. Lantern   Dark adapted   YES/NO   (min)   Mirror   YES/NO

CLOCKWISE					
	5mm		3mm		1mm
W					
LG					
BG					
SG					
SY					
SR					
Y					
DR					

ANTI - CLOCKWISE					
	5mm		3mm		1mm
DR					
Y					
SR					
SY					
SG					
BG					
LG					
W					

SR → SY								
SG → SY								

Tentative opinion.

DA, PA, D, P, Normal.

7.8.6 Real signals

A look at actual signal lights, as it were 'with Daltonic eyes', enables us to see lantern tests in their true context. Many experiments have used simulated traffic lights, flashing laboratory stimuli and signal light 'guns' in addition to those practical situations mentioned elsewhere. Navigation lights present more difficulties than 'pilot lights', such as are used in cockpits (see Sloan and Habel 1955a).

Improvement of flash recognition, as duration and luminances increased, did not invariably follow with certain colours used by Heath and Schmidt (1959). Normals and colour defectives usually improved with white or red (relatively dim) light seen together with the signals but protanopes tended to perform badly. Further distinct disadvantages for protanopes were shown by Cole and Brown (1966) using 'driving conditions', but even protanopes using a long response time could see 'normal optimal luminance'. Signal light gun data for day and night conditions with assorted Daltonics proved to be equivocal and led Steen and Lewis (1972) to express the real need for a test better than 'clinical' types for predicting night-time colour ability. Allyn *et al* (1972) related diode emission functions and anomalous subjects' performance; green GaP diodes were shown to be relatively efficient for all subjects, while the GaAsP (even more than a red GaP) device gives protans difficulty. Even with normal observers there is a complex interaction between signal colour, ambience and background, as found by Reynolds *et al* (1972).

British Standards for signal lights, e.g. BS 1376:1947, BS 1376:1953 and the 1974 edition have had to recognise both the needs of colour deficient persons and the fact that most observers are normal. Thus blue elements tend to be retained in green lights required for 'high recognition' and the reduction in transmission as well as the 'reddening' of filters as they become hot has to be considered. Martin (1976) has summarised reversible and irreversible effects (figures 7.48 and 7.49).

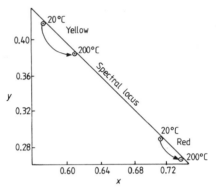

Figure 7.48 Alterations of the characteristics of signal filters, after Martin (1976). The colours are plotted on part of a CIE *xy* diagram.

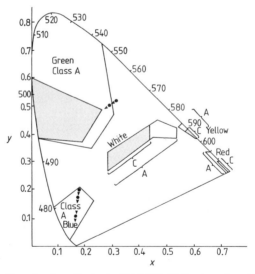

Figure 7.49 Signal colour limits to BS 1376:1974, class A, widest limits, and class C, or most strict limits for high discrimination. The chains of arrowed dots represent changes of chromaticity of two different filters when light-source colour temperature is increased from 2000 to 3000 K.

The size and intensity of red lights involved in traffic compensate for most protanopes' practical difficulties, according to Verriest *et al* (1980a,b) which reinforces the value of different stimulus variables in lantern testing, as explained in §12.2.

7.9   NEUTRAL POINTS—MEASUREMENT TECHNIQUE

If time permits extensive observations, there is value in following the methods of Massof and Bailey (1976); this colour naming method allows considerable consistency. Their observers located the position in a spectrum, provided by a monochromator, where the 'least amount of colour' was present; an average of numerous settings is required to find the 'achromatic point', in the region of 490 nm for protanopes and 495–505 nm for deuteranopes. The intensity of the stimulus appears to have little effect on such measurements. A single method is unlikely to give overlap between protanopic and deuteranopic figures but once the apparatus has been 'calibrated' the distinction should be clear. Although earlier work often involved the use of illuminant B (i.e. a source approximately 5000 K) American investigators showed that a colour temperature of 7500 K (illuminant D) provides the best results. Typical neutral points are thus, for protanopes, 490–492 nm and, for deuteranopes, 494–498 nm.

A distinct bias may result from an individual's ocular (macular, lenticular, etc) pigment; dense pigment could increase the wavelength noted and vice versa. Judd (1944) gave figures relative to illuminant B and it has been shown that aphakia can decrease the wavelength by 2 nm, indicating the magnitude of the lens pigment influence.

In figure 7.50 two Munsell papers (5G 5/8 and 5B 4/8) are 'mixed'. A $2°$ hole in a screen of N5 paper displays the coloured mixture beside a mixture of N4 and N6 as a bipartite field. The mixtures are both adjusted to suit the observer and the true coordinates of the colours involved are plotted. The points representing the mixtures on a chromaticity diagram are joined by a line and its intersection with the spectrum is located. Alternatively a simple mirror box device, after the fashion of the anomaloscope in figure 7.36, is fitted with filters. One side of the field is provided with a neutral (e.g. illuminant D) produced by a blue filter for tungsten light as described in Chapter 6. The other side of the split field is a mixture of tungsten light filtered through 'green' and 'blue–green' screens whose relative contributions can be adjusted. Suitable filters are Wratten 61 'deep green' for illuminant A ($x = 0.246$; $y = 0.699$) and Wratten 44 'light blue–green' for illuminant A ($x = 0.124$; $y = 0.413$).

The 'neutral point', as a special case of the pseudoisochromatic principle, has been considered elsewhere in the text when PIC plates were described; colours 'complementary' to the neutral points in the spectrum will also appear neutral to the dichromat.

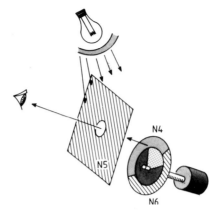

Figure 7.50 A practical 'neutral point' device, after Walls and Heath. Maxwell tops are rotated by a motor and illuminated by blue filtered light; they are seen through a hole in a grey (N5) sheet. The small top comprises 5G 5/8 and 5B 4/8; the larger comprises N4 and N6.

Saturation discrimination studies provide an additional method of estimating the neutral point (see the work of Chapanis (1944) and Cole *et al* (1965)).

7.10 ALTERNATIVE TEST METHODS

Many means of investigating Daltonism, including the usual methods, have been described, but some experienced individuals could claim that they commonly add to these, on account of interest or local conditions. There are so many methods that those now to be described comprise only a selection. In each case it should be evident how far they are valued, or untried, by the authors and the aim is to suggest sources to which the reader could turn.

Those 'clinical' or 'field' tests which are commonplace do leave room for doubt; consequently scientific needs or purposes such as genetic investigation force a choice from a larger armoury. Abney (1913) relied on his 'colour patch' and spectroscopic apparatus in his frequent examination of seamen whose colour perception presented problems. This was somewhat to the disgust of Edridge-Green (1891) who as an ophthalmologist was inclined to feel that physical scientists were 'non-professionals' in the field who could harbour certain prejudices; Edridge-Green favoured his lantern and cards as well as his spectroscope.

Sometimes devices intended for other uses have been applied, for instance the Color Aptitude Test of Dimmick (1956) whose 1964 edition consisted of 48 colour chips which have to be matched. Colorimeters will be considered, with other laboratory devices and tests of field of vision for colours which are being revived. Clarke (1968) was emphasising laboratory methods when he explained the use of wavelength discrimination to distinguish degrees of anomaly; it is weak in separating protan from deutan. He stressed how relative luminous efficiency had value when protans or rod monochromats are suspected, and indicated the use of purity studies in evaluating the extent of anomalies.

Most common tests yield different results in separate hands and under varied conditions, and this diversity can be expected more in the procedures which follow, since field size, intensity, adaptation, mode of presentation, stray light and many other factors differ. If care is used to maintain apparatus each test can be calibrated on subjects whose normality is proved by a battery of other tests. Continuity is improved by a single operator making regular settings, using his own eye as a monitor. Radiometry is required to control stimulus fluctuations from the electricity supply, from lamp aging.

7.10.1 Colorimeters

Terms such as 'trichromatism' and 'dichromatism' became popular in a context of results obtained with colorimeters using three stimuli. Maxwell (in Wilson 1855) showed with his colour top matches how the colour blind differed from normal, and the redundancy of one of the trichromatic stimuli for dichromats is such that Wright (1946) said that their matching is with two controls only. Judd (1943, 1945) noted the requirement of a dichromat to be of mixtures of only two coloured lights in order to achieve his complete gamut of colour experience; also, the fact that just as certain lights of different spectral composition can appear to match in colour for normal subjects, as 'metamers', so a dichromat usually accepts these normal matches and has additional matches, or 'confusion colours', distinguishable by normal people. Such confusions have been considered earlier in the discusion of the anomaloscope and other colour vision tests.

The anomalous trichromat does need all three tricolorimeter stimuli, but in abnormal proportions which are unacceptable to normals. Therefore it is natural that a colorimeter yields information on Daltonism. In a criticism of the 'monochromatic plus white' approach Guild (1926) described how some types of abnormal colour vision could lead to greater errors in colour measurements involving saturation determination than in tricolorimetry.

Important data, obtained with Wright's spectral colorimeter by Pitt (1935), are summarised in figures 1.42 and 3.16 which should be compared with the normal observer.

Another important study of dichromatic colour mixing, using wavelengths 458.7 nm and 570 nm, was reported by Hecht and Shlaer

(1937); this was the subject of comment by Fry (1944), including the disadvantage of altering the monochromatic test stimulus to match different mixtures. Figure 3.18 shows how a protanope was found to give variable results, particularly in regions of the spectrum where his wavelength discrimination is poor; a deuteranope's data were similar. Hecht and Shlaer pointed out how normal trichromats' colour mixture is worse where their wavelength discrimination is worse. Thus only between 480 and 520 nm are the dichromat's mixtures reliable. The mixture data, taken with V_λ for the appropriate dichromat, can predict the spectral neutral point using a method described by Fry, together with a method of differentiating between dichromats and anomalous trichromats. Dichromats would readily match illuminant C with a mixture of the 458.7 nm and 570 nm lights.

A much discussed account of a 'unilateral dichromat' by Graham *et al* (1961), included colour mixing curves resembling those described above. Most colour mixing today is with spectral stimuli but Birch (1973) used a commerical subtractive colorimeter for matches of desaturated test objects, showing matching ranges. The relative imprecision of the method revealed normals with unusually large MacAdam ellipses; deuteranomalous observers gave the expected increases in range.

Anomalous trichromats were examined by Nelson (1938) and by Wright (1946), disclosing considerable variations in trichromatic coefficients, e.g. frequent absence of negative red for protans and greater matching tolerances for yellow parts of the spectrum. Generally the 'normal match' is unacceptable to the anomalous.

7.10.2*   Wavelength discrimination

Colorimeters have often been adapted for this threshold measurement, as by Pitt (1935) or Hecht and Shlaer (1937), and figure 7.51 indicates how the present writers have combined two monochromators for their own studies. Monochromators may be replaced by interference filters which have different transmission features for different wavelengths. Those setting up apparatus would greatly benefit from reading a text by Crawford *et al* (1968), including the examples of construction.

Procedure

The subject must be comfortable, using a chin and head rest and a chair set at a convenient height. The position of his eye is critical, therefore a nose piece is the most useful simple means of avoiding sideways shift; this can be similar to the pad bridge of a spectacle frame or simply two plastic spheres slightly separated. At intervals and ideally between each 'final judgment' the subject should move bodily for a rest, taking 10 seconds look at a neutral surface about as bright as the 2° test field, as a control of adaptation.

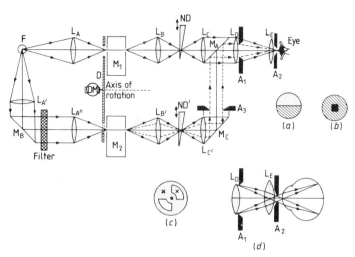

Figure 7.51 Instrument for determination of wavelength discrimination. Two channels are represented each bring rendered 'monochromatic' by a variable monochromator and having a common illuminant, a filament, F. The amount of light is varied by neutral wedges, e.g. ND, or a filter can be inserted. A disc D is rotated, when needed, by a motor DM or acts as an *open* shutter for both channels. The eye sees a Maxwellian view of the part of lens L_D exposed by aperture A_1, viewing A_1 through an artificial pupil A_2 and a lens L_E, as (*d*). By an oblique mirror M_A over the lower half of the beam between L_C and L_D, the lower half of L_D, of A_1 and thus of the field (as (*a*)) is illuminated from M_2 via the mirror M_C. Ideally the beam from L_A onto the slit of M_1 should conform both to the '*f* number' of the monochromator and to the real needs of the field subtense; thus in the figure A_1 and A_3 waste much of the light, for instance the more peripheral parts of the beam passing $L_B{}'$ and $L_C{}'$. If M_A is replaced by a semi-reflecting beam splitter covering the whole area of L_D, and A_3 is reduced to a small square aperture, a circular field of one colour can have superimposed on it a square element of another colour, as (*b*) illustrates. An emmetropic eye has its retinal image size dictated by the ray paths shown by (*d*). (*c*) Shows one form of the disc shutter D with the letters x indicating the optical axes of the two channels for a mode in which the channels can be closed alternately by rotating the disc so that one of the apertures transmits light. Note that the parallel beam reflected from M_B permits neutral or interference filters to be inserted. Stray light filters can be placed between M_1 and L_B or between M_2 and $L_B{}'$.

Procedure is similar to anomaloscope technique. Starting with both half-fields set at, for example, 500 nm, the subject equalises luminosity (manipulating the neutral wedges) and is shown a colour match; one half of the field is then varied in wavelength to show a colour difference and the subject is shown how to equalise the luminosities of the halves of the field. Continue at the 500 nm position, presenting 505, 490, 502, 495, 503,

498 nm, in turn, while the subject ensures equal brightness of the field halves, to ascertain the minimum *difference* detected. In the example, if 505, 490 and 495 nm are rejected as matches but 502, 503 and 498 nm are accepted, the increments 2, 3 and 2 are averaged to give a threshold of 2 nm.

Measurements at 500, 490, 590 and 530 nm are usually of greatest interest but in a few cases additional points may be plotted. The results will be influenced by technique, for example whether a 'match' or a 'just-noticeable-difference' is sought. The criterion should be constant. Some experimenters believe that it is fast and accurate to move from the true position of equality in steps of ± 2 nm until the subject detects a *hue* difference.

Colour defectives' data

Normal curves for wavelength discrimination for $2°$ fields of reasonably high brightness can be compared with the abnormal. There are characteristic maxima of sensitivity (i.e. small increments are detected) at about 444, 495 and 595 nm in the normal data published, for instance by Pitt (1935), McKeon and Wright (1940) and Wright (1946). A very few points of the curve, carefully chosen, give a guide as to the presence of a defect and its extent. Crone (1961) committed himself to a single point in the interest of speed; he attempted grading of the extent of anomaly by measuring $\Delta\lambda$ at 590 nm, at which the normal has a 1 nm threshold. Data at this part of the spectrum compared to other assessments by Fletcher (1976) did not strictly endorse this approach. Any single point between 570 and 620 nm might be used. Tritanopes have dips in their curves in two places (Wright 1952, Voke-Fletcher and Fletcher 1978), corresponding to neutral points (see figure 3.15). The paucity of blue light in a typical apparatus makes measurements in the short wavelengths particularly difficult in the investigation of tritanopia but a minimum is readily shown in the region of 600 nm. The need to ensure that stray light is elminated from apparatus was emphasised by Nelson (1938). Such effects can be controlled by subsidiary filters.

Hue discrimination data for non-spectral colours, give useful information. Thus Clarke (1968) suggested a simple adaptation of two FM 100 Hue tests, mounted on two wheels and presenting a split field for hue discrimination, and Krill and Schneiderman (1964) used the normal FM 100 Hue test to reveal hue discrimination defects in carriers. Nickerson and Granville (1940) gave data for 100 hues (at Munsell 5/5) and 50 hues, of varied Value and Chroma, showing differences from (spectral) wavelength discrimination, particularly near the middle of the spectrum.

Variations of conditions

Curves for normals can be altered if conditions are different. Thus Siegel and Dimmick (1962a) found that a 'constant stimulus' (rather than 'method of limits' or 'adjustment to match by subject') approach was best. Five positions on each side of the wavelength studied were used, the furthest positions appearing distinctly different in hue. Random presentation, many times, for each position gave data including standard deviations. Wright (1946) compared the smaller values of $\Delta\lambda$ obtained by adjustment to a match with the just-noticeable-difference approach, indicating that a reasonably constant difference exists in data obtained by the two methods.

Size and position of the retinal image are potent factors in altering wavelength discrimination and the 'standard' procedure with a $2°$ field of fairly high luminance, surrounded by a dark field, gives 'typical' results only when central fixation is used. Thomson and Trezona (1951) found that as luminance is lowered, with a $1°20'$ field, discrimination worsens; this is most marked between 490 and 620 nm (see also Walraven and Bouman 1966). At the periphery of the retina, both at high low luminances, Weale (1960) established thresholds of wavelength discrimination of about 60 nm in the middle of the spectrum but more like 20 nm at 450 nm.

A most practical issue, 'small field' tritan defects, has been described by a succession of writers form Charpentier (1888), as reviewed by Hartridge (1950), Wright (1946) and Willmer (1946); the defect, a physiological one, is greater at lower luminances. Figure 1.46 indicates how Thomson and Wright (1947) studied the variations of a normal trichromat's central vision. Not least among the applications of this tritan, but normal, phenomenon is the difficulty of recognising pale yellow objects on a blue ground, in the presence of white areas of similar size; dinghies for sea rescue are now orange rather than yellow. Variations in wavelength discrimination at low luminance, as reported by Brown (1951), were described by Walraven and Bouman (1966) as 'pseudo-tritanomaly' since they differed in some respects from pure tritanomaly.

Conclusion

Wavelength discrimination can be used in the study of colour vision defects, inherited or acquired, using the apparatus described above with two monochromators. Experience shows that reasonable results can be obtained even if one or both monochromators are replaced by a wedge-type interference filter which transmits different dominant wavelengths as it is moved across a narrow slit; a half bandwidth between 10 and 20 nm is satisfactory but the more intense the light is, and the narrower the slit, the narrower the waveband that can be used.

7.10.3 Luminous efficiency

Discussion on the characteristics of the relative luminous efficiency in normal trichromats, anomalous trichromats and dichromats is given in

Chapters 1 and 3. In particular the protan features should be noted. Reference should be made to techniques described in Chapter 1 and the work of Wright (1946) and Weale (1960). The value of the V_λ curve as a test method is discussed in this section. Chapter 1 includes reference to appropriate sources for V_λ measurements. Additional guidance can be found in the works of Wright (1946), Weale (1960), Le Grand (1972) and Fry (1981).

A practical method

The optical arrangement shown in figure 7.51 provides a suitable means for examining the V_λ curve although the collection of data is time-consuming. A standard light (white or monochromatic in the region 550 nm) is rapidly alternated with each of a series of spectral lights which are usually emitted from a double monochromator. At about 20 Hz flicker is noticeable and one source is adjusted in luminance (usually by adjusting neutral filters) until the flicker is minimised. Ten carefully selected points in the spectrum provide the basis for an adequate curve; foveal measurements and those involving spectral regions from green to red are of most value in a study of colour vision deficiency.

The reference data of Wright (1946) indicate the wide variations of the V_λ curve under different conditions in both colour normals and Daltonic observers. Calibration of the apparatus is required but provided the main source is maintained at a constant luminance the *relative* performances of many observers can be compared.

*Alternatives to classical techniques of measurement

Detailed measurements of the relative luminous efficiency are time-consuming. A rough estimate to indicate departures from normal, using selected wavelengths, is often adequate. An early alternative suggested by Ives in 1915 involved the determination of a yellow/blue ratio using chemical solutions placed in cells on either side of a direct comparison photometer. Using these values, through interpolation, it is possible to indicate displacements from the standard V_λ curve. A simple three-light comparison test capable of differentiating between the protanopic and deuteranopic V_λ curves has been described by Walls and Mathews (1952). Such an arrangement also indicated normal observers, as Walls and Heath 1956) later showed.

A comparatively simple and rapidly obtained 'luminosity quotient' was shown to be valuable by Crone (1959), who used only two parts of the spectrum at 530 and 650 nm.

An LED (light emitting diode) device is used by one of the present authors (figure 7.52).

The following series of four boys, each 15 years old, illustrates how this test assists the identification of a protan condition.

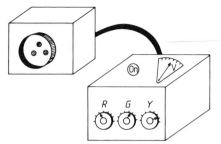

Figure 7.52 Three-light test (Fletcher) using three light emitting diodes of variable intensity. Arbitrary scales are used.

The first (SAE) was referred by a school medical officer for an opinion on his 'fairly severe colour defect'. The boy had never experienced 'real' difficulty with colours but had an interest in an Army career.

(*a*) Ishihara (10th edn) 24/24 protan errors, which could be reduced almost to no errors using a red filter.

(*b*) HRR indicated a 'mild' protan.

(*c*) CUT (2nd edn) showed 7 normal and 4 protan responses.

(*d*) The Nagel anomaloscope indicated a mid-point of 65 with a range of approximately 5, this range being more consistent with 'neutral adaptation'.

(*e*) Giles–Archer aviation lantern revealed:

Large aperture: SR = 'orange'; Y = 'green'; DR = 'red'; W = 'green'

Medium aperture: R = '??'; G = '?'; SY (after SR) = 'green'

(*f*) F_2 plate—green square only seen.

(*g*) The three-light test was used and red set either at medium or maximum intensity. Both green and yellow were set at a minimum intensity but appeared distinctly brighter than red.

The use of tests (*a*), (*c*), (*f*) and (*g*) only would have occupied little time and would have indicated a protan condition of medium severity, sufficient to present likely hazards and occupational limitations. The potential of the filter should be noted.

The second subject (FG) was referred on account of an interest in printing as a career. Mixing colours was difficult for him but as a Scout he had no trouble with maps. The three-light test (in which both red and yellow were named 'orange') indicated a protan condition.

(*a*) 24/24 Ishihara (10th edn) errors indicated protan and could be more than halved by a red filter.

(*b*) CUT (2nd edn) gave 3 normal, 4 protan and 4 deutan results.

(*c*) D.15, with many crossings, was ambivalent.

A third boy (PJW) showed normal brightness matches on the three-light test, also

(*a*) Ishihara (10th edn) 11/24 errors with deutan indications.
(*b*) CUT (2nd edn) 10/10 normal responses.
(*c*) HRR gave 20 normal responses.
(*d*) Giles–Archer aviation lantern:
Small aperture: DR = 'red'; SR = 'red'; SG = 'red'; or 'green'; SY (after R) = 'green' or (after SG) 'red'.
(*e*) Nagel anomaloscope mid-point 24, range 7.
(*f*) FM 100 Hue test gave an error score of 102.

The fourth boy (JC) had experienced some difficulties with art, for example he painted grass red. The three-light test rapidly showed a protan result, with the green light at minimum intensity brighter than the red set to medium.

(*a*) Ishihara (10th edn) 24/24 protan errors, most corrected by a red filter.
(*b*) CUT (2nd edn) 5 normal, 4 protan and 1 deutan response.
(*c*) FM 100 Hue test gave an error score of 180 with large extensions in the $R–G$ areas.
The patient uses a red filter, usually monocularly for preference, glazed in a spectacle frame. This is worn for short periods for assistance with maps, chemistry, table games etc but not for model painting.

When the anomaloscope was considered (in §7.7.2) the variation of 'red' brightness settings was mentioned since the comparison of red and yellow (or of green and yellow) is a practical possibility. This is illustrated by the following three subjects.

(*a*) A man who appeared to be markedly deuteranomalous on several tests tended to obtain a complete Nagel range throughout which he found an intangible 'less than satisfactory match'; neutral adaptation disturbed his matches more and suggested a vague 'deutan' range. He made a brightness match of 11, 13 or 15 against red (scale 73). He had no difficulty with reds in a lantern.
(*b*) A male apprentice in an electrical cable firm made extensive Ishihara errors but protan–deutan differentiation was difficult on this test; other tests (HRR and CUT) indicated a *medium* protan condition, and he made typical errors for red on a lantern. On the Nagel anomaloscope a complete range was elicited but his observational accuracy was considered to be questionable. He had no difficulty in producing these brightness matches: yellow 15 against green (scale 0) and yellow 2 against red (scale 73).
(*c*) A deuteranomalous man (range 32 with mid-point 18) gave brightness matches of yellow 14 against green (scale 0) and yellow 30 against red (scale 73). This observer was only moderately anomalous on most other tests.

The OSCAR test (Medilog, Nieuwkoop, Holland)

A relatively simple screening device for protan and deutan conditions has been produced by Estevez *et al* (1983), which uses light emitting diodes as a flickering colour mixture. The relative proportions are adjusted to a position of minimum subjective flicker, which is dictated by the presence and varieties of the retinal photopigments. Since the name is an acronym of 'objective screening of colour anomalies and reductions' but the test is subjective, the use of the word 'objective' serves to emphasise the objective interpretation of results obtained and the objective knowledge of colour vision applied to the design.

Comparisons of the prototype of the device were made using 118 subjects tested with a variety of colour vision tests. Children between 4 and 6 years of age were evaluated with success. The comparisons were generally satisfactory.

The V_λ in acquired defects

Foveal spectral sensitivity has been investigated in cases of acquired colour disturbance by a number of investigators, notably Francois and Verriest (1957a,b), Marré (1973) and Jaeger and Grützner (1963) although classificiation on this basis alone is difficult. The photopic spectral sensitivity represents a summation of the three basic colour vision mechanisms, and where one colour mechanism is selectively impaired by disease the V_λ curve reflects the reduction in sensitivity accordingly. Marré (1973) noted that in most retinal diseases the V_λ curve indicates disturbances of all three colour channels. Although there is a general reduction in all spectral regions in optic nerve diseases, a distinction can be made between one or other of the two possible sites of the deficiency, because sensitivity in the long wavelength region is affected more in retinal than in optic nerve disease. In senile nuclear cataract Verriest (1970) noted a V_λ shift towards long wavelengths, while shifts towards short wavelengths were found in retro-ocular disease and strabismic amblyopia.

Scotopic luminous efficiency

Under conditions of dark adaptation the sensitivity of the eye to brightness at different wavelengths can be explored using large stimulus fields to allow the activity of rod receptors. Peripheral vision is typically used to facilitate the measurement of rod activity. Direct comparison techniques are usually employed and are easier than for cone vision since no colour differences are present to complicate judgments. The classical curves of Wald (1945) and Crawford (1949) were incorporated into the scotopic standard young observer adopted by the CIE in 1951, although earlier notable measurements were made by König (1894). Typically a peak at around 507 nm is seen. The displacement from the photopic peak at 555 nm is the cause of the Purkinje shift which was described by the Czech physician Purkinje in 1825. When corrected for absorption characteristics of the ocular media (principally the lens, since macular pigment is not involved in peripheral measurements) the V'_λ curve is similar to the action spectrum of rhodopsin, the rod visual pigment.

7.10.4 Saturation discrimination

Hue, saturation and luminosity interact in a colour, and Hecht and Schlaer (1936) appropriately argued how a dichromat's wavelength discrimination is influenced by spectral saturation differences, particularly since the 'neutral point' is a central peak of desaturation. However some separation by experimental conditions can be used and there is a well known difficulty for anomalous trichromats, notably confusions with grey for desaturated colours. Their coloured element is badly discerned, and Chapanis (1944) readily demonstrated some basis for the quantitative scaling of different anomalies using saturation discrimination. His normal subjects easily described when two samples, unlike in brightness, were of the same hue; but his Daltonic observers, who might easily match neutral with a colour by slight brightness adjustment, did not. When two greys of slightly different brightness were compared, dichromats and some anomalous trichromats described the *darker* as 'coloured' and this effect could be reversed by brightening the darker sample. The anomalous subjects studied by Chapanis clearly differed from normal in saturation threshold, protans giving data distinctly shifted from deutans and the amount of 'decrease in saturation' appeared to be related to the degree of colour blindness; the decrease was thus seen to offer hope for a 'really quantitative' scale of defect.

Saturation thresholds are usually expressed in terms of the 'first step from white', the least amount of colour added to white to be discerned. Brightness balance must be maintained. Other steps exist, all the way to the appropriate part of the spectrum. Normal and abnormal characteristic data for photopic fields of about $2°$ have been published, notably by Martin *et al* (1933), Chapanis (1944) and Wright (1946) and are represented in figures 1.47 and 1.48.

In a study of saturation discrimination for normal and Daltonic observers Bouman and Walraven (1957) noted a minimum of 580 nm for both deuteranomalous and protanomalous observers, around a 2848 K white.

Clinical use

Both inherited and acquired defects can be studied by saturation thresholds, with greatest desaturation for Daltonics being related to dominant wavelengths around 500 nm (see figure 3.16). Cole *et al* (1965) described typical tritan characteristics.

Reciprocal threshold purity (or saturation) is commonly plotted against wavelengths as $(L_S + L_W)/L_S$ where L_S is the luminosity of the spectral colour added to a white of luminosity L_W. Onley *et al* (1963) made magnitude estimates of saturation with filtered lights; a difficult measurement. Practical shortcomings of several methods were explained by Martin *et al* (1932). It is ideal to present a standard white field, superimposing the increment of colour on part, but ensuring brightness adjustment. First match two white fields (see Chapanis 1944) then add a distinct amount of colour to one; reduce this colour until both appear white, always with brightness equalisation, and once again increase the amount of colour until threshold is reached.

Using static perimetry with a modified Tübingen perimeter Drum (1976) provided a compromise (13 mL) background luminance which adapts the rods sufficiently but allows colour thresholds to be measured. A practical method used with this device involved successive pairs of targets and the identification of colours; a slow flicker was introduced for testing based on the assumption that spectral sensitivity is related to achromatic sensitivity.

It is possible to distinguish the 'specific' threshold at which a true colour appears and a 'chromatic' threshold where the subject knows there is no match, without identification of the true colour involved, as shown by Schmidt and Bingel (1963) when investigating oxygen deficiency or alcohol effects on saturation. Using experience gained with the anomaloscope they reduced colour adaptation by a 7 s pre-exposure to white. Generally, thresholds were lower when subjects knew in advance what colour to look for, therefore this study raised several points of practical importance.

Frisen's (1973) colour saturation test, a simple 'confrontation' method, compared luminous objects in equally eccentric but 'central' areas of the visual field; this showed how desaturation in acquired defects tends to be for colours which give difficulty in inherited conditions. The test is not an exact one but has some 'clinical' value.

Marré (1979) (see Pokorny and Smith 1979) has measured saturation discrimination thresholds in acquired colour deficiency. In cases such as retinal detachment, *retinitis pigmentosa* and juvenile macular degeneration where the blue channel is predominantly affected, the short wavelength region of the function was abnormal, although a minimum around 580 nm as measured in the normal trichromat was maintained. Where all three colour mechanisms showed disturbances the form of the saturation discrimination curve was distorted compared with the normal, with lower overall sensitivity and minima similar to those of normal observers. For changes to colour vision caused by prereceptoral absorption (e.g. lens opacities) saturation discrimination is impaired throughout the spectrum, most marked effects being noticed in the short wavelength regions.

7.10.5   Objective assessment of colour vision defects

Colour vision is essentially subjective and it is natural to assess its defects by subjective methods, but objective confirmation is important. The investigation of animals, children, non-cooperative adults and malingerers is likely to be enhanced by reliable objective approaches and physiological elements may be indicated. In everyday situations, for most practical purposes, subjective means are simplest and sufficient; the complications and costs of objective tests usually outweigh the benefits. Even in infants 'preferential gaze' and other subjective possibilities compete with the need to 'wire up' the subject for electrodiagnostic approaches, but the non-contact method of optokinetic nystagmus has appeal. This section introduces the methods and gives suitable references for those wishing to use them.

Visually evoked responses

Electrophysiological recordings can indicate the potential changes that occur in the photoreceptors as a consequence of light absorption, and follow the course of neural transmission through the retina and optic nerve to the visual centres of the mid-brain and cortex.

During the past twenty years it has become possible to record gross electical activity from the brain evoked by peripheral sensory stimuli in the fully conscious subject. The visually evoked cortical potential (VECP) or visual evoked response (VER) is one of the many such which can be recorded simply with scalp electrodes. Although strictly a measure of neural function the evoked response reflects the activity of the photopic visual system. Armington (1966) used the VECP to measure photopic spectral sensitivity and showed good agreement with the CIE curve. He used stimuli in different parts of the retina; one position, the optic disc, was related particularly to the important consideration of 'stray light'. He showed how it could produce a VER, a method tending to favour 'photopic' activity, while the ERG predominantly revealed scotopic functions.

One of the disadvantages of these techniques is the strong dependence on the stimulus used and the position of the electrodes for recording; considerable intersubject differences are noted even among normal observers.

The visually evoked cortical potential has been used frequently to study both normal and anomalous human colour vision, but most studies have tended to use flashes of coloured light which have the additional difficulty of some involvement with the luminance visual channel. Riggs and Sternheim (1969) using a square-wave grating stimulus, in which the stripes were of equal brightness but different only in wavelength, showed that as the wavelength difference between adjacent stripes was increased the VECP amplitude increased to a saturation level. Such an approach provides an electrical correlate of wavelength discrimination.

Useful reviews have been provided by Kinney and McKay (1974) and Boynton (1979).

Colour differences alone will induce a VECP, suggesting that the signals from the cones are segregated at least partially before processing by contrast detectors (Regan and Spekreijse 1974). Dichromats produce no VECP to colour-contrast stimuli, indicating that the physiological signals produced by the coloured stimuli are pooled before they reach the first contrast-sensitive neurones.

There is an effective absence of one cone mechanism in dichromats (medium wavelength absence for deuteranopes and long wavelength absence for protanopes) indicated by the measured VECP of spectral sensitivity (Estevez et al 1975).

The clinical potential of the VECP for colour vision examination has been shown by White et al (1977); individual differences in protan observers could be indicated by these techniques, which were superior to the established clinical methods such as the Ishihara test and anomaloscope. Thus colour contrast VECPs do provide a sensitive objective test for the detection and diagnosis of colour vision anomalies. Shipley et al (1968) compared the time and amplitude VER characteristics of various Daltonics with their characteristics on a series of colour vision tests; among their valuable conclusions they emphasised that trained observers were needed for obtaining reliable data. Perry et al (1972) showed repeatability over three years. In addition Regan and Spekreijse (1974) presented a deuteranope with patterned red and green checkerboard objects to find alterations of response amplitudes.

VECPs allow the fundamental response functions of the colour vision mechanisms to be measured high up in the visual pathway, in both colour normals and the colour defective. The absence of one response mechanism in the protanope as shown by this method is in close agreement with psychophysical and other objective measurements. Zrenner and Kojima (1976) found close agreement of VECP measurements on protanopes with the spectral sensitivity curve of the protanope and densitometric measurements of cone pigments, and concluded that the sensitivity function of the protanope appears to be only slightly affected by neural transmission to the visual cortex.

Significant differences were revealed when Kinney and McKay (1974) involved normal trichromats, deuteranopes and protanopes, although the protanopes showed variations among themselves. One tritanope showed distinctive features. VER changes in amplitude and waveform, according to whether blue or red–green mechanisms were involved, agreed with subjective estimates of visual function in studies by Klingaman and Moskowitz-Cook (1979). Yorke (1976) published the variation of VECP components with a latency of 70 ms in deuteranomaly. In the acquired colour vision defects associated with optic neuritis Wildberger and van Lith (1976) could differentiate between the various stages with the VER.

Undoubtedly the expense and difficulties of standardisation of the apparatus for stimulation, recording and interpretation are obstacles to the wider use of the VER in the foreseeable future.

The electroretinogram

The potential difference between the cornea, with respect to the posterior of the eye, which arises after light stimulation, and measured as the electro-retinogram (ERG) since the mid-nineteenth century, has been developed for a variety of clinical uses since the 1930s. In particular it is used as an indicator of retinal function. It represents the total response of millions of retinal cells, although exactly how specific neurones contribute to the total response is still being studied.

Three characteristic elements to the waveform have been identified (figure 7.53) and examination of the specific photopic and scotopic parts has been used since Bernhard (1940) to assess cone function. Under conditions of dark adaptation the 'b' wave ERG response is equivalent to the scotopic sensitivity curve and the absorption spectrum of rhodopsin. Under light adaptation conditions sufficient to eliminate rod contribution a photopic function is seen. The three cone mechanisms can also be separated in the 'b' wave (Mehaffey and Berson 1974); the photoreceptors appear to make a definite contribution to the negative component, the 'a' wave, while the inner nuclear layer contributes to the 'b' wave positive component. All the cells in the inner nuclear layer (the horizontal bipolar, amacrine, and Müller cells) appear to contribute to the ERG, some contributing more than others. The ERG can be used as an index of function of cells only in the outer and inner nuclear layers of the retina, since the ganglion cells appear to play no part. The shape of the waveform, rather than its amplitude is of greatest clinical significance and the exact form depends on the stimulus used with wide variations being shown in the normal value even when standardised techniques are employed.

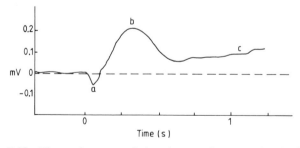

Figure 7.53 Three elements of the electroretinogram (ERG). Different waves and combinations of rod and cone responses, have been selected to convey how a typical recording appears. The magnitudes of reactions and times involved vary according to the stimulus conditions and the condition of the eye.

Armington (1952) and many others have met some success in relating colour vision to the ERG, despite difficulties such as the involvement of rod action and the need for careful control of stimuli. Thus a component of the ERG was demonstrated by Armington which gave maximum response to light of 630 nm wavelength, separate from either photopic or scotopic functions and not only absent in protanopes but also present in other types of Daltonism. Other effects attributable to abnormal colour mechanisms were found when Copenhaver and Gunkel (1959), using a flicker method and interference filters, used the ERG as an objective relative luminous efficiency measure; they supported the view that red—green defects are retinal in origin and showed striking variations, chiefly for protans. Deuteranopes also revealed reductions in sensitivity but the area of retina involved influences the results. Retinal stimulation under such circumstances is complicated by stray light.

The 'b' curves of the ERG follow differences in retinal function and Francois *et al* (1960) deserve credit for demonstrating unusual components and 'culmination time' in protans, with 'typical' achromats giving variations. Most of the 39 achromats investigated by Auerbach (1974) had poor or missing photopic elements in the ERG. Krill (1964) demonstrated the value of using flickering stimuli.

An atypical monochromat with deutan sons were the subjects of retinal function studies, involving the ERG, by Krill and Schneiderman (1966); the monochromat's ERG was normal but different sons gave different results. ERG abnormalities in both protanopes and protanomalous observers and monochromats have been shown by Ponte and Anastasi (1978) but deutans tend to show a normal response.

Sachs (1929) had first shown that the ERG of protanopes is reduced in red light. The 'c' response is not present in the cone-dominated retina.

The early receptor potential is believed to originate from the outer segments of the photoreceptors, being an electrical manifestation of the bleaching process (Brown and Murakami 1964).

ERG changes in carriers of blue cone monochromacy (incomplete achromatopsia where the blue cone mechanism is believed to remain functioning) which might assist in their identification, have been reported by Berson *et al* (1982).

Electrophysiological studies are standard procedure for diagnostic purposes in eye clinics where there is special interest in the causes and locations of retinal malfunctions but it is difficult to imagine that routine applications to inherited forms of Daltonism will become widespread, although the potential for studying acquired defects is considerable.

Optokinetic nystagmus (OKN)

Train passengers move their eyes in jerky attempts to fixate objects moving rapidly into and from their view. Both eyes slowly follow the object,

moving quickly back again, seeking a new fixation. This happens when
there are several changes of stimulus each second and the OKN movements
can be detected by an observer, photographed or displayed by the electro-
oculograph (EOG). The objects must be 'resolved' and will not produce
OKN unless they are large enough to be seen. Black and white patterns or
stripes are sufficient for estimation of visual acuity and these were replaced,
by Moreland and colleagues, with colours which contrast only to normal
subjects. Thus Moreland and Smith (1974), and Moreland *et al* (1976) used
a well controlled, but simple, device to induce eye movements whenever
coloured stripes could be differentiated; this detected deutans successfuly.

The method requires coloured papers or other stimuli carefully equated
to give equal brightness to the appropriate subject and, since it can be
applied even with the use of a telescope to observe the eye, is worthy of
attention by those concerned with examining children. Early studies using
the EOG were carried out by Imaizumi (1966). Figures 7.54 and 7.55 show
how electrodes are placed for recording the EOG and ERG. The electro-
oculograph results in part from the pigment epithelial layer of the retina and
the action of cells in the outer and inner nuclear layers (Arden and Kelsey
1962).

Figure 7.54   Electrodes in the horizontal meridian lateral to the eye, to
record horizontal movements.

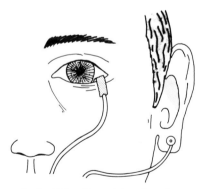

Figure 7.55   Electrode (gold leaf) position for the ERG with the secondary electrode on the ear.

Other methods

It does little justice to some of the alternative objective ways of approaching defective colour vision, to relegate them to this small space; some are elegant but they are such as would not often be attempted.

Pigments in the retinal receptors and their variations have been studied objectively in the living eye by measuring light after it emerges from the eye, having been absorbed selectively. Weale (1959) showed that the fovea of a cone monochromat contained normal cone pigments, and Rushton (1963a) applied his retinal densitometry to the fovea of a protanope; this person was shown by Rushton to lack the pigment erythrolabe while possessing chlorolabe. Rushton made similar measurements in 1965 (Rushton 1965a) to determine a deuteranope's foveal pigment.

The fundus reflection of protanomalous subjects was compared to that of normals by Alpern and Torii (1968c) in an attempt to estimate possible prereceptor filter effects. Finding that in both normal and protanomalous retinae there was no difference in reflection, for red, at the fovea and periphery, they concluded that such a filter is not present. They suggested that the proposed natural filter would make the macula much darker than it is actually seen to be if a red filter is used in an ophthalmoscope.

Damage to selected cones in animal eyes has revealed many features which fit the psychosensory data. The method, as developed by Sperling (1980) really induces, at least temporarily, a variety of acquired colour vision defects but is not for use on patients.

Pupil contraction, when the retina is stimulated by different coloured lights, has been of limited use with Daltonic subjects. Glansolm *et al* (1974) and Hedin and Glansolm (1976) distinguished some variations and Cohen and Saini (1978) showed up the relative insensitivity of a protanope using controlled adaptation and the pupil response.

7.10.6 Colour naming methods

The visible spectral colours look characteristically different to the various types of colour defectives. The relation between colour naming and hue discrimination has been indicated in §7.10.2. By masking narrow regions in turn and confining the subject to five basic colour names it is possible to locate preferred names in different parts of the spectrum. A number of investigators have used this method, as simple apparatus and procedures are involved (see Schiebner and Boynton 1968 and Kalmus and Case 1972). Ingling *et al* (1970) used the method to show small field effects, including foveal tritan tendencies, and Smith (1973) studied tritan observers by this method.

Boynton's colour naming technique (Boynton and Gordon 1965), has been used reliably in colour normals and inherited colour defects (Scheibner and Boynton 1968, Smith *et al* 1973b). Acquired defects were examined by Lanthony (1980). Monochromatic sources are ideal for such investigations although Voke (1976a) used the stable colours provided by Lovibond glass, and surface colours in the form of the coloured bands painted on colour-coded resistors; with glass samples, white was misnamed on 33% of occasions by deuteranopic observers and 40% of occasions by deuteranomalous observers. Some of Voke's results are presented in tables 7.3, 7.4 and 7.5. Typically, desaturated colours cause greater problems to the colour anomalous.

Rubin (1961) presented 'yellow' parts of the spectrum, asking subjects to identify its pure (non-orange, non-green) location; the positions of blue, green, cyan and orange were also found using normals and anomalous trichromats. All colours except blue had mean positions displaced (relative to normal) to longer wavelengths for deutans and to shorter ones for protanomals. Alexander *et al* (1977) concentrated on the 'unique green' of 24 deuteranomalous trichromats and found that 'medium or severe' subjects tended to a location between 500 and 520 nm, while 'mild' observers (fewer in number) tended to move to a position greater than 525 nm; their normals had a distribution (with bimodality for males) around 525 nm.

The unique green approach used by Kalmus (1972b) indicated overlapping results between protans and deutans; dichromats' neutral zones were at shorter wavelengths than unique greens in trichromats (forced choice method) and in some cases desaturation of parts of the spectrum was found.

Table 7.3 Colour naming using the Lovibond colour vision analyser. Eight deuteranopic observers. All figures are expressed as a percentage of possible namings (Voke 1976a).

Number of presentations	Colours as presented	Red	Orange	Yellow	Green–yellow	Green	Green–blue	Blue	Mauve–blue	Mauve	White	Brown	Black	Grey–green
32	Red	62	3			22	3	6			3			
32	Orange	25	12	25	3	22		3				6		3
24	Yellow		33	16		33						16		
8	Green–yellow		37		13	37					13			
16	Green	6	6			62						25		
24	Green–blue					67	8	13		13				
32	Blue					6	3	66		25				
40	Mauve	3			5	5		55		33				
8	White					50		13			37			

Table 7.4 Colour naming using the Lovibond colour vision analyser. Fourteen deuteranomalous observers. All figures are expressed as a percentage of possible namings (Voke 1976a).

		Colours as named												
Number of presentations	Colours as presented	Red	Orange	Yellow	Green–yellow	Green	Green–blue	Blue	Mauve–blue	Mauve	White	Brown	Black	Grey–green	Red–green
60	Red	70	7			10				2	2	4			2
60	Orange	10	27	27		15						13			
45	Yellow		6	10	10	57					2	9			
15	Green–yellow	10		10		80									
30	Green	4	6			86	4								
45	Green–blue	2				71	6			2	7				
60	Blue					6	7	74		8	2				
75	Mauve	1				1		32	3	55	1	1			
15	White	6				33					60				

Table 7.5 Colour naming using the Lovibond colour vision analyser. Three protanomalous observers. All figures are expressed as a percentage of possible namings (Voke 1976a).

		Colours as named											
Number of presentation	Colours as presented	Red	Orange	Yellow	Green–yellow	Green	Blue	Mauve–blue	Mauve	White	Brown	Black	Grey–green
12	Red	75									25		
12	Orange	8	8	25		8					25		
9	Yellow		11	33		56							
3	Green–yellow			33		66							
6	Green			17		83							
9	Green–blue					100							
12	Blue					33	66						
15	Mauve						66		26				
3	White					33				66			

Test conditions, for example the direction of changes in wavelength, can influence results significantly. When subjects view a sequence increasing in wavelength towards the probable position of 'unique hue' they choose a shorter wavelength position, and vice versa. The neutral zones of some deuteranopes depended on order of presentation (Kalmus 1979). Random presentation is most suitable.

7.10.7 Maxwell's spot

The 'dark spot' which Maxwell (1855b) observed most clearly in the blue part of the spectrum was also seen by means of a slowly rotating disc carrying 'sectors of ultramarine and chrome yellow in the proportion of 3 to 1'. Abney (1895) demonstrated the pigment of the yellow spot by looking at a bright white cloud through a layer of chrome alum which 'transmits red and blue–green rays'.

As one of the classical entoptic phenomena, Maxwell's spot is usually demonstrated with a screen such as an x-ray or transparency 'viewing box' (see Miles 1948). The field is viewed alternately through a purple–blue filter and either a neutral or a yellow filter, and a dark spot is seen when purple light is reaching the macula, being absorbed by the yellow prereceptor pigment. Considerable variations exist between observers so that Miles, Isobe (1955), Spencer (1967) and others have described fading and various forms of an outer 'spot' about $3°$ in subtense and pink or red in normal subjects; inside is often a small central spot (about $33'$ in diameter) with a surrounding intermediate ring (see figure 7.56).

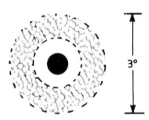

Figure 7.56 Maxwell's spot (after Miles 1954). A central dark spot is surrounded by a pink lighter ring, outside which is a darker purple desaturated annulus subtending some $2.5°$, with a less distinct outer rim.

Walls and Mathews (1952) gave a very detailed description of the phenomenon, including work by many authors since Maxwell, and on the basis of observations on assorted subjects built up a 'receptor distribution pattern' idea of the fovea. They were challenged by Murray (1954) but somewhat supported by Judd (1953) on a theoretical basis. The RDP test, as it became known, indicated that few deuteranomals and very few deteranopes saw Maxwell's spot although Miles (1954) had notice that a bluer 'neutral' phase increased the sighting rate. Walls and Heath (1956) discovered that deuteranopes usually saw a spot with illuminant D but not with illuminant A. Protans tended to see spots as 'dark' or 'blue' with illuminant A. Among the 13 protanopes examined by Walls and Heath 12 saw the 'dark' spot on a bright blue ground under suitable conditions, while most of their 15 deuteranopes saw none. Verriest (1960a) reported results consistent with those of Walls and Mathews, adding that the 'spot' was absent or weak in achromats.

The RDP suggestion may be regarded as now withdrawn and the use of Maxwell's spot in Daltonism remains associated with macular pigment variations (see Bone and Sparrock 1971). Before including the method in any series of tests it is essential to standardise one's own conditions of illumination and filter presentation; it is useful to be able to alter the colour of the 'neutral'. A series of subjects of known characteristics must then be used to establish norms.

A Wratten 34 or Lee 126 (mauve) filter can be used for Maxwell's spot (figure 7.57) since the filters described in the literature are difficult to imitate. Filters described by Spencer (1967) can be used for adaptation to coloured lights, a method which may be profitable in deutans.

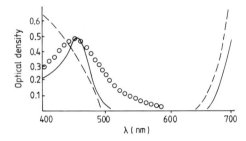

Figure 7.57 Macular pigment densities (circles), compared to transmission of a purple filter used by Miles and Walls (see Judd 1953). The broken line shows the approximate transmission curve of a Wratten 34 filter.

7.10.8 HTRF

A heterochromatic threshold reduction factor, referred to as HTRF, was devised by Boynton and Wagner (1961) for the classification of protans and deutans. It separated normals and 'defectives', went some way towards quantitative assessment of the extent of anomaly and distinguished protans from deutans The method used slide projectors, filters and small motors rotating sector discs, by which a background colour received an increment of another colour for about 50 ms. In some ways (e.g. the continuous conditioning field) this approach resembles Stiles's increment threshold technique; protans showed differences when red and blue lights were paired.

Undeniably, such innovative tests are used only by a few enthusiasts despite the expectation expressed by Boynton (1979) that they could provide superior information. Certainly there is value in a diversity of approach when confidence may be shaken by the fact that well established tests sometimes contradict each other.

7.10.9 Miscellaneous methods of examination

The essential approaches to the examination of the colour sense have been covered in this chapter. The remaining methods are seldom used in a clinical capacity but are outlined here for completeness. Motokawa's phosphene method for obtaining spectral response curves, outlined by Motokawa and Isobe (1955), is not widely used, but provides different data for dichromats. Coloured light pre-exposure of the dark-adapted retina is followed by DC electrical stimulation of the eye region and the study of the phosphenes induced. The method is slow and can be unpleasant for the nervous subject.

The 'hue cancellation' technique proposed by Jameson and Hurvich (1955) was used on anomalous trichromats by Romeskie (1978a). This involves a multiple-channel apparatus in which light of one colour is used to cancel the responses of a particular colour system at different wavelengths and the results gave a correlation with variations in colour discrimination.

A 'rainbow disc' of coloured papers, rotated one way or another about 5 Hz, in work by Aspinall et al (1978), revealed different responses according to normality of colour vision; conditions were critical and the temporal responses of different cones were shown to vary.

Stiles–Crawford effects

The retinal directional sensitivity, Stiles–Crawford effect I describes the variation of sensation as light (dimming) as a small pencil is moved from the centre of the pupil towards the periphery (Stiles and Crawford 1933a). This has been applied to colour vision by measurements in dichromats (Starr 1977) and has been a valuable tool in the investigation of achromatopsia (Alpern et al 1960) and acquired cone dystrophies (Smith et al 1978b) and as an approach to analysis of receptor status (Fankhauser et al 1961), since the effect indicates the physical characteristics of the photoreceptors including photopigment aspects (see §1.21 and figure 1.49).

The small subjective change in colour that is noticed when a monochromatic beam is incident at different angles on the same point of the foveal retina, first described by Stiles (1937) and known as the Stiles–Crawford effect II, shows characteristic changes in colour deficiency (Walraven and Leebeck 1962b, Voke 1974). In the colour normal the hue appears to become redder as the point of entry moves towards the periphery of the eye pupil for wavelengths greater than 560 nm and below 500 nm. Between these wavelengths the hue appears bluer (see Enoch and Stiles 1961). Saturation changes are also observed. Considerable precision is required in positioning the entrance beam. Anomalous trichromats, especially deuteranomalous observers, show larger hue shifts (Voke (1974) as in figure 1.49). Central retinal cones which receive coloured lights at oblique directions from the pupil margins show an effect resembling protanomaly in normal trichromats, as shown by Enoch and Stiles (1961). Such peripheral entry points were also used by Alpern and Torii (1968a), who found no evidence that protanomaly is associated with foveal cone misalignment. A different behaviour of cone pigment according to direction of stimulation appears to be likely.

The critical frequency of flicker has been used to establish Stiles–Crawford effects in protanopes and deuteranopes; the blue-sensitive cones could be removed from consideration by using frequencies over 20 Hz, prompted by work such as that of Green (1969). No selective alignments of dichromats' cones were found. Some disorientation of foveal cones must be involved in certain disturbances of the retina which produce protan tendencies, as noted by Birch *et al* (1980) and by comparison with Stiles–Crawford effects.

Temporal sensitivity curves in Daltonism

Varied temporal investigations have found value in studying anomalous colour vision. A rod monochromat examined by van der Tweel and Sperkreijse (1973) had a normal de Lange curve with low frequency attenuation which could not be due in any way to interactions among mechanisms. Temporal sensitivity measurements on protanopes were made by Estevez and Cavonius (1976) and a later study combined both protanopes and deuteranopes (Cavonius and Estevez 1976). The sensitivity values for the long wavelength mechanism, unaffected in each case, correspond to those of the similar mechanism in colour normal observers. Cicerone and Green (1978) found that the protanope does not show an abnormal sensitivity to green flicker. Evidence suggests that the temporal integrating properties of the three cone systems are different (see for example the work by Krauskopf and Mollon (1971)).

It is thought that flicker detection measurements may be useful in identifying carriers of X chromosomal recessive blue-cone (incomplete) achromatopsia (Starr and Fishman 1982). In one case studied a significant reduction in sensitivity for flicker detection at high frequencies beyond $10°$ from fixation was noted in the carrier, which closely resembled the results from her affected son. Analysis of the spectral and spatial properties of the simple opponent $R-G$ receptive field shows that at low spatial frequencies it sends a difference signal, subtracting, say, R centre cone signals from G surround cone signals. At high spatial frequencies it sends a sum signal, adding centre and surround (Ingling and Martinez 1982). It is believed that low spatial frequency red and green gratings (up to 1 Hz/degree) are determined by the opponent colour system at threshold (Mullen 1982).

Increment thresholds in colour vision defects

Inherited defects. Early use of the technique in a clinical capacity involved standard optical components and was confined to research laboratories. Boynton and Wagner (1961) made the first measurements in groups of Daltonics and found the two-colour method to be reliable for a quantiative and qualitative diagnosis as well as an indication of colour weak observers. No normal observers were misclassified. Wald and Brown (1965) noted that deutan observers showed an abnormal green response, protans showed an abnormal red response and tritans an abnormal blue response, but could not distinguish reliably between the sub-groups of anomalous trichromacy and dichromacy, a point confirmed by Marré (1969). Significant distortion of the green π_4 was noted by Watkins (1969a) for deutans, with changes to most other mechanisms, possibly as a consequence. The π_4 mechanism is characteristically closer to the π_5 (red) function in deuteranopes than in normals but it is noteworthy that the so-called green mechanism π_4 is not always absent. Wider variations than are typically found among normal observers were noted for the deuteranomalous subjects.

Suggestion of a possible fusion of π_4 and π_5 mechanisms was indicated in protanomalous observers by the same author, with a marked loss in sensitivity compared with the normal response of the red mechanism (π_5) at long wavelengths. The secondary lobe of the blue π_2 mechanism is frequently almost completely absent in protanopes and severely reduced in protanomalous observers, as shown by Watkins (1969b). Essentially there can be considered a functional loss of the π_5 mechanism in protans, with some residual activity remaining. Furthermore there is evidence that more than one pigment may be involved as changes to the blue curves representing π_1, π_2 and π_3 are often shown in the red–green defects. Cole and Watkins (1967) noted blue insensitivity in tritan defects, as shown by changes to the π_1, π_2 and π_3 mechanisms. Sperling (1973) confirmed that for anomalous trichromats the main deviations from normal are 'a shape and peak factor'. Thus in deuteranomaly the anomalous green function shows a displaced peak wavelength but retains its green cone shape; in protanomaly the anomalous red sensitivity function lies at a displaced wavelength but retains its normal red cone shape. These findings permit some postulations concerning the character of anomalous cone pigments to be made (see Chapter 1) although such associations must be cautiously interpreted, since the physiological basis of the π mechanisms has never been fully established.

A relatively simple type of tritan increment threshold test has been used by van Norren and Went (1981). Used in a mass screening survey, it appeared to be reliable on retesting subjects. In some 1500 subjects three 'tritan' subjects were found, although two protans were also discovered with the test as well as some possible pathological subjects.

Use as a clinical method in acquired defects. Marré (1969, 1971), using an adaptation of Wald's technique involving less intense adaptation fields, first applied the two-colour threshold technique to diseased eyes. A clear two-stage involvement of the colour vision mechanisms was noted in retinal disease; a blue impairment is typically followed by a disturbance of all three channels, the red and green becoming involved at approximately the same time and with similar degree. Even when visual acuity remains good, Marré frequently measured an impairment of the blue mechanism as the first symptom of optic nerve disease. Eventually both the blue and green channels become affected, unlike the inherited defects. In the most common pattern, reduced sensitivity of all three mechanisms (frequently with differences between the red and green) is shown; such is the pattern in optic neuritis and optic atrophy. Disturbances of the green channel seem to be noted earlier and often give rise to more severe impairment than the red.

Practical considerations. Reliable measurements of this type involve careful consideration of the design features of the parameters in use. In particular:

(*a*) A high background luminance sufficient to ensure photopic vision even in the periphery, and to ensure that the selective chromatic adaptation provided by the bleaching out of two colour mechanisms is as complete as possible.

(*b*) The relative radiances of test objects and background field must be calibrated.

(*c*) The use of narrow-band interference filters for monochromatic sources tends to reduce the intensity.

A $2°$ test field and about $10°$ surround minimum is required. Static or kinetic modes can be used. In order to overcome the local adaptation effect (Troxler's effect) particularly in the periphery, which results in fading of the stimulus, it is usual to provide a flashing test target.

Experiments using a modified Goldmann perimeter to meet the IPS Perimetric standards of 1978 have been performed. A 75 W xenon arc lamp giving 8000 $cd\,m^{-2}$ for the target luminance which approximates to standard illuminant C and a 150 W xenon arc lamp providing around 250–300 $cd\,m^{-2}$ for the background source are used. Stimulus durations of between 50 and 1000 ms are possible and preadaptation of 25 minutes in darkness followed by four minutes at 10 $cd\,m^{-2}$ and 250 $cd\,m^{-2}$ is recommended. Carta *et al* (1980) suggest it is possible to use the two-colour threshold method even with moderate luminance. Typical colour combinations which allow analysis of the individual π mechanisms are:

480 nm on 650 nm background to elicit π_4, π_1, π_3

520 nm on 650 nm background to elicit π_4 and π_4'

650 nm on 650 nm background to elicit π_5 and π_5'

(see Vola *et al* 1980).

Wratten filters (Kodak) numbers W16 Yellow/Orange, W44A Blue/ Green, and W35 Purple provide the background field. When the increment threshold for *monochromatic* test stimuli of varying wavelengths is found response curves peaking at around 440 nm (blue), 540 nm (green) and 570 nm (red) result. With increasing eccentricity the sensitivity of the blue mechanism increases, while that of the red and green decreases, the red being more decreased than the green mechanism (Marré and Marré 1972).

Moreland *et al* (1976) used glass filters—Schott BG12 1 mm for a blue background and OG2 3 mm for the orange background. Considerable energy is needed to stimulate the red mechanism at long wavelengths.

7.11   VISUAL FIELD INVESTIGATIONS

The recent extension of conventional perimetric methods with two-colour increment thresholds has just been described and in Chapter 1 there was consideration of the normal coloured fields of vision. It is now appropriate to deal with the more clinical conventional use of the perimeter and scotometer to investigate abnormal colour vision, although it must be stressed that the apparatus used must be modified or specially constructed for most techniques.

7.11.1   Perimetric investigation of inherited defects

Hansen (1979) noted the weaker response functions, compared to normal, among the colour anomalous, but could not distinguish between dichromats and anomalous trichromats with the two-colour technique. Carta *et al* (1980) examined five protanopes and eight deuteranopes and could not locate the π_4 mechanism in deuteranopes or the π_5 in protanopes. Differentiation between these two forms of dichromats was possible since, with blue adaptation, the sensitivity of π_1 (blue mechanism) was higher for protanopes than deuteranopes and for an orange background the relative sensitivity of the π_1 mechanism was reversed for the two groups (see figures 7.58 and 7.59).

7.11.2   Other applications of colour perimetry

Section 4.10 deals with acquired defects and their investigation by field studies. The possibility of detecting heterozygotes (carrier women) for inherited dichromatism has been explored successfully using a $1°$ test field located up to $10°$ in the visual field. This example deserves further research study.

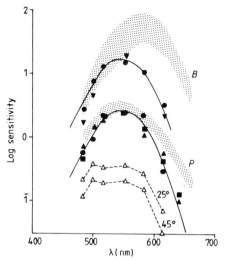

Figure 7.58 Spectral sensitivity of protan observers. Normal variations (mean, plus or minus 1 SD shown shaded, for blue and purple backgrounds. Full curves indicate the 'green primary'. Blue background values are displaced 1 log unit upwards; data for 45° peripheral position displaced $\frac{1}{4}$ log unit downwards (courtesy of E Hansen).

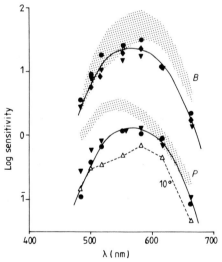

Figure 7.59 Spectral sensitivity of deutan observers. Normal data (mean plus or minus 1 SD) shaded. Blue and purple background values are shown with those for blue displaced 1 log unit upwards (courtesy of E Hansen).

8 Interpretation of records

8.1* INTRODUCTION

Experience is a major factor in the interpretation of colour vision tests and the provision of occupational advice, but the degree of confidence involved will nevertheless vary. It may be noticed that initial timidity and uncertainty in diagnosis soon give way to very distinct views of interpretation after many different degrees and varieties of subject are seen. As more evidence accumulates (often in the form of difficulties reported by subjects) one appreciates alternatives, particularly with regard to advice; this may lead to a review of some earlier decisions. Clearly a policy of constant reappraisal is desirable.

The choice of the particular series of tests applied to study the colour vision of one individual may be dictated by circumstances. At times the arrival of a new test may cause it to be included. The needs of the patient or industry should to some extent dictate what is used, as a test closely simulating a specific job is often suitable for inclusion. Factors such as age and intelligence may be important and the time available frequently constrains the variety of tests possible.

In this chapter a series of carefully selected observers is presented, some showing mixed features. Brief comments on each will sometimes explain the choice of methods. Reading between the lines gives insight into the real practise of colour vision testing. Wherever there is an apparent lack of key information we are reminded of the need for planning ahead, adequate time and the presence of a suitable range of equipment, i.e. an ideal situation; when the subject has left it is often too late for second thoughts. The desired 'perfection' is, unfortunately, seldom achieved.

8.2* SUBJECT A

Typical of those schoolboys who need guidance on which careers are open or closed, on account of a recently discovered slight anomaly, this fourteen year old arrived accompanied by his parents. His mother's father, a medical doctor, sent him for an opinion and his mother recounted the difficulties of one of her brothers and her maternal grandfather. Absolutely no idea of his deuteranomaly had been present until the recent test at school.

Commencing with Ishihara, and followed by CUT, five tests were applied. The desaturated D.15 test confirmed the plates 8 and 10 on CUT and indicated that the many deutan errors with Ishihara should not be taken as an indication of EDA (extreme deuteranomaly). The anomaloscope range of 11, from a mid-point of 25, included some ambiguities which were not worth following in greater detail. Practical performance with a series of 'post-office' wires was good and reassuring to the parents. Consequently an opinion that he could proceed to his intended scientific career, with caution, could be given in the context of a full discussion. Data were obtained in 40 minutes and the parents were invited to write about any queries which arose after they returned home.

COLOUR VISION/DALTONISM RECORD for: ..A......................

Male/~~Female~~ AGE 14

Address: .SUSSEX. ENGLAND...............................

Tel. No: .... Occupation: .SCHOOLBOY...........

History: .FOUND RECENTLY. AT. SCHOOL. MEDICAL. EXAM........

Examined by ...RF.....................

on ..1979..................

Referred by .Medical. practitioner...

~~or~~ ..Maternal. grandfather

'R-G'

Brothers normal

Colour difficulties: NONE. PARENTS HAVE NOT NOTICED ANY REQUIRES GUIDANCE re CAREER IN SCIENCE

since ....

NOTE: Records normally coded R.E. Red, L.E. Blue, B.E. Black Green

DIAGNOSES

ISHIHARA:	H.R.R.:	8/2	D.15: DA Just
T.C.U.: DA (Just)	ANOMALOSCOPE: DA 25MP 11R	LANTERN:	Pass/Fail
100 HUE:	HUE DISCRIMINATION:	REL.LUMINOUS EFFICIENCY CURVE:	
COLOUR MATCHING:	TINTOMETER TEST:		

OTHERS: Farnsworth test — sees one green square Wires etc. accurate

OVERALL DIAGNOSIS SLIGHT DA Unlikely to be a bar to science

ISHIHARA..... 10th ....EDITION, ILLUMINATION. Day . 11am .

TENTATIVE OPINION.......... DA ? EDA ....

12	8	6	29	57	5	3	15	74	2	6	97	45
12	3	5	70	25	2	5	17	21	–	–	25	–

5	7	16	73	–	–	–	–	26	42	35	96	
8	2	28	–	5	2	45	–	2	4	3	9	

ANOMALOSCOPE   NAGEL/~~OTHER~~

Subject's   R 1)   2)   3)   MEAN
first 3
matches   (L)1) 31   2) 16   3) 20   MEAN 33

NEUTRAL PRE-EXPOSURE FOR EACH DECISION?   (YES)/NO

SETTING	1	2	3	4	5	6	7	8	9	10	11	12	13	14	15	16	17	18	19
R																			
L										X								X	✓

SETTING	20	21	22	23	24	25	26	27	28	29	30	31	32	33	34	35	36	37	38
R																			
L	✓✓	X	✓		X	✓				✓✓	✓	✓	X		X				

SETTING	39	40	41	42	43	44	45	46	47	48	49	50	51	52	53	54	55	56	57
R																			
L		X					X												

SETTING	58	59	60	61	62	63	64	65	66	67	68	69	70	71	72	73	74	75	
R																			
L																			

RANGE   R   L   MIDPOINT   R   L

ANOMALQUOTIENT   R   L

LOWER DIAL AT 0 BRIGHTNESS MATCH   R   L

LOWER DIAL AT 75 BRIGHTNESS MATCH   R   L

Tentative opinion ....

D.15 TEST   ILLUMINATION...Day...........

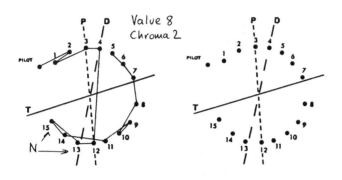

Value 8
Chroma 2

Tentative opinion ..DA......................................
....

THE CITY UNIVERSITY COLOUR VISION TEST

Illumination ('Daylight') Type ...Day.... Level.......

	(PAGE)	SUBJECT'S CHOICE OF MATCH		NORMAL	DIAGNOSIS		
		R | L | BE			PROTAN	DEUTAN	TRITAN
'CHROMA FOUR'	1	B	Ⓑ ▽	R	L	T
	2	R	Ⓡ ◇	B	L	T
	3	L	Ⓛ ◁	R	T	B
	4	R	Ⓡ ◇	L	B	T
	5	L	Ⓛ ◇	T	B	R
	6	B	Ⓑ ▽	L	T	R
'CHROMA TWO'	7	L	Ⓛ ◇	T	R	B
	8	B	R ◇	L	Ⓑ	T
	9	B	Ⓑ ▽	L	T	R
	10	L	T ◁	B	Ⓛ	R
SCORE	At Chroma Four	6/6	/6	/6	/6
	At Chroma Two	2/4	/4	2/4	/4
	Overall	8/10	/10	2/10	/10

Probable type of Daltonism:
P, PA, EPA, D, ⓄA, EDA, Tritan, Mixed

8.3* SUBJECT B

An older boy seeking a career in electronics was referred by the school medical officer. He admitted a typical 'red chalk' difficulty, suggesting a protan condition, which emerged from six selected tests.

A red filter improved his Ishihara performance and the D.15 and CUT passes indicated that his deficiency was minor. The four 'differentiation' plates of Ishihara's test gave the unexpected impression of a deutan, soon dispersed by a well displaced protanomalous range on the Nagel anomaloscope, which was obtained quickly with about 20 settings.

The lantern confirmed the protan defect and the Igaku-Shoin plates showed the mild nature of the anomoly.

He was reassured, but advised of possible dangers and recommended to use a red filter for carefully selected purposes.

COLOUR VISION/DALTONISM RECORD for: *B* ....

Male/~~Female~~   AGE *18*

Address: *BERKSHIRE. ENGLAND* ....

Tel. No: .... Occupation: *SCHOOL* ....

History: *OBJECTIVE — ELECTRONICS WORK* ....

Examined by ....

on .....*1979*................

Referred by *SCHOOL MEDICAL SERVICE*

of ....

Colour difficulties: *NONE — even with HI-FI hobby, except red chalk under poor lighting*

Hobby electronics
since *— no trouble* ...

NOTE: Records normally coded R.E. Red, L.E. Blue, B.E. Black Green

DIAGNOSES

ISHIHARA: *19/24 ?*   H.R.R.:   D.15: *N*

T.C.U.: *'Normal'* ANOMALOSCOPE: *PA*   LANTERN: *? PA* Pass/Fail

100 HUE: HUE DISCRIMINATION: REL.LUMINOUS EFFICIENCY
                                       CURVE:

COLOUR MATCHING: TINTOMETER TEST:

OTHERS: *IGAKU-SHOIN :— 2 screening errors only (7 and 13)*

OVERALL DIAGNOSIS *Slight PA. To use filter for selected tasks*

ANOMALOSCOPE   (NAGEL)/~~OTHER~~

Subject's   (R) 1) 46   2) 57   3) 49   MEAN 54
first 3
matches   L 1)   2)   3)   MEAN

NEUTRAL PRE-EXPOSURE FOR EACH DECISION?   YES/NO MOST

HAS BEST RESULTS IF HE
CONTROLS BOTH KNOBS

SETTING	1	2	3	4	5	6	7	8	9	10	11	12	13	14	15	16	17	18	19
R																			
L																			
SETTING	20	21	22	23	24	25	26	27	28	29	30	31	32	33	34	35	36	37	38
R																			
L																			
SETTING	39	40	41	42	43	44	45	46	47	48	49	50	51	52	53	54	55	56	57
R						X						X							
L																			
SETTING	58	59	60	61	62	63	64	65	66	67	68	69	70	71	72	73	74	75	
R	X	√	√?		√		√	√	√√	X			X						
L																			

RANGE   R 8 L   MIDPOINT   R 63 L   PA
ANOMALQUOTIENT   R   L
LOWER DIAL AT 0 BRIGHTNESS MATCH   R   L
LOWER DIAL AT 75 BRIGHTNESS MATCH   R   L
Tentative opinion ....PA...............

G.A. Lantern   Dark adapted YES/(NO) (min)   Mirror (YES)/NO

CLOCKWISE								
	5mm			3mm			1mm	
W							W	
LG							G	
BG							G	
SG							R	
SY							Y	G
SR							R	
Y							Y	
DR							NIL	R	NIL

ANTI - CLOCKWISE								
	5mm			3mm			1mm	
DR								
Y								
SR								
SY								
SG								
BG								
LG								
W								

SR → SY	G							
SG → SY								

Tentative opinion.

(DA, PA,) D, P, Normal.

THE CITY UNIVERSITY COLOUR VISION TEST

Illumination ('Daylight') Type ...Day.... Level.200.lux

	(PAGE)	SUBJECT'S CHOICE OF MATCH R ┊ L ┊ BE	NORMAL	DIAGNOSIS		
				PROTAN	DEUTAN	TRITAN
'CHROMA FOUR'	1		Ⓑ ▽	R	L	T
	2		Ⓡ ▷	B	L	T
	3		Ⓛ ◁	R	T	B
	4		Ⓡ ▷	L	B	T
	5		Ⓛ ◁	T	B	R
	6		Ⓑ ▽	L	T	R
'CHROMA TWO'	7		Ⓛ ◁	T	R	B
	8		Ⓡ ▷	L	B	T
	9		Ⓑ ▽	L	T	R
	10		Ⓣ △	B	L	R
SCORE	At Chroma Four	/6	/6	/6	/6
	At Chroma Two	/4	/4	/4	/4
	Overall	10/10	/10	/10	/10

Probable type of Daltonism:

P, PA, EPA, D, DA, EDA, Tritan, Mixed   'Pass'

D.15 TEST   ILLUMINATION. Day. 200.lux..

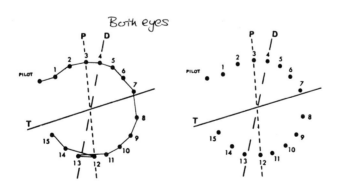

Both eyes

Tentative opinion ....
....

ISHIHARA.....10th....EDITION,   ILLUMINATION. Day...2 pm
TENTATIVE OPINION....19/24. Errors....?DA..........

12	8	6	29	57	5	3	15	74	2	6	97	45
12	8	5	20	35	2	5	17	21	—	—	87	—
12	8	6	29	57	8	3	15	74	—	—	—	—

Both eyes (row 1), Filter 164 (row 2)

5	7	16	73	-	-	-	-	26	42	35	96	
—	—	10	18	5	2	15	—	26	42	35	96	
5	7	16	73	—	—	—	—	28	42			

BE (row 2), 164 (row 3)

?DA on last 4 plates

8.4 SUBJECT C

On the reasonable assumption that she was unlikely to have had an inherited condition of the blue colour sense, this patient's mainly tritan 'errors' may be seen as acquired. She was known to have rapidly developing multiple sclerosis but reasonable visual acuity.

The CUT 'desaturated' plates showed a variety of errors and 100 Hue results, plotted on the Chromops device, gave an 'anarchic' display. The remaining data are self explanatory and the Ardern grating tests showed a variation between the two eyes of a monochrome sine wave grating discrimination loss.

Note that the 100 Hue test was used with the right eye only but other tests were used binocularly, chiefly in an attempt to minimise the time and thus the patient's fatigue.

COLOUR VISION/DALTONISM RECORD for: ...C..........................

~~Male~~/Female   AGE 28

Address: .....LONDON...........................

Tel. No: .... Occupation: .Science..............

History: .MULTIPLE SCLEROSIS......................
No Daltonism in family

Examined by ....

on ......1981...............

Irish descent   Referred by ...Clinic.............

of ....

Colour difficulties: Hardly any. Colour 'taken for granted'
'Likes' blue and browns

since ....

NOTE: Records normally coded R.E. Red, L.E. Blue, B.E. Black Green

DIAGNOSES

ISHIHARA: 3 errors   H.R.R.: 2 errors   D.15:

T.C.U.: DESTAT. MIXED   ANOMALOSCOPE: LANTERN: Pass/Fail

100 HUE: ANAR.   HUE DISCRIMINATION: REL.LUMINOUS EFFICIENCY CURVE:

COLOUR MATCHING: TINTOMETER TEST:

OTHERS: VELHAGEN — 12 errors   MOSAICS — TRITAN
         IGAKU — 6 errors
         3 LIGHT - brightness anomalies
      Arden — BE = 3:15; 4:18
OVERALL DIAGNOSIS RE = 3:17 others nil } High ~~~
                   LE = 3:19 others nil }   loss

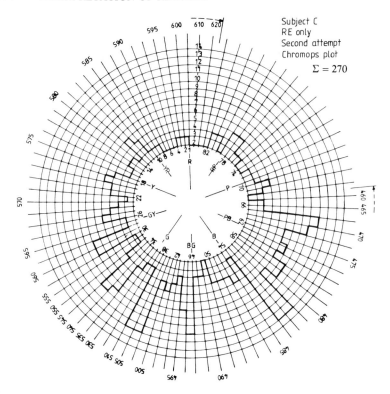

Subject C
RE only
Second attempt
Chromops plot

$\Sigma = 270$

THE CITY UNIVERSITY COLOUR VISION TEST

Illumination ('Daylight') Type .Day...... Level Average

	(PAGE)	SUBJECT'S CHOICE OF MATCH R | L | BE	NORMAL	DIAGNOSIS		
				PROTAN	DEUTAN	TRITAN
'CHROMA FOUR'	1	B	Ⓑ ▽	R	L	T
	2	R	Ⓡ ◊	B	L	T
	3	L	Ⓛ ◊	R	T	B
	4	R	Ⓡ ◊	L	B	T
	5	L	Ⓛ ◊	T	B	R
	6	B	Ⓑ ▽	L	T	R
'CHROMA TWO'	7	LTR	Ⓛ ◊	Ⓣ	Ⓡ	B
	8	LR	Ⓡ ◊	Ⓛ	B	T
	9	RB	Ⓑ ▽	L	T	Ⓡ
	10	TLB	Ⓣ ◁	Ⓑ	Ⓛ	R

SCORE	At Chroma Four	/6	/6	/6	/6
	At Chroma Two	/4	3/4	2/4	1/4
	Overall	/10	/10	/10	/10

Probable type of Daltonism:

P, PA, EPA, D, DA, EDA, Tritan, (Mixed)

ISHIHARA.....10th....EDITION,   ILLUMINATION..Day......
TENTATIVE OPINION...?..3/24 errors

BE	12	8	6	29	57	5	3	15	74	2	6	97	45
	12	8	6	28	57	3	3	15	71	2	6	—	45
	5	7	16	73	-	-	-	-	26	42	35	96	
	5	7	16	13	-	-	-	-	26	42	35	96	

AO H.R.R. ..2....Edition.   Illumination...Day.........

(FIRST 4 PLATES, DEMONSTRATION, DO NOT SCORE)
Ask 'How many symbols?', then 'What?', then 'Where?'.
If no errors 1-6 screening (Errors 1 or 2, go on to 17-20; errors
3-6, go on to 7-16).

SCREENING SERIES   DIAGNOSTIC SERIES

B-Y			R-G				7	8	9	10	11	12	13	14	15	16		17	18	19	20
1	2	3	4	5	6	Protan	0		Δ	0	X	Δ	0	Δ	X	0	Trit	Δ	X	0	Δ
X 0		Δ 0	Δ X	Δ 0	0	X Deutan	Δ	X		X	0	0	Δ	X	0	Δ	"Tet"	X	0	Δ	X
R						R											R				
L						L											L				
BE X 0		Δ 0	-	-	0	X BE	0 Δ	X		Δ X	0 0	X	0 Δ	Δ 0	X 0	Δ 0	BE	Δ X	X 0	0 Δ	

↑   ↑   ←Mild   Defects↔Medium ↔Strong→   Medium↔Strong→

Errors 1 or 2, go on to 17-20
Errors 3 or 6, go on to 7-16

Tentative opinion   TYPE:....R.G.?.........................
                    EXTENT:.....Mild.........................

8.5   SUBJECT D

This precocious young man had a well 'documented' family of Daltonics to report, plus his earlier rendering of grass in red paint. The Local Health Authority medical officer was puzzled about his test results and the boy presented unusual insight and maturity. At the same time he was not an easy person to test, volunteering some information which could be confusing to the unwary.

He appeared to be strongly protanomalous and the subsequent prescription of plastics dyed corrective lenses produced a mixed response. Generally he found benefit for selected tasks; the filters made red 'lighter' and blue darker but did not greatly aid television viewing. Map reading was improved and his 'wargames' pieces could be identified more easily, although he found that the tint reduced the brightness of paints when model making. When one lens fell out of his spectacles, he found the monocular filter useful in binocular vision. He 'enjoyed' the extra attention and comments aroused by his red filters.

COLOUR VISION/DALTONISM RECORD for: ..D..........................

Male/~~Female~~ AGE IS

Address: ....LONDON................................

Tel. No: .... Occupation: ..School..............

History: ..PASSED SCREENING - PRIMARY SCHOOL.. Tested himself in
....Science Museum

Examined by ....

on ....

Referred by ..Local Health Authority..
since 'odd answers on
~~of~~ ....test'

Colour difficulties: With art. 'red grass'

since ...?...............

NOTE: Records normally coded R.E. Red, L.E. Blue, B.E. Black Green

DIAGNOSES

ISHIHARA: 24 errors PA   H.R.R.:   D.15:

T.C.U.: PA   ANOMALOSCOPE: LANTERN: Pass/Fail

100 HUE: ? PA DA ?   HUE DISCRIMINATION: REL. LUMINOUS EFFICIENCY
         E180   ~~CURVE:~~ 3 lights distinct
COLOUR MATCHING: TINTOMETER TEST: PROTAN.

OTHERS: *To try red tint   (Mother indicates
                                   as protan carrier)

OVERALL DIAGNOSIS Medium to heavy PA

   *3 months trial - used for confusing colours
                     red/brown
                  - make modern art 'different'
                  - indicators in chem. some better
                     some harder
                  - maps improved
        Prefers one tint, one clear

THE CITY UNIVERSITY COLOUR VISION TEST

Illumination ('Daylight') Type .... Level.......

	(PAGE)	SUBJECT'S CHOICE OF MATCH R | L | BE	NORMAL	DIAGNOSIS PROTAN	DEUTAN	TRITAN
'CHROMA FOUR'	1	B	(B) ▷	R	L	T
	2	B	R ◊	(B)	L	T
	3	R	L ◊	(R)	T	B
	4	R	(R) ◊	L	B	T
	5	L	(L) ◊	T	B	R
	6	LT	B ▽	(L)	(T)	R
'CHROMA TWO'	7	L	(L) ◊	T	R	B
	8	L	R ◊	(L)	B	T
	9	B	(B) ▽	L	T	R
	10	L	T △	B	(L)	R

SCORE		NORMAL	PROTAN	DEUTAN	TRITAN
	At Chroma Four	/6	/6	/6	/6
	At Chroma Two	/4	/4	/4	/4
	Overall	5/10	4/10	1/10	/10

Probable type of Daltonism:

P, (PA, EPA,) D, DA, EDA, Tritan, Mixed

ISHIHARA.....10.th...EDITION, ILLUMINATION...Day......
TENTATIVE OPINION.......PA................................

	12	8	6	29	57	5	3	15	74	2	6	97	45
BE	12	3	5	70	35	2	5	17	21	−	−	−	−

Filter 164		8	6	29	57	5	3/6	15	74	−	−	112	45
	5	7	16	73	−	−	−		26	42	35	96	

BE	−	−	−	−	5	8	45	78	6	2	5	?8
164	5	7	16	73	−	−	−	−	26	40?	30	96

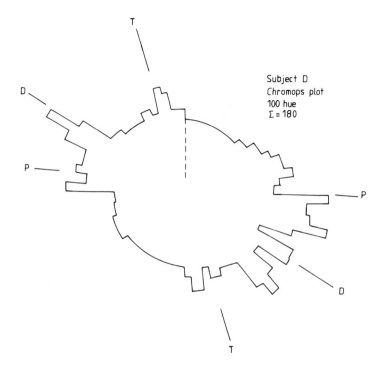

Subject D
Chromops plot
100 hue
$\Sigma = 180$

8.6   SUBJECT E

A small boy of seven, given a limited series of tests was sufficiently identified as slightly deuteranomalous by his Ishihara errors and largely normal CUT performance. The parents' query could be dealt with simply and the provision of a magenta filter for discretionary use was much appreciated.

His intelligent mother found benefit from a discussion of occupations likely to be barred to him and the fact that his anomaly need not be considered to be extreme.

COLOUR VISION/DALTONISM RECORD for: ..E...................

Male/~~Female~~   AGE 7

Address: ...MIDDLESEX...................

Tel. No: .... Occupation: ..SCHOOL...............

History: ....

Examined by ....

on ....

Referred by ...Optometrist...........

of ....

V. YOUNG
? □

Colour difficulties: Picked up at school with Ishihara
Parent - 'is it significant and how?'

since ....

NOTE: Records normally coded R.E. Red, L.E. Blue, B.E. Black Green

DIAGNOSES

ISHIHARA: DA   H.R.R.:   D.15:

T.C.U.: Slight DA   ANOMALOSCOPE: LANTERN: Pass/Fail

100 HUE: HUE DISCRIMINATION: REL.LUMINOUS EFFICIENCY
                                           CURVE:

COLOUR MATCHING: TINTOMETER TEST:

OTHERS: Child mosaics = Daltonic (R-G)
             113 filter provided for use
             at parents' discretion — advised parents
             to have other son examined

OVERALL DIAGNOSIS   SLIGHT DA.

THE CITY UNIVERSITY COLOUR VISION TEST

Illumination ('Daylight') Type .... Level.......

	(PAGE)	SUBJECT'S CHOICE OF MATCH R | L | BE	NORMAL	DIAGNOSIS PROTAN	DEUTAN	TRITAN
'CHROMA FOUR'	1	B	(B) ▽	R	L	T
	2	R	(R) ◊	B	L	T
	3	T	L ◊	R	(T)	B
	4	R	(R) ◊	L	B	T
	5	L	(L) ◊	T	B	R
	6	B	(B) ▽	L	T	R
'CHROMA TWO'	7	L	(L) ◊	T	R	B
	8	B	R ◊	L	(B)	T
	9	B	(B) ▽	L	T	R
	10	T	(T) △	B	L	R

SCORE		NORMAL	PROTAN	DEUTAN	TRITAN
	At Chroma Four	≤5/6	/6	1/6	/6
	At Chroma Two	3/4	/4	1/4	/4
	Overall	/10	/10	2/10	/10

Probable type of Daltonism:

P, PA, EPA, D, (DA) EDA, Tritan, Mixed

ISHIHARA....10th.....EDITION,   ILLUMINATION..Office....

TENTATIVE OPINION.....................................

12	8	6	29	57	5	3	15	74	2	6	97	45
	3	5	70	35	2	5	17	21	—	—	—	—
5	7	16	73	—	—	—	—	26	42	35	96	
—	—	—	—	5	8	45	70	2	4	3	9	

Lee 113 Improves

8.7   SUBJECT F

This protanope has overcome his colour vision defect and left eye ambly-opia sufficiently for many aspects of his career in electrical engineering. Both anomaloscope and hue discrimination confirm his condition. The 100 Hue result shown is of strong protan tendency, though not as distinct as might be expected. His lantern performance with red lights should be noted. HRR plates show a 'strong protan' tendency as does Ishihara, followed by CUT although this record is not shown. The differences between the two eyes with Ishihara's test are to be expected, as it is very sensitive.

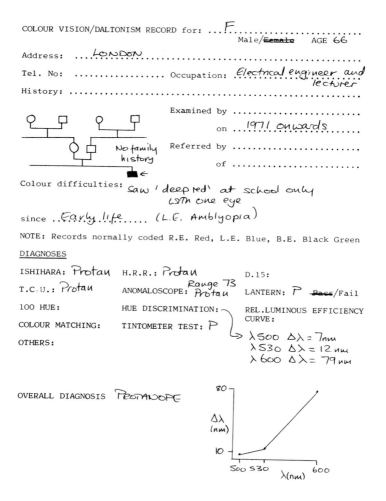

COLOUR VISION/DALTONISM RECORD for: ...*F*............................

Male/~~Female~~   AGE 66

Address: ...*LONDON*....................................

Tel. No: .... Occupation: *Electrical engineer and lecturer*

History: ....

Examined by ....

on ...*1971 onwards*......

No family history

Referred by ....

of ....

Colour difficulties: *Saw 'deep red' at school only with one eye*

since ..*Early life*..... (*L.E. Amblyopia*)

NOTE: Records normally coded R.E. Red, L.E. Blue, B.E. Black Green

DIAGNOSES

ISHIHARA: *Protan*   H.R.R.: *Protan*   D.15:

T.C.U.: *Protan*   ANOMALOSCOPE: *Rouge 73 Protan*   LANTERN: *P* ~~Pass~~/Fail

100 HUE: HUE DISCRIMINATION: REL.LUMINOUS EFFICIENCY CURVE:

COLOUR MATCHING: TINTOMETER TEST: *P*

OTHERS: ↳ *λ500 Δλ = 7nm*
λ530 Δλ = 12 nm
λ600 Δλ = 79 nm

OVERALL DIAGNOSIS *PROTANOPE*

ANOMALOSCOPE   NAGEL/~~OTHER~~

Subject's   R 1) 28   2) 66   3)   MEAN
first 3
matches   L 1)   2)   3)   MEAN

NEUTRAL PRE-EXPOSURE FOR EACH DECISION?   ~~YES~~/NO

SETTING	1	2	3	4	5	6	7	8	9	10	11	12	13	14	15	16	17	18	19
R	✓	✓	✓	✓	✓	✓	✓	✓	✓	✓	✓	✓							
L																			

SETTING	20	21	22	23	24	25	26	27	28	29	30	31	32	33	34	35	36	37	38
R																			
L																			

SETTING	39	40	41	42	43	44	45	46	47	48	49	50	51	52	53	54	55	56	57
R																			
L																			

SETTING	58	59	60	61	62	63	64	65	66	67	68	69	70	71	72	73	74	75	
R												✓	✓	✓	✓	✓	✓	✓	
L																			

RANGE   R 73 L   MIDPOINT   R   L
ANOMALQUOTIENT   R   L
LOWER DIAL AT 0 BRIGHTNESS MATCH   R 35 L 34
LOWER DIAL AT 75 BRIGHTNESS MATCH   R 2 L 2.5
Tentative opinion ....P...

TINTOMETER (LOVIBOND) COLOUR VISION ANALYSER

Quantitative diagnosis. After determining type of deficiency, the
figure for the range (extent of deficiency) is derived from the
   SUM OF THE TWO HIGHEST THRESHOLD SATURATION SETTINGS
(on opposite sides of the circle) each minus 10 (the normal threshold).

~~RD/LD~~/BOTH

Filters chosen N	14	1	2					
Threshold saturation chosen	TOTAL	TOTAL	TOTAL					
Neutral setting	0.7	1.0	1.1					

TYPE............ RANGE.........
Tentative opinion ....

Actual (random) order of colours
N, 20, 9, 13, 24, 4, 1, 14, 19, 16, 21, 11, 8, 17, 23, 2, 18,
12, 15, 6, 25, 10, 7, 22, 3, 5, 26, (then N again)

G.A. Lantern   Dark adapted YES/NO (min) Mirror YES/NO

CLOCKWISE								
	ALL 5mm			3mm		1mm		
W	G	G	G	G				
LG	Y	Y	Y					
BG	G	G						
SG	W	G						
SY	?	Y	?					
SR	G	R						
Y	Y	Y	Y	Y				
DR	—	—	R					

ANTI - CLOCKWISE								
	5mm			3mm		1mm		
DR								
Y								
SR								
SY								
SG								
BG								
LG								
W								

SR → SY								
SG → SY								

Tentative opinion.

DA, PA, D, (P,) Normal.

ISHIHARA....10th.....EDITION, ILLUMINATION..MacBeth..

TENTATIVE OPINION.....P................................

12	8	6	29	57	5	3	15	74	2	6	97	45
RE	12	3	5	20	55	—	5	17	—	—	—	—	—
LE	12	3	5	10	6-	—	5	17	—	—	—	—	—

	5	7	16	73	-	-	-	-	26	42	35	96
RE	—	—	—	—	5	2	—	—	-6	-2	-5	-6
LE	—	—	—	—	—	—	—	—	-0	-2	-6	-6

D.15 TEST

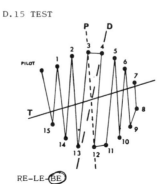

RE–LE–(BE)

AO H.R.R. ...2....Edition.   Illumination....MacBETH....

(FIRST 4 PLATES, DEMONSTRATION, DO NOT SCORE)

Ask 'How many symbols?', then 'What?', then 'Where?'.

If no errors 1-6 screening (Errors 1 or 2, go on to 17-20; errors 3-6, go on to 7-16).

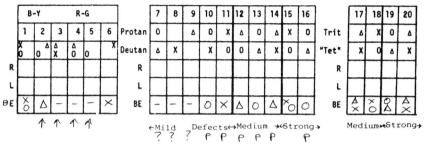

SCREENING SERIES

	B-Y			R-G		
	1	2	3	4	5	6
Protan	X O	O	Δ X	Δ O	O	X
R						
L						
BE	X O	Δ	−	−	−	×

↑ ↑ ↑ ↑

DIAGNOSTIC SERIES

	7	8	9	10	11	12	13	14	15	16
Protan	O	Δ	O	X	Δ	O	Δ	X	O	
Deutan	Δ	X		X	O	O	Δ	X	O	Δ
R										
L										
BE	−	−	−	O	X	Δ	O	Δ	X O	O

←Mild   Defects↔Medium →←Strong→
? ? ?   P P P P   P

	17	18	19	20
Trit	Δ	X	O	Δ
"Tet"	X	O	Δ	X
R				
L				
BE	Δ ×	× O	O Δ	Δ ×

Medium←Strong→

Errors 1 or 2, go on to 17-20
Errors 3 or 6, go on to 7-16

Tentative opinion   TYPE:....PROTAN.........................

                    EXTENT:..................................

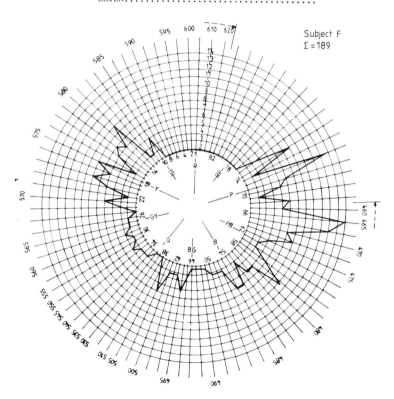

Subject F
Σ = 189

8.8 SUBJECT G

This eighteen year old woman was only slightly deuteranomalous but she experienced difficulties, probably on account of a careful approach to colours, in everyday life. The Ishihara errors she made should be compared with those of other subjects. Other tests showed that her anomaly is slight, while the rapid use of the Ishihara test on its own could be misleading.

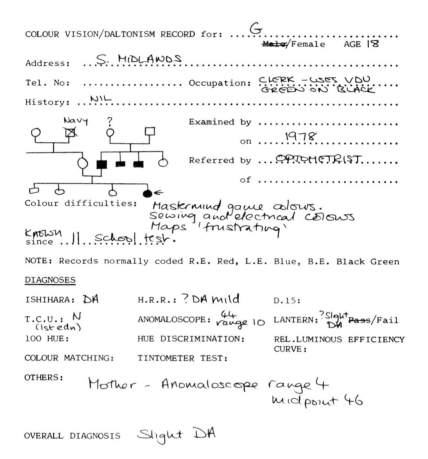

COLOUR VISION/DALTONISM RECORD for: .....G............

~~Male~~/Female   AGE 18

Address: ..S. MIDLANDS...............

Tel. No: .... Occupation: CLERK – USES VDU
GREEN ON BLACK

History: ..NIL........

Examined by ....

on ....1978.......

Referred by ...OPTOMETRIST.......

of ....

Colour difficulties: Mastermind game colours.
Sewing and electrical colours
Maps 'frustrating'

known
since ..II...School test.

NOTE: Records normally coded R.E. Red, L.E. Blue, B.E. Black Green

DIAGNOSES

ISHIHARA: DA   H.R.R.: ?DA mild   D.15:

T.C.U.: N   ANOMALOSCOPE: range 10 (44)   LANTERN: ?Slight DA ~~Pass~~/Fail
(1st edn)

100 HUE: HUE DISCRIMINATION: REL.LUMINOUS EFFICIENCY
CURVE:

COLOUR MATCHING: TINTOMETER TEST:

OTHERS: Mother – Anomaloscope range 4
midpoint 46

OVERALL DIAGNOSIS Slight DA

SUBJECT G

ANOMALOSCOPE   NAGEL/OTHER

Subject's   R 1) *52*   2) *42*   3) *46*   MEAN *48*
first 3
matches   L 1)   2)   3)   MEAN

NEUTRAL PRE-EXPOSURE FOR EACH DECISION?   YES/~~NO~~

SETTING	1	2	3	4	5	6	7	8	9	10	11	12	13	14	15	16	17	18	19
R																			
L																			
SETTING	20	21	22	23	24	25	26	27	28	29	30	31	32	33	34	35	36	37	38
R																X			X
L																			
SETTING	39	40	41	42	43	44	45	46	47	48	49	50	51	52	53	54	55	56	57
R		✓		✓					✓	✓		X							
L																			
SETTING	58	59	60	61	62	63	64	65	66	67	68	69	70	71	72	73	74	75	
R																			
L																			

RANGE   R *10* L   MIDPOINT   R *44* L
ANOMALQUOTIENT   R   L
LOWER DIAL AT 0 BRIGHTNESS MATCH   R   L
LOWER DIAL AT 75 BRIGHTNESS MATCH   R   L
Tentative opinion ....

AO H.R.R. ....Edition.   Illumination.................

(FIRST 4 PLATES, DEMONSTRATION, DO NOT SCORE)
Ask 'How many symbols?', then 'What?', then 'Where?'.
If no errors 1-6 screening (Errors 1 or 2, go on to 17-20; errors
3-6, go on to 7-16).

SCREENING SERIES   DIAGNOSTIC SERIES

B-Y			R-G				7	8	9	10	11	12	13	14	15	16		17	18	19	20
1	2	3	4	5	6	Protan	0		Δ	0	X	Δ	0	Δ	X	0	Trit	Δ	X	0	Δ
X 0	0	Δ X	Δ 0	0	X	Deutan	Δ	X		X	0	0	Δ	X	0	Δ	"Tet"	X	0	Δ	X
R	X 0	0 Δ	−	Δ	0	X	R	Δ	X	Δ	O X	X O	O Δ	Δ O	Δ X	X O	Δ O	R				
L							L											L				
BE							BE											BE				

↑ ↑   (←Mild) Defects↔Medium →←Strong→   Medium→←Strong→
                     Λ

Errors 1 or 2, go on to 17-20
Errors 3 or 6, go on to 7-16

Tentative opinion   TYPE:.....Mild DA...................
                    EXTENT:.................................

G.A. Lantern   Dark adapted   YES(NO) (min)   Mirror (YES)/NO

CLOCKWISE							
		5mm		3mm		1mm	
W			−			Y	
LG						G	
BG							
SG						R	W
SY						G	
SR						R	
Y						Y	
DR						R	

ANTI - CLOCKWISE							
		5mm		3mm		1mm	
DR							
Y							
SR							
SY							
SG							
BG							
LG							
W							

SR → SY									
SG → SY									

Tentative opinion.
(DA) PA, D, P, Normal.

ISHIHARA....10th.....EDITION, ILLUMINATION. MacBeth

TENTATIVE OPINION....Mild DA........................

	12	8	6	29	57	5	3	15	74	2	6	97	45
BE	12	3	5	70	35	2	5	17	21	-	-	-	-

5	7	16	73	-	-	-	-	26	42	35	96
-	-	-	—	5	8	45	-	2-	49	33	9-

Deut figs less distinct

8.9   SUBJECT H

The large range on the anomaloscope tends to upset the general impression that this subject is only slightly deutan. He could identify coloured wires well but the lantern was the most convincing test used; despite this he found it difficult to accept that he would really be handicapped if he tried flying duties.

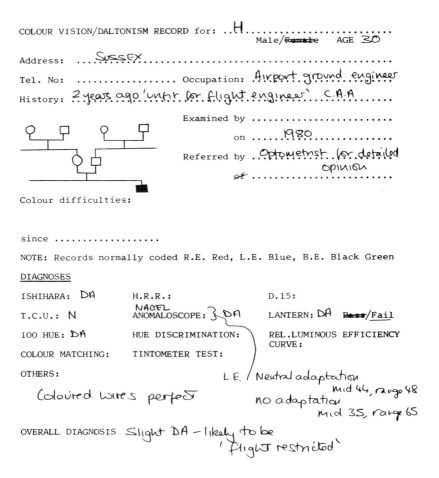

COLOUR VISION/DALTONISM RECORD for: ..H...............................
                                          Male/~~Female~~   AGE 30

Address: ...Sussex...

Tel. No: .... Occupation: Airport ground engineer

History: 2 years ago 'unfit for flight engineer` C.A.A.......

                              Examined by ....

                                on ......1980...............

                        Referred by ..Optometrist for detailed
                                                    opinion
                            ~~of~~ ....

Colour difficulties:

since ....

NOTE: Records normally coded R.E. Red, L.E. Blue, B.E. Black Green

DIAGNOSES

ISHIHARA: DA   H.R.R.:   D.15:

T.C.U.: N   NAGEL
                      ANOMALOSCOPE: } DA   LANTERN: DA ~~Pass~~/Fail

100 HUE: DA   HUE DISCRIMINATION: REL.LUMINOUS EFFICIENCY
                                           CURVE:
COLOUR MATCHING: TINTOMETER TEST:

OTHERS: L.E. / Neutral adaptation
                                                        mid 44, range 48
    Coloured wires perfect   no adaptation
                                                      mid 35, range 65

OVERALL DIAGNOSIS Slight DA — likely to be
                          ' Flight restricted`

G.A. Lantern   Dark adapted YES (NO) (min)   Mirror (YES)/NO

CLOCKWISE	5mm			3mm			1mm		
W							W	W	G
LG							G	Am	W
BG									
SG							G	?	?
SY							R	G	Am
SR							R	R	
Y							G	Am	
DR							R	-	

ANTI - CLOCKWISE	5mm			3mm			1mm		
DR									
Y									
SR									
SY									
SG									
BG									
LG									
W									

SR → SY	G							
SG → SY								

Tentative opinion.

(DA,) PA, D, P, Normal.

ISHIHARA.......10th.....EDITION, ILLUMINATION. Office day.

TENTATIVE OPINION......................................

12	8	6	29	57	5	3	15	74	2	6	97	45
12	3	5	20	35	2	5	17	21	-	-	-	-
Lee148	3or8	5or6	20									

5	7	16	73	-	-	-	-	26	42	35	96	
-	-	-	-	-	-	45	-	30	42	3-	96	
Lee148							-	26	42	35	96	

THE CITY UNIVERSITY COLOUR VISION TEST

Illumination ('Daylight') Type ...Office day.. Level.......

	(PAGE)	SUBJECT'S CHOICE OF MATCH R | L | BE	NORMAL	DIAGNOSIS		
				PROTAN	DEUTAN	TRITAN
'CHROMA FOUR'	1	B	(B)⟷	R	L	T
	2	R	(R)◊	B	L	T
	3	L	(L)◊	R	T	B
	4	R	(R)◊	L	B	T
	5	C	(L)◊	T	B	R
	6	B	(B)⟷	L	T	R
'CHROMA TWO'	7	C	(L)◊	T	R	B
	8	R	(R)◊	L	B	T
	9	B	(B)⟷	L	T	R
	10	T	(T)⟷	B	L	R
SCORE	At Chroma Four		/6	/6	/6	/6
	At Chroma Two		/4	/4	/4	/4
	Overall		10/10	/10	/10	/10

Probable type of Daltonism:

P, PA, EPA, D, DA, EDA, Tritan, Mixed 'Pass'

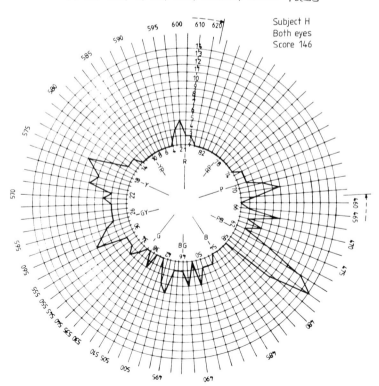

Subject H
Both eyes
Score 146

8.10   SUBJECT I

A moderately deuteranomalous boy, this subject was not easy to examine. He accepted the magenta filter provided for his trial under practical conditions.

COLOUR VISION/DALTONISM RECORD for: ....I.........................
                                      Male/~~Female~~   AGE 10'/2

Address: ...LONDON..

Tel. No: .... Occupation: ...School...................

History: .Almost. died. from. infection. at. birth... Reading.......
                                                          difficulty

Examined by ....

on ....1980.................

Referred by .Family. doctor......

of ....

Colour difficulties:
        Errors on wallpaper colours etc.

since ..only. recently. noticed difficulties

NOTE: Records normally coded R.E. Red, L.E. Blue, B.E. Black Green

DIAGNOSES

ISHIHARA: SA   H.R.R.: M-H DA   D.15: MIX

T.C.U.: MILD   ANOMALOSCOPE: LANTERN: Pass/Fail

100 HUE: HUE DISCRIMINATION: REL.LUMINOUS EFFICIENCY
                                               CURVE:
COLOUR MATCHING: TINTOMETER TEST:

OTHERS: Advised filter to aid

OVERALL DIAGNOSIS   Medium DA

AO H.R.R. ....Edition.   Illumination. Office day....

(FIRST 4 PLATES, DEMONSTRATION, DO NOT SCORE)
Ask 'How many symbols?', then 'What?', then 'Where?'.
If no errors 1-6 screening (Errors 1 or 2, go on to 17-20; errors
3-6, go on to 7-16).

SCREENING SERIES   DIAGNOSTIC SERIES

	B-Y		R-G					7	8	9	10	11	12	13	14	15	16			17	18	19	20
	1	2	3	4	5	6	Protan	O		Δ	O	X	Δ	O	Δ	X	O	Trit		Δ	X	O	Δ
	X O	Δ O	Δ X	Δ O	O	X	Deutan	Δ	X		X	O	O	Δ	X	O	Δ	"Tet"		X	O	Δ	X
R	X O	O Δ	–	–	–	–	R	–	–	–	–	O	O	Δ O	X	O	O Δ	R		Δ X	X O	O Δ	X Δ
L							L											L					
BE							BE											BE					

←Mild   Defects↔Medium ↞Strong→   Medium↞Strong→

Errors 1 or 2, go on to 17-20
Errors 3 or 6, go on to 7-16

Tentative opinion   TYPE:.....Mild - Medium DA..........
                    EXTENT:................................

ISHIHARA.....10th.....EDITION,   ILLUMINATION.Office day.
TENTATIVE OPINION.....Slight DA.....................

	12	8	6	29	57	5	3	15	74	2	6	97	45
	12	3	6	70	35	2	5	15	21	–	–	–	–
	5	7	16	73	–	–	–	–	26	42	35	96	
Filter Straud 13 assists	–	–	–	–	5	8	45	?	2–	4–	3?/36	9–	

D.15 TEST   ILLUMINATION. Office day......

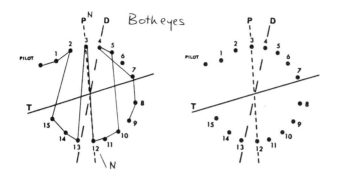

Both eyes

Tentative opinion ....
....

THE CITY UNIVERSITY COLOUR VISION TEST

Illumination ('Daylight') Type Office day. Level.......

	(PAGE)	SUBJECT'S CHOICE OF MATCH R ⌐ L ⌐ BE	NORMAL	DIAGNOSIS		
				PROTAN	DEUTAN	TRITAN
'CHROMA FOUR'	1	R	B ▽	(R)	L	T
	2	L	R ◊	B	(L)	T
	3	L	(L) ◊	R	T	B
	4	R	(R) ◊	L	B	T
	5	B	L ◊	T	(B)	R
	6	T	B ▽	L	(T)	R
'CHROMA TWO'	7	L	(L) ◊	T	R	B
	8	L	R ◊	(L)	B	T
	9	T	B ▽	L	(T)	R
	10	L	T △	B	(L)	R
SCORE	At Chroma Four	/6	/6	/6	/6	
	At Chroma Two	/4	/4	/4	/4	
	Overall	/10	/10	/10	/10	

Probable type of Daltonism:
P, PA, EPA, D, (DA,) EDA, Tritan, (Mixed)

8.11 SUBJECT J

Confusions of colours of wires confirmed the real difficulties this fifteen year old was likely to experience. Time was particularly short and other circumstances of this examination were difficult, therefore the tests used were confined to those shown. This boy also obtained interesting results with filters and he took a sample for further trials.

COLOUR VISION/DALTONISM RECORD for: ....J.......................

Male/~~Female~~   AGE 15

Address: .....*KENT*...

Tel. No: .... Occupation: ..*School*..............

History: ...*Mother has colour difficulties*.................

ex-Navy

Examined by ....

on ....

Referred by ..*Local health authority*
            medical officer

of ....

Colour difficulties:
*Cleaned shoes with brown instead of black
polish. Paints models. Cricket ball seems
                                    brown.*

since ....

NOTE: Records normally coded R.E. Red, L.E. Blue, B.E. Black Green

DIAGNOSES

ISHIHARA: *D or EDA* H.R.R.:   D.15:

T.C.U.: *? EDA*   ANOMALOSCOPE: LANTERN: Pass/Fail

100 HUE: *D or EDA* HUE DISCRIMINATION: REL.LUMINOUS EFFICIENCY
                                              CURVE:

COLOUR MATCHING: TINTOMETER TEST:

OTHERS: *Hopeless with
        Coloured wires*   *Filter supplied*

OVERALL DIAGNOSIS   *D or EDA*

ISHIHARA ....10th..... EDITION, ILLUMINATION ..Office....
TENTATIVE OPINION ....D or DA................

12	8	6	29	57	5	3	15	74	2	6	97	45
12	3	5	70	35	2	5	17	21	-	-	-	-
	8	6	29	57	5	3	15	74	5	-	89	45

Filter Strand 12

5	7	16	73	-	-	-	-	26	42	35	96
-	-	-	-	5	-	45	75	2-	4-	3-	9-
5	7	16	73	-	-	-	-	26	42	35	96

Strand 12

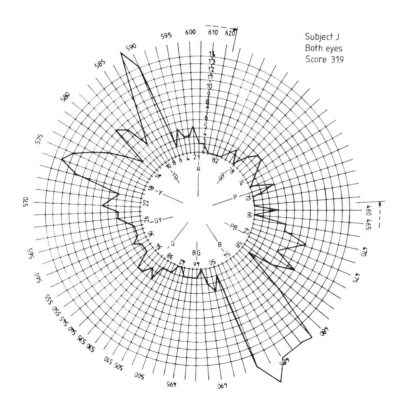

Subject J
Both eyes
Score 319

THE CITY UNIVERSITY COLOUR VISION TEST

Illumination ('Daylight') Type ..*Office*.... Level.......

	(PAGE)	SUBJECT'S CHOICE OF MATCH R | L | BE	NORMAL	DIAGNOSIS		
				PROTAN	DEUTAN	TRITAN
'CHROMA FOUR'	1	L	B ▽	R	(L)	T
	2	R/L	(R)	B	(L)	T
	3	T	L ◁	R	(T)	B
	4	B	R ▷	L	(B)	T
	5	B	L ◁	T	(B)	R
	6	T	B ▽	L	(T)	R
'CHROMA TWO'	7	R/R	(L)	T	(R)	B
	8	B	R ▷	L	(B)	T
	9	T	B ▽	L	(T)	R
	10	T/L	(T)	B	(L)	R

SCORE	At Chroma Four	/6	/6	/6	/6
	At Chroma Two	/4	/4	/4	/4
	Overall	/10	/10	/10	/10

Probable type of Daltonism:

P, PA, EPA, (D, DA, EDA,) Tritan, Mixed

8.12 SUBJECT K

This physics graduate sought advice relating to employment but had insufficient time for a full investigation of his colour vision. Both CUT performance and failure on a rapid lantern test led to advice that his condition
was sufficiently severe to be likely to give many occupational difficulties;
also that he might consider help from a suitable filter. A rapid examination
such as is illustrated by this case is often all that can be done initially if time
is limited and points the need for more extensive tests at a later date. Nevertheless the information gained from two tests, carefully selected, can form
the basis of suitable advice.

COLOUR VISION/DALTONISM RECORD for: .... *K* ....

Male/~~Female~~   AGE 22

Address: .... *HERTS* ....

Tel. No: .... Occupation: *Milkman at present –*
physics graduate

History: ....

Examined by ....

on .... *1978* ....

Referred by *Just arrived, no appointment*
rapid tests only possible

of ....

Colour difficulties: *As he approaches them colours less*
difficult. Thinks 'red–brown' blind. Requires
estimate of defect

since ..... *10 years* ...

NOTE: Records normally coded R.E. Red, L.E. Blue, B.E. Black Green

DIAGNOSES

ISHIHARA: *DA*   H.R.R.:   D.15:

T.C.U.: *DA*   ANOMALOSCOPE: LANTERN: ~~Pass~~/*Fail*

100 HUE: HUE DISCRIMINATION: REL.LUMINOUS EFFICIENCY
                                              CURVE:
COLOUR MATCHING: TINTOMETER TEST:

OTHERS: *Note restricted test.*
Informed ' fairly high DA'. Proposed filter
aid trial later

OVERALL DIAGNOSIS *EDA likely*

THE CITY UNIVERSITY COLOUR VISION TEST

Illumination ..*Office*....   *Both eyes*

(PAGE)	SUBJECT'S CHOICE OF MATCH	NORMAL	DIAGNOSIS		
			PROTAN	DEUTAN	TRITAN
1	R	B ⬙	(R)	L	T
2	R	(R) ◊	B	L	T
3	R	T ⬠	B	(R)	L
4	B	R ◗	T	(B)	L
5	T	L ◖	R	(T)	B
6	B	R ◗	L	(B)	T
7	T	L ◖	(T)	B	R
8	B	L ◖	R	(B)	T
9	T	B ⬙	L	(T)	R
10	R	L ◖	T	(R)	B

1/10   2/10   7/10

Tentative Opinion: ..DA...................

ISHIHARA.....10th....EDITION, ILLUMINATION..MacBeth.
TENTATIVE OPINION...DA...............................

12	8	6	29	57	5	3	15	74	2	6	97	45
12	8	6	70	35	2	5	17	21	—	—	—	5
5	7	16	73	—	—	—	—	26	42	35	96	
—	—	—	—	—	—	45	28	2-	4-	3-	9-	

8.13 SUBJECT L

This medium protanomalous trichromat was able to identify many common object colours but he gave typically 'bad' results on the Ishihara test. CUT (2nd edn) results just show the protan confusions and HRR test results are as expected from the above. The single 'deutan' errors on CUT and Igaku-Shoin should be noted; also the nil response on HRR plate 9 and the mixed plot from the 100 Hue test. The F2 plate worked well as a screening approach and the way in which he required a bright red to match a relatively dull yellow on the three-light test was characteristically protan. A red filter (monocularly) was likely to be helpful at times.

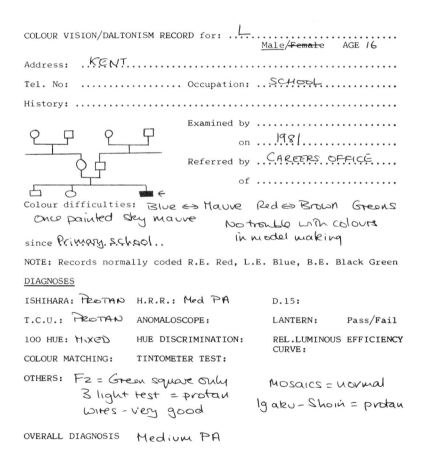

COLOUR VISION/DALTONISM RECORD for: ...L...........................

Male/~~Female~~   AGE 16

Address: .KENT..

Tel. No: .... Occupation: ..SCHOOL..............

History: ....

Examined by ....

on ....1981................

Referred by ..CAREERS OFFICE.....

of ....

Colour difficulties: Blue ⇔ Mauve   Red ⇔ Brown   Greens
Once painted sky mauve   No trouble with colours
since Primary school..   in model making

NOTE: Records normally coded R.E. Red, L.E. Blue, B.E. Black Green

DIAGNOSES

ISHIHARA: PROTAN   H.R.R.: Med PA   D.15:

T.C.U.: PROTAN   ANOMALOSCOPE: LANTERN: Pass/Fail

100 HUE: MIXED   HUE DISCRIMINATION: REL.LUMINOUS EFFICIENCY
                                         CURVE:
COLOUR MATCHING: TINTOMETER TEST:

OTHERS: F2 = Green square only   MOSAICS = normal
         3 light test = protan
         wires - very good   Igaku-Shoin = protan

OVERALL DIAGNOSIS   Medium PA

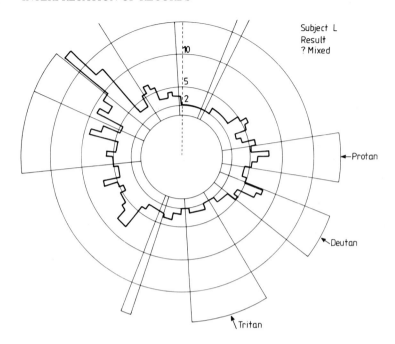

Subject L
Result
? Mixed

←Protan

←Deutan

Tritan

ISHIHARA....10+h.....EDITION,   ILLUMINATION. Office. day.

TENTATIVE OPINION...P.A.................................

12	8	6	29	57	5	3	15	74	2	6	97	45
12	3	5	70	35	2	5	11	21	–	–	51	–

Filter Lee 164

12	8	6	20	37	8	5	15	74	–	–	–	–
5	7	16	73	–	–	–	–	26	42	35	96	
–	7	–	–	5	8	15	75	-5	-2	-5	-6	
5	7	16	73	–	–	–	–	2-	4-	3-	9-	

AO H.R.R. ....Edition.   Illumination.................

(FIRST 4 PLATES, DEMONSTRATION, DO NOT SCORE)
Ask 'How many symbols?', then 'What?', then 'Where?'.
If no errors 1-6 screening (Errors 1 or 2, go on to 17-20; errors
3-6, go on to 7-16).

SCREENING SERIES   DIAGNOSTIC SERIES

B-Y			R-G				7	8	9	10	11	12	13	14	15	16		17	18	19	20
1	2	3	4	5	6	Protan	0		Δ	0	X	Δ	0	Δ	X	0	Trit	Δ	X	0	Δ
X 0	Δ 0	Δ X	Δ 0	0	X	Deutan	Δ	X		X	0	0	Δ	X	0	Δ	"Tet"	X	0	Δ	X
R						R											R				
L						L											L				
BE X 0	0 Δ	—	0	—	—	BE	O O	—	—	O	X O	Δ	O Δ	Δ	X O	O Δ	BE	Δ X	X O	Δ O	X Δ

←Mild   Defects←(Medium) →←Strong→   Medium→←Strong→

Errors 1 or 2, go on to 17-20
Errors 3 or 6, go on to 7-16

Tentative opinion   TYPE:.........PROTAN..................
                    EXTENT:.......MEDIUM....................

[Standard Pseudoisochromatic Plates IGAKU-SHOIN]

Name _____ L ____ (M)F ___ Age _16_ Date _____

                                    Examiner _____

Screening Series

Plate No.	Normal	Protan/Deutan
5	3	(8)
6	2	(9)
7	37	(illegible)
8	7	(4)
9	(8)	(7)
10	(4)	(3)
11	2	(4)
12	7	(5)
13	58	(illegible)
14	3	(6)
Total		

Classification Series

Plate No.	Protan	Deutan
15	(8)	3
16	(5)	7
17	(4)	8
18	(9)	(4)
19	(3)	5
Total		

Result :   Normal
           (Protan)
           Deutan
           Others

THE CITY UNIVERSITY COLOUR VISION TEST

Illumination ('Daylight') Type *office day.* Level.......

(PAGE)	SUBJECT'S CHOICE OF MATCH (R L BE)	NORMAL	DIAGNOSIS PROTAN	DEUTAN	TRITAN
'CHROMA FOUR' 1	ß	(B)→	R	L	T
2	L	R ◊	B	(L)	T
3	R	L ◊	(R)	T	B
4	R	(R)◊	L	B	T
5	L	(L)◊	T	B	R
6	ß	(B)→	L	T	R
'CHROMA TWO' 7	L	(L)◊	T	R	B
8	L	R ◊	(L)	B	T
9	ß	(B)→	L	T	R
10	T	(T)→	B	L	R

SCORE		NORMAL	PROTAN	DEUTAN	TRITAN
	At Chroma Four	4 /6	1 /6	1 /6	/6
	At Chroma Two	3 /4	1 /4	/4	/4
	Overall	/10	/10	/10	/10

Probable type of Daltonism:

P, (PA), EPA, D, DA, EDA, Tritan, Mixed

8.14 SUBJECT M

The variation in responses to Ishihara plates and the complete 'pass' on both HRR and CUT should be noted. The anomaloscope range was small but displaced well to the deutan side of normal and small lantern stimuli gave well understood difficulties such as the successive contrast miscallings of *SY*. A career in 'biology' in which he expressed interest should present less difficulty from his colour vision point of view than police or airline work, especially if a magenta filter is used with discretion.

COLOUR VISION/DALTONISM RECORD for: ...M...............................

Male/~~Female~~ AGE IS

Address: ...MIDAX...

Tel. No: .... Occupation: .SCHOOL....................

History: .Seeks ideas re careers - Police, Pilot, Biology......

Examined by ....

on ...1980.......................

Referred by .School medical service

of ....

Colour difficulties: Indefinate results with school tests at various times.

Never keen on colours or painting

since ....

NOTE: Records normally coded R.E. Red, L.E. Blue, B.E. Black Green

DIAGNOSES

ISHIHARA: DA 11/24   H.R.R.: N   D.15:

T.C.U.: N   ANOMALOSCOPE: DA 24 range 7   LANTERN: DA ~~Pass~~/Fail

100 HUE: ? DA but 270 HUE DISCRIMINATION: REL.LUMINOUS EFFICIENCY CURVE:

COLOUR MATCHING: TINTOMETER TEST:

OTHERS: 3 light test = not protan

OVERALL DIAGNOSIS   Slight DA

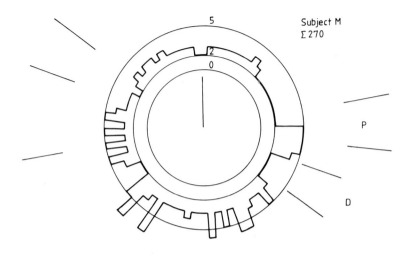

Subject M
Σ270

G.A. Lantern   Dark adapted YES/(NO) (min) Mirror (YES)/NO

	CLOCKWISE							
	5mm		3mm		(1mm)			
W								
LG						Y	G	
BG						G		
SG						R	G	R	G
SY						Y		
SR						R		
Y						Y		
DR						R		

	ANTI - CLOCKWISE						
	5mm		3mm		1mm		
DR							
Y							
SR							
SY							
SG							
BG							
LG							
W							

SR → SY	G							
SG → SY	R							

Tentative opinion.

DA, PA, D, P, Normal.

ANOMALOSCOPE (NAGEL)/OTHER

Subject's	R 1) 26	2)	3)	MEAN
first 3				
matches	L 1)	2)	3)	MEAN

NEUTRAL PRE-EXPOSURE FOR EACH DECISION? (YES)/NO

SETTING	1	2	3	4	5	6	7	8	9	10	11	12	13	14	15	16	17	18	19
R										X									
L																			
SETTING	20	21	22	23	24	25	26	27	28	29	30	31	32	33	34	35	36	37	38
R	X	√	√			√	√	√	X	X						X			X
L																			
SETTING	39	40	41	42	43	44	45	46	47	48	49	50	51	52	53	54	55	56	57
R		X																	
L																			
SETTING	58	59	60	61	62	63	64	65	66	67	68	69	70	71	72	73	74	75	
R																			
L																			

RANGE   R 7 L   MIDPOINT   R 24 L

ANOMALQUOTIENT   R   L

LOWER DIAL AT 0 BRIGHTNESS MATCH   R   L

LOWER DIAL AT 75 BRIGHTNESS MATCH   R   L

Tentative opinion .....P.A.......................................

AO H.R.R. ...2....Edition.   Illumination......day.......

(FIRST 4 PLATES, DEMONSTRATION, DO NOT SCORE)
Ask 'How many symbols?', then 'What?', then 'Where?'.
If no errors 1-6 screening (Errors 1 or 2, go on to 17-20; errors
3-6, go on to 7-16).

SCREENING SERIES   DIAGNOSTIC SERIES

	B-Y			R-G				7	8	9	10	11	12	13	14	15	16		17	18	19	20
	1	2	3	4	5	6	Protan	0		Δ	0	X	Δ	0	Δ	X	0	Trit	Δ	X	0	Δ
	X 0	0	Δ X	Δ 0	0	X	Deutan	Δ	X		X	0	0	Δ	X	0	Δ	"Tet"	X	0	Δ	X
R							R											R				
L							L											L				
BE	X 0	0	Δ X	Δ 0	0	X	BE	0 Δ	X	Δ	X 0	X 0	0 Δ	Δ 0	Δ X	X 0	0 Δ	BE	Δ X	X 0	0 Δ	Δ X

←Mild   Defects↔Medium ↔Strong→   Medium↔Strong→

Errors 1 or 2, go on to 17-20
Errors 3 or 6, go on to 7-16

Tentative opinion   TYPE:.....NORMAL...................
                    EXTENT:................................

ISHIHARA....10th.......EDITION, ILLUMINATION..Day.......
TENTATIVE OPINION......DA................................

12	8	6	29	57	5	3	15	74	2	6	97	45
12	8	6	20	53	8	3 9/8	15	21	8	–	97	45
5	7	16	73	–	–	–	–	26	42	35	96	
5	7	16	29	5	8	45	24	26	42	35	95	

Filter
Straud 12
helps

8.15   SUBJECT N

The condition had probably been stable for some years before this examination was made and the acquired defect can be regarded as superimposed upon a red–green defect. The record is not clear in some respects. Visual acuity was remarkably good and the results of the illness and perhaps of the treatment appear to be confined to a colour vision deficiency. Although it would be difficult to prove that no malingering was in operation there is no reason to suspect such malingering. Both D.15 and 100 Hue test results were variable; there was no clear axis with the 28 Hue test and the RE anomaloscope data probably suffered on account of frequent reports of brightness mismatching. CUT results are not included but data from a preliminary model of the test in use in 1974 showed a tendency to give 'normal' matches, mixed with protan and tritan confusions, with inconsistencies on retesting. The observer frequently said that 'these colours look similar, except in brightness'. He tended to use comparisons with familiar object colours, such as 'pillar box red'. His naming of surface colours showed confusions between blue objects and green objects but he could identify orange; yellow was called either 'green' or 'yellow'.

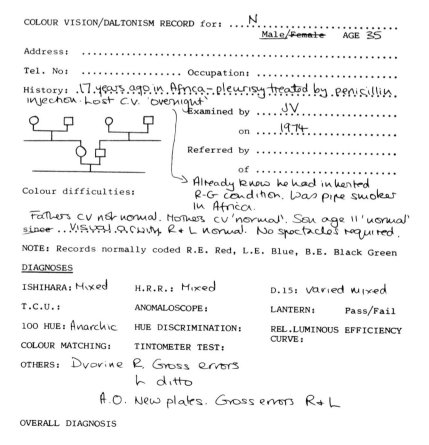

COLOUR VISION/DALTONISM RECORD for: ....N.............................

Male/~~Female~~   AGE 35

Address: ....

Tel. No: .... Occupation: ....

History: 17 years ago in Africa - pleurisy treated by penicillin injection. Lost C.V. 'overnight'

Examined by ....JV................

on ....1974.................

Referred by ....

of ....

→ Already knew he had inherited R-G condition. Was pipe smoker in Africa.

Colour difficulties:

Fathers CV not normal. Mothers CV 'normal'. Son age 11 'normal' since ...Visual acuity R+L normal. No spectacles required.

NOTE: Records normally coded R.E. Red, L.E. Blue, B.E. Black Green

DIAGNOSES

ISHIHARA: Mixed   H.R.R.: Mixed   D.15: varied mixed

T.C.U.:   ANOMALOSCOPE: LANTERN: Pass/Fail

100 HUE: Anarchic   HUE DISCRIMINATION: REL.LUMINOUS EFFICIENCY
                                          CURVE:
COLOUR MATCHING: TINTOMETER TEST:

OTHERS: Dvorine R. Gross errors
                  L ditto

         A.O. New plates. Gross errors R+L

OVERALL DIAGNOSIS

ANOMALOSCOPE   NAGEL/~~OTHER~~

Subject's   R 1) 42/21   2) 32/14   3) 37/22   MEAN 37
first 3
matches   L 1) 36/48   2) 44/30   3) 43/20   MEAN 41

NEUTRAL PRE-EXPOSURE FOR EACH DECISION?   ~~YES~~/NO

SETTING	1	2	3	4	5	6	7	8	9	10	11	12	13	14	15	16	17	18	19
R	✓	✓			✓														
L	✓																		

SETTING	20	21	22	23	24	25	26	27	28	29	30	31	32	33	34	35	36	37	38
R																			
L																			

SETTING	39	40	41	42	43	44	45	46	47	48	49	50	51	52	53	54	55	56	57
R																			
L																✓			

SETTING	58	59	60	61	62	63	64	65	66	67	68	69	70	71	72	73	74	75
R			✓											✓		✓		
L	✗		✗					✗								✗		

RANGE   R   L   MIDPOINT   R   L
ANOMALQUOTIENT   R   L
LOWER DIAL AT 0 BRIGHTNESS MATCH   R 34 L 32
LOWER DIAL AT 75 BRIGHTNESS MATCH   R 55 L 28
Tentative opinion ......?...

ISHIHARA....10th......EDITION, ILLUMINATION..MacBeth..

TENTATIVE OPINION.....Acq. MIxED...................

	12	8	6	29	57	5	3	15	74	2	6	97	45
RE	12	–	—	–	—	1	–	–	–	–	–	–	–
LE	12	–	–	–	–	–	–	–	–	–	–	–	–

	5	7	16	73	–	–	–	–	26	42	35	96	
RE	–	–	–	–	–	–	–	–	–	–	–	–	
LE	–	–	–	–	–	–	–	–	–	–	–	–	

AO H.R.R. ...2...Edition.   Illumination...MacBETH....

(FIRST 4 PLATES, DEMONSTRATION, DO NOT SCORE)
Ask 'How many symbols?', then 'What?', then 'Where?'.
If no errors 1-6 screening (Errors 1 or 2, go on to 17-20; errors
3-6, go on to 7-16).

SCREENING SERIES   DIAGNOSTIC SERIES

B-Y			R-G				7	8	9	10	11	12	13	14	15	16		17	18	19	20
1	2	3	4	5	6	Protan	0		Δ	O	Δ	O	Δ	X	O	Trit	Δ	X	O	Δ
X 0	0	Δ X	Δ 0	0	X	Deutan	Δ	X		X	O	O	Δ	X	O	Δ	"Tet"	X	O	Δ	X
R —	—	—	—	—	—	R	—	—	—	—	—	—	—	—	X	O	R	—	—	O	—
L —	—	—	—	—	—	L	—	—	—	—	—	—	—	—	O	—	L	—	—	O	—
BE						BE											BE				

←Mild   Defects↔Medium ←Strong→   Medium←Strong→

SYMBOLS SEEN BY CONTRASTING
'BRIGHTNESS' NOT COLOUR

Errors 1 or 2, go on to 17-20
Errors 3 or 6, go on to 7-16

Tentative opinion   TYPE:.....ACQUIRED MIXED.........
                    EXTENT:.........GROSS................

G.A. Lantern   Dark adapted YES/NO⃝ (min) Mirror YES⃝/NO

BOTH EYES	CLOCKWISE										ANTI - CLOCKWISE									
	5mm			3mm			1mm					5mm			3mm			1mm		
W	G	O									DR									
LG	R										Y									
BG											SR									
SG	B	G	Y								SY									
SY	W	Y									SG									
SR	B	B									BG									
Y	G										LG									
DR	B										W									

SR → SY									
SG → SY									

Tentative opinion. ?
DA, PA, D, P, Normal.

D.15 TEST   ILLUMINATION........MacBeth.........

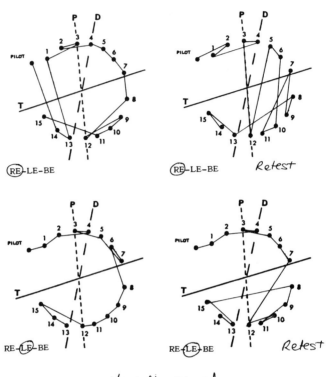

Tentative opinion .....Variable mixed.....................
....

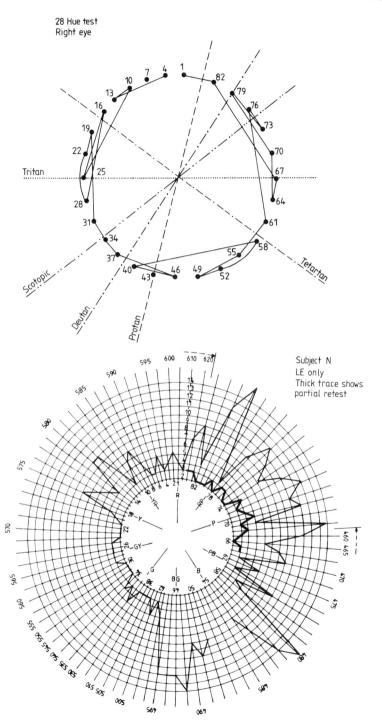

28 Hue test
Right eye

Tritan

Tetartan

Scotopic

Deutan

Protan

Subject N
LE only
Thick trace shows
partial retest

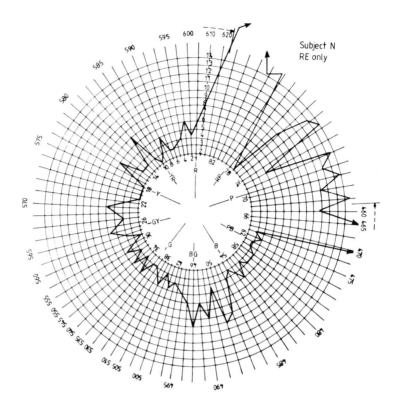

Subject N
RE only

TINTOMETER (LOVIBOND) COLOUR VISION ANALYSER

Quantitative diagnosis. After determining type of deficiency, the
figure for the range (extent of deficiency) is derived from the
        SUM OF THE TWO HIGHEST THRESHOLD SATURATION SETTINGS
(on opposite sides of the circle) each minus 10 (the normal threshold).

RB/LE/BOTH

Filters chosen	N	N	1	26	N	8	7			
Threshold saturation chosen	80	55	55	45	45	45	45			
Neutral setting	1.5	1.5	1.5	1.5	1.5	1.5	1.5			

TYPE............ RANGE.........

Tentative opinion ....

Actual (random) order of colours
N, 20, 9, 13, 24, 4, 1, 14, 19, 16, 21, 11, 8, 17, 23, 2, 18,
12, 15, 6, 25, 10, 7, 22, 3, 5, 26, (then N again)

Colour specification on the CIE (1960) chromacity diagram

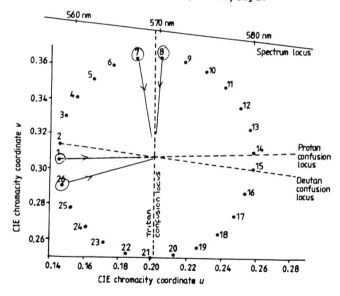

TINTOMETER (LOVIBOND) COLOUR VISION ANALYSER

Quantitative diagnosis. After determining type of deficiency, the
figure for the range (extent of deficiency) is derived from the
SUM OF THE TWO HIGHEST THRESHOLD SATURATION SETTINGS
(on opposite sides of the circle) each minus 10 (the normal threshold).

(RE)/~~LE~~/~~BOTH~~

Filters chosen	N	N	N	N	6	2	1	N	2	N	3
Threshold saturation chosen	80	70	60	55	45	45	45	45	40	40	40
Neutral setting	1.5	1.5	1.5	1.5	1.5	1.5	1.5	1.5	1.5	1.5	1.5

TYPE.............   RANGE.........

Tentative opinion ....

Actual (random) order of colours
N, 20, 9, 13, 24, 4, 1, 14, 19, 16, 21, 11, 8, 17, 23, 2, 18,
12, 15, 6, 25, 10, 7, 22, 3, 5, 26, (then N again)

Colour specification on the CIE (1960) chromacity diagram

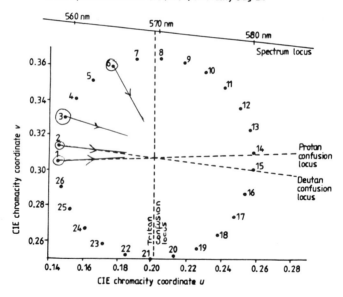

9 Assistance for colour vision defects

9.1 INTRODUCTION

A person with an inherited anomaly of colour vision, Daltonism, cannot be 'cured' but it is usually unfair to ignore the possibility of assistance. It is important to stress possible dangers from the condition but most subjects have learned some tricks to minimise the 'disability' and these can be encouraged. Design of signals to include cues other than colour is often useful and the Daltonic must be encouraged to make use of position, shape and other clues.

A hundred years ago the literature contained reports of attempts to 'educate' Daltonics and to 'treat' Daltonism; Favre, in France, between 1874 and 1878, Delboeuf in Belgium and Magnus and Kalischer in Germany all contributed. Giles (1950b) considered that for a few boys colour perception 'developed' late in childhood and since the advent of colour vision tests some Daltonics have tried to improve their guesswork combined with discernment, using brightness and other aids. Blinking the eyelids at a rapid rate is sometimes used as self-help. Some use coloured glass or plastics filters as an aid, a technique which has long history, for example the device invented by Maxwell in 1854, involving spectacle lenses split into two coloured parts, red and green, enabling Daltonics to alternate their vision between the two (figure 9.1). Many rediscoveries appear in the literature.

9.2 THE USE OF FILTERS

If correctly applied, coupled with cautious advice against over-confidence, and stressing the dangers of tiredness and adverse conditions, filters can be prescribed to assist in many cases. Schmidt (1976) published a valuable summary at a time when interest was being shown in aiding the colour defective. Her account was particularly rich in examples from the German and American literature and concentrated on the uses of filters and of illumination. Naming three groups of colours, as follows, Schmidt indicated that patients would benefit from help in differentiation within each group. Group 1: red, yellow, green; Group 2: blue and purple; Group 3: red colours, grey, and blue–greens.

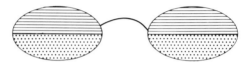

Figure 9.1   Maxwell's spectacles. One half of each lens is red, the other half green

Some of Schmidt's quotations included practical ideas to help protanopes with Wratten (Kodak) yellow filters, 8, 9, 12 or 15, or a magenta number 32 (see figure 9.2); the last choice is endorsed by the present writers. Deuteranopes could be assisted by the above filters, except 32.

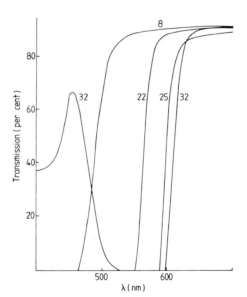

Figure 9.2   Wratten filters numbers 8, 22, 32 and 25 with approximate transmissions, after Kodak data. Note that 32, a magenta, has two peaks of transmission.

It is therefore important that persons dealing with Daltonic subjects should consider the possibilities and the aids now available, bearing in mind the wide range of attempts in this field, including special lighting and some little publicised approaches. Firstly the principles of the use of filters must be considered and some case records will illustrate the real help sometimes possible with these methods.

9.2.1 How filters assist

A view of the world through coloured filters is altered in several respects; the 'brightness' of coloured objects is apparently altered and the chromaticity may also be changed. The change in chromaticity, or 'chromaticness' is by virtue of a manipulation of saturation and of dominant wavelength. The three basic features of colour—brightness, hue and saturation ('hue, value and chroma' in the terms and order of Munsell)—tend to be altered interdependently when filters are involved. The successful use of filters is possible without an understanding of the principles, but a consideration of the effects of the underlying principles can be helpful.

The transmission curves of a selection of filters are shown in figure 9.3. An orange filter is shown which passes yellow and red parts of the spectrum in addition to orange; a person looking through the filter when viewing the spectrum naturally would find that the green part of the spectrum is reduced in brightness by the filter, relative to the red part. Daltonics could use this information to distinguish red from green under some circumstances.

Calculations showing the way in which filters 'induce' colour changes which can be helpful to dichromats have been described by Richer and Adams (1984a,b). These authors made trials with protanopes and deuteranopes, showing that D.15 colour placement could be altered with filters in the ways predicted by their calculations. Protanopes appeared to derive less benefit than deuteranopes from the information related to luminous features which was given by X-Chrom lenses.

Everyday colours and signal lights are seldom 'monochromatic'. A 'green' object colour can be imitated by a variety of component colours. If an object normally appears 'white' in daylight it looks a 'desaturated green' if it is illuminated by two lights mixed, one a nearly spectral blue–green and the other a close spectral orange (see figures 9.3 and 9.4). If the orange component dominated, the 'green' would tend towards yellow–green, to a normal observer. An anomalous trichromat (for example a deutan), looking at the 'green' object described above through an orange filter held close to the eye, would be under the impression that the object had gained in 'saturation', or purity. He would also interpret variation in brightness, particularly relative to the appearances of other objects and their colours, as a function of the 'green' object.

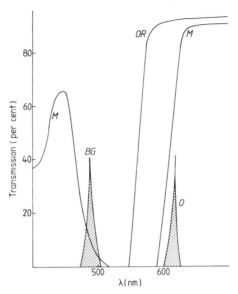

Figure 9.3   Transmission curves for four filters. *M* is a magenta, passing red and blue light; *OR* is a 'broad band' orange filter. *BG* and *O* represent 'narrow band' filters which isolate, respectively, blue–green and orange parts of the spectrum.

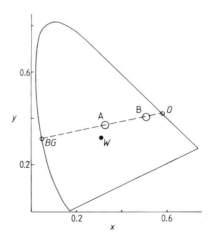

Figure 9.4   Colour diagram showing white (*W*) and the two monochromatic lights *BG* and *O*. A and B represent mixtures of the lights; A when *BG* and *O* are almost equal and B when *O* predominates.

The principles can be expressed by an example using familiar objects. On a table a small white cotton handkerchief is held vertically in front of two projectors; one projector transilluminates the screen with a nearly monochromatic 'blue–green' light and the other sends 'orange' light which mixes with the first. On the table in front of the screen are many varied objects, illuminated by daylight. A normal trichromat would describe the handkerchief as a desaturated green; an anomalous trichomat would probably have difficulty in naming the colour, but if he looked through an orange filter this would probably enhance the saturation of the object (figure 9.5). If, relative to the foreground objects, the filter produced an increase in brightness he would tend to identify the colour as 'reddish' or 'yellow'; a decrease in brightness would probably be interpreted as indicating that the colour should be called 'green'. If the Daltonic person uses a magenta filter instead, there being two peaks of transmission for this filter, his reception of blue–green light is spoilt as a consequence, and the chromaticity of the overlapping projector lights, the mixture on the screen, would tend to move towards the spectral locus. Although no brightness is impaired by the magenta filter it is likely to be described as having 'enhanced chromaticness'. Magenta spectacles would therefore darken green leaves in the foreground and a ripe tomato would be as bright as before.

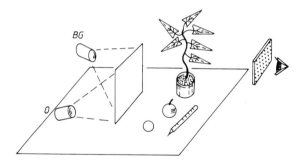

Figure 9.5 A translucent diffusing screen (a white handkerchief) illuminated from behind by two monochromatic lights (*BG* and *O* of figure 9.3) Several familiar objects on the table are illuminated by the normal room lighting and the scene is observed with or without a filter held in front of the eye.

Some protan observers may find magenta filters helpful but the deuteranomalous subject is most helped by this aid. Bizzare sunglasses are commonly worn today and deep magenta lenses may be mounted in a frame without great comment; a clip-on full or half lens can be provided or a loose piece of filter can be less ostentatious if carried discreetly by the subject. The material can be inconspicuously mounted in the plastics cap of an eye dropper bottle, using the aperture for viewing while the cap is held in the loosely closed fist. The method of dyeing an organic resin spectacle lens can be used, provided a good control of transmission is achieved.

9.2.2   Previous success with filters

Wilson (1855) quoted Seebeck as proposing, in 1817, the use of a vessel containing coloured liquid to aid Daltonism, viewing objects through the liquid. Wilson evidently tried such devices with many observers, concluding that yellow or blue filters gave the best results but reporting success with orange glasses in addition. According to Jennings (1896), Delboeuf had earlier overcome his own dichromatism by using a fuchsin solution as a filter to assist him in distinguishing between red and green. More recently Lyons (1958) reported assistance for a rod monochromat with photophobia by a scleral contact lens tinted deep blue, combined with a Crookes B2 spectacle lens. Ohara and Akutsu (1956) showed how a patient could pass a colour vision screening test for railway workers using coloured contact lenses or filters and Stonebridge (1968) explained how red filters can help some patients. McKay Taylor (1978) has described how filters (red in the left eye, green in the right eye) enabled Daltonic yachtsmen to differentiate red and green lights at night. He favoured glazing binoculars with the colours. These writers proposed that a yellow filter could be substituted for red for 'partial' defects.

A German approach in about 1930 involved 'Neophan' glass which incorporates neodym oxide; this was recommended as an aid for colour discrimination and for use in foggy weather. It has maximal absorption (about 50%) at 589 nm. Birch-Hirschfeld (1932) was an anomalous trichromat who described his success in colour distinctions using gelatin filters or neodym glass. The transmission of the glass is shown in figure 9.6.

Very predictable results were obtained by Phillips and Kondig (1975) in using a series of observers including defectives to investigate the practical effect of filters. A magenta filter made a red signal invisible and rendered amber traffic lights 'red' and 'blue'. Brightness changes and apparent colour changes were as expected. Thus colour normals and three types of Daltonics found that certain magenta filters made a green signal hard or impossible to see; a different magenta filter however made only a small difference to the apparent brightness of a red traffic signal. Clearly there are real potential dangers in using coloured lenses, including contact lenses, and these must be stressed to patients who seek such assistance. Night driving is likely to provide the greatest danger. Richards and Grolman (1962) stressed the wisdom of the correct choice of lens transmission according to the individual requirements; more recently the variations in transmissions of photochromic lenses have promoted discussion. A study by Farnsworth (1945), however, showed how 'Calobar' lenses did not 'seriously impair' colour perception although a yellow Novial and a Rose-smoke lens each reduced performance on the FM 100 Hue test. A rose-coloured filter provides greatest benefit for deuteranomalous observers. Everson (1973) discussed how such a filter provided good contrast between sky and foliage, noting how some ski experts consider the 'DOC' tint to improve viewing of snow surface contour. The transmission curve is shown in figure 9.7.

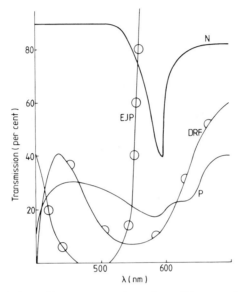

Figure 9.6 Various filters: N, Neodym glass; EJP, a reddish magenta used by subject EP; DRF is the filter used by patient DF and P is Polaroid plum filter

Fletcher (1980c) has experimented for approximately twenty years with filters which have assisted patients. Fletcher and Nisted (1963) reviewed the visual performance achieved when coloured contact lenses are used, noting that colour normals suffer no effects upon pseudoisochromatic plate test performance when red, blue, 'black', brown and green corneal lenses are worn, with the exception of a slight loss of the desaturated HRR characters with blue and red lenses. No practical improvement was elicited for a group of colour defective observers and reduced performance was caused by brown and red corneal lenses in some cases. In a more recent study a series of coloured corneal lenses was used by Harris and Cabrera (1976) on eight observers with normal colour vision; none of the lenses were found to alter significantly the colour vision as assessed with the FM 100 Hue test.

Good results were obtained by Fletcher in 1959–61, in aiding tobacco-curing workers with varying types of Daltonism by the use of three Ilford gelatin filters; these were held between afocal lenses with the coloured filter in the lower half of a spectacle. Filters found to be of assistance included a pale magenta, 502 micro 7, used for slight deuteranomaly, a deep magenta 503 minus green, for other deutan observers and 302, a pale turquoise, for protans (see figure 9.7). All filters were prescribed by the optometrist B Rivron in conjunction with the present writers. Although filter 302 assisted four protans to achieve better performance on PIC plates no practical aid in the judgment of tobacco leaf colours resulted in a colour range from green to brown. Filter 502 assisted three out of 22 slight deutans in leaf quality discrimination; for 10 out of 14 'severe' deutans, filter 503 enabled green lines and spots to be recognised as required. The need to overcome difficulties in discerning slight traces of green in an otherwise yellow-brown leaf was involved. Magenta filters assisted several men to judge correctly when the colour was properly 'fixed' during curing since the leaf rib showed up as a 'grey' line. One anomalous trichomat found that the contrast filter enabled him to detect traces of green with greater accuracy than colleagues with normal colour vision.

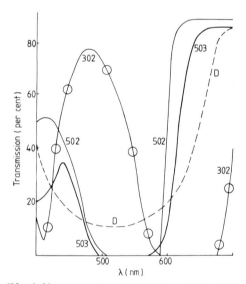

Figure 9.7 Ilford filters numbers 302, 502 and 503. Redrawn transmissions after Ilford data. D is the Doc filter described by Everson in 1973.

9.3* METHOD OF CHOICE OF A FILTER

The prescription of filters to assist colour defective patients can be attempted in a simple way or can be taken very seriously. For best results care, patience and time are required. A full colour vision assessment of type and severity of defect is appropriate, and the patient's needs should be discussed sympathetically, taking care to learn by what means he already tries to overcome his colour vision disabilities. Full records should be kept. A restrained approach is essential since some have heard of 'wonder lenses' and have misinformed impressions as to what is possible. Patients are thus encouraged to hope for slight improvements, if any. It is ideal practice to commence with, or to introduce early, some filters which are unlikely to be of any use at all. A range of magenta, red, orange and purple filters of different densities is required. Samples of Wratten and Ilford filters are useful, but gelatins must be safeguarded by glass covers. Theatrical stage light filters, e.g. those supplied by Lee Filters of Walworth, Andover, are robust and useful and plum Polaroid material, polarising and non-polarising, is worthy of inclusion. The present writers use a series of dyed resin lenses, graded in different mixtures of red and blue dyes. Each filter should be coded.

About six filters are introduced in turn for experimentation, moving towards a magenta series for deutans and reddish colours for protans, but switching back to earlier choices for comparison. Common coloured objects are viewed with daylight and tungsten light for assessment of the effects and results can be recorded on tables as shown below. Usually monocular viewing is used and detailed assessment is recorded, e.g. performance at certain PIC plates. In PIC books, plates of different types of colour combinations are used and two or three of these must be reviewed repeatedly as filter after filter is compared. The remarks of the subject are often worth recording. It is useful to allow the patient to borrow three samples, including a 'bad' one, for a week, after which he returns for a brief discussion on the basis of which prescription, if desired, is determined. If resin lenses are possible, transmission curves of the sample and of the dyed result should be compared.

Notes on filters used by   from   to

Please write short notes.   ⌐ can be used for good ⌐ ⌐ for very good
                            x can be used for useless
                            ? for uncertain.

Code for filter	Method of use e.g. 'clip on' or 'held in hand'	Both eyes used or one only?	What was the 'task'? Name important objects e.g. sorting tomatoes or painting a portrait	What sort of light was used? Daylight, fluorescent bulb	Did the filter help?

Extra comments: Please note good and bad points. Any ideas for improvement of method of use, including mounting or concealing the 'aid'?

9.4 SELECTED CASE STUDIES

9.4.1. Case AG (DA + PA)

A healthy young male was extensively tested since he is a good observer.His errors are more akin to deuteranomalous errors although some protan errors are seen as shown below.

(*a*) Ishihara (10th edition) used normally revealed PA and DA errors but a magenta filter (Lee) enabled him to overcome the PA errors.

(*b*) CUT test. Under normal conditions he gave 6 protan, 2 deutan, 1 normal and 1 mixed responses: these errors persisted when viewing through (Lee) 107 light rose but he gave completely normal results when using (Lee) 113 magenta. Using a plum Polaroid (polarising) (figure 9.6) he produced 3 protan results, no deutan and 2 'N + P'.

9.4.2 Case BC (DA)

Referred by his optometrist, at age 33, he had been aware of his colour vision difficulty since the age of 6. His occupation as a butcher was affected by inability to detect meat and poultry changes noticed by colleagues.

(*a*) CUT test. 8 deutan, 1 protan, 1 normal (both eyes).

(*b*) Ishihara (10th edition). 23 errors.

(*c*) HRR and FM 100 Hue. Marked deuteranomaly.

(*d*) Anomaloscope. Mid-point 30, range 60.

Trials indicated that a magenta filter could assist and he was prescribed a light purple pair (figure 9.8) of distance lenses for optional use. A few days after using this appliance he expressed great satisfaction. 'I have seen things that I knew existed but could not see before, e.g. traces of blood in veins... poultry showing green blemishes and slight marks on poultry caused by fat.... I can see different shades of green in trees and grass... clouds are no longer grey. We have a colour television and I can now see and appreciate more'.

After almost one year he reported that he wore the tinted lens constantly with benefit. He considered that 'driving glare' was reduced but the likelihood of his being safer driving at night with untinted lenses was stressed.

The tint he was wearing reduced his Ishihara errors by five. Filters 502 or 13 enabled him, with these (enhanced purples) filters, to read almost all Ishihara plates correctly and a new pair of spectacles was ordered, urging the wise choise of non-tinted lenses or a particular tint according to need. Figure 9.8 indicates the respective transmissions of the most important filters involved. His new spectacles enabled him to appreciate colours more fully.

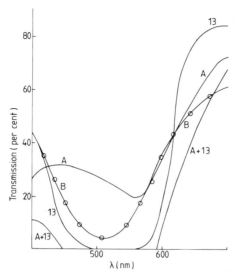

Figure 9.8 Filters used by patient BC, A being the initial tint and B representing the dyed improvement. 13 is Cinemoid 13 Magenta and A + 13 is the transmission of A and 13 combined. B was the result of seeking a dyed imitation of 13 in a prescription lens.

9.4.3 Case CB (DA)

A medium to strong deuteranomalous male aged 16 with hobbies of electronics and painting claimed that he had no difficulties with colour codes. A variety of tests showed his anomaly but he could demonstrate his 'practical' skills in colour identification; his CUT test record was 4/10 deutan.

Ishihara (10th edition) deutan errors of 24/24 were unaffected by a Neophan 50 filter but a 13 filter reduced these errors to 4/24.

9.4.4 Case DF (P)

A male protanope aged 35 years, gave typical results on a variety of tests, including a complete anomaloscope range. He complained of difficulties when using coloured inks, etc.

Different filters gave varying results, as indicated below, and resulted in the prescription of a deep rose tint in CR 39 prescription lenses. He found Wratten 25 (red) helpful for some colour decisions but inconvenient in the gelatin form. His preferred method is to alternate vision with and without a filter; thus he identifies a green object by its 'darkened' appearance with the filter (see figure 9.6).

	No filter	Filter 24 dark green	Filter 35 deep amber	Filter 6 primary red
10 selected Ishihara plates	10/10 errors	10/10 errors	3/10 normal	9/10 normal
CUT test	7/10 protan	6/10 normal 1/10 deutan and protan 1/10 tritan		8/10 normal 1/10 tritan

9.4.5 Case EP (PA; ?DA)

At the age of 15 this young man was referred by an industrial medical officer. He had been successful with electrical hobbies but showed significant anomaly, with both protan and deutan results. Samples of the test results follow.

 (a) Anomaloscope. Deutan mid-point 32; range of 62.
 (b) CUT test. 5 normal, 4 protan, 1 deutan.
 (c) Ishihara (10th edition). 14/24 errors, 3 protan, 1 deutan.
 (d) HRR. 9 errors, protan.
 (c) Giles–Archer lanterns. PA errors including inability to see dark red at smallest size.

A pair of 'deep reddish magenta' filters was prescribed, as CR39 dyed lenses (figure 9.6) and was worn for about a year. The subject experienced some selfconsciousness wearing the lenses and had variable success in use. The spectacles enabled him to read all but two Ishihara plates correctly. He decribed his experiences as follows. 'I had the pleasure of really seeing autumn tints on leaves for the first time.... I could differentiate between reds, browns and greens on a Mahonia bush. It made little real difference ... when trying to pick out colour-coded resistors... thanking you, especially for the autumn colours'.

9.5 SUCCESSIVE FILTERS

Seebeck, and later Maxwell, used red and green filters which the patient looked through alternately. Daltonics noted how objects of different colours became brighter or darker, particularly red and green colours. Loose filters of different shapes were used, or they were glazed into spectacle frames, either one red lens with one green or in bicolour lenses, each lens having the upper half one colour with the other colour below. Mechanically it is easy and cheap to produce such devices. Split lenses can be cemented with modern adhesives.

Kessler (1977) in an interesting but non-detailed account, outlined the way in which red and green filters can be alternated at about 2 Hz. This method may be expected to combine the advantages of steady filtering with those of Bartley's alternation of illumination, and Fletcher has long considered the possibility of a motorised disc built into a pair of 'spectacles' to optimise the alternation of colours and occlusion. The results obtained by Kessler do not appear to have been satisfactory.

In another approach Rosenstock and Swick (1974) described a simple adaptation of the left lens of a pair of spectacles in which the centre of the left lens was 'transparent', while a vertical strip of red filter (about 10 mm wide) was incorporated into the nasal portion of the left lens. Figure 9.9. indicates this arrangement in which the temporal portion of the left lens has a similar strip of *green* filter. The wearer shifts his head when he wishes to interpret the sequential dark and light appearances of coloured objects.

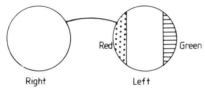

Figure 9.9   Successive filters applied to left eye by Rosenstock and Swick (1974) where the nasal part of the left lens is red and the temporal part is green.

Hope (1972) attributed another proposal to Piringer of Austria and to the National Patent Development Corporation, Delaware (BP 1 291 453). The approach attempts to overcome the likelihood that adaptation will reduce the aid of filters. A mosaic of hexagonal filters is built into these spectacles, using transmissions which are 'suitable' for the individual. The small segments are used in a rapid alternation brought about by head and eye movements.

9.6   MONOCULAR CONTACT LENS CORRECTION

During the last decade Zeltzer (1971, 1973), in the USA, has used chromaticity 'enhancement' in one eye with the chance to compare 'brightness' differences between the two eyes. A monocular contact lens, usually described as red, is worn on suitable occasions, usually in front of the non-dominant eye. The term X-Chrom lens is used. Transmission characteristics and thickness and power are described in the following paragraphs.

The 1975 edition of the X-Chrom Manual describes the characteristics of this corneal contact lens as transmitting substantially in the red part of the spectrum, from approximately 590 to 700 nm. Schmidt (1976) showed how the transmission ranges from some 2% to 570 nm to 75% at 625 nm, flattening out from 650 nm to the red end of the spectrum at about 80%; in addition this author showed a secondary peak of transmission reaching a maximum at 400 nm and going to zero at about 480 nm (figure 9.10).

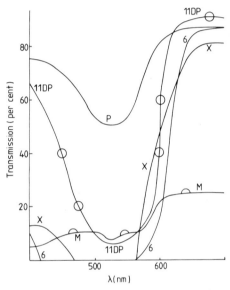

Figure 9.10 Various filters. X, approximate transmission by Zeltzer's X-Chrom corneal lens, from the literature, M being a red corneal lens supplied in the UK (Madden). 6, Cinemoid number 6 Red filter; 11DP is Cinemoid Dark Pink. P is a pink CR39 resin lens used by a deuteranomalous patient with approbation.

Lens fitting is usually arranged to avoid 'edge flare' with a centre thickness of 0.12 mm, although thinner or thicker lenses can be acceptable. A 'chromatic aberration' correction of approximately +0.25 DS is usual. Attention has been given to the possibility of a spectacle correction being combined with a contact lens, a combination which could improve anisoeikonia, if a clear contact lens is not worn in the other eye, or to enable power changes to be made to the contact lens in the interests of transmission changes. Some optical practitioners are prepared to permit restricted wearing of a 'steep' lens dictated by 'colour' considerations.

9.6.1 Prescribing

Protanomalous trichromats are most likely to find assistance from the X-Chrom lens, although reports include some success with deutans and even dichromats. Initial trials with monocular spectacle-type red filters are usual, establishing whether the patient can thereby increase his score on a PIC test. A corneal lens is then introduced, avoiding any distinctive impression of darkening of the filtered visual impression. A clear, or brown for cosmetic preference, corneal lens is usually considered for the other eye. A training regime, involving the discussion of coloured work and hobbies, as well as outdoor observations, is a valuable adjunct. Zeltzer (1973) has outlined the procedure for the device. In the UK several light red materials are obtainable.

9.6.2 Subjective reports

The most striking report by patients is that the X-Chrom lens provides additional lustre or glitter particularly to objects viewed. Bartley (1941) has described how binocular rivalry produces a glittering effect for flickering lights between 6 and 10 Hz. Such lustre is frequently seen when a difference in illumination of the two retinal images exists. Changes in the apparent brightness and chromaticity of objects must be interpreted as part of a learning and relearning process in the early use of the device.

There is a strong possibility that the Pulfrich stereophenomenon will be evident, giving anomalies in depth perception—at least until perception adjustments can be brought into play.

9.6.3 Prospects of success

Mixed reports of success and failure appear in the literature and those of la Bissoniere (1974), Schmidt (1972) and Zeltzer (1973) are worthy of comparison. Evidently performance on the Ishihara plates is often susceptible to improvement and many colour deficient observers are enthusiastic and appreciative of the X-Chrom lens. Further description of X-Chrom-type contact lenses was given by Kemmetmüller et al (1980) with calculations as to the confusion colour contrasts likely to be found by protanopes or deuteranopes. These authors considered that anomalous trichromats could gain more benefit than dichromats from such an aid.

9.7 ILLUMINATION

9.7.1 Quality

In the same way that aspects of the illumination, in particular the quality of illumination, influence tests for colour vision, it is well established that colour defective observers can benefit from certain variations in lighting. It is easier for anomalous trichromats, particularly deuteranomals, to make colour judgments (as provided by PIC tests) under the relatively red illuminant A than with the relatively bluer illuminant C, as described by Hardy *et al* (1946), Volk and Fry (1947), Farnsworth and Reed (1948) and Schmidt (1952). A test pass may be possible in tungsten light with PIC plates designed for 'daylight', by an observer who would fail when the test is carried out under the correct recommended illumination. Yellow filters clearly have a role in assisting some colour defectives, and over a century ago Wilson (1855) described the experiences of Daltonics who found gaslight or candle illumination helpful in distinguishing between some 'confusion colours'.

Boyce and Simons (1977) used a range of lamps to show considerable variations in FM 100 Hue performance among normal trichromats. Thornton (1974) made proposals for illumination, by a mixture of blue, green and red spectral lights, balanced to enhance the colour discrimination of colour normals and defectives.

9.7.2*   Quantity

A slight benefit from high intensities of illumination is noticed for 'colour blind' observers, particularly deuteranomalous trichromats, who have been shown by Schmidt (1952) to make fewer PIC plate errors at 1100 lux than at 270 lux. Boyce and Simons (1977) using normal subjects under different types of illumination, assessed by the FM 100 Hue test, showed that increasing illuminance from 400 to 1200 lux improved 'hue discrimination' for older persons but not for young ones; some such benefit for older anomalous trichromats may apply (see figure 9.11).

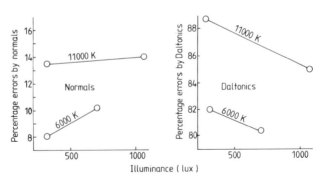

Figure 9.11   Variation of errors on PIC plates, for normal and Daltonic subjects, as illuminance is altered (after Schmidt 1952).

A wide range of luminances for a colorimeter field of just over one degree was used by Thompson and Trezona (1951) for colour matching experiments. A normal trichromat and a protanomalous observer were used to show the deterioration in colour discrimination as luminance is reduced.

9.7.3 Duration

Building on the initial work of Walter (1956) and others, whereby flickering
light had been used to modify normal colour vision, Ball and Bartley (1967)
exposed protan and deutan observers to HRR plates with temporally
modulated illumination conditions; the pulse to cycle fraction (PCF) was
varied using illuminant C. Further deterioration in 'colour deficiency' was
apparently produced by some conditions, although one deutan observer was
helped to achieve better performance using a PCF of $\frac{3}{4}$ at between 1 and
4 Hz. This manipulation of colour vision is readily made by interrupting the
light with a sector disc of the type shown in figure 9.12, ideally avoiding
alternating current for the illuminant. Ball and Bartley found no
blue–yellow vision effects; nor have the present writers found any.

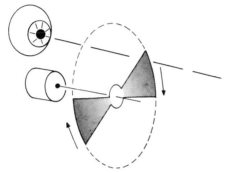

Figure 9.12 Sector disc rotating once a second to give a pulse to cycle
fraction of $\frac{3}{4}$ at 2 Hz while the subject views through the rotating
apertures.

Such an arrangement can be used conveniently by the simple closure of
the eyelids. Sector discs or stroboscopic lights are impractical, clumsy and
objectionable for most everyday conditions. For some colour anomalous
observers colour perception can be improved during short spells of enforced
blinking at a frequency which they can choose by trial. Advice can thus be
given to encourage patients to experiment with blinking or with fanning a
small ruler in front of one eye, the other eye being shut, to enhance their colour discrimination.

9.8 ELECTRICAL STIMULATION OF THE COLOUR SENSE

In the mid 1960s a novel electrical device was produced in Japan and
described by the Hayakawa Electric Company of Osaka as the 'Sharp
Sanivista' colour perception restorer. Electrodes were applied to each of the

subject's temples providing stimulation from two 9 V batteries, using diodes and transistors to pulse a square wave form at 42.5 and 77 Hz, alternating every three seconds.

Training sessions of twenty minutes daily were recommended and this was to be continued for up to one year. The present writers and colleagues applied the device over many months to several Daltonic subjects but the slight variations in their colour perception which were determined were indefinite and not suitable for publication. It is therefore not surprising that little more has been heard of the method, despite the fact that at the time this 'treatment' was linked with Motokawa's resonance frequency approach to the phosphenes which are seen when the eye is electrically stimulated. The fullest description, which gives a very limited account in English, is by Imamura (1964).

9.9 ALTERNATIVES

A journalistic account of the Tokyo 'treatment for colorblindness' has been given by Rogers (1982), in which Yamada is described as having improved colour discrimination in many cases. The method appears to require the location of small 'critical points' on the face; two of these are stimulated electrically, using an oil to assist conduction, often using electrodes on the ears. No explanation is offered and data such as test results do not appear.

10 The Daltonic child

With the increased use of colour in all forms of education, particularly in the primary sector, the Daltonic child must receive special understanding. Early detection of anomalous colour vision is clearly desirable so that parents and teachers can be informed and the child is helped to overcome his defect as far as is possible, from an early age, and later to avoid a few unsuitable careers.

10.1 EARLY DEVELOPMENT OF COLOUR VISION

The investigation of the visual capacities of human neonates is itself in its infancy, but recent electrophysiological and psychophysical techniques have been successfully applied to give greater insight into the colour discrimination of infants and their visual resolution capacity. Objective studies using the visually evoked cortical responses have established that by six months a child can achieve 6/6 vision (Sokol 1978), with the macula and fovea reaching maturity at about four to five months (Duke-Elder 1963). Accommodation is well established by two to three months (Haynes *et al* 1965) and colour discrimination (as measured by photopic spectral sensitivity, psychophysically and electrophysiologically) at two months is almost identical to adult capacity (Peeples and Teller 1978, Teller *et al* 1978). Many psychological studies have established full colour sense at an early age, even at fifteen days (Chase 1937). Reasonable ability in the naming and matching of colours is developed at two years (Cook 1931) with increased accuracy at four years. Infants as young as two months can discriminate blue, blue-green, orange, red, reddish-purple and bluish-purple from white (Peeples and Teller 1975) and at eleven to twelve weeks can discriminate red from green (Schaller 1975). A slight super-sensitivity to blues has been noted among infants between three and fourteen weeks (Dobson 1976 and Moskowitz-Cook 1979). A most extensive literature survey by Verriest

(1981) cited about 100 references and drew two main conclusions; the early detection of colour deficiency was seen to be advantageous and different tests came into their own for different ages.

There is some recent evidence, by Verriest and Uvijls (1977a), that foveal sensitivity is less in the young although they have some advantage peripherally. Furthermore, marked individual variations in the sensitivity of the short wavelength colour mechanisms might lie behind the other colour vision mechanisms during early post-natal development.

10.2   COLOUR EDUCATION

In recent years extensive use has been made of colour naming, coding and matching in the teaching of reading and elementary mathematics. The Cuisenaire rods system involves blocks of material of different length and colour representing different quantities allowing the four rules of arithmetic to be taught by simple selection of the rods. The colour factor is critical, for increasing magnitude is represented by an increase in the depth or strength of the colour. The tendency towards colour printing in recent years has promoted the use of colour to assist in the teaching of reading. Dr Cattengo's words in colour' for example, use forty colour shades representing different sounds. In secondary education colour becomes important in the teaching of geography, geology, biology, chemistry, electronics, art, craft and needlework.

10.3*   AN EDUCATIONAL HANDICAP?

Few comprehensive studies have been conducted, but preliminary reports suggest that although the colour defective child may be confused and perhaps even anxious by the colour emphasis at school, he is usually unlikely to suffer educationally from his defect. Equally surprising, and despite the increased use of colour in primary education, infant teachers seldom report cases of difficulty with colours among children. Either the problem has been overlooked completely, or there is no problem. Studies by Mandola (1969) and Lampe *et al* (1973) found no relationship between colour anomalies and elementary school achievement.

In England, Bacon (1971) gave a series of colour tests of varying complexity to children of all school ages and consulted with teachers of all types of schools, to assess likely difficulties. Greater confusion in using Cuisenaire rods which had been cut down to standard length was shown by colour defective children, but the normal children also had some difficulty.

It is instructive to listen to the youngsters who come for advice on account of their 'colour blindness'. They usually range from eleven to six-

teen years of age and are frequently partially informed as to what careers
are closed to them. Few appear to have suffered from accidents or in their
education but about one third express the disadvantages encountered in
terms of teachers' criticisms and the need to overcome the disability. They
have often disliked discussion of difficulties. In instances where learning or
reading difficulties are found Daltonism may require particularly sympathetic attention.

Reactions of children to the 'handicap' of a colour vision defect were
studied by Carpenter (1983) in Wiltshire schools. This account was
summarised by Voke (1984). Difficulties such as unattractive food colours,
involvement in accidents, poor recognition of car rear lights and lowered
enjoyment of colour television were probed. The older the child the more
reticent they were about their condition. It was usually a school nurse who
explained the condition to these children. In the study it was shown how
readily mistakes were made in art and how coloured chalks on blackboards
caused difficulty.

10.4 EARLY DETECTION ADVISABLE

Although mild defects rarely pose a major occupational handicap (see
Chapter 11) sympathetic guidance to the young Daltonic, to help him avoid
careers which rely heavily on colour recognition, is clearly advisable. There
is therefore a strong case for early detection. In a study of the career choice
of 850 Daltonic children aged ten to eighteen, Taylor (1971) noted that 62%
had chosen one which was unsuitable or incompatible with their handicap,
by the time the test was given in early adolescence. Seventy per cent of the
150 contacted four years later had changed their career aspirations after
consultation (Taylor 1977). Sloan (1963), Waddington (1965), Gardiner
(1972), Voke (1978b,c) and Fletcher (1979) all favour testing at an early
age. Many educationists and physicians consider the optimum age to be at
initial school entry (for example Gallagher and Gallagher 1964 and Bacon
1971). Thuline (1972) recommended a pre-school examination. Sloan (1963)
pointed out that assessment of later possible acquired defects of the colour
sense is aided by early screenings for inherited disturbances.

10.5 HISTORICAL PERSPECTIVE AND CURRENT PRACTICE

The desirability of examining the colour vision of children as a matter of
routine in schools in Great Britain has been recognised for some considerable time but the need to test at an early age has received far less attention.
In 1934 it was recommended that the results of tests given at school should
be passed on to the Youth Employment Service. The Minister of Education

assured the House of Commons in 1944 that steps had been taken to see that the needs of the colour defective child were brought to the attention of doctors through school teachers and nurses. By January 1965 routine testing of colour vision was established in most local authorities in England and Wales; of the 33 authorities who did not test routinely 19 could make arrangements for special cases on request. Some form of test was administered in all but five authorities in England and Wales in 1971/2 and by 1976 all authorities in England gave some form of test (Voke 1976a). Deliberate encouragement to test is given by the Department of Education and Science and there is a specific space on the official school medical card to record the result, although Voke (1976a) found that this was seldom used. In addition 27 authorities repeated the test at some stage in a child's school career. Testing was included in Scottish schools as part of the school medical examination by 1962. But as far as examination at an early age is concerned, in 1964 no Welsh local education authority tested under the age of ten years, and only two English authorities in that year tested at five. Voke (1976a) noted that 14 authorities (also 14%) tested between the ages of four and six, and a further 14 between seven and nine. The most popular age was between ten and twelve (64% of cases).

Without doubt girls should be included in routine screening despite the low incidence of colour deficiency among females. Bacon (1971) takes this view and Voke (1976) noted that in 69 authorities in England (69% of cases) the test is given to both sexes. It is pleasing to note that where Daltonism is identified, teachers and parents are usually informed (Voke 1976a). A few authorities send a standard explanatory letter to parents explaining the anomaly. In 13 authorities it was the practice to inform the child. The child's general practitioner was informed routinely by two authorities. Fletcher has promoted considerable interest in the UK by means of one day courses on 'colour blindness' in London, at which teachers, school medical officers and careers specialists have readily exchanged views. Greater awareness of the range of tests available has stimulated frequent referral of problem cases for a second opinion and at this stage it has often been worthwhile to provide a tentative 'grading' of the anomaly; usually advice on palliative methods (see Chapter 9) can be given.

10.6   METHODS FOR CHILDREN'S TESTS

Several attempts have been made to provide colour vision tests specifically for children, such as those by Waddington (1965), Bacon (1971) and Gardiner (1972); all suffer from severe disadvantages. Alexander (1975) considered no test to be entirely suited for children and indicated that in most cases this was on account of cognitive demands beyond the capability of young children. The task involved is evidently an important feature of test method. Bacon's (1971) approach was to evaluate success or failure in discriminating a bar of contrasting colour from a series of other coloured bars. Gallagher and Gallagher (1964) have noted that a five year old cannot normally cope adequately with a test which depends on the recognition of numbers; shapes seem more appropriate. Thus Gardiner (1972) provided plastics cut-out figures for the young child to place on confusion plates adapted from the Ishihara series. However no diagnostic assessment can be made with this screening test. Sassoon and Wise (1970) reported favourable results with three and four year olds using the D.15 diagnostic test and indicate that the 'game' of arranging the coloured discs in order appeals to many children. Verriest (1981) reported that the 'ranking' task of the D.15 was unfamiliar to Belgian kindergarten children. Undoubtedly tritan errors are often found as false positives. The 'nearest match' criterion, forcing a choice between four alternatives against a standard colour has been developed by Fletcher (1975, 1978b) in the City University Test which can be used with five year olds without the need for physical contact that can seriously soil the colours. The use of dominoes and lotto games in coloured card, described by Verriest (1981), gave variable results. His conclusions distinctly support suitably designed 'games' for use in colour vision testing. There is a special (1970) Ishihara edition which uses patterns of paths, circles and squares.

'Tracing paths' which forms part of several pseudoisochromatic tests is not always reliable. Among 141 boys and 129 girls tested at pre-school age with the Ishihara 'tracing paths' Cox (1971) noted that 40% of Daltonics were missed; when examined two and a half years later with eight Ishihara numeral plates 25% were classed as anomalous. Medical officers responsible for child health frequently report the difficulty of the severity of the Ishihara test and the false positives it often creates (personal communication to authors). Special children's plates by Kojima-Matsubara and a version of the TMC plates are in the Japanese idiom of pictures. Polak's plates contained some pictures, such as a chicken. The illiterate E, used commonly on visual acuity charts, has been used as a means for examining colour vision in children as young as four (Velhagen 1980). Using Ishihara plates, Krekling and Andersen (1973) found about 1% of girls and 6% of boys to be 'colour deficient' in both special and ordinary Norwegian schools. Performance of boys, between 3 and 11 years, was evaluated by Hill et al (1981) using four PIC tests and an anomaloscope. The probability of successful use of each test was determined.

The practical difficulty of applying standard tests to very young children has been daunting. Millodot and Lamont (1974) used Dvorine and some HRR plates to test French-Canadian children (8–18 years). Among 758 boys and 779 girls 6.7% and 0.9%, respectively, revealed red–green deficiencies. The D.15 test was also used and an important suggestion arose, namely that the next generation of boys may have more Daltonism; schools therefore can take note of this. Richards (1975) described a boy's improvement from 'atypical achromate' at age 10, to deutan with poor discrimination at $11\frac{1}{2}$. By age 14 considerable improvement to 'deuteranomalous' had emerged and Richards clearly accepted that improvements (up to adolescence) could be recognised.

10.7 VARIATIONS IN TEST RESULTS

Some idea of the diversity among the percentages of 'failures' when children of different ages are tested by different people at different sites emerges from the following experiences. In a certain area of England, within the last decade, the 'failures' among boys were reasonably constant, between 3.25% and 4%, whereas the figures for girls ranged from 0.18% to 1.2%. The area will not be identified. As the somewhat 'rounded' figures show there was great variation and a large proportion of doubtful results. It is possible to read much into these results and to conclude that greater standardisation of approach, with a better range of methods, is indicated.

(*a*) Boys
Age about 5 years. Sample of approximately 4300. Failures on Ishihara 'tracks' were 3.25% overall (variations between testing sites such as no failures, 2% or 9% failures; half as many 'doubtful' responses recorded as failures).
Age about 10 years. Sample of approximately 5200. Failures on Ishihara numerals were 4.00% overall (site variations 2%, 4%, 4.5% and 7%; with three times as many failures recorded as 'doubtful').
Age about 15 years. Sample of approximately 5000. Failures on Ishihara numerals were 3.8% overall (site variations 1%, 3%, 5% and 7%; with six times as many failures recorded as 'doubtful').

(*b*) Girls
Age about 5 years. Sample of approximately 2000. Failures on Ishihara 'tracks' were 1.2% overall (variations between testing sites such as no failures, 0.5% or 2.5% failures; as many 'doubtful' responses recorded as failures).
Age about 10 years. Sample of approximately 3300. Failures on Ishihara numerals were 0.33% overall (site variations, nil, 0.8% up to 1.1%).
Age about 15 years. Sample of approximately 4400. Failures on Ishihara numerals were 0.18% overall (site variations, nil, 0.2%, 0.3%, and 0.5%).

10.8 MOSAICS IN PLASTICS MATERIALS

Fletcher (1979, 1980d) has approached the need for 'child-proof' tests of special design using selected plastics laminates in one centimetre square units; these are illuminated not by daylight but by tungsten light, on the grounds that this is universally available for easy standardisation. Various 'games' are presented. A coloured unit is presented and the colour most nearly matching this has to be chosen from a selection. The simplest form (figures 10.1 and 10.2) can be to fill a gap in a wall. A set of small dominoes can be used (figure 10.3) in which successive matching errors can be revealed. Children as young as 4 years accept this mode of presentation. Alternatively if the concept of 'difference of colour' is feasible the child can point to contrasting units (figures 10.4 and 10.5).

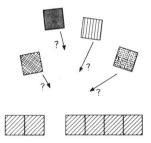

Figure 10.1 The task is to complete a 'wall' of coloured 'bricks' composed of flat miniature coloured tiles. One of the loose alternatives having the *nearest* colour from four choices has to be selected and inserted into the gap.

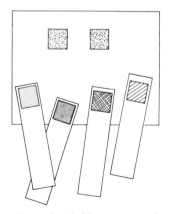

Figure 10.2 Construction of a bridge or 'stepping stones'. Two fixed squares of identical colour are to be bridged by the best choice from the loose colours, which are all on small holders to enable them to be moved easily.

Figure 10.3 Plastics dominoes, each comprising two colours with one used as a starting point which is the same each end. The correct confusion colours are included to reveal colour vision defects.

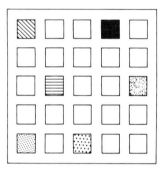

Figure 10.4 A mosaic comprising a background colour, itself a confusion colour for one of the squares which are seen as contrasting by normals. The mosaic can be rotated for retesting.

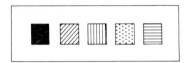

Figure 10.5 A set of five coloured squares from which the subject has to identify at least one which is obviously a different colour. Confusion colours are included, with one very contrasting colour.

Children often respond to the challenge of noting the difference between pictures, and screening for 'red–green defects' is possible along the following lines. In figure 10.6 two pseudoisochromatic colours A and C would be confused by a Daltonic while N is nearly a normal colour match for C. A Daltonic says that N is the 'different' colour and a normal says that A is different, assuming he accepts C and N as being sufficiently similar to each other. Protans could be differentiated from deutans if (as in figure 10.7) a 'first mode' of presentation is used, with P and D confused, respectively, with C; protans and normals choose D as different but deutans say P is different. Figure 10.8 shows a 'second mode' where protans choose D and deutans identify P as different from the other two colours. Table 10.1

indicates how six 'plates' provide a reasonable chance of detecting a protan defect (*D* chosen as different) or a deutan (by consistent choice of *P*). Normals would be expected to choose *D* the same number of times as *P*.

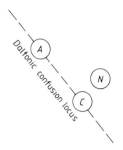

Figure 10.6 Colours *A* and *C* are positioned on a colour diagram which forms the imaginary background for the figure and are confusion colours for someone with a red–green defect.

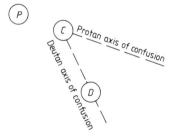

Figure 10.7 A 'first mode', to separate deutan subjects from others.

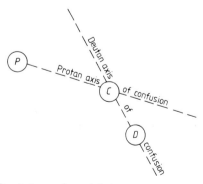

Figure 10.8 A 'second mode', to separate protans from others.

Table 10.1 Use of 'plates' in detecting protan and deutan defects.

| | Colour seen different by | | |
	Normal	Protan	Deutan
Plate 1, Mode I	D	D	P
Plate 2, Mode I	D	D	P
Plate 3, Mode I	D	D	P
Plate 4, Mode II	P	D	P
Plate 5, Mode II	P	D	P
Plate 6, Mode II	P	D	P

These principles have been incorporated into a practical test, produced by Hamblin Instruments, London, as the Fletcher–Hamblin Simplified Colour Vision Test (Fletcher 1984). Three 'units' are provided, each constructed of a mosaic of 16 washable squares. Units 1 and 3 detect 'red–green' defects and the third unit is for tritans. Units 1 and 2 involve detection of colour contrasts by normals, while selected colour confusions are possible in order to fail the subject. Unit 3 shows an isolated 'model' colour with four others, one of which has to be selected as the best colour match for the 'model'. With this test most 'significantly defective' subjects can be detected and in some cases relatively slight conditions are revealed (see figures 10.9 and 10.10). Tungsten light is recommended for the test.

Figure 10.9

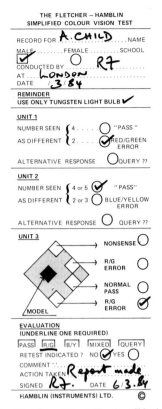

Figure 10.10 The Fletcher–Hamblin simplified colour vision test.

10.9 A REFLECTION METHOD

Munsell papers are useful as components of colour tests which can be specified to fairly close tolerances but their drawback is that they are easily spoiled and therefore difficult to use with children. A method used by Fletcher (1980, unpublished) overcomes the difficulty to some extent, while encouraging a child to use touch as a means of identification of his choice of the 'different' colour in a set of three Munsell papers. The samples are mounted upside-down in the roof of part of the simple device but are seen by reflection in a mirror below them; the appearance of three shaped apertures (each with a tactile reinforcement) is produced by using a perforated plastics sheet over the mirror, in which there are three apertures. One set of three apertures resembles three balls, another three flowers and fish, motor cars and cats are also used in order to stimulate response, such sets being changed rapidly.

A tungsten lamp illuminates the samples and the apertures. The child views the apertures and his own hand through an eyepiece so that his line of sight is directed correctly and a blue filter in the eyepiece effectively modifies the quality of the illumination, as shown in figures 10.11 and 10.12. Few children have difficulty in peeping through the eyepiece and in making the series of 'difference' choices.

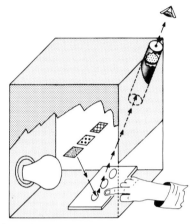

Figure 10.11 Coloured objects, which cannot be touched or seen directly, are reflected in separate areas of a mirror. In the viewing tube is a blue filter since tungsten illumination is used.

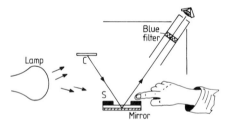

Figure 10.12 Light from a lamp is reflected from colour C and then from the mirror. A tactile surround S locates the finger when the child has chosen the colour which appears to be different from the others. The blue filter is positioned as shown.

10.10* LIMITATIONS OF PRESENT PRACTICE

While it is pleasing that there is such widespread use of colour vision tests among the school population, it is clear that the method of testing leaves much to be desired. There is, at present, no standardised procedure or guidance on this matter given to specialists in community medicine responsible for child health, working with the local authority; it is the responsibility of individual medical specialists to make their own choice. Very little thought appears to have been given to the choice of test. Very often it was reported to have been 'inherited from predecessors' and its use continues without further consideration of the possible alternatives. 'Historical reasons' accounted for the present policy in a third of all authorities. In only one authority had a deliberate examination of the alternative test methods been made (Voke 1976a).

The Ishihara test is used almost universally. The Keystone screener, which has poor facilities for colour vision testing (only one or two 'Ishihara-type' plates), is used by a third of all authorities, but in some cases the complete Ishihara tests are supplemented in possible cases of abnormality. These PIC tests give poor differentiation between protan and deutan and no idea of severity of defect (see Chapter 7). Clearly a more complete diagnosis of defect should be made so that some idea of the potential handicap can be formed. Great confusion is evident over the question of whose responsibility it is to provide career guidance for colour defective children; few child health specialists consider it to be their role. With limited time and resources the task must be to screen and to refer the Daltonic for more precise examination. Thus a two-tier system of testing would be ideal. A screening test would identify those children who were likely to have a colour defect, and a further diagnostic test for failures would indicate the severity and the likelihood of handicap in leisure and educational activities and future careers. A series of tests, possibly coupled with a practical assessment of colour ability (trade tests) and a sympathetic interpretation by a person with experience is most helpful in a referral capacity. Advisory clinics for the Daltonic (in particular, children whose career choices may conflict with their handicap, and adults who experience trouble with their work because of their anomaly) are needed in at least the major cities of the UK. The facility for such consultation has been available at the Heathfield Hospital, Ayr, Scotland under the direction of consultant ophthalmologist W O G Taylor for more than ten years. In the first five years of its functioning more than 900 children were examined; 70 proved to be colour normal. These figures alone show the need for and value of detailed examination and consultation of this nature. A clinic has recently been introduced at the Oxford Eye Hospital.

10.11* PRACTICAL RECOMMENDATIONS FOR CHILD TESTING

As in any situation of this type attempts must be made to achieve standardisation of illumination, test materials and mode of presentation, with carefully determined protocol and well trained staff. A rapid screening test can be assembled locally by extracting the most useful of the Ishihara plates (e.g. plates 6, 8 and 5 from the 1978 edition) plus plates 1, 2 and 5 from the City University test (first edition) or the first three plates from the second edition of the City University test. Children with less than perfect responses can be retested, preferably by another person and with alternative tests.

The very young child can be tested if cooperative and the following approach can be used for four or five year olds. A quiet room, familiar to the child and providing suitable illumination, should be used and one child should be brought in at a time with someone he or she trusts, such as an adult layworker, teacher or parent, to sit at a suitably low table with a neutral coloured surface. A table-cloth, white or black, can be used. Behind the table, on another or on a chair, various tests are kept until, one at a time, they are brought forward. A record should be kept. After brief kindly pleasantries, two identical red 'building brick' cubes are placed before the child and he is prompted to agree that they are the same colour; two others, different in colour, are next shown and their difference is agreed. Simple colour naming can be elicited and praised at this stage. Finally three cubes are presented, two the same and one different, for confirmation by simple discussion that the child can point to the 'two the same' and 'the one which is different'. Cards with colours can be used but they must be perfect matches as some children have a keen sense of slight differences. The preliminaries can lead on to chosen tests, selected from a range. The mosaic test, or other method of presenting 'difference' identification, is a useful start and one of the matching games should follow. If PIC plates are used it is best to place a set of mixed black card or plastics numerals in front of the child; the PIC plate should be further away, not to be touched, and, starting with a 'dummy' plate, the child should be asked to touch the loose numerals to indicate his choice. With care, superimposition of the numeral onto the plate may be permitted. Frequently a child names the number after one or two silent matches.

A D.15 test can be attempted in many situations. Wherever possible a variety of approaches should be provided, without exceeding the time which can be spared. Most children can be assessed within five minutes if the organisation is adequate. The aim is to identify a significant degree of Daltonism and to arrange suitable follow up.

Filters and other palliative approaches described in Chapter 9 should be introduced with the parents' cooperation, for use in a selective manner. The parents should be informed as fully as possible and steps should be taken to ensure that at least the extent of the difficulty is known to teachers, medical advisers and the family. Whenever possible, a simple explanation of the cause and effects of the condition should be offered to the child.

10.12 APPLICATIONS TO MENTAL HANDICAP

Mental patients often have child-like characteristics and colour vision tests produced for children can be applied to the mentally handicapped. Since such patients frequently receive drug treatment and may display syndromes where colour vision is useful as a 'marker', there are good reasons why inherited or acquired colour vision deficiency should be investigated in such cases. Although reliable results are not always easy to obtain, the incidence of Daltonism seems to be similar in the mentally retarded population and the normal population (Schein and Salvia 1969, Salvia and Shugerts 1970).

10.12.1 The incidence of Daltonism in these patients

The mentally handicapped often have other handicaps, including visual troubles, and a survey by Ellis (1979) supports this contention, with references to many studies. Although it would be expected that mental patients might be slightly more liable to defective colour vision the proposals of Archer (1964) and Krause (1967) are surprising. They proposed figures of 30.9% and 22.3% for the incidence of colour vision deficiency for male mentally handicapped 'children and adolescents', their figures for females being 26.1% and 21.5%. Earlier, Kratter (1957) and O'Connor (1957) had proposed figures of 10% and 13% for male 'feeble minded' adults. Courtney and Heath (1971) used the HRR plates in a special manner on over 100 'mentally retarded' and 107 'emotionally disturbed' children and found between 4.3% and 9.7% of male colour defectives.

10.12.2 The influence and choice of tests

Results obtained are evidently dependent upon the choice and method of use of tests. This is substantiated by Salvia and Ysseldyke (1972) whose sample of 90 boys with a mental handicap was examined with five different tests. These were the D.15, HRR, Ishihara and Dvorine tests as well as a filter anomaloscope. Percentages of colour deficiency ranged from 11% (anomaloscope) to 40% using Ishihara numerals and estimates were made of false positives and false negatives. It was concluded that all tests gave substantial false positives and than none of the tests was 'sufficiently accurate' for this purpose. This was in accord with earlier data, obtained with certain HRR plates by Salvia and Ysseldyke (1971).

Four tests were given to 32 mentally disordered men by Verriest et al (1980a); the 100 Hue test possibly proved excessively hard or tedious and there was over a 50% failure rate with D.15, showing worst results among non-active patients. Younger subjects tended to perform relatively badly. This study was linked to aspects of alcoholism in other patients and, while showing how difficult it is to interpret the cause of poor performance, brought out the value of colour vision tests in this context. The defects could be classified as 'acquired'.

A variety of suitable tests is as important for mental patients as it is for children, since some will work with a particular individual and others will be rejected. Ill-treatment of tests must be anticipated. Assuming that a patient (accompanied by a nurse who is well known) sits at a table with controlled illumination, it is best to start with obvious colour sorting. Two red and one green bricks (or cards) are presented and the patient's cooperation and ability to name colours is probed. Identification of 'similar' and 'different' colours should be possible with the majority of adult Down's syndrome patients resident in a hospital.

In one (unpublished) series one of the authors applied several different tests to about fifty patients with the variations of success shown in figure 10.13. In 146 separate attempts only 5 failed through lack of cooperation by the patient but a total of 47 attempts (each with a different patient) failed to be categorised as 'acceptable' or better. The important point is that usually one mode of approach will work, even if others fail.

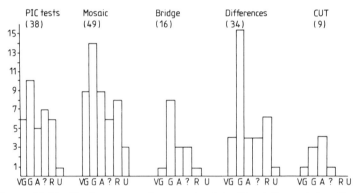

Figure 10.13 Performance of approximately 50 adult mental patients with different colour vision tests. Local conditions and patients' likely abuse of tests accounted for the variety of sample numbers. VG, very good; G, good; A, acceptable; ?, uncertain response; R, random, U, uncooperative.

One seriously handicapped man could perform satisfactorily when invited to identify 'differences' between confusion colours, while failing completely to be engaged by other methods. Another (retested over a period of months) consistently revealed deuteranomaly but only on laboriously presented PIC tests and the children's 'bridge' (matching) tests; he appeared 'normal' by using plastics mosaic and gave equivocal results with the 'differences'. This man, with Down's syndrome, was not receiving any treatment with drugs.

11 Vocational and industrial aspects

11.1* INTRODUCTION

The impact of colour on society, and in particular the world of work, has increased steadily in recent years; thus numerous occupations, careers and professions require some degree of colour identification. The widespread use of colour coding (often where safety is important), aids identification of similar objects; it conveys information quickly and has permeated every aspect of life, particularly manufacturing industry. Increasing industrialisation has increased competition. More discerning customers require colour tolerances of consumer goods to be small. The observer with normal colour discrimination has a remarkable sensitivity to hue and brightness differences so that industry must maintain the highest standards in colour control.

11.2 A LONG-STANDING OCCUPATIONAL PROBLEM

These problems are not altogether new, for long before industry became organised on the scale we see today the Romans displayed an awareness of the need for uniformity in the design of their coloured mosaics. Monks of the Middle Ages made a careful selection of colours for their illustrated manuscripts. The dyeing of cloth and yarn, one of the earliest industrial activities, required careful colour matching. Tann (1967) noted that the cloth was inspected on arrival in London in the late sixteenth century and that clothiers were fined if the colour of goods was faulty.

Little thought was given in the past to the vocational and industrial problems of Daltonism or to the social consequences. Texts dealing specifically with occupational aspects of vision seldom devote adequate discussion to colour vision problems at the practical level which industry requires. Yet isolated examples have appeared in the literature and it is noteworthy that one of the earliest accounts of defective colour vision involved the colour problems experienced by Harris, a shoemaker from Cumberland, described by Huddart (1777). The following year Whisson (1778) mentioned the occupational difficulties of a shoemaker, Scott. Jeffries (1883) reported cases of numerous individuals who were frequently in doubt about colours in their work; a bookbinder, an architect, a weaver, a physician, a post office clerk and a farmer were included. He noted that 'volumes might be written on the subject if all the embarrassment to which they gave rise were cited'.

Some activities in the work environment rely heavily on colour assessment or critical judgment; these include the obvious colour industries of dyestuffs, textiles, ceramics, paints, inks, photography, printing and electronics where inability to identify basic colours might cause embarrassment or reduced efficiency. Other less obvious ones include food production/colouration, brewing, cosmetics and pharmaceutical manufacture, tanning, brick and furniture industries and the repair of car exteriors. Buyers and sales people of every kind frequently have to make colour judgments, from the butcher to the jeweller; storekeepers and packers are also involved. A hairdresser might make the wrong selection of hair dye, post office clerks could have difficulty in identifying postage stamps and the tanner, upholsterer or tailor may confuse coloured threads. A waiter might mistake a bitter lemon drink for Coca Cola!

It is surprising therefore that the potential difficulties of the 8% of men who have some degree of colour deficiency have been largely neglected, and that, in general, industry has been content to learn by its mistakes. An error in the weaving of a carpet, unripe tomatoes picked too early, a day's production of colour-coded resistors that had to be discarded, green sweets in place of red ones—all these mishaps occurred recently in this country, costing the companies involved some degree of loss. All had a common cause—a *single* employee with faulty colour vision.

Even in the early days of industrial medicine Pearce (1934) and MacGillivray (1947) emphasised the *slowing down of industry* by the colour defective employee and Pearce also mentioned the financial consequences of industrial errors made by the colour anomalous worker. Cavanagh (1955) wrote of the possible dangers, but Norman (1948) noted that there were no statistics available of accidents in which the defective colour vision of an employee may have been responsible.

It is a useful exercise to group into categories work activities in which it is desirable or vital for operatives to have normal colour vision, provided it is taken as a general listing and is not used as a rigid guideline. Few authors have attempted to do this in the past, perhaps because such a list could never be fully comprehensive, and many jobs fall into several categories, for there are often different activities with a specific trade, profession or occupation. However, drawing on published skeleton lists compiled by various organisations and individuals, with additions, a fairly comprehensive listing is given in tables 11.1, 11.2 and 11.3.

Table 11.1 Jobs, careers and industries where defective colour vision is a handicap and in which important consequences might result from errors of colour judgment. From Voke (1980b).

Air traffic controller	Cotton grader
Buyers—textile, yarns, tobacco, food e.g. fruit, cocoa, timber	Coroner
Car body resprayer, retoucher	Forensic scientist
Cartographer	Market gardener e.g fruit
Ceramics—painter/decorater of pottery	Meat inspector
Ceramics—inspector (quality control)	
Chemist and chemicals laboratory analysis	Oil refining
food chemist	
teacher of chemistry	Paper making
manufacturer of chemicals and polishes and oils	Pharmacist
	Plastics
Colour printer, etcher, retoucher	Paint maker and distributor
Colour photographer	
Colour TV technician	Restorer of paintings/works of art
Coloured pencils/chalks/paints manufacturing	Safety officer
Colourist/colour matcher in paints, paper, pigments, inks, dyes, wallpaper	Tanner
	Tobacco grader

Table 11.2 Jobs and careers where good colour vision is desirable but in which defective colour vision would not necessarily cause a handicap. From Voke (1980b).

Accountant	Builder/Bricklayer
Anaesthetist	Buyer—textiles, yarn, tobacco, food e.g. fruit, cocoa, timber
Architect	
Artist—graphic, commercial, advertising	Carpenter
Auctioneer	Carpet/lino fitter/planner
	Chiropodist
Barmaid/barman	Clothes designer
Bacteriologist	Cook or chef
Baker	Coroner
Beautician	Confectioner
Botanist	Cosmetics director (stage, film, TV)
Brewer	
Butcher	

Table continued

Table 11.2 *Continued*

Dental surgeon and technician	Nurse
Draughtsman	
Dressmaker	Optometrist/ophthalmologist
Driving instructor	Osteopath
Driver in public services e.g. bus	
	Painter
Engineer (various)	Pharmacist assistant (counter service)
	Physician
Farmer	Physiotherapist
Fishmonger	Post Office counter assistant
Florist	Potter
Forester	
Furrier	Salesman/woman (fabrics, drapery, yarns, wool, carpets)
	garments/footwear
Gardener and landscape gardener	china and glass
Geologist	linen
Gemmologist e.g setting stones, diamond grader	cosmetics/toiletries
	jewellery
Grocer	confectioner
	stationer
Hairdresser	storekeeper
Horticulturist	Shoe repairer
	Surgeon
Illuminating engineer	
Interior decorator/designer/planner	Tanner
	Tailor
Jeweller	Telephone switchboard operator
	Theatre/stage props manager
Librarian	
Lighting director (stage, film, TV)	Veterinary surgeon
Manicurist	Waiter
Metallurgist	Window dresser
Milliner	
Miner	Zoologist

11.3* TRANSPORT INVOLVEMENT

Those entrusted with the safety of passengers (pilots, seamen, train crews) should clearly have normal colour vision or only *very* mild abnormalities; efforts to maintain such standards extend back to the last century after a number of accidents by rail and sea were found to be caused by faulty colour recognition. The fear of widespread disasters has led to the enforcement of stringent standards and the Armed Services are similarly concerned. The first practical colour vision test to be introduced was Holmgren's wool test in 1887, following a railway accident in Sweden, probably caused by defective colour vision.

Table 11.3 Jobs and careers requiring perfect colour vision. From Voke (1980b).

Armed forces—certain grades (see Chapter|13)

Civil aviation

Colour matcher in dyeing, textiles, paints, inks, coloured paper, ceramics, cosmetics

Carpet darner/inspector, spinner, weaver, bobbin winder

Electrical work
  electrician
  electronics technician
  colour TV mechanic
  motor mechanic
  telephone installer

Navigation—pilot, fisherman, railways

Police—certain grades

Radio—telegraphy

Occupation where defective colour vision may be an asset
Camouflage detection.

The history surrounding the introduction of colour vision requirements for the Armed Services and Merchant Navy makes fascinating reading (Topley (1955) gives details and see Chapter 12). Farnsworth (1957a) felt there was little justification for disqualifying the mild defective from occupations such as pilot, locomotive engineer and bus driver.

Rigid colour vision requirements have also been challenged by Hoogerheide (1959), and the colour vision requirements in different flying operational roles have been the subject of international discussion through AGARD (1972). There is a growing feeling that mild anomalous trichromats are acceptable for flying and naval duties, a belief which led Farnsworth to develop his dichotomous D.15 test, to separate the 'safe' from the 'unsafe' individual. Cavanagh (1955) was correct in pointing out that any anomalous trichromat who would not be accepted for certain duties in the Armed Forces might be considered safe for work in civilian life.

11.4 INDUSTRIAL CONSEQUENCES

Industry and commerce have evidently felt the consequence of Daltonism for many years but the absence of guidelines and the lack of suitable consultation has left attitudes confused. The large public corporations (such as the Civil Service and the Gas and Electricity Boards and the Post Office)

gave preliminary consideration to colour vision requirements; their medical departments compiled categories and grades within their operations which required satisfactory colour vision (see Chapter 13). Seldom, however, was this based on trials to discover how capable the colour defective was in performing the required colour tasks. Research associations frequently advise individual member companies, but the small individual company is most at risk and has few sources to approach for consultation. The Employment Medical Advisory Service seeks, through its itinerant industrial medical advisers, to assist those companies who are too small to warrant employment of their own medical officers and there is a case for regarding the matter as part of occupational health care.

Clearly the industrial hazard of defective colour vision must be avoided if possible. It is in the interest of both employer and employee that defects of colour vision should be identified before engagement so that where there are slight differences from normal the possible consequences can be assessed in the light of the individual situation.

The importance of good colour discrimination among operatives and employees varies widely from industry to industry and from profession to profession. The consequence of faulty colour perception also differs and the financial consequences are of most concern to industry. It is inefficient to employ someone who may be slow, indecisive or inaccurate when judging colours in industries such as textiles, paint manufacture or colour printing.

Social consequences tend to be treated less seriously but the humiliation felt by an individual among his work-mates associated with the simple error of mistaken colour identification may be considerable. It is a strange, but dangerous, fact that those with faulty colour perception are often loath to admit to their problem and will go to great lengths to conceal it.

The personal disappointment felt by a youngster who has set his or her heart on a specific vocation, only to be refused on the grounds of Daltonism, is a saddening experience for all involved and could so easily be avoided if extensive testing facilities and appropriate counselling were open to children at suitable ages. Yet this approaches the problem from only one angle. Employers, trade unions, associations, professional bodies and recruiting agencies need to be aware of the extent of the handicap imposed by colour vision defects in specific fields, setting their standards accordingly. Too much arbitrary setting of standards has been seen in the past, based neither on experience nor on proper evaluation. Different testing methods have been neglected leaving those most closely involved in the technical, personnel or medical areas unable to offer appropriate advice. Often the colour defective individual has been the victim of unfair discrimination.

It is unfair to exclude the Daltonic from professions and occupations where performance would only be slightly impaired, and where they might, given a chance, prove to be able to overcome their minor disability. Nevertheless industry is seldom prepared to accept mild colour defectives for employment in colour spheres—the risk is considered to be too great. Only when an individual is highly attractive for employment on other accounts will this belief be questioned.

Many authorities accept that mild anomalous trichromats can be entrusted with some jobs involving colour provided they are aware of their possible handicap and they realise their limitations (Hardy *et al* 1954, Farnsworth 1957a, Walraven and Leebeck 1958, Cole 1964 and Voke 1976a). However a moderately or severely colour defective person must never be placed in a situation in which failure of colour judgment or recognition would jeopardise persons or property. The problem is usually how to recognise 'unsafe' individuals.

One of the present authors has completed a study on the effects of faulty colour perception on industry (Voke 1976a), the first major study of its kind. Twenty major colour industries were examined in detail and an overall impression was gained from numerous visits to inspect the jobs in progress. Discussions with relevant personnel, both employees and employers, and the comments of a number of colour defective individuals working in these areas proved to be valuable. Several industrial medical advisers working within the Employment Medical Advisory Service (EMAS) presented helpful opinions through personal discussion. A field trial was undertaken in a paper mill to determine the performance of paper colour matchers at standard 'clinical' colour vision tests. Five hundred copies of a questionnaire were circulated to individual companies to investigate the role of human colour judgment in order to establish the incidence of testing. Industrial errors attributable to Daltonism were investigated. In addition a number of industrial jobs involving fine colour judgment were simulated, using 100 colour defective volunteers and a group of control normals to give an indication of the potential handicap of the colour defective in industry. An analysis was made to compare performance at these tasks with scores on clinical colour vision tests. Thus informed recommendations could be made, and a comprehensive list of jobs and professions which required some recognition of colours was compiled, with specific groupings. The interested reader is further referred to a summary of these finding in a substantial document of recommendations to industry and the professions written by Voke (1980b). Individual occupations are treated in detail with the role of the observer in each of the colour jobs outlined, together with indications of the current testing methods employed, if any.

11.5 EXTENT OF INADEQUATE TESTING

The severity of the defect is of great importance in the occupational and industrial context. Yet where testing methods are used in industry today they are almost always restricted to the screening type which lack diagnostic ability. The use of such tests, which fail the entire 8% of colour anomalous, rather than diagnostic tests which grade individuals, has continued unquestioned by those responsible for personnel selection. Frequently their use is confined to the pre-employment stage with little opportunity for examination at a later date. Futhermore such tests rarely indicate the blue function so have little value for acquired cases due to disease. Perhaps most important of all is evidence that failure at these chart screening tests does not correlate well with actual colour ability (Voke 1976a). Industrial users frequently report them to be too severe for their purposes and that many who fail can cope adequately with a colour job. Jolly (1954) and Jordinson (1961), who were both involved in critical colour industries, showed this earlier, and the numerous contacts made by the authors, together with the systematic simulation of industrial colour jobs in the laboratory and an evaluation of the ability of colour defective volunteers, confirm this view. Nevertheless industry has been slow to investigate the alternative testing methods now available, which would suit their needs and provide a more realistic measure of performance.

11.6*   REQUIREMENT OF A REALISTIC APPROACH

Those who are in a position of responsibility to make decisions on medical or technical grounds concerning personnel placement and recruitment should be assisted to a more realistic approach to colour vision examination, suited to the needs and requirements of the individual trade, profession or industry; frequently this role is underestimated. At the outset a number of questions must be considered: is the purpose of testing to eliminate all Daltonics or should a less stringent standard be applied in order to qualify a larger number of applicants? Are we dealing with selection of the 'safe' and rejection of the 'unsafe'? A judgment must be made to set an acceptable pass/fail level which is appropriate for the situation. This involves a consideration of whether any form of anomalous colour vision is compatible with the requirements, by estimating the likely handicap for a specific job. Testing by the layman in the occupational environment is therefore not an easy matter.

Colour vision is most suitably investigated with at least two tests, seeking both the nature and the extent of any defect. Sometimes such a procedure may not be practicable in the industrial setting. It is unreasonable to expect a small company to have more than one standard colour vision test, and in some cases a single diagnostic test will suffice; it can be used as a simple screening test as well as providing more detailed information. An assessment can be made effectively by the appropriate choice of test for the circumstances. An unambiguous scoring method, clear instructions and an individual record sheet will aid the inexperienced tester to reach the correct assessment at the possible risk of over-simplification, which in such circumstances can be safely tolerated.

Where a clinical test is also a close simulation of a specific job or task, there is some justification for making a diagnosis solely on the basis of the performance at this one test. The examination of a pilot or a train driver by a lantern test, where coloured lights must be named, is an obvious example. Where possible a test should be chosen to represent a level of difficulty which is consistent with the colour requirements of the job in question. Thus an individual who is daily concerned with small colour differences in the textile or paper-making industries, or in grading gems, should be examined, where possible, by a sophisticated colour matching or discrimination/colour blindness test. The FM 100 Hue test involving the task of arranging carefully produced coloured paper samples in a colour sequence, is ideal for this purpose, as is the Lovibond colour vision analyser. 'Trade tests' which simulate closely the job in question, have a useful function for they can indicate, in some circumstances, the actual handicap; but since they are difficult to standardise, and give no general information on the characteristics of the individual's colour sense, they should always be used alongside a clinical test.

The approach to testing in the occupational sphere cannot therefore be divorced from the material covered elsewhere in this book and the reader is particuarly advised to consider again comments made in Chapter 5 on the approach to testing. Familiarity with the range of defects and testing methods is also assumed as outlined in Chapters 3, 4, 7 and 8. Chapter 8 is of particular value; it deals with interpretation of results in borderline cases and with decisions and applications to specific needs, with extensive examples. With such knowlege the examiner is in an excellent position to embark on the often difficult task of setting a dividing line between the acceptable and unacceptable.

11.7 INDUSTRIAL EXAMPLES

Despite increasing automation the eye continues to play a vital role in colour assessment in a wide variety of industries. A few of the more important ones will be discussed to give examples of critical colour jobs and the consequences of defective colour vision.

11.7.1 Textiles and dyeing

Despite the introduction in recent years of spectrophotometers and computer match prediction techniques, particularly among the larger manufacturers, the eye continues to remain the final arbiter. The ultimate assessment of acceptability must be a visual detection by the discerning customer and it is unlikely that instrumental methods will completely replace the human eye. One of the main obstacles to automated colour analysis and correction is the lack of agreement between objective methods of colour analysis and visual discrimination.

In the textile and dyeing industries there are few jobs where some degree of colour judgment is not required. Although men form the bulk of those working in the initial processes, women make up the majority of machinists and sales staff in textiles. Dressmaking, tailoring, retail clothes sales and haberdashery and carpet sales are jobs frequently occupied by women, who will be at less risk than their male counterparts of inheriting a colour vision defect. But critical jobs such as technical managers, chief chemists, carpet weavers and 'bobbin boys', textile chemists, bleachers and dyers, quality control staff and laboratory personnel, are often male-dominated, and many of these men will never have been screened for Daltonism.

These activities call for the highest degree of sensitivity to colour differences. The job of colour matcher, colourist or dyer in the textile world is one of the most exacting, ideally calling for perfect, even above average, colour perception. A great many people are additionally involved in minor aspects of colour identification during production.

The dyer or colourist is normally required to dye yarn or fabric to match, as accurately as possible, a small sample provided as a standard or master. Drawing on previous experience and recipes of previously dyed samples, a match must be produced which is 'commercially acceptable' to the customer. This judgment is almost always made visually, under a standard illuminant as set out in BS 950 (Part I) usually by a dyer working in isolation. In arriving at a proper match three factors have to be taken into consideration—the shade or hue of the colour, the strength or depth of colour or saturation and the lightness of the colour. The procedure is approached in stages, because of the obvious difficulties of removing dye once it has been added. Additional problems arise if the standard sample is made of a different material or is of a different weight, weave or knit from the required goods. Apparent colour changes can be introduced by texture differences, by the colour of the background and by a line of demarcation, so that standard viewing conditions are hightly desirable. The matching of stockings in pairs and the detection of faults calls for a high degree of colour sensitivity since visual assessment is used exclusively for this activity. The final use to which the material must be put must often be considered, as must the degree of fastness required; a good dyer is able to produce an appropriate dye formula or recipe within a few minutes. The commerical colour match is one which is as near to the standard as is possible under the circumstances, deviating only slightly from the standard by an amount that probably will not be detectable to most inexperienced observers. Tolerances in colour difference vary from manufacturer to manufacturer and customer to customer. Higher standards of matching are required in the dyeing of cloth compared to printed textiles because the print hides small colour differences; very close matches must be maintained for plain-dyed cloth. When large quantities of fabric must be dyed the most important colour problem is the accurate repetition many times over of a standard colour or pattern; shade variations must not occur within the width and length of the fabric. Additional problems arise in the making up of garments using components which are made of different fabrics since different materials take up dyestuffs to different extents.

Whites pose special difficulties since most textile fabrics, particularly natural fibres are slightly yellow. Colour traces have traditionally been removed by chemical bleaching or by adding a small amount of blue dye; more recently fluorescent or whitening agents have become popular. Great care is taken in the industry to ensure that metameric pairs (samples which match in colour under all illuminants) are used together.

Almost every job in carpet manufacture relies on critical colour judgment, from the designer to despatch staff who frequently select goods from a warehouse store on the basis of written instructions referring to colour. Buyers of raw materials and employees who receive and sort them must detect differences in quality often judged by colour. Occasionally dyeing takes place on the same premises, but often carpet manufacturers buy ready-dyed materials which may or may not be required to be wound on to bobbins; similarly dyed yarn must obviously be wound on each bobbin! The colour range is usually large—one company reported using over 300 shades—and frequently close shades are used together. 'Bobbin boys' select yarns according to instructions in order to set up a loom, usually in isolation; up to 12 000 bobbins of up to 32 different shades may be required on one loom. One company contacted thought that at least 200 of the 350 employees engaged in colour decisions could make errors which could not be checked. Up to 5000 people may be involved in one company. The financial consequences of mistaken colour identification could be considerable. One of the most exacting visual tasks is the inspection of finished carpets and the insertion by hand of missed threads.

Testing policies

Large textile and dyeing establishments have recognised the need to restrict employment to colour normals for some time, although no central policy operates in the industry and typically the relevant research associations seem unable to advise adequately. Smaller manufacturers are less aware of the need to examine for colour vision defects and a number of important errors have been reported in addition to the frequent disagreements in shade matching between normal observers. Since almost every employee in the textile world is concerned daily with decisions based on visual colour appraisal, every effort should be made to restrict employment to colour normals. The testing of key personnel with a sophisticated colour discrimination/matching test such as the FM 100 Hue test or the Lovibond colour vision analyser is recommended.

The need to recognise eye changes

The gradual deterioration in colour perception with age attributed to an increase in opacity and density of the ocular crystalline lens and possible macular changes (Lakowski 1962) is largely unrecognised by textile and printing manufacturers. Such changes affect the matching of whites and near whites in particular (Wright 1946) and lead to a tendency for older colourists to match with a red bias. Warburton (1954) argued that since the age differences can be compared to putting a yellow filter in front of the eye, the effect in practical colour matching is very similar to that of varying the colour temperature of the light source by as much as 4000 K. For example a change of illuminant from the standard illuminant A source of red bias (tungsten) to one of a higher colour temperature approaching standard illuminant C (daylight) with its bluer bias. Standardisation of the illuminant for colour matching is of very little value unless the observer is standardised around the average pigmentation value. McLaren (1960) has described a simple test for detecting such differences but doubts the practical significance of age changes (personal communication 1976). In his experience colour matching by older observers is no less reliable than for young observers, but this can be disputed. Experience is presumably a valuable factor in any colour assessment and the trade-off between this factor and deteriorating colour perception on account of increasing age is a most difficult decision.

11.7.2 Printing and paper industries

In common with dyeing activities, printing and allied manufacturers rely heavily on visual colour judgment, although objective analysis of plain paper and electronic scanning techniques to assess colour print are rapidly being introduced. The unaided eye is generally superior to instrumental methods in the detection of differences, particularly in the near-white and desaturated (pale) range. Colour printing is a highly competitive industry and a large proportion of personnel is concerned with colour appraisal in some capacity; ink mixers, compositors, engravers, etchers, printers, bookbinders and especially colour retouchers. Few women are involved. Printing is still considered a craft-based industry with the average retoucher seeing himself as a minor artist whose job it is to make the best result possible visually using a standard original picture as a guide. Since printing pigments are imperfect and their spectral sensitivity characteristics are not identical to the sensitivity curves for the eye's photopigments, colour corrections must be made visually by skilled craftsmen.

The need for accurate colour reproduction is felt particularly by the postal sales organisations, who rely on coloured catalogues for the display of their merchandise, and who are now controlled by the Representation of Goods Act. In the printing of postage stamps and the manufacture of wallpaper great care has to be taken to maintain uniformity of colour; the eye is quick to spot small colour differences. Shade guides for paints and cosmetics rely on careful attention to colour reproduction.

Frequent difficulties arise when printers or publishers deal with advertising agencies, and the quality of coloured material has to be judged visually by both parties in isolation since most communication is by telephone or in writing. Interpretations are likely to differ if the illumination conditions are not standardised and if one of the observers has faulty colour perception. Recently a degree of cooperation has been achieved and standard illuminants are now used by many advertising agencies, but widespread colour vision testing of these personnel has yet to be established.

Colour matchers in the paper industries must be able to detect minute colour additions to white or coloured paper and recommend changes if the production sample does not fall within the accepted tolerance; the job is as demanding as the textile dyer or colour chemist. Paper mills and printing businesses, like dye works, are often small establishments with a minimum of checking possible on each shift, requiring individuals to take major responsibilities for colour acceptability.

Testing requirements

Despite the obvious need for adequate colour perception among all operatives concerned with coloured paper and printing, manufacturing industry has been slow to introduce clinical testing and apply colour vision standards. Voke's 1976 survey of eleven paper mills indicated four who as yet administer no colour vision test. Errors in colour judgment are commonly reported and only then, it seems, have management personnel been prepared to consider the need for a systematic approach. There are instances in this field, as indeed in other critical colour industries, where Daltonics are working in positions where they are actually responsible for the most important colour decisions in the whole operation; often they were engaged before colour screening was introduced.

A red defective, promoted to colour foreman, would admit to few problems in practice when interviewed, but his manager attributed his success to the fact that everyone knew of his defect and he was sensible enough to seek advice if he was in any difficulty. Among twenty colour matchers in one mill where routine testing was not made, Voke found two to have deuteranomalous vision. This incidence is consistent with the national average and confirms Richter's (1954) observation, based on textile mills, that the proportion of unscreened colour defectives working in industry is the same as the national average.

While it is unreasonable to deny the mild Daltonic employment in specialist colour areas if he is able to show adequate competence at the job, by compensation (the two deuteranomalous matchers mentioned above were aware of no obvious disadvantage) it is still in the interests of both employer and employee to have full details of a colour vision abnormality so that both can be prepared for the possibility of errors. In large companies where work is indirectly checked at many stages of production, a skilled colour defective with a mild defect may perform adequately in the team, particularly if he knows the extent of his own limitations, and is not afraid to draw on the second opinion of a colleague. More usually, however, the colour defective will try to conceal his anomaly, and when screening measures are not taken he is in a good position to pass unnoticed, perhaps for a number of years. In a paper mill where only ten men were concerned routinely with colour analysis it was four years before the anomalous colour perception of a worker came to light on account of an important error. Undoubtedly even a mild defect will limit the confidence of most colour workers. A deuteranope involved in the production of formica-laminate paper-based materials asked to be moved from the subjective elements of quality control to a more scientifically-based job because he could not cope with wood-grain patterns.

It is strange that in industries such as paper-making and textiles, where great efforts have been made to standardise the illuminant for colour work, so little attention should be given to the visual characteristics of the colour matcher himself.

The importance of screening for acquired defects cannot be over-emphasised in the printing and paper industries, for the typical impairment of the blue function leads to confusion between white, yellows and greys—the critical colours in paper-making. Some production staff may be heavy smokers and drinkers—features which add to the risk. Cases of acquired defects have been reported in the printing world, with the obvious consequences—too much dye was added by one individual resulting in an incorrect match which appeared correct to the man involved. Methods of examination currently used in industry fail to screen for these defects; recommended tests include such an evaluation, and it is advisable to repeat testing at three yearly intervals particularly for employees over 45 years of age.

Role of professional associations

The paper, printing and allied industries are heavily involved with the trade unions and there are many organisations which support the employee and assist in apprentice selection. Since a high proportion of entrants to the printing world pass through apprentice schemes this would seem an ideal medium through which to establish initial screening for colour perception; it would ensure that those with severe anomalies who would be unsuited to colour work are channelled into other more appropriate areas.

A general awareness of the need to test the colour perception of workers in the industry is acknowledged by the various organisations operating nationally and regionally and a colour vision test (screening only) is sometimes part of an apprentice selection scheme. However there is confusion over who should be responsible. Such bodies are uncertain of where to go for advice. As a result there is no national policy for colour vision examination; individual companies in every case must make their own decisions whether to accept or reject a colour anomalous applicant. Small companies are particularly at risk for they seldom have access to testing methods or the consultation of medical advisers.

11.7.3 Engineering industries

The extensive use of colour coding for a wide variety of individual purposes is a considerable source of risk to the Daltonic electrician, engineer or factory worker, his/her colleagues and all who subsequently use the hardware. These potential hazards have been largely overlooked by the bodies in the United Kingdom who are responsible for establishing colour codes (Voke 1980b). This subject is dealt with in detail in Chapter 12—Colour coding and safety—but it is pertinent to discuss briefly some aspects here.

In the electrical industry a whole range of responsible jobs requires careful colour identification, from the painting of colour codes on electronic components to the job of rack wire-men and cable joiners, telephone exchange and TV chassis assemblers. The human element is unlikely to be replaced by automation for some time. Naming is frequently involved, posing considerable difficulties for the colour defective individual; for example the joining of a red wire to a space marked 'red' on the circuit board, or using a number code, for example red to area 3, blue to area 4. Efforts have been made by electrical bodies to reduce the number of colour combinations to a minimum, but particularly in the activities of the telecommunications industry the colour range is wide and striped wires are a particular source of confusion to the colour defective. Although the new European code for flexible domestic wiring using brown, blue and yellow/green stripes, is less confusing for the colour defective (especially protans), the choice was not made by considering the needs of the colour defective individual (see Voke 1976a and 1980b for detailed discussion). Tables 11.4 and 11.5 give typical confusions.

The correct setting up of a colour TV is an almost impossible task for even the mild anomalous trichromat. Grey-scale adjustments and balancing the channels to give a white will undoubtedly be different for each Daltonic.

Table 11.4 Common errors made in pairing single wires. From Voke (1976a).

Deuteranomalous	Deuteranopes	Protanopes
Blue/violet	Blue/violet	Blue/violet
Blue/brown	Red/green	Orange/green
	Red/brown	

Table 11.5 Confusions in pairing double-twisted wires. From Voke (1976a).

Deuteranomalous	Deuteranopes	Protanopes
(red + blue) + (brown + violet)	(violet + green) + (violet + brown)	(blue + red) + (violet + brown)
	(orange + red) + (orange + green)	(brown + black) + (green + red)
	(yellow + brown) + (yellow + green)	(yellow + green) + (yellow + orange)
	(green + yellow) + (green + orange)	(violet + yellow) + (violet + green)
	(red + green) + (red + grey)	(red + orange) + (red + green)
	(white + brown) + (white + green)	(white + green) + (white + orange)
	(white + green) + (white + orange)	(violet + orange) + (violet + green)
	(green + red) + (green + brown)	
	(red + brown) + (red + green)	
	(yellow + orange) + (yellow + green)	

While mild anomalous trichromats are able to identify colour-coded resistors and capacitors with some degree of success, dichromats are particularly at risk. Errors made by mild Daltonics amounted to 7.7% (suggesting one likely error in the identification of twelve resistors) while dichromats misnamed 20% of the total. A whole range of errors is involved but the confusion arising when red or green appears on a resistor of brown background is particularly common. Complete failure to identify the presence of a band was also more frequent for dichromats than anomalous trichromats (see tables 11.6, 11.7 and 11.8 and figure 11.1).

In metallurgy superficial colour recognition by most operatives is required. Chemical colorimetric and microscopic analysis may require fine colour discrimination. Small traces of pinks, browns, oranges, yellows and greys must be seen in sectioned material, and, in common with electronic components and cables, it is unfortunate that this range spans the typical confusion colours of Daltonics. Some inspection procedures rely on colour clues for defects and inclusions in stock, and colour is an important criterion in the grading of steel and aluminium compounds. Before the introduction of industrial pyrometers the temperature of steel was estimated largely by a subjective colour assessment of the emitted radiation.

Throughout industry the presentation of alphanumeric and graphical data on the phosphor screen of a cathode-ray tube has become a popular means of retrieval and display. Displays are usually in monochrome—green being the most common at present—but technological advances are adding a wide range of colours rapidly. The human factors group of one large nationalised industry consulted one of the authors about the likely problems for a Daltonic. The reduced colour and brightness contrasts seen with such a presentation undoubtedly makes the task more confusing for the colour anomalous than for instance a multicoloured graph on paper, and care to restrict use to colour normals is desirable.

Table 11.6 Resistor colour confusions made by deuteranopes (16 observers). Percentages are given to the nearest per cent. The following colours were completely missed by deuteranopes: brown not seen on brown, 3%; red not seen on red, 4%; red not seen on brown, 3%; green not seen on brown, 3%; grey not seen on brown, 3%. The percentages in tables 11.6 and 11.7 do not add up to 100 because some colour bands were missed altogether. From Voke (1976a).

Colours as named

Colours as presented	White	Black	Grey	Brown	Red	Orange	Yellow	Green	Blue	Violet	Silver	Gold
White	100											
Black		100										
Grey			63	2	6				8	5	14	
Brown		1		48	13	1		35	1			
Red		1		27	53			9	2			
Orange				1	3	82	11	2				
Yellow							100					
Green				4	22	7		2	62			
Blue									100			
Violet			8					1	51	40		
Silver											100	
Gold												100

The human eye shows a degree of chromatic aberration and a variety of colours displayed simultaneously can lead to problems in accommodation (focusing); this is more common when different distances are also involved. A wiring operative manipulating a coloured cable at one depth in space and at the same time required to read the relevant code or pathway in another colour at a different distance perhaps several centimetres away from the cable position, will clearly experience visual problems in the form of headaches and fatigue. If he has faulty colour vision also, his work task could be a near impossibility.

Table 11.7 Resistor colour confusions made by deuteranomals (22 observers). Percentages are given to the nearst per cent. The following colours were completely missed by deuteranomalous trichromats: brown not seen on brown, 1%; grey not seen on light brown, 1%; grey not seen on brown, 1%; red not seen on red, 1%; red not seen on brown, 2%; green not seen on brown 1%. From Voke (1976a).

Colours as named

		White	Black	Grey	Brown	Red	Orange	Yellow	Green	Blue	Violet	Silver	Gold
	White	100											
	Black		100										
	Grey			83	2				2		11		
	Brown			1	79	5			15				
	Red				4	91	2		1				
	Orange					3	96	1					
	Yellow							99	1				
	Green				4	2			94				
	Blue									94	6		
	Violet		4						2	15	79		
	Silver											99	1
	Gold												100

Colours as presented

Testing procedures

It seems likely that the numbers of severely colour anomalous individuals employed in the electrical assembly industry are small on account of a degree of self-selection and the large number of women operatives. Boys who discover their colour deficiency early are also encouraged to avoid colour-orientated careers; the determined few, who despite the possible handicap continue with their chosen career, usually make the higher grades on the basis of their academic qualifications, and can often rely on objective measurements or a second opinion. A number of deuteranopic observers who cope adequately in this way have been encountered. Nevertheless it is disturbing that only half of the sixteen electrical manufacturers surveyed screen for colour vision defects. Frequently, large numbers of people within a single company are involved. In one cable factory 1200 employees were daily required to identify colours, yet comments such as this summarise the familiar attitude: 'The ability to distinguish colours is required in about half of our employees, but apart from asking each individual when he or she joins the company, whether they are able to distinguish colours I cannot say that we give them a test'.

Table 11.8 Colours most frequently misnamed by deutans in resistor and capacitor trials. Grouped colours show similar percentages of errors. From Voke (1976a).

	Resistors	Capacitors
Deuteranopes	Violet	Violet
	Brown ⎫ Red ⎭	Brown ⎫ Grey ⎭
	Grey ⎫ Green ⎭	Red
Deuteranomalous trichromats	Violet ⎫ Brown ⎭	Violet
	Grey	Brown
	Red	Red ⎫ Orange ⎭

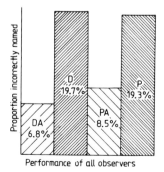

Figure 11.1 Errors made in naming colours of electrical resistors. From Voke (1976a)

Trade tests have a particular value in the electrical industry; a carefully chosen series of wires or colour-coded resistors presented to a borderline candidate for identification can usually indicate his capabilities clearly, but there is a real danger and temptation to restrict testing to a brief practical test of this sort. The candidate who has practised his colour naming beforehand may well succeed for the occasion, especially if he is an anomalous trichromat, and the lighting is good and the samples clean. He might well falter at a later date when required to repair an installation underground by pocket torchlight after a frustrating day.

11.7.4 Medical and chemical analysis

In the medical field there are many occasions where a correct colour change must be recognised for diagnostic, analytical or identification purposes. The successful staining of biological materials requires care in choosing a wide range of chemicals, many of which are coloured. A deuteranopic medical technician known to one of the authors had to choose a counterstain very carefully in order to be able to distinguish bacteria stained by Gram's method. A protanopic zoologist mentioned identification problems with stained sections and their preparation.

Fetter (1963) compared the accuracy with which colour normal observers and ten colour defectives performed colorimetric medical laboratory tests covering a wide range of colours; these involved a number of analytical procedures to urine samples, ketone bodies, sugar, glucose and protein proportions and pH. Surprisingly the colour defective observers he used performed well at evaluating an increase in colour intensity as well as colour differences. A deuteranopic nurse found urine testing confusing; in particular the matching of the colour of the Billi-labstix reagent with a coloured chart as the chemical reacts with the urine. It is likely that this is a common problem among diabetic individuals who, when the disease is advanced, suffer from marked acquired blue defects which makes the discrimination of yellows also difficult. The administration of insulin is frequently gauged from a colour-coded chart (see Chapter 4).

Acid–base titrations, using coloured indicators, a standard procedure in industrial and medical laboratories, are the main problems for the colour defective. Screened methylorange poses a common difficulty for deutan observers, both dichromats and anomalous trichromats. The problems concern the immediate point of change from green to grey (the end-point) before the solution further changes to violet. The colour change from purple to blue associated with an assessment of water hardness as the magnesium and calcium react with the reagent ethylenediaminetetraacetate (EDTA) is not easily discriminated by deutan observers.

Within the medical profession colour coding of patients' charts and records may cause problems to the colour defective. Bottles of blood are sometimes colour coded to identify the additive. Red, black and greens should not be used together to avoid possible confusions.

Medical diagnosis

Colour discrimination plays a very important role in the recognition of pathological conditions in an organ or tissue; where other clues give an incomplete picture, a colour change can particularly aid diagnosis. The severely colour defective physician, surgeon or nurse is likely to be at a great disadvantage in this respect and, especially where the defect remains undetected, the potential risk to the patient may be significant.

The similarity of yellow pus in infected wounds to blood, was mentioned by a deuteranopic female nurse who also found blood-stained sputum to be indistinguishable from plain infected sputum. The presence of blood in vomit had occasionally been missed, and she reported a general difficulty in describing the colour of vomit and faeces. Face discolouration, particularly in jaundice, often went unnoticed by her on account of the confusion between pink and yellow skin.

The red defects (protanopia, the more severe form, and protanomaly) often cause greater practical problems than the more common green (deutan) anomalies of colour vision. A protanomalous anaesthetist admitted that he placed greater reliance in his mechanical monitoring devices to give an indication of the proportion of gases in the anaesthetised patient, than the changes in patient colouration, which he found difficult to identify. A protanopic ophthalmologist also used his other faculties and diagnostic techniques to overcome his handicap. To his knowledge he did this successfully, but admitted that some assessments in which varying shades of red were involved, for instance retinal lesions, did cause him trouble. A protanopic optometrist admitted to great difficulty in identifying some changes in fundus appearance as a result of his reduced sensitivity to red. He also noted difficulty in identifying tinted lenses and, perhaps understandably, in administrating a colour vision test to a patient!

It is interesting that the greatest problem for the Daltonic individual in medical identification appears to concern the presence of blood. John Dalton, the atomic chemist, who first gave a full description of the colour confusions resulting from his own protanopia a century ago (Dalton 1798) also remarked on the similarity of blood to a 'bottle green colour'. The confusion of reds, yellows, dark orange or brown with greens, is typical for all red–green defectives and it is perhaps unfortunate that body fluids generally fall within this colour gamut.

Endoscopic examination is becoming more widely used in preventative, as well as diagnostic, medicine. In these circumstances colour is a vital clue because stereoscopic vision is lacking and the handling of tissues is impossible. A physician at a London teaching hospital, with a medium deuteranomalous colour vision defect, consulted the authors when he became aware of a different interpretation in examination from his colour normal colleague. Despite his very considerable experience, he realised, after many years, the possible handicap that his defect might pose (see Fletcher 1978a).

Finally, dental surgeons, dental technicians and assistants must be able to match artificial teeth to natural ones with a fair degree of accuracy which one would imagine requires very fine colour perception.

11.7.5 Examples of other occupations

Turning to the less obvious areas where a colour vision anomaly might give rise to daily problems the jobs of carpenter, carpet or lino planner or fitter, upholsterer, tanner, brushmaker and shoemaker and repairer must be considered. In the manufacture, assembly and sale of furnishings careful matching of woodgrains is important, and for upholstery materials an inappropriate choice of threads and trimmings could lead to goods being sold as rejects or seconds. The demands on human colour discrimination are similar to those required of the tailor and dressmaker though usually less critical.

The tanner, brushmaker and shoemaker have a more exacting task, one which normally calls for good colour discrimination. Several thousand people are involved in these activities in the United Kingdom. The shoemaker and shoe repairer must also be able to identify colours and on occasions make exact colour matches.

In the timber trade the carpenter and finisher must be able to make an appropriate choice of polish, colourant or preservative for the wood grain. Colour is the most obvious method of identifying timber lengths in the timber yard. Both the Furniture Industry Research Association and the Timber Research and Development Association, recognise the desirability of normal colour discrimination in operatives involved in these activities, but colour vision testing is not carried out among personnel in these organisations and rarely in the individual companies they represent.

The grading and appraisal of gems relies upon highly critical visual colour judgment by the jeweller or gemmologist. Colour is one of the major characteristics of a gemstone and often determines the quality and price. Emeralds, sapphires and rubies vary greatly in the green, blue and red ranges respectively, confusion colours for Daltonics. The most valuable diamond is colourless, but generally small traces are seen due to impurities and absorption effects, involving the colours blue–white, green, yellow, brown and red. Yellow traces are seen in 90% of diamonds, and mixtures of yellows and browns and occasionally yellow and green occur, but green and red varieties are rare. A blue bias is the most precious.

Since minute colour differences must be identified for the selection, purchase, grading and resale of gemstones, the necessity for colour perception cannot be over-emphasised in this trade, and above average ability with colours is hightly desirable for all those involved. Of all the occupations, trades etc surveyed this area appears to be the most exacting and demanding in terms of human colour judgment. The job of diamond grader, especially, calls for the highest degree of sensitivity to colours. Approximately 100 000 people in the UK are involved in the colour grading, cutting, polishing and retailing of gemstones including diamonds. The financial consequences of faulty colour judgment could be disastrous in this trade since imitations are often identified by their colour characteristics alone. The most significant colour mistakes are likely to occur with diamonds. These stones account for over 90% of the total value of the world's trade in gems. The EEC have agreed a standardisation procedure for diamond colour grading based on master stones held in each member country.

Since gemstone colours, and the colour range for diamonds, typically fall within the colour confusion range of red–green defectives, a colour vision defect of even the mildest variety is likely to be a very considerable handicap for all concerned, particularly when the stones are small and pale. The tendency towards tritanopic vision (blue defects) among colour normal observers when objects subtend small visual angles (small field tritanopia) may lead to difficulty in identifying yellow/white, blue/white and blue/yellow differences. The impaired blue discrimination of the elderly is an important element which can easily be overlooked. Graders should be young, healthy individuals with good accommodative power, free from binocular defects and all colour vision abnormalities. Distinguishing between pink topaz stones, tourmaline, mauve spinel, pale amethyst, kunzile and pink beryl by colour is often difficult. Filters are frequently used by gemmologists to reveal underlying differences in colour, and are particularly valuable for discriminating between emeralds and their imitations.

One of the world's leading gemmologists, a British man who has deuteranopic vision, has had to develop his career with a physical/scientific emphasis because of his handicap. He confuses pale green and pink stones readily and to mistake an emerald for a ruby of similar brightness would be quite possible.

In view of this strong dependence on visual colour assessment it is highly surprising that no evidence of colour vision testing among those engaged in the diamond trade could be established from the extensive enquiries made in London's Hatton Garden. Discussions with editors of the gem and jewellery magazines confirmed that this matter is not taken seriously in other branches of the trade either. Efforts must be made to impress upon all those concerned with gemstones, in whatever capacity, the necessity of good colour vision, for it is likely that many disagreements on the pricing of stones and diamonds result from individual variations in colour perception.

In the ceramics industry careful colour control is required for bathroom units made in different factory locations. The eye is quick to notice small colour differences in a row of tiles and because small tolerances must be upheld visual colour matching between production and standard samples has been maintained. Colour vision testing is widespread among sanitary-ware manufacturers but no precautions are taken by the majority of tile establishments even though they take great pains to maintain colour control through temperature regulation. The close network of ceramics/porcelain manufacturers in the NW Midlands of England makes some degree of cooperation for colour vision testing feasible.

In the pottery industries colour is assessed entirely by eye and although a fairly wide tolerance is accepted for routine tableware, individual pieces are matched to standard samples requiring a high degree of sensitivity to colour differences.

11.8*   RECOMMENDATIONS

The ultimate responsibilty to detect and eliminate those with severely faulty colour perception from colour work must lie with the individual companies. Pre-vocational testing should be encouraged by trade unions, professional bodies and apprentice selection boards. It is surely their responsibility also to eliminate the sense of distress and waste which the discovery of a colour vision defect must certainly cause to individuals who have settled on a chosen career. When appropriate tests are administered, preferably in consulation with experts in the field, costly errors will be avoided and applicants will be treated realistically and fairly. Clearly, colour vision defects are seldom treated with the seriousness that is typical of other medical conditions. Industrial medical officers are often ill-equipped to advise in the occupational setting; there is confusion over the question of whose responsibility it is to take decisions or give advice. Optometrists and medical practitioners must be ready to assist industrial managers, personnel staff, careers advisers and individuals in this often difficult matter.

12 Safety aspects

12.1 COLOUR CODING AND DALTONISM

Colour is used extensively in industry, commerce, navigation and military situations to code information. In many cases the correct identification of a colour code is essential for safe and efficient work. Colours have become associated with specific meanings summarised below:

White/black	Traffic directions
Red	Fire/danger/stop
Orange	Danger
Yellow	Caution
Green	Safety, first aid
Blue	Caution
Purple	Radiation hazard.

The industrial uses are particularly numerous—to reinforce hazard warnings, to control the movement of vehicles in transport (signals and flags), colours for building purposes and to code contents of pipes and cylinders, gas and medical containers, capacities of pipettes, pressure ratings, values of electronic components and electric cables and wires.

12.1.1* Optimum colours

Numerous studies have been carried out on the optimum use of colour in coding systems (Green and Anderson 1956, Conover and Kraft 1958, Jones 1962, Smith 1962, Reynolds 1972). Under optimal viewing conditions between 5 and 12 colours can be distinguished reliably, and there is evidence that colour is better than shape in tasks which involve locating displayed data (Conover and Kraft 1958). As many as 30 categories can be identified if three dimensions of colour (hue, brightness and saturation) are used (Bishop and Crode 1961). Green and Anderson (1956) found that when observers know the colour of the target the search time is approximately proportional to the number of target colours.

The most effective colours for 'stimulus lights', as measured by the speed of detection and the accuracy of identification, go in the order (fastest to slowest) red, green, yellow, white. For errors in colour naming (least to most) the order was green, red, white, yellow (Reynolds 1972). If the signal-to-background contrast is low, it was shown there was a marked advantage in using a red signal with green, yellow or white following, in that order. Orange is known to be seen best at greatest distances and care must be taken to avoid colours which will be confused by colour normal observers when the angle subtended is 20′ or less, due to small field tritanopia.

12.1.2   Problems for the colour defective

The special difficulty the Daltonic has with colour codes has not been forgotten. Pitt (1942) mentioned the problem, and it was his hope that 'one day the various responsible authorities will cease to colour code in the so-called confusion colours'. The suggestion that colour codes should be ancillary to pattern codes and the colours chosen should be saturated ones was made by Wright (1953). Wilson (1960) described symbols which have been recommended for use in conjunction with safety colours 'for the sake of those with defective colour vision'.

12.1.3   Attempts to aid the anomalous

Efforts have been made to select colours for some codes which will cause the least difficulty to colour defectives. The American Standards Association has developed a safety colour code for marking physical hazards and the identification of certain equipment in connection with accident prevention, with special regard to anomalous colour vision (Conover and Kraft 1958). Strict specifications for the purple and grey limits for 'safety red' and the yellow limit for 'safety green' are documented. When BS: 1376 (Colours of Light Signals) was revised, a bluish-green hue was selected which is more reliably recognised than a yellowish-green by colour normals and colour defective observers.

12.1.4   How able are they?

Cole (1964) reported American and Australian experimental studies of the ability of colour defective observers to recognise colour codes; dichromats were found to be unable to respond to a 'three category' colour code, although attempts have been made to devise colour codes of three or more categories which the dichromat could appreciate (Judd 1952). Cole (1964) recommended that dichromats should not be employed in operations requiring reliable recognition of colour codes involving three or more categories. It is possible to choose at least nine different colours which even the colour defective should not confuse. Taylor (1975) and Conover and Kraft (1958) recommend red, orange, yellow, purple, blue, grey, white, black and buff for use with the colour anomalous. The potential hazards of the colour defective have largely been overlooked by the bodies in the United Kingdom who are responsible for establishing colour codes, for example in electrical components such as resistors and capacitors (Voke 1976a). Despite careful specification of the colours used in the British Standard code no tolerance limits are given and there is no legal enforcement of colour control. Lack of standardisation of the background colour leads to wide variations in the colour of coding bands. The use of a standard white background would eliminate this problem; a transparent coating could be applied to minimise soiling. Good contrast is essential and the addition of fluorescent substances to heighten conspicuousness could be considered. Many colour defectives find a glossy finish more helpful than a matt finish. The colours at present recommended by the British Standards Institution (BSI) are those which cause great confusion to the colour defective individual (see Chapter 11 and figure 12.1).

Studies directed at establishing the likely consequence of a colour vision defect in those concerned with identifying colour codes of electronic components all agree that the colour defective is likely to make multiple confusions, especially if the illumination is poor and his defect marked (see Hardy et al 1954, Walraven and Leebeck 1958, von Ricklefs and Wende 1966 and Voke 1976a). Similarly, anomalous trichromats, and more especially dichromats, have extreme difficulty in identifying and pairing colour-coded cables and wires (Voke 1976a). In a study in which 55 colour anomalous observers identified the colour codes of 27 resistors and 18 capacitors of varying sizes under controlled illumination conditions Voke found a high correlation between the colour-coded bands which were actually confused in the trial and the confusions which could be predicted on the basis of the positions of the band colours on the CIE chromaticity chart, relative to the confusion loci for dichromats.

The departure from the British Standard colours in practice is particularly noticeable for brown, red and yellow bands (on resistors) which appear more desaturated (paler) than the recommended colours and a large spread is typical for blues and violets, all of which are confusing for the colour anomalous. A luminance difference of around 17% is necessary to enhance the colour differences sufficiently to allow successful recognition by the Daltonic.

Voke found that anomalous trichromats made an average of 7.7% of errors (6.8% for the deuteranomalous and 8.5% for the protanomalous), suggesting one likely mistake in the identification of 12 resistors. Dichromats show a much greater risk; incorrect recognition of the code for resistors could occur in 20% of cases (19.7% for deuteranopes and 19.3% for protanopes). Dichromats can thus be expected to place incorrect values on one out of every two resistors (total eight bands). Complete failure to identify the presence of a band was more frequent for the dichromat than anomalous trichromat.

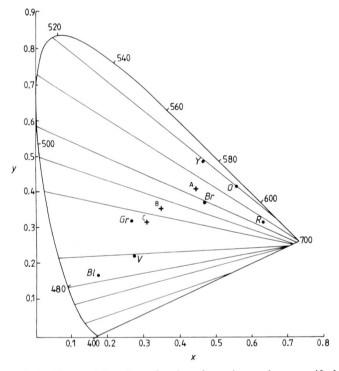

Figure 12.1 Chromaticity plots of resistor/capacitor code as specified by BS 318C 1964 Suppl. no 1 (1966) with protanopic confusion loci. A, B and C are the standard illuminants.

12.1.5 Cables

Considerable efforts have been made to reduce the number of colours involved and the colour combinations for coloured cables. Although Daltonics were not considered when the new brown, blue, yellow/green striped code for domestic wiring was introduced, it does cause fewer problems to the anomalous than did the old red, black, green code (Voke 1976a).

12.2* COLOUR SIGNALS AND THE COLOUR DEFECTIVE

Colour signals, a special category of colour coding, are used widely in conveying information to operators of land, sea or air vehicles. Colours are associated with specific instructions, either by displayed information as in self-luminous navigation signals, or in association with a verbal or symbolic instruction as in non-luminous road signs. The danger associated with misinterpretation demands that very special attention must be given to the choice of colour, particularly in view of possible confusion by the colour anomalous group.

Traditionally red and green have been associated with danger and safety; the use of yellow as a warning signal came much later. Research indicates that red signals give rise to a faster response than any other colour (Reynolds 1972). In a study on luminance requirements for hue perception of small-signal targets (subtending visual angles from $1°$ to $21''$ diameter and exposure duration 44 and 700 ms) Conners (1968) noted thresholds to be lower for red than for blue or green; thus red is identified more frequently than other hues of similar luminance. Reynolds found the speed of detection and recognition to be in the order (fastest to slowest) red, green, yellow, white.

Experience has shown that if precise meanings are to be understood the colours must be strictly defined within narrow limits. The establishment of internationally agreed standards requires national standards for a framework; many countries have no national standards for signal colours and in some countries where standards exist enforcement is lax because there is no legislative control. Even today colour signal codes for railway use show wide variation between countries. In the 1940s the fact that the preferred signal colours varied very widely between countries and for different signal usage was indicated by the choice of the French lighthouse 'green' as a blue–green, while the German railway 'green' was a yellow–green. Uniformity is clearly most desirable.

12.2.1 Signal colours for specific usage

The history of coloured signals for use in navigation, their choice and corresponding meaning is long and complex, particularly in the United Kingdom where Government legislation has established many colour standards. Stringent requirements for recognition are generally enforced for operators of moving vehicles in view of the dangers of anomalous colour vision. The first colour vision test for maritime officers was introduced in the United Kingdom in 1877. Flashing coloured lights are a particular category of a signal which have found widespread application in maritime, aviation and road transport.

12.2.2 Maritime signal navigation

Flags, in use from about 1650, were the first means of conveying signals on board ship; initially the position on the mast was the coding element. Gradually coloured flags were introduced involving red, blue, yellow, white and black colours with a distinctive two-colour flag being most often used. Coloured signal lights then followed, principally red and white sources arranged vertically on the mast. An 1839 Parliamentary Report on the carriage of lights noted that ships were carrying coloured lights but there was no set order in them. In 1848 the standard red–port, green–starboard, white–masthead, specification was laid down by an Act of Parliament for ships' navigation lights, following recommendations produced in 1840. Clearly it is essential that two ships approaching each other should be able to see each other's red light and the angles of emission of light were specified to avoid confusion. This appears to be the first time that instructions on ships' lights had been issued anywhere in the world (Holmes 1981).

Maritime signals are often restricted to red, green and white because light signals of these colours and the corresponding possible luminances ensure safe recognition over a long range. Blue is an unsuitable colour for self-luminous signals. Intermittent emission of a signal gives greater scope for coding and is used widely in water transport often to enhance an acoustic signal. Thus flashing signals are used as aids to marine navigation both on sea and inland waterways. Lighthouse beacons on land and buoys and lightships on the water mark obstacles such as reefs, wrecks, shallow water etc; flashing lights on ships indicate dangerous loads, e.g. chemicals, radioactive materials or ammunition. Typically a flashing signal from, or in connection with, marine craft (e.g. that from a lighthouse) is white because filters absorb too much light. A blue light is used by police craft.

12.2.3 Aviation signal navigation

In air transport speeds are faster than on land or sea, safety is more critical and therefore prompt recognition is essential. In many instances only part of a signal display may be visible, for example runway lights indicating whether the angle of descent and direction is correct. In both maritime and aviation signalling detection at least is desirable at the maximum range; at small angular subtences this can be a most demanding requirement. Flashing lights have been routinely used for airborne traffic, particularly before the widespread use of radio and radar control were associated with specific signals. The first night-flying during the 1930s in the USA made use of flashing red and white beacons located ten miles apart. Red, green and white lights are used by air traffic control towers for both air and ground traffic control. The colour code for aviation lights is shown in table 12.1.

Table 12.1 The colour code for aviation signals. From Gibbons and Lewis (1969).

Colour and type of signal	Meaning with respect to aircraft on the surface	Meaning with respect to aircraft in flight
Steady green	Cleared for take off	Cleared to land
Flashing green	Cleared to taxi	Return for landing (to be followed by steady green at proper time)
Steady red	Stop	Give way to other aircraft and continue circling
Flashing red	Taxi clear of runway in use	Airport unsafe—do not land
Flashing white	Return to starting point on airport	Not applicable
Alternating red and green	Exercise extreme caution	Exercise extreme caution

Navigation lights on aircraft involve a red light on portside and green on starboard with a white light on the aircraft tail. These lights thus indicate whether an aircraft is receding or approaching. Anti-collision lights on aircraft are flashing red or white.

In military aviation colour signals play a dominant role for coding. For ground-to-air signals red, green, yellow, blue lights are used. Red and green are used in search and rescue, and smoke markers used by the Army to indicate targets can be red, green, white, yellow and blue. Airfields have characteristic coloured beacons—a flashing red beacon indicating a military airfield, a flashing green beacon indicating a civil airfield. Runway lights are also colour-coded; on a civil aerodrome the threshold lights across the end of the runway are green and the stop lights at the other end are red. Low-intensity approach lights are red formed in a characteristic pattern (Holmes 1981). On a military airfield the lights which indicate the lead-in path to the runway are red or white, the threshold and end of runway lights are transversely positioned red or green, and runway edges are indicated by white lights. The lights on the centre line of modern runways are usually green although older runways may use white lights. The most modern fields however have alternate red and white lights. Red lights indicate obstacles on buildings of a certain height near an airfield. Warning signals in the cockpit involve red lights to indicate hazards demanding immediate attention, amber lights indicate a lesser malfunction but which if left unattended could lead to a major hazard, while blue, green or white lights confirm that a service is functioning. Normally audio signals accompany immediate hazards, and aircrew appear to rely to a great extent on the position and temporal modulation of lights, rather than the actual colours, to indicate problems (Brennan 1972). Maps are used extensively for air navigation, and occasionally must be viewed in red cockpit illumination. Typically woods are coded green, rivers blue, roads yellow, railways black and towns brown. With the development of larger, faster, higher flying, instrument-navigated aircraft, most authorities recognise that the role of colour vision should decrease in importance (Tredici *et al* 1972).

Features of coloured navigation lights on aircraft have been considered by Bowman and Cole (1981), showing how visual range, atmospheric transmissivity and signal intensity interact; also considered were the likely patterns of coloured lights to be seen when aircraft are placed on different relative courses. The need to recognise both red and green is evident and the likely difficulties of normal and Daltonic aviators were stressed.

12.2.4   Railway signal navigation

Semaphore signals were used generally from the latter part of the eighteenth century even until very recently—a horizontal bar indicating 'stop' and a vertical position for 'all clear'. The earliest light signal for railway navigation was a candle burning in the window of a point watchman's house, it having been agreed with the driver that the presence of a light meant that he had to stop! This developed into a code on the railway line whereby a red light indicated 'stop' and a white one 'go'. Eventually the white was replaced by green because of the danger of confusion with non-signal lights in the vicinity of the track. A yellow signal was introduced on one line in 1905 and was universally adopted about 1925. A fairly complex modern code has now developed. Red conveys a command to 'stop', two yellows indicate 'proceed expecting a single yellow at the next signal', one yellow means 'proceed to the next signal' and green 'line is clear to the next signal'. Rationalisation of signal colour codes has only been achieved comparatively recently in the UK, around 1948 when the railways were nationalised. For example on the Stockton to Darlington line in 1930 red meant 'stop', green 'proceed with caution' and white meant 'line clear'. A three-aspect signal with yellow indicating 'be ready to stop' was installed at Neasden in 1923 and the Southern Railway introduced four aspects in 1926 with the double yellow indicating 'caution—proceed at medium speed'. There appears to be no international coordination on the colours of railway signals at the present time.

12.2.5*   Road traffic signals

Coloured light signalling in road transport is a much more recent introduction having been in use in the UK since about 1930. Signals involve the red–stop, amber–caution and green–go signals. In the USA traffic control signals and signs initially used four colours: black, white, yellow and green. In 1929 red was added for parking control coding and blue was introduced for pedestrian regulations. The red–stop light was not used until 1954 and orange to act as a wait signal was introduced in 1961 (Robinson 1967). In addition flashing vehicle direction indicators and hazard warning lights (yellow to indicate service and blue to indicate emergency vehicles and police), amber Belisha beacons at pedestrian crossings and red warning lights at automatic level-crossings are found typically on roadways.

Colour is the dominant feature for recognition of traffic signals but other cues such as position of the signal, brightness differences between colours, movement of other traffic and shape provide secondary indications although these may not always be present and seldom are at sea. Switzerland has made use of shape-coded traffic signals although this has the disadvantage of reducing signal intensity. In Canada the red traffic signal is larger than the amber and green (Cole 1970).

12.3 SPECIFICATION AND STANDARDISATION OF COLOURED SIGNALS

12.3.1 Self-luminous signals

Self-luminous signal lights are frequently seen as very small coloured sources, either in very bright surroundings (which impairs detection and recognition) or in the dark. Both sets of conditions affect the apparent brightness of the signal on account of simultaneous contrast effects. Thus coloured signal lights are normally described in terms of colour values alone, without reference to intensity or apparent brightness. The CIE chromaticity specification, allowing a precise numerical value, is an appropriate and established means for indicating signal characteristics.

The BSI published some of the earliest standards on signal colours—three separate documents in the 1930s for road traffic signals, lighting in aerodromes and airways and railway signals. This followed general recommendations by the CIE for traffic light signals in 1931, but the colour limits were wide being a compromise between a reliable recognition of colour, reasonable significance and practical accuracy from a manufacturing point of view. Specific British research in the late 1930s then sought to coordinate the BSI coloured light signal standards to use the same specification, establishing optimum chromaticities and maximum safe tolerances for principal signal colours at various intensity levels, with the minimum values corresponding to absolute threshold. The resulting document BS: 1376 *Colours of Light Signals* first published in 1947 and revised in 1953 to include the properties of coloured glass filters was the first attempt in any country to specify chromaticity ranges for coloured light signals for general use, and was valid for over 20 years. The CIE produced international recommendations in 1959 drawing heavily on the BSI document and taking into account the needs of Daltonic observers (CIE 1959).

In the USA attempts were first made to standardise traffic control signs in 1927 with revisions in 1939, 1948 and 1954. Judd (1952) outlined specifications for red, green and blue pilot lights which the colour anomalous could use, one of the first attempts to consider this minority group.

Since those early recommendations a wide range of improved light sources have been developed, along with more stable plastics coloured filters But modern signal equipment tends to be more compact than earlier designs and the heating effects of the source can change the colour of the signal considerably, even in a short timespan. This is particularly critical for yellows and reds which tend to become darker with a colour shift towards longer wavelengths. Associated transmission losses can also occur on this account. Greater use is being made of coloured lights, many for non-signal purposes, and lights are generally more intense. This poses a potential distraction from signal recognition. These changes, together with the non-acceptance of the 1953 revised BS: 1376 document by the railway authorities demanded reassessment of signal specifications.

A second revision of the BS: 1376 document now combines in one the colour requirements for light signals for all purposes by specifying ranges of chromaticity of the light emitted by a signal, indicating which ranges are appropriate to each of the signalling services, and defining the colours of glass and plastics filters that can be used with a variety of lamps to achieve these requirements. Coloured filters can be measured with one standard light source (CIE illuminant A at 2856 K) despite being used with sources which range from an oil lamp to a tungsten-halogen lamp. Guidance is given on the performance and stability of lamps and, filters and on photometric test methods.

The CIE have a continuous interest in recommending specifications for light signals through their committee on visual signalling.

Typically, mariners have been required to discriminate three colours in use in maritime light signalling. These colours, red, green and white were defined by IALA (International Association of Lighthouse Authorities) in 1968 with strict chromaticity limits. In 1977 separate colours for the lights of special marks were established (IALA 1976) incorporating yellow as a fourth colour. Yellow signals have been increasingly used to indicate special functions at sea, e.g. hovercraft in motion, powered vessels towing another craft, vessels fishing with pure seine nets, and on buoys.

In setting out new recommendations for a four-colour system (IALA 1977) the needs of the colour defective have been considered. New recommendations for surface colours in use at sea have been made (IALA 1980). As in other such recommendations the colours have been defined on the basis of achieving a very high probability of correct recognition for colour. A minimum luminance factor for red has been specified (0.07). Orange is considered to be the 'best ordinary colour for conspicuity against the sea' and it is suggested that its use should be reserved for objects for which detection in the water is more important than recognition of their colours, such as emergency equipment, life-jackets and rafts. For recognition purposes orange is not a suitable colour when used in a colour code which employs red, for at very small angles orange and red or orange and yellow can easily be confused.

Similarly it is indicated that discrimination between yellow and white is difficult at a small angular subtence so it is highly recommended that these two colours should not be used together. Green does not show up well at sea but in the case of red, yellow, green and orange greater conspicuity can be obtained with fluorescent additions to the colour. Blue is recommended particularly for use on inland waterways, estuaries and harbours where the colour is seen at close range.

12.3.2   Non-luminous signals

Surface colours and highly efficient retroreflecting materials (which reflect most of the incident light in the direction from which it came) are used widely for non-luminous signs where high conspicuity is not required. Retroreflecting signs are particularly valuable for night use. Much attention has been given to an optimum choice of colour codes for safety and specifically for highway use in the USA. In the latter case the colour anomalous have been considered. Twelve colours have been accepted by the National Joint Committee on Uniform Traffic Control Devices of the USA and the American Standards Association Safety Colour Code (Robinson 1967). Specifications have been set by both national and international standardising bodies such as the CIE to limit the colour boundaries of such indicator signals or signs. Surface signal colours are usually assessed by subjective comparison with a standard colour sample or small range of samples. Matching against colour cards or standards such as are set out in BS: 381C *Colours for Identification Coding and Special Purposes* is most appropriate. The use of fluorescent materials to aid conspicuity is now popular but their lifetime is short on account of poor withstanding of environmental conditions.

12.4*   FACTORS AFFECTING SIGNAL RECOGNITION

The detection of a coloured signal involves a situation of absolute threshold; accordingly normal factors limiting perception under such conditions must be
(*a*) Signal size and luminance.
(*b*) Signal colour.
(*c*) Signal presentation time—flashes less than 0.25 s tend to impair recognition.

(*d*) Signal to background contrast, both luminance contrast and colour contrast.

(*e*) Texture of background and presence of distracting 'clutter' (objects in close proximity).

(*f*) Observer to signal distance, or the visual angle subtended by the signal at the eye.

(*g*) Atmospheric attenuation of signal on account of fog, smoke, haze etc.

(*h*) Observer characteristics such as:
  (i)   colour vision;
  (ii) state of adaptation;
  (iii) pupil size and ocular transmission factors which regulate the intensity of the signal;
  (iv) psychological factors such as motivation, stress and experience, familiarity etc. These are particularly important in recognising the colour. The experience factor is important—air, sea, and railway signals tend to be interpreted by trained personnel, whereas road signals must be capable of interpretation by any driver.

12.4.1 Chromatic and luminance contrast

Simultaneous and successive contrast effects can have a profound influence on the appearance and recognition of a coloured signal. Simple theories of chromatic contrast effects suggest a selective change in sensitivity of visual mechanisms. Hence a physically achromatic area may appear green in the presence of a strong red surround owing to a reduction in the sensitivity of the red mechanism in the central as well as the surrounding area. These ideas are compatible with the known behaviour of visual pigments (their bleaching and regeneration) and are essentially based on lateral adaptation concepts.

12.4.2 Possible visual mechanisms involved in signal recognition

Opponent processing theory has now become a favoured explanation, postulating that stimulation of a specific retinal area induces an opponent or opposite response in adjacent retinal regions. The theory implies that the induced response that occurs in conditions involving simultaneous brightness and colour contrast involves only the response of members of that opponent pair, and is independent of activity in other mechanisms. Thus the brightness contrast produced when a coloured field lies adjacent to a stimulus should be related only to its own brightness and should not be affected at all by the wavelength, except where wavelength affects luminance, since chromatic and achromatic processing are considered to be separate. In practice, however, some complex interactions have been observed which are not consistent with this theory. It is well known that the apparent brightness and saturation (in addition to hue) of a coloured stimulus can be markedly affected by an adjacent coloured stimulus or surround.

The role of both chromatic and luminance processing channels in the visual pathway has been studied in relation to the detection of coloured signals or targets. King-Smith and Carden (1976) suggest that in some circumstances (for instance the detection of coloured stimuli subtending small angles) the chromatic processing channel is superior in sensitivity to the luminance system and plays the major role in detection. This might well be the case for coloured signals detected at long range at sea. Threshold experiments reveal that coloured test stimuli, of all wavelengths except yellow, are as easily discriminated from white as they can be detected (i.e. discriminated from a blank), supporting the suggestion that the opponent colour system is used for their detection. The view is consistent with a number of other observations—in particular the fact that temporal integration is greater in the opponent colour system than in the luminance system, and the suggestion, based on psychological evidence, that spatial integration may also be greater in the colour channels. King-Smith and Carden suggest that a stimulus will be detected if it exceeds the threshold of either the luminance or chromatic system, and the threshold will be determined by whichever system is most sensitive for the conditions that are operating. They further indicate, on the basis of experimental results, that the colour system makes the dominant contribution to detection for relatively long durations of stimuli (200 ms) presented on white backgrounds. By reducing either the stimulus duration (to 10 ms) or the stimulus size, or by eliminating light adaptation, the contribution of the luminance system becomes much more evident. Kerr (1974) confirmed the role of the colour mechanisms in detection. In a similar argument to that of King-Smith, Kerr hypothesised that if the colour mechanisms are operating at threshold, then the stimuli should be readily identifiable as coloured.

12.4.3 Target size and luminance requirements

It has been suggested the total luminous flux is the critical factor in absolute threshold detection. This is described by three well known summation laws: Ricco's law (area × luminance = a) for small angles, Piper's law ($\sqrt{\text{area}}$ × luminance = b) for larger angles and Bloch's law (area × time = c) for small angles and short stimulus durations, where a, b and c are constants. When the visual angle subtended by the target is very small (less than 20′) the eye tends to become dichromatic and tritanopic (blue blind) i.e. small field tritanopia results (see Chapter 1). This is one reason why blue is an unsuitable colour for signals. Hofman (1975) confirmed other reports that with an increased target luminance the effect of angular size on recognition is less pronounced. Kaiser (1968) investigated by colour naming the colour recognition thresholds for very small (12′) monochromatic stimuli varying in retinal illuminance and duration. As predicted by Bloch's law, when retinal illuminance × time was constant observers gave consistent colour names; when this relationship was not upheld colour naming responses varied, particularly for yellow and white. The fact that the degree of temporal summation in the fovea is less for red than for blue indicates that for very brief exposure durations the red receptors play a dominant and possibly the only role in detection. Thus it is not surprising that two independent psychophysical studies have shown that red is perceived sooner than blue of equal brightness (Piéron 1932, Scott and Williams 1958).

12.4.4 Effect of adaptation

Hofman (1975) investigated the effect of dark and light adaptation on colour signal recognition. Observers were asked to give one of five colour names (red, yellow, green, blue or white) to 2 s presentations of a 1′ point source target of luminance 10^3 cd m^{-2}. For such a small stimulus size small field tritanopia would be expected to operate. For a 90% confidence level the state of adaptation did not influence the perceptibility of red. Yellow was perceived with a better than 90% frequency only when the eye was dark adapted. When light adapted, yellow–green sources were frequently named white (characteristic of small field tritanopia) and confusion of blue and green occurred more often when the eye was light adapted. Adaptation state was shown to affect the recognition of blue, particularly for very short wavelengths, where for light adaptation conditions blue was frequently named white.

   Clearly these findings are of great importance when considering the optimum design of signal lights and any display units which employ small coloured stimuli.

12.5* HOW HANDICAPPED IS THE COLOUR DEFECTIVE?

The difficulty which the colour anomalous experience with coloured signals has long been recognised. The possibility that colour vision defects might be the cause of rail and sea disasters attracted the attention of Professor Wilson of Edinburgh in 1855 and stimulated the development of colour vision tests. Crude colour vision tests similar to those developed by Seebeck in 1837, based on the sorting of coloured objects, were applied to drivers of the then Great Northern Railway. In 1877 Professor Holmgren in Sweden devised his famous Wool Test, following a serious railway accident caused by the failure to recognise a red signal correctly, and by 1885 Donders had developed one of the first lantern tests, a crude simulation of signal lights, to be followed by a variety of lantern modifications in the following century (see Chapter 7).

During the Second World War the need to separate the 'safe' from the 'unsafe' defective was realised in the USA, particularly by Farnsworth, on account of the loss of some 8% of possible recruits to the services on the grounds of anomalous colour vision. Interest then became intensified in lantern tests and the ability of the colour defective to recognise coloured signals.

One of the first considerations of the potential problems of Daltonics with self-luminous colour codes, and the desire to assist them, was made by Judd (1948) in his choice of coloured filters for instrument panel design. Realising that dichromats have an equal tendency to call yellow as red, Judd restricted his code to red, green and blue in the belief that red would then be correctly named. The possibility of deuteranopes confusing a green signal with a white one was eliminated by not including white in the choice of colours. Judd recognised the necessity of ensuring adequate luminance and angular subtence for the blue signal code.

12.5.1 Point source recognition

Sloan and Habel (1955a) found that 'the majority' of colour defectives could use Judd's code successfully when presented with point colour sources, but some protanopes experienced difficulty and it was Sloan and Habel's recommendation that protanopes should be excluded from tasks involving recognition of signals. The difficulty was found to vary with the intensity of the stimulus and correlated with the severity of colour vision defect. Heath and Schmidt (1959) extended this study to include colour signals flashed for duration times of 0.1 and 0.25 s, presented to colour normals and Daltonics, and investigated contrast effects—signal recognition in the presence of neighbouring secondary signal sources. Increasing the signal brightness and/or flash duration did assist recognition generally by all groups although some added confusion did occur—namely colour normals and some Daltonics tended to call blue–green 'blue' more often at higher luminances than low levels and the colour defectives called yellow–green signals 'red' more often at higher luminances than at low ones. The presence of either a white or red distracting light in the field of view at a lower luminance than the signal colour slightly improved the ability of all observers to identify the signals correctly, this being most effective for green recognition. The colour anomalous group were assisted in this way even more than the colour normal observers. Both groups of researchers emphasised that a three-colour code is unsafe for signals of small angular subtence, a situation which frequently arises in maritime and, to some extent, aviation fields.

A real-life maritime circumstance involving Naval midshipmen, among them 81 Daltonics, required colour recognition of typical signals at a distance of one, two and three miles. Kinney *et al* (1979) made observations which indicated a greater number of errors by the colour anomalous group, and large individual variability. Although on average the mildly anomalous were more able than the other colour defective men, no systematic degradation of performance with increasing degree of defect could be established, so that a general prediction of possible handicap by any one class of defectives has proved to be impossible. Consequently the inclusion of colour defectives for signal duties demands individual assessment in every case. Such a conclusion was also reached by Steen and Lewis (1972) for aviation signals, especially at night, and by Nathan *et al* (1964) for road traffic signal recognition. Although *in general* those with mild defects are less handicapped than the medium or severe Daltonics, there are always isolated cases of even protanopes and deuteranopes who pass a signal test correctly. Performance on clinical colour vision tests can thus never be taken as a fool-proof indication of practical ability, a conclusion which was also reached by Voke (1976a) who found little correlation between clinical test performance and ability at industrial colour tasks. Kinney *et al* (1979) found that white and green signals presented the greatest problem to the colour anomalous group, with protans experiencing more difficulty than deutans.

Performance became significantly retarded with increasing distance from the source (a smaller angular subtence); at a distance of two miles colour normals could identify correctly 95% of the time while the worst performance for the anomalous group was 50% correct. At three miles only 80% of signals were correctly recognised by colour normals, the majority of errors being failure to see a signal at all. The colour anomalous groups both confused colours and failed to perceive them at this distance. Red was incorrectly recognised on 56% of the occasions by moderate deutans who also failed to see it 17% of the time. Fewer than 20% of the red lights were reported correctly by any of the groups of protans and 61% of protans *never* saw even one of the most distant red lights. These errors indicate the considerable scale of the colour recognition problem for the colour anomalous and confirm that dichromats experience most difficulties.

12.5.2*   Road traffic signal recognition

The presence of the yellow/orange/amber road traffic signals (these names are used for the same signal colour) in conjunction with a red and green can be expected to cause difficulty to the red and green anomalous because these three colours lie along the confusion loci of protanopes and deuteranopes. A number of experimental studies of colour signal recognition, often involving simulations, indicate that all red−green defectives are significantly slower and make more errors than normal observers, especially for low intensity signals at short observation times and considerable distances (see

Ganter 1955, 1956, Hager 1963, Nathan *et al* 1963, Allen 1966, Cole and Brown 1966). In order to reduce the chance of faulty recognition the CIE recommended in 1975 that the green signal should have a blue bias (its range being restricted on the yellow side) and the red signal chromaticity should be restricted on the purple side and on the extreme red side. Verriest *et al* (1980a,b) consider these restrictions are insufficiently severe, for even with the boundaries specified deuteranopes still cannot reliably recognise the signals from the colour clue, although brightness differences which normally exist in practice between the yellow and red signals tend to enhance recognition. The fact that colour defective observers are much slower and make more errors in identifying road traffic signal lights was shown by Nathan *et al* (1964). Verriest *et al* (1980a,b) extended this observation by measuring the mean minimum distance, d, of the traffic signals at which recognition, was possible, for at least one normal, one protanomalous, one protanope, one deuteranomalous and one deuteranope. When the positional cue was present as in the usual vertical presentation of the signals red, yellow, green, the value d for correct identification of the red lights was about 55% of the normal values for protans and 73% of that for deutans. It was 41% of the normal distance for protans and 33% for deutans when signals were presented singly on a post giving no positional cue to the subjects. Deutans rely heavily on the positional cue. Recognition of the yellow signal was at a distance correponding to 69% of normal for all Daltonics when the positional cue was present and was 69% of normal value for the green signal for protans and 61% for deutans, showing that the green signal presents greater difficulty to the deutan group. A slight difference in performance was noted between Belgian and Austrian signal colours, the Belgian orange being better recognised than the Austrian one, but the Austrian green being more easily identified than the Belgian one. When the positioning was randomised (i.e. red in the middle, orange above) performance by the colour anomalous group was very poor—being only 10% of normal values.

The main danger in traffic signal recognition lies with the protan group who are particularly handicapped in recognising the red signal. This is accounted for by the confusion of red and white/neutral by the protanope's confusion locus which lies along the red—white—blue—green axis, and by the protan's brightness response to reds which causes them to apear very dark. The 'optimum' intensity of red traffic signals (which would allow recognition with a minimum time-lag) was investigated for both colour normals and protans by Cole and Brown (1966) who concluded that protanopes require four times the intensity needed by normal observers. In their opinion the red signal at the optimum luminance for normal drivers is likely to be seen by protanopes but reaction time may be longer; also recognition of traffic signals by all groups is more difficult on a sunny day than in overcast conditions.

12.5.3   Road sign and car light recognition

Both protan and deutan defectives do have impaired recognition for road signs, such as red stop signals, and car lights, such as red rear brake lights. Verriest *et al* (1980a,b) found that the mean perception distances for both groups together were reduced to about 53% of the normal mean for the detection of the stop signal (a red octagon with a white rim and the words 'stop' in white letters) and 85% of the normal for the detection of the danger signal (a triangle with a red rim and a corner pointing upwards). Protans have particular difficulty at night detecting red rear lights on bicycles moving at 40 km h^{-1} (perception distance being 60% of the normal value, while deutans achieved 90% of the normal value). Red reflectors on a variety of vehicles gave considerable difficulty to protans, who achieved only 50% of the normal range; the value for deutans was between 78% and 89% of the mean normal figure. In daylight the perception of red rear stop lights on cars and motorcycles depended considerably on the meteorological conditions, particularly for the colour anomalous; in some conditions the deutan performance was as good as or better than normal. The overall protan detection distances varied from between 39% and 75% of the normal ones and in sunny conditions some protans did not see the stop lights at 50 m. Experience does assist the survival of the Daltonic.

12.5.4*   Traffic accident frequency among the colour anomalous

Although one author is personally acquainted with a protanope whose failure to notice a red traffic signal caused a fatal road accident and many examples of accidents are quoted, studies have shown no definite statistical indication that the colour anomalous have more traffic accidents than colour normals (Norman 1960, Gramberg-Danielson 1961, Sachsenweger and Nothaas 1961, Zehnder 1971, Verriest *et al* 1980a,b). The most recent study involved modern cars and surveyed the incidence of colour vision defects among 2058 drivers involved in accidents. The drivers were traced by means of police records. Their incidents of defects was 8.41%, the same figure as is typical among Caucasian populations. However certain types of accidents did show some connections with colour deficiency, as confirmed by statistical analysis. Protans had more rear-end collisions and accidents brought about by failure to notice red rear, stop and warning lights, than all other drivers (43% of accidents caused by protans, compared with 26% caused by normals). Deutans had twice as many accidents involving traffic lights as did normals, i.e. 4.55% of all accidents in this study compared with 2.1% in normals; they also tended to have accidents when changing their driving direction to the left. (Note that the study is a Continental one, so driving is on the right.) Protans caused more accidents on wet and slippery roads than other drivers. The lighting conditions (daylight, dusk, darkness) showed no influence on the accident behaviour of protans, nor did personal factors such as age, profession, visual acuity, refractive condition or number of driving years.

The authors consider that a psychological compensation on the part of the colour anomalous and more vigilance in driving explains why their accident rate is not more serious.

12.5.5 Should licensing restrictions be imposed on the colour anomalous?

The practical question of whether licensing restrictions should be imposed on the colour anomalous is a difficult one in the light of research which shows no definite proof of increased accidents on account of colour deficiency but nevertheless shows evidence of clear confusion. Experts are divided on this issue and many believe that protanopes should not be allowed licences as professional drivers or professional drivers of passenger vehicles. The WHO recommendations of 1956 suggest no restrictions should be imposed on the colour anomalous, but different countries enforce their own standards and requirements (see Chapter 13). Verriest (in Verriest *et al* 1980a,b) considers restrictions to be unnecessary, but Marré (in Verriest *et al* 1980a) recognising a potential risk, prefers to exclude protanopic and protanomalous observers from professional driving. Neubauer (in Verriest *et al* 1980a) concludes that protanopes, deuteranopes and achromats should be excluded from professional conveyance of passengers and dangerous materials. Cole (1970) suggests exclusion for deuteranopes, protanopes and protanomalous observers. Most agree that a simple colour vision examination is desirable for all candidate-drivers in order that the anomalous should be aware of a possible handicap.

12.5.6 Improvements which would assist the Daltonic

Engineers of every kind involved in road, vehicle and lighting construction should design installations so as to improve the visibility and conspicuity of road safety signs and signals thus reducing the potential handicap to the colour defective driver. The following suggestions could make a significant difference:

(*a*) universal increased size of red traffic signal compared with green and yellow signals;

(*b*) high intensity for traffic signals and rear vehicle brake lights;

(*c*) additional markings in the vicinity of rear brake lights;

(*d*) compulsory use of high intensity fog lamps;

(*e*) choice of rear red vehicle lights with an orange bias.

12.6 NAVIGATION ACCIDENTS ATTRIBUTABLE TO DEFECTIVE COLOUR VISION

The first published reports of Daltonism appeared towards the end of the eighteenth century (Huddart 1777) yet almost another century went by before it was suggested that faulty colour vision might be responsible for fatal accidents that had resulted involving ships and trains (Wilson 1855).

When one looks into the early literature it is perhaps surprising that the first documented cases of casualties involving marine and rail transport actually occurred in or round about the same year, 1869. Two ophthalmic surgeons Drs Joy Jeffries, an American, and E Nettleship, ophthalmologist at St Thomas' and Moorfields Eye Hospital, London, extensively documented such accidents and near accidents from both their own countries and overseas which had occurred near the time of their writing (Jeffries 1879 and Nettleship 1913). The following survey draws heavily from these materials, since many of the original sources are poorly referenced.

12.6.1 Railway accidents

The importance of brightness is well illustrated by the case of a Swiss railway engine driver described by Nettleship in 1913. This man had no serious trouble until a new fireman was detailed to work with him. Their engine then over-ran signals both at stations and at level-crossings, fortunately without any serious accident. It was found that they were both colour defective. The red lamp gave the driver the most difficulty, the colour not being identifiable until he was too near to stop the engine. He managed to distinguish the green from the white by the greater brightness of the latter and he maintained that he could imitate all the three signal colours by turning the wick of his oil lamp more or less up or down!

Nettleship outlined ten cases of accidents which occurred, and one instance where a potential disaster was avoided, because of defective sight, the majority of incidents involved defective colour perception. Two railway cases were documented in Europe before the well-known Swedish accident publicised by Holmgren (1877), which led to the widespread introduction of his wool selection test as a screening method for potential railway employees. Jeffries (1883) notes that Dr A Favre medical officer to the Paris-Lyons Railroad, had, by 1855, examined about 5000 candidates and rejected more than fifty for being red-blind. The first documented railway accident attributable to defective colour vision was said by Favre (1873) to have been in England 'several years before' but no details were given. Three years before this report (i.e. in or around 1870) a colour defective pointsman in Westphalia was responsible for an accident which injured twenty people. The collision in 1875 of two trains near Logerlunda, Sweden, which resulted in nine fatalities has received greatest attention historically since it was believed by Holmgren that colour blindness was the principal cause.

Nevertheless no indication of the source or nature of the data which led him to this conclusion was presented. Nagel (1907a,c) clearly believed one of the engine drivers in question to have been colour blind, but no tests were made because both drivers were killed. Nettleship outlines the case in detail; apparently one driver, although previously recognising a green caution signal, later failed to stop at a red light. In 1876 another collision followed in Finland. This was documented by Gintle (1878) and involved a Daltonic pointsman who had held up the green instead of the red light to the approaching train.

Acquired defects were clearly affecting a good number of railway employees at this time, principally on account of tobacco amblyopia due to the smoking of 'shag'. Nevertheless there is no evidence that anyone other than a few of the sufferers were aware of such potential difficulties and hazards. Walton (1877) writing in *The Times* described the case of an engine driver who after an accident in which he missed a red signal confessed to his awareness of a gradual loss of red perception, which had previously been shown by examination to be perfect.

Retrobulbar optic neuritis resulted in the loss of colour vision in one eye to a young level-crossing attendant who was uncertain about the colour he had set the lamp (Nuel 1879).

In the USA, too, the early consequences of Daltonism were felt on the railroads. An engine driver lost a leg because of an accident attributable to a colour defective employee, and received considerable damages (Carter 1890).

12.6.2   Accidents at sea

Numerous early maritime incidents involving faulty colour vision are documented by Nettleship (1913). These incidents involved some fourteen actual casualties and many potential difficulties. Some of the early major accidents are summarised here. A Dr Feris, a French medical officer in the maritime service, showed early interest in the examination of seamen for colour vision defects. The screening of 775 seamen revealed 10% with difficulty in colour recognition of whom 19 totally confused red with green (Jeffries 1879); Feris described three early accidents.

The first accident at sea which involved casualties on account of Daltonism occurred off the coast of Brittany in 1869 when a French lugger confused a white flashing light with a white and red intermittent signal. Each signal indicated a different island. The same year a Swedish vessel made a similar mistake in identifying land between Calais and Dunkirk. Feris's third case, which came to him secondhand from the literature, occurred in 1871 and involved an English steamer which hit reefs off Marseilles, having mistaken the green light of a pier for the red light of a ship in her vicinity.

In 1875 a collision between two steamships off the North American coast cost the lives of ten people and was clearly due to the misinterpretation of red and green signals by the tug-master who was later shown to have a colour defect (see Nettleship (1913) who quotes US Government sources). Compulsory testing of pilots and would-be pilots of steam vessels for colour vision was instituted in 1880, the year in which the tug-master involved in the accident was examined twice and shown to be colour anomalous. Long gaps in time such as this between accidents and colour vision examination are highlighted by Edridge-Green (1911) who deplored the refusal of authorities to act more promptly.

Armstrong (1888) reported a case of the master/owner of a steamboat on the Mississippi who though 'proved to be colour blind' several years ago and aware of his deficiency was nevertheless not prevented from carrying out shipping activities. The master in question had even previously caused the loss of another vessel on account of his failure to identify coloured signals correctly.

In the Caribbean in 1877 a further confusion between coloured lights on a ship with white harbour lights led to an accident and two years later mistaken identity between the white light of a house and the red harbour light at the end of the quay also caused a major shipping accident in the same region (Nettleship 1913).

In 1879 British waters were the scene of further accidents to vessels on account of Daltonism. These accidents were widely reported by Bickerton (1887, 1888, 1900) in the British medical press. In particular he highlighted the inadequacy of the official Board of Trade examination methods and the wool test. One case is quoted of a first mate who mistook the green light of an approaching ship for a white one. Had it not been for the captain, who by chance spotted the error, a collision would have occurred. The officer was *retested* by the Board of Trade and considered to be fit to go to sea again, which he did, although the officer freely confessed that he could never distinguish a green light from a white one with certainty. The vessel involved was sailing in the North Sea at the time of the incident. The other example Bickerton quoted was an actual collision which occurred in the same year in the south Channel.

Some early accidents involved considerable loss of life. A collision on the River Elbe in Germany in 1902 involving the loss of 107 lives was shown by a colour examination, performed five years later, to be due to faulty colour vision involving the confusion of the red and green lights on the ship (Guttmann 1907).

Acquired defects

In 1881 the loss of a steamer in a harbour off the coast of Florida (which cost the owners £40 000) was apparently attributable to defects resulting from tobacco excess. The ship's pilot mistook the colour of the buoys marking out the harbour channel. This example, quoted by Edridge-Green (1911) represents one of the first cases of a ship's loss on account of acquired colour deficiency. Even the slight and temporary disturbances to colour vision from toxic amblyopia were sufficient to have a serious shipping consequence, as reported by Nagel (1907a,c). In 1906 the collision occurred between Danish and Finnish vessels off the south coast of Sweden. The steersman's colour vision was shown to be abnormal but at this and a later re-examination he made only slight mistakes.

Nettleship's thorough bibliography of early accidents by rail and sea includes eight examples of near-accidents involving colour deficiency as a cause. In many cases the prompt intervention of the captains involved led to the removal of the faulty officer who had displayed his lack of colour perception whilst on board ship. One case, reported originally by Jeffries in personal correspondence dated 1913, involved two defective men on duty on the same ship simultaneously.

The accidents speak for themselves, clearly indicating the need for careful colour vision examination of persons entrusted with the guidance of moving vehicles dependent on coloured light signals. Examination should take place before an employment agreement is reached. Other aspects emerge from these historical examples; one, highlighted by the Committee on Colour Blindness of the Royal Society (1892), concerns the requirement that witnesses giving evidence concerning coloured signals or lights should themselves be tested for colour vision. A second is the need for regular examination of key personnel for acquired disturbances and a third is the desirability of prompt colour vision examination of those involved in accidents or near-accidents following the incident. Persons found to be at risk should be transferred quickly to jobs where colour recognition is not a responsibility.

Although it is not likely that all such precautions are taken in *every* case arising today, those concerned with transport take the responsibility for colour vision examination more seriously than in the past. In the UK, for example, an estimated 200 000 employees of British Rail (1975 figures) are thought to have satisfied the colour vision standard and a periodic re-assesment (usually at five-yearly intervals) is enforced for all drivers and operators involved with moving vehicles. Although there is no record of a major accident which resulted or was said to have arisen from defective colour vision in the last fifty years, mistakes due to colour anomalies are nevertheless reported occasionally, emphasising the real need for caution. Holmes, whose expertise on signal colours spans some fifty years, did not know of any aircraft accident in whch defective colour vision had been established as a contributory cause (personal communication 1980).

13 Official colour vision standards in the UK

The armed services and public bodies concerned with transport (e.g. British Rail) have been aware of the need for colour vision standards for many years.

The history surrounding the introduction of official colour vision requirements for the armed services makes fascinating reading, and it is perhaps surprising to discover that the subject had even been considered by both Houses of Parliament in 1912, and before that by a committee of the Royal Society in 1892. Colour vision standards within the national and public services were briefly considered by the Committee of the Colour Group (GB) of the Physical Society responsible for a report on defective colour vision in industry in 1946. More recently the subject came up for discussion at a meeting of the Royal Society of Medicine (Ophthalmology and Occupational Medicine sections) held in London in February 1975.

13.1 TRANSPORT

The first tests for defective colour vision were based on the matching of coloured wools or beads after Seebeck in 1837. Holmgren produced his wool test as a result of a major train crash in Sweden in 1877, and it was popular for use among railway employees in Europe and the USA until the early twentieth century. In most countries there is no colour vision restriction for possession of a driving licence for a private car, goods vehicle, heavy goods vehicle or public service vehicle.

13.1.1 British Rail

All persons who apply to join the staff of British Rail are given a colour vision test and the result is recorded. An estimated total of 200 000 people throughout the UK are required to satisfy the colour vision standards. The tests performed are the Ishihara test and the Edridge-Green lantern test (identification of red, green, yellow and white lights), the Ishihara test being the primary test.

All staff concerned with lines controlled by signals and jobs requiring normal colour vision must pass the Ishihara test. Such jobs include foot-platemen (drivers), drivers' assistants and trainees, operators, traction trainees, some supervisors in the traffic section, signalmen, lookout-men, guards, shunters, patrolmen, track maintenance staff, those concerned with signal and telegraph wiring and all signal and telecommunications and overhead line staff.

All drivers, operators and some supervisors in the traffic section have a reassessment of colour vision at five-yearly intervals at age 45 until the age of 60 and thereafter at two-yearly intervals. Statistics are available concerning incidents in which misreading of signals may have been responsible for a rail accident. It is only since 1919 in Great Britain that accidents attributable to the failure of enginemen to obey danger signals have been separately recorded and in the six years prior to 1924 there were 23 such accidents; there is no record of the numbers of these accidents which may have been due to defective colour vision of the drivers. Mistakes due to suspected colour vision anomalies are known by British Rail medical officers—at least one case of the passing of a red signal and the mistaking of coloured signals.

13.1.2 London Transport

All persons who apply to join the staff of London Transport are given a colour vision test and the result is recorded. Most people who fail the colour vision test are refused employment, except for graduates who go into railway or signalling jobs, and in this case their supervisors are advised of their disability. There are over 9000 motormen and 2000 guards employed by London Transport. The Ishihara test and Edridge-Green lantern test are performed. A candidate can be refused employment for failing one plate of the Ishihara test. The Edridge-Green lantern is used with two apertures, three colours and a filter to simulate rain and fog. Both tests are performed by a doctor. The present tests have been used since the 1940s. Until 1932 the Holmgren wool test was used.

Drivers, signal engineers, signalling staff, shunters, guards and certain track workers must pass the colour vision tests. All drivers and operators concerned with moving stock are re-examined with the lantern test at ages 50, 55, 60 and 63.

Two cases of acquired colour vision defects resulting from tobacco amblyopia have been reported recently in persons who previously passed the colour vision requirement. Since Norman (1960) failed to show any difference in driving ability between 150 Daltonic bus drivers and a control group the colour vision requirement for bus drivers was dropped.

Trains have an automatic stopping mechanism for red signals on the tracks, but not in sidings. No accidents or incidents reported in the last 25 years have been attributed to defective colour vision although an accident on the Circle line in 1938 was said to have been caused by the incorrect wiring of signals.

13.2   CIVIL AVIATION

Applicants to the Civil Aviation Authority must 'demonstrate their ability to perceive readily those colours the perception of which is necessary for the safe performance of duties'.

The following tests are used: Ishihara plates illuminated by illuminants C or D. Failures can be assessed as fit provided they pass a colour lantern test (Martin or Giles–Archer lantern, or Holmes–Wright lantern). See the Appendix for further details.

All flying personnel (airline transport pilot, flight navigator, flight engineer, radio officer), air traffic controllers, private pilots and most engineering apprentices must pass the colour vision tests.

In the 1946 Report on Defective Colour Vision in Industry it is noted that defective colour vision is not necessarily a bar to safe flying and if the applicant for a private licence is able to distinguish the coloured lights used in air navigation he may be awarded a licence for a day and night flying. If he fails to identify correctly the coloured lights his licence is endorsed, and he is restricted to flying only between sunrise and sunset. Recent studies (Hoogerheide 1959, AGARD 1972) emphasise that some anomalous trichromats can be considered safe for flying duties.

13.3   ROYAL AIR FORCE

The Royal Air Force has three categories for colour perception.

(a) CP2 perfect colour vision. No errors on Ishihara test when carried out in daylight or equivalent artificial illumination.

Persons who must achieve this standard: RAF and WRAF officers—engineer, marine, photographic interpreter. RAF and WRAF other ranks—photographic personnel, safety and surface personnel, electrical, radio and some aircraft engineers, electronic engineers, some general engineers and some marine personnel and aerospace operators.

(b) CP3 'colour vision safe'. Errors on the Ishihara permitted but ability to recognise the colours used in aviation as presented by the Holmes–Wright lantern (the only recognised test).

Persons who must achieve this standard: RAF and WRAF officers—fighter control. RAF and WRAF other ranks—physical training instructors, air traffic controllers, RAF regiment, MT drivers, police.

(*c*) CP4. Unable to pass standards above. This involves all other personnel.

The Appendix should be consulted for further details.

13.4   ARMY

Before the Second World War colour deficiency was recorded but did not constitute a ground for rejection from services. Today the Army does not specify direct requirements except that 'complete colour blindness' may restrict choice of certain specialist Arms or Corps, as an officer.

Officers must be red/green safe for commissions in the Household Cavalry, Royal Engineers, Royal Signals, Special Air Services, Royal Corps of Transport, Ordnance Corps, REME, Military Police, Intelligence Corps and flying duties. The Youth Employment Service Bulletin mentions also entry to Sandhurst.

'Complete colour blindness' restricts choice of employment in WRAC but is acceptable in QARANC.

The Appendix should be consulted for further details.

13.5   ROYAL NAVY

A special MRC Report on Colour Vision Requirements in the Royal Navy was published in 1933 (Report No 185) at the request of the Admiralty. At that time the regulations stated that persons entering all branches of the Navy should have normal colour vision and re-examination was forbidden. Now three categories have been established as in the Royal Air Force. All are first tested on Ishihara plates.

(*a*) CP1. This standard requires a pass on Ishihara and correct recognition of coloured lights using small paired apertures on the Martin lantern at 6 metres in a darkened room.

Persons who must achieve this standard: seamen and seamen officers, air crew officers, pilots and observers, hydrographic surveying officers.

(*b*) CP2. Pass 13 out of the first 15 plates of Ishihara test (24-plate 1969 edn) shown under artificial daylight in random order.

(*c*) CP3. Correct recognition of coloured lights using large paired apertures on the Martin lantern at 6 metres in a darkened room.

Persons who must achieve this standard: seamen/officers operations branch, missile aimers, seamen/officers tactical sub-branch, air traffic controllers, naval airmen and air crew, diver adquals, hydrographic surveying recorders, Royal Marines officers, landing craft commando air crew, clerks and cooks, divers and photointerpreters.

(*d*) CP4. Correct recognition of colours used in relevant trade situations (e.g. wires, resistors, stationery tabs etc).

Persons who must achieve this standard: other grades than in CP1–3, Naval air mechanics, communications personnel, Royal Marines SD list persons, signallers, musicians, divers, engineering branch officers and ratings, medical and dental officers and ratings, WRNS, QARNNS and instructors.

Colour vision testing is carried out before transferring employment to a group involving a different standard and in cases of doubt. Retests are carried out by a naval eye specialist. The Appendix should also be consulted.

13.6   MERCHANT NAVY

The Ishihara test and Holmes–Wright lantern are used, either together or separately.

Deck officers and ratings (involving look-out duty) are regarded as having normal colour vision if they pass plates 1, 11, 15, 22, 23 of Ishihara's charts, or pass the Holmes–Wright lantern. Failure on the lantern comprises mistaking red for green or vice versa. In doubtful cases (if white is called 'red' or vice versa or white and green are confused) the case is referred to the principal examiner.

Engine-room officers, cadets and ratings, electricians and radio officers must pass 'the modified chart colour tests'.

It is stressed that 'ability to reach the required visual standard when first going to sea does not guarantee an ability some years later to pass the lantern test when seeking admission to an examination for a certificate of competency'. Furthermore it is stressed that 'Candidates other than new entrants who have previously failed to pass the sight test either locally or at a special appeal test conducted by the Principal Examiner of Masters and Mates or his Deputy may, provided aids to vision have not been worn at any of these previous tests, apply to take the test again locally with aids on payment of a further fee'. The Appendix should be consulted for further details.

13.6.1   History

A variety of tests have been used in the Merchant Navy over the years. Topley (1959) noted that before 1845 anyone could sail as a master or mate, no visual standards being applied. In 1877 a primitive form of colour vision test was introduced using coloured glasses and cards. Candidates had to distinguish the principal colours, but there was no uniform method of testing. Soon one of the examiners reported to the Board of Trade that a candidate who could not name the coloured glasses correctly in daylight made no mistake if they were displayed in front of an oil lamp flame. This led to the introduction in 1885 of an oil lamp which was supplied with a removable ground glass screen for use with an improved set of coloured glasses and cards. In 1892 a committee of the Royal Society under the chairmanship of Lord Rayleigh recommended that the card and glass test should be discontinued in favour of Holmgren's wool test. This method was still far from satisfactory, however, as the famous case of Mr Trattles indicates. Trattles had been at sea for several years and became a second mate in 1902 passing the Board of Trade colour test. When examined a year later, as part of the test for first mate, he failed the Holmgren wool test and a later appeal test which included a lantern, and was asked to surrender his ticket. The gentleman persisted and between 1904 and 1907 he was examined six times (three times he failed and three he passed). He was seen also by Sir William Abney, physicist and adviser to the Board of Trade, and Dr Edridge-Green. Both experts produced evidence; one would have passed Trattles and the other failed him. The Board of Trade was severely criticised and it suffered a great deal of adverse publicity. The Holmgren wool test was discontinued and the Board of Trade lantern was introduced in 1913 as a fairer test of practical ability at sea. The case indicates the extreme difficulty which sometimes arises in placing an individual on the proper side of the safety line. The need for standard viewing conditions and a universal technique or procedure associated with the testing of colour vision, is also indicated. At the February 1975 meeting of the Royal Society of Medicine on Visual Standards it was pointed out that the luminous intensity of the Board of Trade lantern and the Martin lantern is so near the colour threshold value that the test presents some difficulty to a number of colour normal people. The need for a colour vision standard to be based on 'reasoned principle and applied with a due sense of proportion' was stressed. The introduction of the Holmes–Wright lantern to replace the Board of Trade lantern may help in future (see Chapter 7).

A very recent case which reinforces this point came to the notice of the present authors. A young man who had passed the examination for a yacht master's (ocean) certificate was not awarded the certificate because he had failed the Board of Trade colour lantern. He attended for a second colour vision test at the recommendation of his optometrist and was found to be virtually perfect on the Ishihara plates and the HRR plates and perfect at the D.15 test. In the first box of the FM 100 Hue test, he made five minor transpositions, all other boxes being perfect. On the Giles–Archer lantern he had a slight difficulty with green on the very small aperture and with the dark red, but many colour normals have this difficulty.

13.7   POLICE FORCE

The ability to distinguish the principal colours is all that is required for constabularies. For further details consult the Appendix.

13.8   CIVIL AND PUBLIC SERVICE

For a number of grades within the Civil and Public Service satisfactory colour vision is required, but the testing is left to the individual departments and they use a method which is most suited to their requirements. This may involve the Ishihara test or may only require an answer to the question 'Do you have difficulty in recognising colours?'†

The grades for which a special slip 'Colour Vision' should be attached to the health questionnaire are listed below.

All departments
   Illustrator
Ministry of Agriculture, Fisheries and Food
   Agriculture Development and Advisory Service (science specialists)
   Assistant Veterinary Investigation Officers (Grade B officers)
   Science Group (all grades)
   Veterinary Research Officers
Ministry of Defence (Air Force Department)
   Telecommunications Technical Officers
   Telecommunications Technical Officers in the Meteorological Office
   Professional and Technology Officers Grade IV (Instruments and Electrical)
   Radio Technicians
   Examiners in the Quality Assurance Service (RAF)
Ministry of Defence (Army Department)
   Telecommunications Group BAOR
Ministry of Defence (Navy Department)
   Professional and Technology Officers Grade IV (L)
   Professional and Technology Officers Grade III and IV
   Professional and Technology Officers Grade IV (Aircraft)
   Professional and Technology Officers (RNSS)
   Cartographic Assistants
   Cartographic Draughtsmen
   Civil Hydrographic Officers
   Chart Depot Assistants
   Reproduction Grade A

†Personal written communication to J Voke from Dr R Oliver, Senior Medical Officer for the Civil Service, December 1973.

Ministry of Defence (Procurement Executive)
Science Group (all grades)
Professional and Technology Officers and engineering graduates
Professional and Technology Officers
Government Communications Headquarters
Telecommunications Technical Officers
Home Office
Science Group (all grades) in the forensic science laboratories
Former Ministry of Posts and Telecommunications
Telecommunications Technical Officers
National Coal Board
Surface and Underground locomotive drivers and electricians are required to pass the Ishihara test.
Meteorological Office
The Ishihara test is given to electricians and technical staff concerned with meteorological instrument development and maintenance.
BBC
Engineering and Technical Operators including film editors are given the Ishihara test and FM 100 Hue test in doubtful cases.
Regional Electricity Boards
Although no national policy exists for colour vision examination, the Ishihara test is widely used. Apprentices and control engineers are involved.
Regional Gas Boards
A great many gas fitters and electricians are involved in colour identification. The Ishihara test, often with a wire identification trade test is widely used.
British Telecom
Colour vision tests are given to engineers. There is no national policy but the Ishihara test and a wire identification trade test are widely used. The possibility of introducing a colour vision test for switchboard telephonists has recently been under consideration.

13.9 INTERNATIONAL STANDARDS

A selection of international standards for colour vision is given in table 13.1.

Table 13.1   International standards of colour vision. A dash indicates that data are not available for that category. All standards are for 1972.

Country	Private car	Lorry/Bus/taxi
Austria	Protanopes and achromats excluded	Protanopes and achromats excluded
Hungary	Protanopes and achromats excluded	Normal
Finland	No limitations	Lorry—no limitations. Bus and taxi—achromats excluded
Sweden	No limitations	Lorry and taxi—no limitations. Bus—normal (likely to be no limitations in future)
Norway	No limitations	No limitations
Denmark	No limitations	No limitations
France	No limitations	No limitations
Netherlands	No limitations	No limitations for lorry and taxi. Bus drivers, lock-keepers, bargee and bargehands on boats must be able to discriminate R and G signal lights. Normal also ambulance drivers.
West Germany	No limitations	Mild deutans, tritans and protans accepted by 2 plate tests and anomaloscope
South Africa	No limitations	No limitations
Canada	No limitations	Normal also ambulance drivers, fire engine drivers and chauffeurs. Firemen mild R/G confusion accepted
Israel	No limitations	No limitations
Belgium	No limitations	Lorry—no limitations. Taxi and buses—protanopes and achromats excluded

Table continued

Table 13.1 *Continued*

Country	Railway	Aviation (civil and military)
Austria	Protanopes and achromats excluded	Protanopes and achromats excluded
Hungary	Normal (always use Nagel)	—
Finland	1–5 category normal Cat. 6 no limitations	Normal
Sweden	Normal for those with signals	Normal or PA/DA without increased exhaustion and for military sharp matching range
Norway	Normal	Normal but some mild/moderates allowed
Denmark	Normal for those with signals	Normal except for flight engineers. Mild DA/PA pass
France	Normal 3rd class—coloured lights only	Normal for private pilot and navigator. Helicopter pilot only pass on lantern
Netherlands	Normal	Mild deutan accepted
West Germany	Normal	Professional—normal. Private—some PA/DA allowed
South Africa	Some accepted as safe if pass red signal light identification but fail Ishihara	Civil pilots only. Some accepted as safe if pass red signal light identification but fail Ishihara
Canada	13/16 correct on Ishihara (16 plate edition)	*Civil*—22/24 correct on Ishihara (24 plate edn). Day private pilot—red flashing light seen only. *Commercial*—13/16 correct on Ishihara (16 plate edn)
Israel	Some must be normal	Air crew and flight supervisor normal
Belgium	Some must be normal. Category E no limitations	*Civil*—normal on Ishihara for professional. Normal on Beyne lantern for private

Table continued

Table 13.1 *Continued*

Country	Marine/Navy	Army	Tests used
Austria	—	—	1 Stilling, Velhagen or Ishihara 2 Nagel anomaloscope in cases of doubt
Hungary	—		1 Stilling or Rabkin or Ishihara 2 Nagel anomaloscope
Finland	Normal	Normal for aviation and transport. No limitation for militia	Velhagen for rail and marine Ishihara for army and aviation
Sweden	Normal		1 Bostrom *and* Bostrom–Kugelberg 2 Anomaloscope in cases of doubt
Denmark	Normal	No restriction	1 Ishihara, Bostrom or Bostrom–Kugelberg must pass all plates on 1 test 2 Nagel anomaloscope in cases of doubt
France	—	No restriction except machine gunners, navigators etc normal	1 Ishihara coloured lights, Beyne's lantern used for aviation and army at 5 metres
Netherlands	Normal for airman, driver, marine engineer, sailor, assembler HRR mild/med accepted	—	1 Ishihara or HRR 2 Anomaloscope in cases of doubt
West Germany	Normal	Normal	1 Ishihara, Bostrom or Stilling or Velhaven 2 Anomaloscope in cases of doubt
South Africa	—	—	1 Ishihara 2 Signal red
Canada	No red/green confusion allowed as assessed by lantern		1 Ishihara 2 Coloured lantern
Israel	Deck normal other categories severe deficiencies excluded		1 Ishihara 2 Signal light recognition
Belgium	Normal on Ishihara	Militia: achromats excluded. Protanopes and achromats excluded from usual categories	1 Ishihara 2 Panel D.15 used as well for taxi and bus drivers and army 3 Holmgren wools for some railway categories

14 Recommendations

A decision to recommend a colour vision examination is often made by persons who will not be involved in the testing procedures. Serious thought should be given to this initial recommendation for screening for detailed tests or for referral. The following questions help to identify the objectives.

What stimulus has prompted the decision to examine colour vision?
Who should conduct the tests?
Which tests should be selected?
Should any special procedures be adopted?
If deficiencies are identified, what actions should result?

No two situations present identical circumstances. Variations must therefore be recognised and accepted in defining the objectives and the methods by which they are to be achieved.

Throughout this book guidance has been offered both as to the choice of testing methods and on the analysis of results. Aspects such as safety and the occupational and industrial consequences of colour vision deficiencies have been highlighted. Such considerations must influence those responsible for making decisions concerning the desirability of permitting Daltonic observers access to certain types of employment.

This short chapter provides additional guidance of a deliberately practical nature, to assist those with responsibility for examining colour vision.

14.1* PROTOCOL

Those responsible for administering colour vision tests will not necessarily interpret the results; cooperation between the 'administrator' and the 'interpreter–controller' is essential. Roles and responsibilities should be clearly understood. A written statement explaining the requirements for testing and the testing methods selected should be available to all those involved. This statement should outline:

(a) the purpose of colour vision testing;
(b) the reasons for the choice of such testing methods;
(c) recommendations for a regular review of (a) and (b).

Likewise, a written 'protocol' or system for administering the tests should be provided. Part of this document can be communicated to those whose colour vision is to be examined, as a means of explanation and reassurance. Some testees may be puzzled, confused and uncertain about what is required of them unless this is clearly stated. For example it may help to explain that they may have to identify coloured lights in a dimly lit room, or to rearrange coloured samples into a colour sequence. Those responsible for the tests should give full explanations and demonstrations to those actually administering the tests or recording the results. The reasons for a precise regime must be made clear to all involved. It will not be in the interests of observers if they are prompted unfairly, nor should they be scolded on account of errors. It is advisable to ensure that personnel involved in administering the tests are well-orientated and frequently updated on progress in the field, by means of suitable texts. They too should be subject occasionally to the tests, administered by different colleagues.

14.2   AIDS TO THE PROTOCOL

Simple, typed, introductory notes can be given to subjects to read while they await the tests. In some cases these should be agreed with trade unions, professional bodies or colleagues, well in advance. Alternatively, or in addition, carefully worded instructions should (literally) be read to the subject, to ensure uniformity of methods and of responses.

For example, a transport driver can be told 'we have to discover how good you are now at detecting signal colours and coloured codes and if your colour vision has possibly altered'. He would then be prompted to sit in the correct place, to use his most suitable optical correction, not to touch certain parts of a test, etc.

There is some latitude to most methods, but it must be made clear how limits are to be set when subjects hesitate, stumble over the intitial stages, or race carelessly through a series of responses. Recording methods must be adequately standardised and should include the date and the name of the examiner or administrator. The pattern of response data should be such that it can be understood by others who view the record sheet. Extra notes may be added on a basis previously explained. In isolated cases a language barrier may be encountered which will call for a retest with an interpreter. The variations of results outlined in §§7.4 and 10.8 give additional guidance.

14.3 DEMANDS

Despite the most rigorous standardisation of procedures, individual responses will undoubtedly vary. Lakowski (1969a) has stressed the need for rapport, the recognition of the complexity of such a psychophysical task, and the fact that several cognitive demands are made on the observer. Those who administer tests must be preconditioned to exercise sympathetic, just and informal procedures.

14.4 APPARATUS

Invariably those establishing themselves in a new field require to know how they should be equipped. A provisio is often added that a limited budget is available. Guidance on the selection of tests, recommendations for equipping a consulting room and proposals for specific 'test batteries' have been considered elsewhere in this text. While the expensive anomaloscope can be fully justified in many cases and the relatively expensive Lovibond colour vision analyser will be invaluable in other situations, there is a natural reluctance to invest in the more expensive devices initially. Special needs such as 'trade tests' or means of investigating the colour perception of children, must be considered; these call for additions to the range of basic apparatus which may be adequate for the average situation. It is well recognised that it would be poor preparation to be without four or five different test methods, even if only two or three of these are in regular use.

14.4.1*　Basic equipment

Two PIC and at least one 'sorting or confusion' tests form the minimum selection. The most suitable artificial illuminant (with spare bulb) which can be provided is a necessary complement to such tests. The authors' preference for simple filtered tungsten light is economical and should appeal to those commencing to gather apparatus. Provision for 'red–green' and 'blue–yellow' testing is always required. In Chapter 7 it has been indicated that a *selection* of individual plates (even as few as six) can suffice for some screening purposes. It may be worthwhile to extract the most reliable plates and to mount them in a separate folder, producing a special record sheet for the purpose. The back or lower part of such a record should have printed spaces for extra tests, which may be compiled in another folder from the remaining plates as a 'second attempt' set. A simple form of lantern should be included whenever possible, on account of its high 'face value'. The appropriate sections of Chapter 7 give adequate justification for this suggestion.

14.4.2 Recommendations for screening tests

Selections of tests have been proposed to provide an optimum variety of approaches for screening purposes, for example those described by Paulson (1973) and Fletcher (1981). Use of a single screening test should not be encouraged for there is no one single ideal colour vision test and variations between editions of plate tests can be misleading. Nevertheless the demands of an excessive series of tests are daunting and confuse the lay person required to screen for colour vision defects. A selected choice of suitable tests or parts of tests is presented below, based on the experience of the authors and with the following objectives in mind: detection of all (or the majority of) red—green variations from normal; detection of significant blue defects; provision of data to guide professional judgment as to the extent of any anomaly.

(*a*) Appropriate daylight illumination.
(*b*) The F_2 plate or a substitute to detect blue defects. (The system described by Taylor (1975b) can be adopted, perhaps substituting short lines or angles for squares.)
(*c*) Three plates from Ishihara's test.

Number	Normal sees as	
6	5	
8	15	for red/green defects only
5	57	

(*d*) Three plates from the City University test.

Number		
1	2nd edn or 1 1st edn	for red/green
2	2nd edn or 2 1st edn	and blue/yellow
7	2nd edn or 7 1st edn	defects

(*e*) Where additional variety is thought necessary, or as an alternative to (*c*), three plates from the Dvorine test are useful.

Number	Normal sees as
1	48
9	74
2	67

Such screening may result in false positives but they should be referred for more detailed examination. A few Daltonics may 'pass', usually on account of the mode of administration rather than the test battery but this is inevitable with screening.

14.4.3 Protocol for a lay person conducting screening

(a) Preliminary discussion with person(s) responsible for (i) setting the 'standards' for 'pass' or 'fail' and (ii) making decisions about those who 'fail' are recommended. Ensure that you have clear instructions about indicating to the individuals you test how well or badly they appeared to perform. Those you 'fail' may best be told they should have a 'second test'.

(b) Ensure that *you* have normal colour vision or that you know errors you make. In either case, keep to a rigid system of questions, display of the tests and recording; this assists uniform standards and fair comparisons.

(c) Familiarise yourself with the procedure and tests privately, well in advance and with the help of someone who is not one of the official subjects. Repeat the procedure and seek advice from this person as to any obvious difficulties which may be caused by local conditions.

(d) Assisting a person to 'pass' is unlikely to be the kindest approach. All 'failed' subjects will be given careful and fuller testing and it is expected that some 'false positives' will be 'passed' at later retesting. Encouragement through remarks such as 'thank you' and 'you are responding helpfully' in a slightly positive but non-commital manner, is often supportive. Mark the record immediately and legibly.

(e) Avoid a queue of subjects which permits the next in line to hear or see the performance of other subjects. Use an assistant to marshall the oncoming subjects to ensure that they arrive promptly.

14.5 ADDITIONAL EQUIPMENT

Definite guidance must be provided for personnel engaged chiefly for 'screening' if they are to be allowed, or encouraged, to extend the tests they give. A good reason could be the added interest involved, provision for the satisfaction of specially concerned parents or subjects, or perhaps the collection of extra data for a new approach to testing. It is important to realise, however, that the use of a 'special' device, especially one with which there is less than full skill available in its operation, can produce misleading information. Furthermore, a complicated or expensive piece of apparatus may carry a self-importance of its own in the eyes of the testee, despite the fact that an advanced device in the wrong hands can be worse than useless. It is nevertheless valuable to have a modest set of more complex devices available for use in retesting, in cases of doubt, or if comparison with data from elsewhere is of value.

In the low cost range there is ample justification for one or more of the matching tests involving Munsell colours, particularly in desaturated (low chroma) form. A simple filter anomaloscope can be valuable, particularly one which permits variable luminance to the red and yellow which can then be matched in brightness. At the other extreme, where a laboratory is to be equipped with V_λ and/or hue discrimination apparatus, or even an evoked cortical response analyser, many thousands of pounds sterling will be involved. More modest laboratory apparatus in a middle cost range is seldom obtained ready-made and must be constructed individually.

14.6   *AD HOC* CONSTRUCTION

Apparatus of the type described in §7.10.3 is commonly constructed using optical benches and a selection of achromatic or plastics lenses mounted in commercial holders. Car bulbs or halogen lamps can be used as light sources. Inexpensive heat filters (or 'cold mirrors'), a series of second-grade neutral density filters, together with devices to provide polarised light and spectral sources (such as interference filters), provide scope for suitable construction of apparatus costing a few hundred pounds. Such equipment can be readily modified to fulfil a variety of functions. Optical benches can be imitated by resting lenses and filters in U- or V-shaped channels, if alignment is carefully checked. Maxwell's classic spectral apparatus was of a simple design. A constant deviation spectroscope can be produced with low-cost prism or grating methods. Small monochromators which are readily available are most versatile. It is possible to construct a wide range of apparatus for less than a thousand pounds sterling (at the time of publication). Such a range could include a simple colorimeter, capable of use as an anomaloscope, as well as simpler increment-threshold measurements, and provide both spectral hue discrimination and V_λ measurements. A single device used for too many purposes, and used by more than one person, however, is often difficult to keep in reasonable calibration. Stray light must be controlled. Few 'instrument makers' are willing to construct such apparatus as a whole, but some may construct certain parts. A design and practical guidance is best obtained through one of the university departments interested, usually on a consultancy arrangement. Some such departments will be prepared to carry out the construction and/or the calibration. Alternatively, one of the industrial concerns involved in lighting or in colour measurement may be prepared to undertake such work. A retired technician or academic will often find such a task, including the inevitable 'shopping around' for components, most rewarding. Such an approach may well be the most suitable arrangement to make. Calibration is important. Standardising bodies such as the National Physical Laboratory can undertake such a service but this would be expensive.

In order to indicate the potential in real terms, figures 14.1, 14.2, 14.3 and 14.4 show plans of devices actually produced and used by the authors for different purposes.

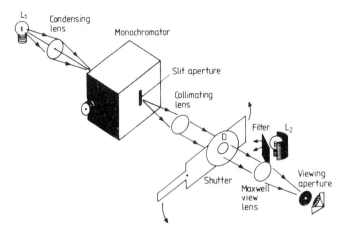

Figure 14.1 Apparatus for the naming of 'unique green' and capable of being used for spectral neutral points with dichromats. D represents the disc of white material.

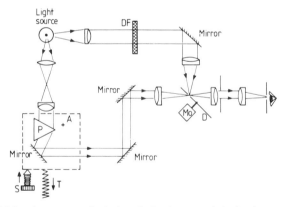

Figure 14.2 A constant deviation (prism) spectral device incorporating a simple flicker photometer for V_λ determination. P = dispersing prism; A = axis of rotation of prism unit, moved by S, a screw acting against tension T; DF = daylight filter for white beam, focused onto disc D of flicker photometer, rotated by motor Mo.

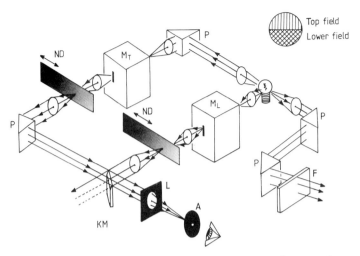

Figure 14.3 Wavelength discrimination apparatus using two low-cost monochromators with surface-silvered mirrors and lenses mounted with glue on metal pillars. Provision is made for different field sizes and for pre-exposure. P = reflecting prism; M_T = monochromator for top field; M_L = monochromator for lower field; F = daylight filter; ND = wedge neutral density filter; KM = knife-edge mirror; L = lens combined with aperture; A = aperture which may be combined with a lens. The inset shows top and lower fields of view.

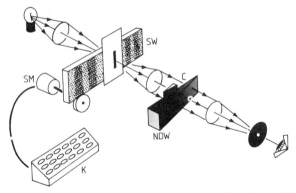

Figure 14.4 Device using a wedge-type interference filter, SW, as a simplified monochromator. A stepper motor, SM, and keyboard control, K, are readily added. The collimated coloured light is adjusted in intensity by a neutral density wedge, NDW, and compensator, C, giving uniform intensity over the beam.

Appendix: Standards of vision, including colour vision, for various occupations

The following information on visual standards etc, is Crown Copyright, and is reproduced with the permission of the Controller of Her Majesty's Stationery Office. (The dates indicate the last revision quoted.)

NAVAL EYESIGHT STANDARDS (APRIL 1981)

Colour perception (CP)

There are four standards of colour perception graded as follows.

   1 The correct recognition of coloured lights shown through the small paired apertures on the Martin lantern at 6 metres (20 feet) distance.

   2 The correct recognition of 13 out of the first 15 plates of the Ishihara test (24-plate abridged edition 1969) shown in random sequence at a distance of 75 cm under standard fluorescent lighting supplied by an artificial daylight fluorescent lamp (British Standard 950: 1967).

   3 The correct recognition of coloured lights shown through the large paired apertures on the Martin lantern at 6 metres (20 feet) distance.

   4 The correct recognition of colours used in relevant trade situations, and assessed by simple tests with coloured wires, resistors, stationery tabs, etc.

Method of testing using the Holmes–Wright lantern

The Holmes–Wright lantern is constructed to simulate in controlled conditions the critical visual task of seamen.

The test must be carried out at a distance of 6 metres (20 feet) in a completely darkened room. The candidate may wear spectacles if he wishes and may be 'dark adapted' if necessary. The colour pairs may be changed by rotating the colour setting flange at the rear of the lantern, the colour pairs presented being indicated by the code number visible in windows on each side and at the rear of the lantern.

The code numbers represent:

1	R2	2	G2	3	W	4	G2		
	W		R1		W		R2		
5	R1	6	R1	7	W	8	G1	9	G1
	R2		G1		G2		R2		G2

The intensity of the lights presented may be varied by the filter change lever at the rear of the lantern, the setting being:

> DEM   for demonstration only
> HIGH BRIGHTNESS
> LOW BRIGHTNESS

In order to reduce errors the examination method and instructions to the examinee should be followed exactly in each case.

(*a*) The examinee is to be seated with the lantern apertures at eye level.

(*b*) Connect the lantern to a 230/240 volt supply and switch on with the rotary switch at the rear of the lantern. No warming up period is necessary.

(*c*) Turn the filter change lever to DEM and the colour setting flange to Code 1.

(*d*) Say to the examinee 'this is a test to find out whether you can readily recognise the colours of red, green or white lights. The colours are shown in pairs one above the other in any combination of red, green, or white. Name both colours calling the one on top first. The top colour you see now is red'.

(*e*) Turn the colour setting flange to Code 2. Say to the examinee 'the top colour you see now is green'.

(*f*) Turn the colour setting flange to Code 3. Say to the examinee 'the top colour you see now is white'.

(*g*) Turn the filter change lever to HIGH or LOW BRIGHTNESS as appropriate. Turn the colour setting flange to Code 4, 6, 8 or 2 (i.e., any red, green, combination). Say to the examinee 'Start now, naming first the top then the bottom colour. Do not use any words other than red, green or white. You will be given 5 seconds to name the colours'. If the examinee uses any colour name other than red, green or white he is to be reminded that only these words will be used. No other comments are to be made by the examiner.

(h) Show each colour pair to the examinee in consecutive order. Each response must be given within 5 seconds. The lantern is not to be opened except for routine annual servicing, at which time the lamp will be changed.

Procedure in testing

Ishihara plates are used as a screening test for all entries. Those who pass the Ishihara test are graded CP2 and require no further testing except for those whose critical visual task requires a categorisation of CP1. Those who fail the Ishihara test are categorised CP3 or CP4 according to requirement.

Retesting of colour perception

Colour perception does not normally change significantly throughout life. It will require retesting, therefore, only in certain circumstances: (i) before employment in a specialisation requiring a different colour perception standard as shown in the regulations, and (ii) if there is any doubt concerning the existing grading. Retest will be carried out invariably by a Naval eye specialist at the request of a Medical Officer.

Application of eyesight and colour perception standards

Specialisation, branch, or sub-branch	Standard VA	CP	Specialisation, branch, or sub-branch	Standard VA	CP
Seamen specialisations and operations branch			Air traffic control	2	3
Seamen officers			Ratings		
GL(X) and SL(X)	1A	1	Naval airmen	2	3
PWO(N)	1A	1	Naval air mechanics	2	4
SD(X)	1A	1	Aircrewmen	2	3
Operations branch			Missile aimers	1	3
Seaman group	2	3	SAR and SAR diver		
Missile aimers	1	3	adquals	1	3
Communications	2	4			
tactical sub-branch	2	3	*Hydrographic Surveying*		
			Service		
Fleet Air Arm			Officers		
Aircrew officers			GL(X) and SL(H)	1	1
Pilots GL(X) and SL	1A	1	Ratings		
Observers GL(X) and SL	1A	1	Surveying recorders	1	3
Officers					
SD (AV)	2	3			

Table continued

Specialisation, branch, or sub-branch	Standard VA	CP	Specialisation, branch, or sub-branch	Standard VA	CP
Royal Marines			*Medical and Dental Branches*		
Officers	1	3	Officers and Ratings	2	4
SD List	2	4			
Other ranks			Instructor	2	4
General duties	1	4	Chaplain	2	4
Landing craft	1	3	Recruiting Officer	2	4
Commando aircrewman	1	3	Royal Corps Naval		
Tradesmen, clerks, cooks			Constructors and		
musicians, drivers,			Royal Naval		
signallers	2	4	Engineering Service	2	4
			WRNS	2	4
Engineering Branch			QARNNS	2	4
Officers and ratings	2	4			
			Sub-Specialisations		
Supply and Secretariat			Regulating	2	4
Branch			Diver	2	3
Officers and Ratings	2	4	Photo-Interpreter	2	3

Extract from qualifying notes on eyesight and colour perception standards

Spectacles and contact lenses

There are in general no restrictions to the wearing of spectacles or contact lenses to improve visual efficiency provided that the standards are met. Spectacles will be provided if required for the efficient performance of duties.

(*a*) *Seamen specialist officers* with bridge watchkeeping responsibilities require to reach a minimum safe level of visual acuity if deprived of spectacles in an emergency, although they may wear spectacles at any time to improve visual efficiency. This safe level (a serving standard of 6/12 in the better eye and 6/24 in the worse eye) is not a desirable standard but is the lowest compatible with safety requirements, and is designed to apply only to fully trained and experienced seaman officers whose visual acuity has deteriorated from the entry standard. As this can occur in the case of incipient myopes, officers entered with a myopic error of refraction require to have an annual eyesight test, as distinct from the usual four-yearly Pulheem eyesight test.

(*b*) *Naval aircrew.* Spectacles in general are incompatible with aircrew duties and are not permitted except in special circumstances decided by an eye specialist.

(*c*) *Royal Marine officers and other ranks (GD)* are not permitted to wear spectacles in the field as they are incompatible with efficient job performance. At other times they can be worn if they improve visual efficiency, and there is no restriction to the wearing of spectacles by officers on the SD list or by musicians, tradesmen, clerks and cooks.

(*d*) *Subspecialist groups.* The nature of the visual tasks in some sub-specialist groups precludes the use of spectacles, and therefore a certain level of unaided visual acuity is required.

Serving and re-engagement standards

In general there is no separate lower serving or re-engagement standard (except for seamen officers Note 1a). The criterion is a degree of visual efficiency required for adequate job performance and this is reflected in the entry standard. Limits apply only on entry and thereafter will be considered in relation to the functional visual efficiency.

Below standard

Personnel falling below standard will be referred to a Naval ophthalmic specialist who will assess each case individually and will indicate in the report if retention in the Service is possible as an exception to regulations.

THE ARMY (JANUARY 1981 AND MARCH 1981)

Visual standards applicable for acceptance in the Army, (including WRAC and QARANC).

Minimum visual standards

(*a*) Candidates for the Royal Military Academy, Sandhurst, Army Scholarships and University Cadetships require a minimum visual acuity with spectacles of at least 6/12 in each eye or 6/6 in one eye and not less than 6/36 in the other. Colour Perception Standard 2 or 3 is necessary.

(*b*) Most Regiments and Corps require a corrected visual acuity of 6/12 in each eye or 6/6 with the right eye and 6/36 with the left eye. Failure to achieve Colour Perception Grade 3 will restrict employment to certain trades.

(*c*) Myopia exceeding 7 dioptres in any meridian in either eye or hypermetropia exceeding 8 dioptres precludes acceptance even if vision is correctable to the required standard.

Diseases of the eye

Any pathological condition is liable to be a cause of rejection of military service.

Colour perception (CP) standards

Three classifications are applied:

(*a*) CP2. No errors made on Ishihara plates in daylight or using artificial light source of equivalent quality.

(*b*) CP3. Ability to recognise signal colours on the approved lantern test. This is normally the Holmes–Wright lantern.

(*c*) CP4. Inability to achieve Grade 3.

These notes are for guidance only. Each case must be judged on its merits and the final decision as to a candidate's fitness will be made by the appropriate Army Medical Board.

ROYAL AIR FORCE (APRIL 1981)

Visual standards applicable for acceptance for flying and non-flying personnel.

The following are the minimum visual standards for acceptance for service in the Royal Air Force. These standards are under active review and are subject to alteration without notice and no responsibility for consequences arising as a result of these changes can be accepted by the Ministry of Defence (Air). The decision as to the individual's fitness is the prerogative of the Medical Board which examines him.

Flying personnel [appropriate extract]

Require colour perception CP2 or CP3 which are as defined below.

CP2. No errors are made using Ishihara plates in daylight or artificial light of equivalent quality. Tests carried out under normal tungsten or fluorescent lighting are not acceptable except where the ADLAKE lamp is used.

CP3. Although errors are made using Ishihara plates the candidate is readily able to recognise the colours used in aviation. At present the Holmes–Wright lantern is the only recognised test.

CP4. Unable to pass Standard 3.

Non-flying personnel

The minimum uncorrected acuity for entry for all non-flying personnel may be less than 6/60, 6/60, provided that it is correctable to 6/9, 6/9, and

(*a*) the fundi are normal;

(*b*) no other ophthalmic pathological condition is present;

(*c*) considering each eye separately, the spherical correction lies between the range of $+8$ and -7 dioptres, and the astigmatic correction is not greater than 6 dioptres.

RAF and WRAF officers

(*a*) *Visual acuity.* Correctable to 6/9, 6/9, is acceptable for the majority of ground branches.

(*b*) *Colour perception.* CP4 with the following exceptions: engineer, marine and photographic interpreter CP2; physical education (including parachute instructor), RAF Regiment, aircraft control and fighter control CP3.

RAF and WRAF other ranks

(*a*) *Visual acuity.* Correctable to 6/9, 6/9, is acceptable for the majority of ground trades.

(*b*) *Colour perception.* CP4 with the following exceptions: electrical, radio, some aircraft engineering trades, electronic engineering (air) and (ground), some general engineering trades, aerospace systems and movements operators, safety and surface trades, photographic trades and some marine trades require CP2; physical training instructors (RAF), air traffic controllers , RAF Regiment, MT drivers and police require CP3.

CIVIL AVIATION AUTHORITY (MAY 1981) [appropriate extract]

Visual requirements for licences applicable to professional pilots, licensed air crew, air traffic control officers, student and private pilots.

Colour perception

The applicant shall be required to demonstrate his ability to perceive readily those colours the perception of which is necessary for the safe performance of his duties.

The applicant should be tested for his ability to correctly identify a series of pseudoisochromatic plates (tables) in daylight or in artificial light of the same colour temperature such as that provided by illuminant C or D as specified by the International Commission on Illumination (ICI). An applicant obtaining a satisfactory score as prescribed by the licensing authority should be assessed as fit. An applicant failing to obtain a satisfactory score in such a test may nevertheless be assessed as fit provided he is able to readily and correctly identify aviation coloured lights displayed by means of a recognised colour perception lantern. Martin, Giles–Archer and Holmes–Wright Lanterns. (The requirements detailed are those approved by the International Civil Aviation Organisiation (ICAO) and implemented March 1973.)

Professional pilot and aircrew licences

The initial medical examination for these licences is arranged only through CAA Medical Department, CAA House, Kingsway, London WC2B 6TE, to whom applications should be made in writing or by telephone 01-379 7311 Ext. 2800/2801. This examination is invariably held at our Central London Medical Examination Centre on the 2nd Floor of CAA House. Appointments for the routine periodic medical examinations may be obtained at our Authorised Centres in CAA House or in the Tower Block, London (Heathrow) Airport or from authorised medical examiners (AME). Lists of AMEs are available on application to Medical Department—Ext. 2800 and 2801.

Students and private pilot licences

Applicants may obtain their initial and subsequent medical examination appointment at the AME most convenient to their home, business or flying club address.

Lists are available on application to the CAA Medical Department, CAA House, Kingsway, London WC2B 6TE. Ext. 2800, 2801 and 2807 or the Air Owners and Operators Association (AOPA), 50a Cambridge Street, London SW1V 4QQ. Tel: 01-834 5631.

Definition: *Accredited medical conclusion* is the conclusion reached by one or more medical experts acceptable to the licensing authority for the purposes of the case concerned, in consultation with flight operations or other experts as necessary.

MERCHANT NAVY (6 JULY 1981 NEW ENTRANTS)
(17 March 1981 other departments)

Eyesight standards for the Merchant Navy [appropriate extract]

Colour vision for deck officers and ratings may be regarded as normal if plates 1, 11, 15, 22, and 23 in Ishihara's charts are read correctly, or if the Department of Trade lantern test is passed.

Deck department

Unaided vision. The use of spectacles, contact lenses, or glasses of any kind, or any other artificial aid to vision is not allowed.

*Sight tests (Merchant Shipping Notice No M. 896 as amended by M.918).
This Notice supercedes Notice No M.758*

SECTION I

*This section applies to young persons and new entrants about to embark on
a seagoing career in the Merchant Navy or fishing industry.*

1 Every person who wishes to go to sea with the object of becoming a
deck officer in the Merchant Navy or the skipper or second hand of a fishing
vessel should realise that whenever applying to take an examination for a
Department of Trade Certificate of Competency he/she must first pass the
Department's sight test in order to become eligible to enter that
examination.

2 It is important that every candidate, especially a young person who
contemplates becoming a deck officer, should realise that ability to reach
the required standard when first going to sea does not guarantee an ability
some years later to pass the letter and lantern tests when seeking admission
to an examination for a certificate of competency; nor will the required
standard of visual acuity necessarily be maintained throughout a person's
working life. It is possible that there may be certain latent defects of
eyesight which may cause a deterioration in vision as the person grows
older. These defects can only be discovered by a more searching examination than the Department's sight test and it is very desirable, therefore, that
before embarking upon a sea-going career in a deck capacity, every young
person should, as a matter of personal interest, undergo a thorough examination in both form and colour vision by an ophthalmologist.

3 The purpose of the Department's test is to ensure that a candidate's
eyesight is sufficiently good to enable him/her to pick up and identify correctly the lights of distant ships at sea. Experience has shown that for this
purpose a person must be able to reach certain minimum standards both of
form and colour vision.

4 A sight test comprises a letter and lantern test. The letter test is a test
of form vision only and is conducted on Snellen's principle by means of
sheets of letters viewed indirectly through a polished mirror so as to place
the letters at a virtual distance of 6 metres from the eye. Each sheet contains
7 lines: every candidate will be required to read correctly down to and
including line 7 with the better eye and down to and including line 6 with
the other eye. The lantern test is a test of form and colour vision combined.
A special lantern and a mirror are provided for the test which is conducted
in a room darkened so as to exclude all daylight. A series of red, white or
green lights will be shown either singly through a large aperture or two at
a time through small apertures, when they will appear side by side. When
the small apertures are used any combination of the three colours may be
shown or they may be of the same colour. To pass the lantern test the
candidate will be required to name the colours correctly as they appear.

5 Candidates are cautioned that, when taking the lantern test, they should not attempt to hurry. They should satisfy themselves as to the colours of the lights shown before reporting them to the examiner. The results of the test can only be judged by what the candidate reports. Carelessness may cause failure and thus prevent the candidate from pursuing a chosen career at sea in a deck capacity.

6 A candidate should be in good health when taking the test otherwise his/her vision may be impaired.

7 During the examination in the sight test, candidates will not be allowed to use spectacles, contact lenses or glasses of any kind, or any other artificial aid to vision. Treatment designed to improve form vision temporarily should not be undertaken shortly before the test.

SECTION II

This section applies to sight tests to be taken by candidates before admission to examinations for a Certificate of Competency, and to sight tests taken voluntary other than those taken by new entrants to the Merchant Navy or fishing industry.

1 Candidates for certificates of competency as masters, deck officers, skippers or second hands and sight test applicants (other than new entrants) may take the sight test with or without aids (conventional spectacles or contact lenses) at their option but candidates who opt to use aids for the letter or lantern test or both must additionally reach a minimum standard in each eye without aids.

2 Every candidate seeking admission to an examination for a first certificate of competency, as a deck officer, must hold a sight test certificate showing that he/she has passed in both letter and lantern tests within the six months period preceding the date of the examination. For certificates of competency subsequent to the first a sight test certificate fulfilling these conditions in respect of the letter test only will be required.

3 Candidates will be asked before the test begins whether they wish to use artificial aids for the test or any part of it. Where aids are used the following conditions will apply:

(*a*) In the letter test the candidate will first be tested without aids, each eye being tested separately. He/she will be required to read correctly down to and including line 5 with the better eye and down to and including line 3 with the other eye.

(*b*) The candidate will then be tested with aids, each eye again being tested separately. He/she will be required to read correctly down to and including line 7 with the better eye and down to and including line 6 with the other eye.

4 Candidates not using aids will be required to read down to and including line 7 with the better eye and down to and incluidng line 6 with the other eye.

SECTION III

General

1 A fee (plus VAT at the standard rate, except when the sight test is taken in conjunction with the examination for a statutory marine qualification at an inclusive fee), payable to the Superintendent of a Mercantile Marine Office, is charged for conducting a sight test at one of the Department of Trade's Offices.

2 Candidates other than new entrants who have previously failed to pass the sight test either locally or at a special appeal test conducted by the Principal Examiner of Masters and Mates or his Deputy may, provided aids to vision have not been worn at any of these previous tests, apply to take the test again locally with aids on payment of a further fee plus VAT.

3 Candidates who have taken a previous test without aids and have failed in the test but did not proceed to the lantern (or having proceeded to the lantern, passed the lantern test) may be re-examined locally with aids after a period of not less than one month.

4 Candidates who have taken a previous test with aids and have failed in the letter test but did not proceed to the lantern (or having proceeded to the lantern, passed the lantern test) may be re-examined locally with aids after a period of not less than one month.

5 A list showing the ports where sight tests are held is available on demand. Applications for appointments should be made to the Department of Trade Sight Examiner of the Mercantile Marine Office or the Marine Survey Office at the port concerned, as appropriate.

6 Unless otherwise indicated, sight tests are conducted between 9.30 AM and 12.30 PM on the listed days. A candidate who lives a distance from the port and cannot attend during the hours at which sight tests are normally conducted at the port should apply in writing to the examiner at the sight test centre for a special appointment.

THE WOMEN'S TRANSPORT SERVICE (FANY) (MARCH 1981)

Normal vision for distant and near is all that is required.

RAILWAY SERVANTS (JUNE 1981)

Footplate grades, i.e. drivers, secondmen and traction trainees and cleaners. On entry into the service as traction trainee and cleaner the following standard arrangements apply: 6/6, 6/6, 6/6 with fogging test; colour vision normal. Re-examinations are carried out at regular intervals and a special scheme is available for providing drivers and secondmen with corrective glasses.

Colour vision

For all candidates for employment the colour vision test will form part of the routine examination on entry and the result thereof will be recorded on their service histories.

Normal colour vision will be required in all cases for employment in grades the duties of which take the staff concerned on or about the running lines (i.e. lines controlled by a system of signalling) or otherwise demand accurate colour sense. By way of example this provision will apply to all permanent way grades, staff employed on flagging duties, lookout men, hand-signalmen and all signal and telecommunications and overhead line staff.

Normal colour vision will also be required by staff not necessarily working on the running lines, whose work requires an accurate appreciation of colour, e.g. signal and telegraph wiring and electrical maintenance in other departments where it is necessary to distinguish the colours of insulated coverings, etc.

A departmental officer, in conjunction with the medical officer, may, at his discretion, having regard to such factors as the safety of operation and of the member of the staff concerned, allow a man with defective colour vision to occupy an appropriate post, the duties of which do not take him on or about the running lines or require accurate colour sense of their performance. Each case of this nature shall be reconsidered before the man is transferred to other work.

Where a man is found to be colour defective he is to be advised of this immediately and the resulting possible restriction upon his future promotion is to be explained to him.

METROPOLITAN POLICE FORCE (MARCH 1981)

Requirements for Constables

Form vision—6/6, 6/12, 6/6, with or without optical aid, subject to minimum 6/18 unaided in each eye. Colour vision—'Ability to distinguish the principal colours.

POSTAL AND TELECOMMUNICATIONS SERVICES (8 JUNE 1981)

Post Office and National Girobank

Candidates for appointment: 'The standards of vision required depend primarily upon the type of occupation for which the candidate has applied. Any visual defect is therefore considered in relation to this, to its cause and its prognosis.'

British Telecom

Candidates for appointment: 'The standards of vision required depend primarily upon the type of occupation for which the candidate has applied. Any visual defect is therefore considered in relation to this, to its cause and its prognosis.'

CIVIL SERVICE (MARCH 1981)

Candidates for appointments in the Civil Service

Except for a few special posts, no precise standards of vision are prescribed, each case of visual defect being considered by the Civil Service Commissioners on its merits. The Commissioners have regard not only to the degree of visual acuity but to the nature and cause of the defect, the health and structural condition of the eyes, the nature of the duties of the post and the likelihood of the sight remaining adequate for the effective performance of those duties for at least five years. Provided that these points were satisfied, the loss of sight in one eye would not in itself be cause for the rejection of a candidate for a situation for which no special sight standard had been prescribed.

Candidates for some appointments of a special character would be rejected for colour blindness, but for most appointments in the Home Civil Servce and for the Diplomatic Service it is not in itself a disqualification.

TEACHING PROFESSION (MARCH 1981)

Visual standards for candidates for the teaching profession

Circular 11 78 (Appendix 2, section 5) requires that students and persons on entry to teacher training and to teaching meet the standards as under.

(*a*) If the best vision for distance, with glasses if worn, does not attain 6/12 Snellen in at least one eye, a report should be obtained from a consultant ophthalmologist on the nature and extent of the defect, whether it is progressive and whether it is likely to interfere with efficiency as a teacher. It may be possible to accept as suitable for training and for teaching a person who has a severe visual defect but who is otherwise in good health. Details of such candidates should be referred to the Department of Education and Science for advice regarding acceptance.

(b) For intending specialists in physical education who have myopia, a report should be obtained from a consultant ophthalmologist on visual acuity for distance with and without glasses, their strength, whether contact lenses can be provided and an opinion on the probability of retinal detachment or other damage to the eye as a result of strenuous physical exertion.

(c) For intending specialist teachers of art where there is a colour vision abnormality, a report should be obtained from a consultant ophthalmologist on the nature and extent of the condition.

References

Aarnisalo E 1980 Effects of reduced illumination on the results obtained with some diagnostic colour vision tests in subjects with congenital red–green defects *Acta Ophthalmol. Suppl.* **142** (Copenhagen: Scriptor)

Abney W de W 1891 On the examination for colour of cases of tobacco scotoma and of abnormal colour blindness *Proc. R. Soc.* **49** 491–506

—— 1895 *Colour Vision* (London: Sampson Low, Marston)

—— 1913 *Researches in Colour Vision* (London: Longmans Green)

Abramov I and Gordon J 1977 Colour vision in the peripheral retina. I: Spectral sensitivity *J. Opt. Soc. Am.* **67** 195–202

Adachi-Usami E, Heck J, Gavriysky V and Kellermann F J 1974 Spectral sensitivity function determined by the visually evoked cortical potential in several classes of colour deficiency (cone monochromatism, rod monochromatism, protanopia, deuteranopia) *Ophthalmol. Res.* **6** 273–90

Adam A 1980 Polymorphisms of red–green vision in some populations in South Africa *Am. J. Phys. Anthropol.* **53** 339–46

Adam A, Doron D and Modan R 1967 Frequencies of protan and deutan alleles in some Israeli communities and a note on the selection–relaxation hypothesis *Am. J. Phys. Anthropol.* **26** 297–306

Adams A J and Rodic R 1981 Use of desaturated and saturated versions of the D.15 test in glaucoma and glaucoma-suspect patients in *6th Int. Symp. Int. Research Group on Color Vision Deficiencies, Berlin, September 1981* (The Hague: Junk)

Adams A J and Zisman F 1980 Colour vision in juvenile diabetics *ARVO Meeting, Abstracts Document* (St Louis: Mosby)

Adams A J, Zisman F, Rodic R and Cavender J C 1981 Chromaticity and luminosity changes in glaucoma and diabetes in *6th Int. Symp. Int. Research Group on Color Vision Deficiencies, Berlin, September 1981* (The Hague: Junk)

Adrian E D and Mathews R 1927 The action of light on the eye. Part I. The discharge of impulses in the optic nerve and its relation to the electric charges in the retina *J. Physiol.* **63** 378–414

AGARD (Advisory Group for Aerospace Research and Development) 1972 *Colour Vision Requirements in Different Operational Roles* AGARD Conference Proceedings 99 (Neuilly sur Seine: AGARD)

Ahmed M M 1971 Ocular effects of anti-freeze poisoning *Br. J. Ophthalmol.* **55** 854–5

Ainley R G 1970 The F-M 100 Hue test in tobacco amblyopia *Trans. Ophthalmol. Soc. UK* **90** 765–72

Aitken J 1873 On colour and colour sensation *Trans. R. Scott. Soc. Arts* **8** 375–418

Alexander J A 1976 A report form for colour vision tests *Aust. J. Optom.* **59** 234–8

Alexander J A, Leason L L and Clancy J 1977 The locus of unique green in deuteranomalous trichromats *Am. J. Optom.* **54** 567–72

Alexander K R 1975 Colour testing in children; a review *Am. J. Optom.—Physiol. Opt.* **52** 332–7

Alken R G 1982 Drug-induced colour vision deficiencies; from side effects to clinical pharmacology *Doc. Ophthalmol. Proc.* Series 33 467–76

Alken R G, Miller A, Semjan R and Rietbrock N 1980 Colour vision deficiencies detected by Farnsworth's 100 Hue test in patients under long-term treatment with digitalis *Colour Vision Deficiencies V* ed G Verriest (Bristol: Adam Hilger) pp 320–6

Allen M J 1966 Vision, vehicles and highway safety *Highway Res. News* **25** 57–62

Allyn M R, Dixon R W and Bacharach R Z 1972 Visibility of red and green electroluminescent diodes for colour-anomalous observers *Appl. Opt.* **11** 250–4

Alpern M 1974 What is it that confines in a world without colour? *Invest. Ophthalmol.* **13** 645–74

—— 1976 Tritanopia *Am J Optom.* **53** 340–9

Alpern M, Falls H F and Lee G B 1960 The enigma of typical total monochromacy *Am. J. Ophthalmol.* **50** 996–1011

Alpern M, Lee G B and Spivey B E 1965a π_1 cone monochromacy *Arch. Ophthalmol.* **74** 334–7

Alpern M and Moeller J 1977 Red and green cone visual pigments in deuteranomalous trichromacy *J. Physiol.* **266** 647–75

Alpern M and Pugh E N Jr 1977 Variation in the action spectrum of erythrolabe among deuteranopes *J. Physiol.* **266** 613–46

Alpern M, Thompson S and Lee M 1965b Spectral transmittance of visible light by the living human eye *J. Opt. Soc. Am* **55** 723–6

Alpern M and Torii S 1968a The luminosity curve of the protanomalous fovea *J. Gen. Physiol* **52** 717–37

—— 1968b The luminosity curve of the deuteranomalous fovea *J. Gen. Physiol.* **52** 738–49

—— 1968c Prereceptor colour vision distortions in protanomalous trichromacy *J Physiol.* **198** 549–60

Alpern M and Wake T 1977 Cone pigments in human deutan colour vision defects *J. Physiol.* **266** 595–612

Ambler B A 1974 Hue discrimination in peripheral vision under conditions of dark and light adaption *Percept. Psychophys.* **T5** 586–90

Amsler M 1953 Earliest symptoms of diseases of the macula *Br. J. Ophthalmol.* **37** 521–37

Anon 1961 Simple colour vision test for graders *Rhod. Tob. J.* (Nov) 1957

Archer R E 1964 Unpublished data cited by D Ellis 1979

Arden G B and Kelsey J H 1962 Changes produced by light in the standing potential of the human eye *J. Physiol.* **161** 189

Arias K 1981 Spectral sensitivity measurements performed with the Goldmann perimeter *Acta. Ophthalmol. Suppl.* **148** 1–58

Ariel B 1979 Colour vision and Stargardt's disease *The Optician* **178** 13–16

Armington J C 1952 A component of the human electroretinogram associated with red color vision *J. Opt. Soc. Am.* **42** 393–401

—— 1966 Spectral sensitivity of simultaneous ERG and occipital responses *Clinical Retinography* (Oxford: Pergamon) pp 225–33

Armstrong S T 1888 *Br. Med. J.* Jan 23 183

Aspinall P A 1974a An upper limit of non-random cap arrangements in the Farnsworth Munsell 100 Hue test *Ophthalmologica* **168** 128–31

—— 1974b Inter-eye comparison on the 100 Hue test *Acta Ophthalmol.* **52** 307–16

Aubert H 1857 *Physiologie der Netzhaut* (Breslau)

Aubert H and Foerster R 1857 Untersuchungen über den Raumsinn der Retina *Graefes Arch. Ophthalmol.* **3** 1–37

Auerbach E 1974 Electroretinographical and psychophysical studies in achromats *Mod. Probl. Ophthalmol.* **13** 169–75

Auerbach E and Merlin S 1974 Achromatopsia with amblyopia I. A clinical and electroretinographic study of 39 cases *Doc. Ophthalmol.* **37** 79–117

Aulhorn E and Harms H 1956 Untersuchungen über das Wesen des Grenzkontrastes *Ber. Dtsch. Ophthalmol. Ges.* **60** 7–10

—— 1972 *Visual Perimetry* Handbook of sensory physiology vol VII/4 (Berlin: Springer) pp 102–45

Austin D J 1974 Acquired colour vision defects in patients suffering from chronic simple glaucoma *Trans. Ophthalmol. Soc. UK* **94** 880–3

Babel J and Stangos N 1972 Intoxication rétinienne par la digitale *Bull. Soc. Belg. Ophthalmol.* **160** 558–66

Bacon L 1971 Colour vision defect—an educational handicap? *Med. Off.* **125** 199–209

Baghdassarian S A 1968 Optic neuropathy due to lead poisoning *Arch. Ophthalmol.* **80** 721–3

von Bahr G 1946 The visual acuity in monochromatic lights and in mixtures of two monochromatic lights *Acta Ophthalmol.* **24** 129–46

Bailey J 1980 Colour perimetry—a review *Am. Acad. Optom.* **51** 843–7

Baker H D and Rushton W A H 1965 The red-sensitive pigment in normal eyes *J. Physiol.* **176** 56–72

Balaraman S, Graham C H and Hsia Y 1962 The wavelength discrimination of some color blind persons *J. Gen. Physiol.* **66** 185–201

Ball G V 1972 Acquired anomalies of colour vision *Br. J. Physiol. Opt.* **27** 1–6

—— 1974 Terminology of colour deficiency *Mod. Probl. Ophthalmol.* **13** 354–62

Ball R J and Bartley S H 1967 The induction and reduction of color deficiency by manipulation of termporal aspects of photic input *Am. J. Optom.* **44** 411–18

Barca L and Vaccari G 1977 On the impairment of color discrimination in diabetic retinopathy *Attiv. Fond. G. Ronchi* **XXXII** 635–40

—— 1978 Colour discrimination in old people: A report of 30 cases *Attiv. Fond. G. Ronchi* **XXXIII** 279–83

Barlow H B and Mollin J D (ed) 1982 *The Senses* (Cambridge: Cambridge University Press)

Bartleson C J 1980 Colorimetry in *Optical Radiation Measurements II* ed F Grum and C J Bartleson (London: Academic) p.128

—— 1981 On chromatic adaptation and persistence *Col. Res. Appl.* **6** 153–60

Bartley S H 1941 *Vision, a Study of its Basis* (London: Macmillan)

Baylor D A, Fourtes M G F and O'Bryan P M 1971 Receptive fields of cones in the retina of the turtle *J. Physiol.* **214** 265–94

Baylor D A and Hodgkin A L 1973 Detection and resolution of visual stimuli by turtle photoreceptors *J. Physiol* **234** 163–98

Beare A A 1963 Colour name as a function of wavelength *Am. J. Psychol.* **76** 284–6

Becker O 1879 Ein Fall von angeborener einseitiger totaler Farben Blindheit *Graefes Arch. Ophthalmol.* **25** 205–12

Bedford R E and Wyszecki G 1958 Wavelength discrimination for point sources *J. Opt. Soc. Am.* **48** 129–35

Belcher S, Greenshields K and Wright W D 1958 A colour vision survey *Br. J. Ophthalmol.* **51** 355–9

Bell J 1926 Colour blindness *Treasury of Human Inheritance* vol II, part II (Cambridge: Cambridge University Press) pp 125–268

Bender B 1973 Spatial interactions between the red and green sensitive colour mechanisms of the human visual system *Vision Res.* **13** 2205–18

Bergmans J 1960 *Seeing Colours* (Eindhoven: Philips Technical Laboratory)

Bernhard C G 1940 Contributions to the neurophysiology of the optic pathway *Acta Physiol. Scandinavia* **1** Suppl. 1

Berson E L, Sandberg, M A, Bromley W C and Roderick T W 1982 Electroretinograms in carriers of π_1, cone monochromatism *ARVO Meeting, Abstracts Document* (St Louis: Mosby)

Berson E L, Sandberg M A, Rosner B and Sullivan P L 1983 Color plates to help identify patients with blue cone monochromatism *J. Opt. Soc. Am* **95** 741–7

von Bezold W 1873 Uber das gesetz der Farbenmischung die physiologizchen Grundfarben *Ann. Phys.,Lpz.* **150** 221–47

Bhargava S K 1973 Tobacco amblyopia and acquired dyschromatopsia anomaloscope tests *Acta Ophthalmol.* **51** 822–8

Bhasin M K 1967 The frequency of colour blindness in the Newars of Nepal Valley *Acta Genet.* **17** 454–9

Bickerton T H 1887 *Br. Med J.* **ii** 498, 505

—— 1888 *Br. Med. J.* **i** 188 and **ii** 1038

—— 1900 *Br. Med. J.* **i** 621

—— 1903 *The Practitioner* reprinted pamphlet p.11

Biersdorf W 1977 The Davidson and Hemmendinger Color Rule as a color vision screening test *Arch. Ophthalmol.* **95** 134–8

Birch D G, Birch E E and Enoch J M 1980 Visual sensitivity, resolution and Rayleigh matches following monocular occlusion for one week *J. Opt. Soc. Am.* **70** 954–8

Birch J 1973 Dichromatic convergence points obtained by subtractive colour matching *Vision Res.* **13** 1755–65

—— 1974 Colour vision testing and advice *Br. J. Physiol. Opt.* **29** 1–29

—— 1975 Use of the Giles–Archer lantern for the examination of defective colour vision *Ophthalmol. Opt.* **15** 1106–8

Birch-Hirschfeld 1932 *Neodym Glass* (Berlin: Karger)

Bishop H and Crode M 1961 Absolute identification of colour for targets presented against white and coloured backgrounds *USAF WADC Report* 60–611

la Bissoniere P (ed) 1974 *Int. Contact Lens Clin.* **1/4** 48–55

Blackwell H R and Blackwell O M 1957 Blue mono-cone monochromasy *J. Opt. Soc. Am* **47** 338–9

—— 1961 Rod and cone receptor mechanisms in typical and atypical congenital achromatopsia *Vision Res.* **1** 62–107

Boettner E A and Wolter J R 1962 Transmission of the ocular media *Invest. Ophthalmol.* **1** 776–83

Boles-Carenini B 1954 Del comportamento del senso chromatico in relazione all'eta *Ann. Ottalmol.* **80** 451–8 (quoted by Young 1974)

Boll F 1876 Zur Anatomie und Physiologie der Retina *Monatsber. Preuss. Akad. Wiss. Berlin* 783–7

Bone R A and Sparrock J M B 1971 Comparison of macular pigment densities in human eyes *Vision Res.* **11** 1057–64

Bonting S L (ed) 1976 *Transmitters in the Visual Process* (Oxford: Pergamon)

Bouma P J 1971 *Physical Aspects of Colour* 2nd edn (London: Macmillan)

Bouman M A and Walraven P L 1957 A study of normal and defective colour vision *Visual Problems of Colour* vol II NPL Symposium no 8 (London: HMSO) pp 463–73

Bouman M A, Walraven P L and Leebeck H L 1956 Another colorimeter for studying color vision *Ophthalmologica* **131** 179–93

Bowmaker J K 1973 Spectral sensitivity and visual pigment absorbance *Vision Res.* **13** 783–92

Bowmaker J K, Dartnall H J A, Lythgoe J N and Mollon J D 1978 The visual pigments of rods and cones in the rhesus monkey *Macaca mulatta J. Physiol.* **274** 329–48

Bowmaker J K, Dartnall H J A and Mollon J D 1980 Microspectrophotometric demonstrations of four classes of photoreceptor in an old world primate *Macaca fasicularis J. Physiol.* **298** 131–43

Bowmaker J K and Mollon J D 1980 Primate microspectrophotometry and its implications for colour deficiencies *Colour Vision Deficiencies V* ed G Verriest (Bristol: Adam Hilger) pp 61–4

Bowmaker J K, Mollon J D and Jacobs G H 1982 Microspectrophotometric measurements of old and new world species of monkey *International Conference on Colour Vision, Cambridge, UK, August 23–27 1982* (Proceedings published by Academic Press, 1983)

Bowman K J 1973 The Farnsworth dichotomous test—the panel D-15 *Aust. J. Optom.* **56** 13–24

—— 1978 The effect of illuminance on colour discrimination in senile macular degeneration *Mod. Probl. Ophthalmol.* **19** 71–6

—— 1980a The clinical assessment of colour discrimination in senile macular degeneration *Acta. Ophthalmol.* **58** 337–46

—— 1980b The relationship between colour discrimination and visual acuity in senile macular degeneration *Am. J. Optom.* **57** 145–8

Bowman K J and Cole B L 1980 A recommendation for illumination of the Farnsworth–Munsell 100-Hue test. *Am. J. Optom.* **57** 839–43

—— 1981 Recognition of the aircraft navigation light color code *Aviat. Space Environ. Med.* (52) **11** 658–65

Bowman K J, Collins M J and Henry C J 1983 The effect of age on performance on the Panel-D-15 and desaturated D-15. A quantitative evaluation paper presented to *IRGCVD Geneva Congress* (transactions to be published by Karger, Berlin)

Boyce P R and Simons R H 1977 Hue discrimination and light sources *Light. Res. Technol.* **9** 125–40

Boycott B B and Dowling J E 1969 Organization of the primate retina. Light microscopy *Phil. Trans. R. Soc. B* **255** 109–84

Boyle R 1688 Some uncommon observations about vitiated sight appended to *A Disquisition about the Final Causes of Natural Things* reprinted 1772 in *Works* vol V (London: T Birch) pp 445–52

Boynton R M 1960 Theory of colour vision *J. Opt. Soc. Am.* **50** 929–44

—— 1979 *Human Color Vision* (New York: Holt, Rinehart and Winston)

—— 1982 Spatial and temporal approaches for studying colour vision *Docum. Ophthalmol. Proc.* Series 33 1–15

Boynton R M and Gordon J 1965 Bezold–Brücke hue shift measured by colour naming technique *J. Opt. Soc. Am.* **55** 78–86

Boynton R M, Hayhoe M M and MacLeod D I A 1977 The gap effect: chromatic and achromatic visual discrimination as affected by field separation *Optica Acta* **24** 159–77

Boynton R M, Ikeda M and Stiles W S 1964a Interactions among chromatic mechanisms as inferred from positive and negative increment thresholds *Vision Res.* **4** 87–117

Boynton R M, Kandel G and Onley J W 1959 Rapid chromatic adaptation of normal and dichromatic observers *J. Opt. Soc. Am.* **49** 654–66

Boynton R M, Schafer W and Neun M E 1964b Hue wavelength relation measured by colour naming method for three retinal locations *Science* **146** 666–8

Boynton R M and Scheibner H 1967 On the perception of red by 'red-blind' observers *Acta Chromatica* **1** 205-20

Boynton R M and Wagner M 1961 Two-colour threshold as a test of colour vision *J. Opt. Soc. Am.* **51** 429–40

Brennan D H 1972 in *Colour Vision Requirements in Different Operational Roles* in AGARD Conference Proceedings 99 (Neuilly sur Seine: AGARD)

Brewster D 1826a On the invisibility of certain colours to certain eyes *Edinburgh J. Sci.* **86** (as cited by Sherman 1981)

—— 1826b Account of two remarkable cases of insensibility in the eye to particular colours. *Edinburgh J. Sci.* **86** 153–9 (as cited by Sherman 1981)

—— 1844 Observations on colour blindness *Phil. Mag. Ser.* 3 25

Brindley G S 1953 The effects on colour vision of adaptation to very bright lights *J. Physiol.* **12** 332–50

—— 1954a The summation areas of the human colour receptive mechanisms at increment threshold *J. Physiol.* **124** 400–8

—— 1954b The order of coincidence required for visual threshold *Proc. Phys. Soc. B* **67** 673–6

—— 1970 *Physiology of the Retina and Visual Pathway* 2nd edn (London: Edward Arnold)

Brindley G S and Rushton W A H 1955 Detection of a visual pigment in living human cones *J. Physiol.* **128** 59

Brindley G S and Willmer E N 1952 The reflexion of light from the macular and peripheral fundus oculi in man *J. Physiol.* **116** 350–6

Brodmann K 1905 Beitrage zur histologischen Lokalisation der Grosshirnrinde *J. Psychol. Neurol., Lpz.* **4** 176–226

Bronte-Stewart J and Foulds W S 1972 Acquired dyschromatopsia in vitamin A deficiency *Mod. Probl. Ophthalmol.* **11** 168–73

Brown J L, Kuhns M P and Adler A E 1957 Relation of threshold criterion to functional receptors of the eye *J. Opt Soc. Am.* **47** 198–204

Brown K T and Murakami M 1964 A new receptor of the monkey retina with no detectable latency *Nature* **201** 626–8

Brown P K and Wald G 1963 Visual pigments in human and monkey retinas *Nature* **200** 37–43

—— 1964 Visual pigments in single rods and cones of the human retina *Science* **144** 45–52

Brown W R J 1951 The influence of luminance level on visual sensitivity to colour differences *J. Opt. Soc. Am.* **41** 684–8

—— 1952 The effect of field size and chromatic surroundings on colour discrimination *J. Opt. Soc. Am.* **42** 837–44

von Brücke E T 1878 Über einige Empfindungen im gebiete des Sehnerven *SB Akad. Wiss. Wien Abt.* 111 **77** 39–71

Burch G 1898 On artificial temporary colour blindness with an examination of the colour sensation of 109 persons *Phil. Trans. R. Soc.* **191** B 1–34

Burggraf H, Hellner K A and Burggraf H 1981 Erprobung und Beurteilung eines neuen Tests zur prüfung des Farbensinns *Klin. Monatsbl. Augenheilkd.* **179** 204–13

Burian H M and von Noorden G K 1974 *Binocular Vision and Ocular Motility* (St Louis: Mosby)

Burnham R W 1951 The dependence of colour upon area *Am. J. Psychol.* **64** 521–33

—— 1952 A colorimeter for research in color perception *Am. J. Psychol.* **65** 602–8

Burnham R W and Clark J R 1955 A test of hue memory *J. Appl. Psychol.* **39** 164–72

Burnham R W, Hanes R M and Bartleson C J 1963 *Color, a Guide to Basic Facts and Concepts* (London: John Wiley)

Burns C A 1966 Indomethacin reduced retinal sensitivity and corneal deposits *Am. J. Ophthalmol.* **66** 825–35

Butler J H 1910 On the futility of the official tests for colour blindness *Br. Med. J.* Feb 5 316–17

Cameron R G 1967 Rational approach to colour vision testing *Aerosp. Med* **38** 51

Campbell F W and Gubisch R W 1967 The effect of chromatic aberration on visual acuity *J. Physiol.* **192** 345–58

Campbell F W and Rushton W A H 1955 Measurement of the scotopic pigment in the living human eye *J. Physiol.* **130** 131–47

Carlow T J, Flynn J T and Shipley T 1976a Colour perimetry parameters *Docum. Ophthalmol.* **14** 427–9

—— 1976b Color perimetry *Arch. Ophthalmol.* **94** 1492–6

Carpenter D V 1983 An examination of the difficulties encountered by colour vision defective pupils in a Wiltshire school *MEd Dissertation* Bristol University

Carr R E, Gouras P and Gunkel P D 1966 Chloroquine retinopathy *Arch. Ophthalmol* **78** 171–8

Carr R E, Henkind P, Rothfield N and Siegel I 1968 Ocular toxicity of anti-malarial drugs *Am. J. Ophthalmol.* **66** 738–44

Carta F, Maione M, Lettieri S and Bondi L 1980 Stiles' mechanisms in clinical ophthalmology *Colour Vision Deficiencies V* (Bristol: Adam Hilger) pp 122–5

Carta F, Vincinguerra E and Barrera E 1967 A study of colour sense in patients with cirrhosis of the liver *Ann. Ottalmol.* **93** 350

Carter R 1890 *Nature* **xiii** 55–61

Caspersson T 1940 Methods for the determination of the absorption spectra of cell structures *J. R. Microsc. Soc.* **60** 8–25

Cavanagh P 1955 The Ishihara test and defects of colour vision *Occup. Psychiatr.* **29** 43–57

Cavonius C R and Estevez O 1975a Contrast sensitivity of individual colour mechanisms in human vision *J. Physiol.* **248** 649–62

—— 1975b Sensitivity of human colour mechanisms to grating and flicker *J. Opt. Soc. Am.* **69** 966–8

—— 1976 Flicker sensitivity of the long wavelength mechanisms of normal and dichromatic observers *Mod. Probl. Ophthalmol.* **17** 36–40

—— 1978 π mechanisms and cone fundamentals in *Visual Psychophysics and Physiology* ed J C Armington, J Krauskopf and B R Wooten (London: Academic) pp 221–33

Chamberlin C J and Chamberlin D G 1980 *Colour, its Measurement, Computation and Application* (London: Heyden)

Chapanis A 1944 Spectral saturation and its relation to colour-vision defects *J. Exp. Psychol.* **34** 24–44

—— 1948 A comparative study of five tests of colour vision *J. Opt Soc. Am.* **38** 626–49

—— 1949 Simultaneous chromatic contrast in normal and abnormal colour vision *Am. J. Psychol.* **62** 526–39

Chapman V A 1953 *Trans. Sect. Ophthalmol. Am. Med. Assoc.* 301

Charpentier A 1888 *La Lumiere et les Couleurs* (Paris)

Chase W 1937 Colour vision in infants *J. Exp. Psychol.* **20** 203–22

Chioran G M and Sheedy J E 1983 Pseudoisochromatic plate design – Macbeth or tungsten illumination *Am. J. Optom.* **60** 204–15

Chisholm I A 1968 The dyschromatopsia of tobacco amblyopia paper to *2nd Scottish Symp. on Colour, Edinburgh* (unpublished)

—— 1969 An evaluation of the Farnsworth-Munsell 100 test as a clinical tool in the investigation and management of ocular neurological deficit *Trans. Ophthalmol. Soc.* **80** 243–50

—— 1972 The dyschromatopsia of pernicious anaemia *Mod. Probl. Ophthalmol.* **11** 130–5

—— 1979 in *Congenital and Acquired Colour Vision Defects* (New York: Grunne and Stratton) pp 291–300

Chisholm I A, Bronte-Stewart J and Awduche E O 1970 Colour vision in tobacco amblyopia *Acta Ophthalmol.* **48** 1145–56

Chisholm I A and Kearns J A 1981 Pattern of visual recovery after relief of chiasmal compression *Docum. Ophthalmol. Proc.* Series 33 453–6

Chisholm I A and Pinckers A J L G 1979 in *Congenital and Acquired Colour Vision Defects* (New York: Grunne and Stratton)

Chung Y, Yun-Chin C, Pao-Huo F and Nai-Hua Y 1958 Colour blindness and the Chinese *Chin. Med. J.* **76** 283–4

Cicerone C M and Green D G 1978 Relative modulation sensitivities of the red and green colour mechanisms *Vision Res.* **18** 1593–8

CIE 1959 *Colours of Light Signals* CIE Publication no 2 (Paris: CIE)

Clark B A J 1968 The effects of tinted ophthalmic media on the recognition of red traffic signals *Proc. 4th Conf. Aust. Road Res. Board* **4** 898–930

Clark E C 1973 *Illumination in Colour Vision Testing* unpublished special study report, The City University, London

Clarke F J J 1968 Laboratory measurement of colour vision *Proc. Symp. AF-NRC Comm. on Vision* (Washington, DC: National Research Council) pp 86–90

Clements F 1930 Racial differences in colour vision *Am. J. Phys. Anthropol.* **14** 417

Coates 1970 *Neoset Letterpress Inks* (London: Coates Brothers)

Cobb S R 1980 Red-green colour defect among the Roman Catholic and non-denominational population of Glasgow in *Colour Vision Deficiencies V* ed G Verriest (Bristol: Adam Hilger) pp 247–50

Cobb S R and Shaw F G 1980 The effect of industrial exposure to lead *Colour Vision Deficiencies V* ed G Verriest (Bristol: Adam Hilger) pp 343–5

Cohen G H and Saini V D 1978 The human pupil response as an objective determination of colour vision deficiency *Mod. Probl. Ophthalmol.* **19** 197–200

Cohen J D 1975 Temporal independence of the Bezold–Brücke hue shift *Vision. Res.* **15** 341–51

Cole B L 1963 Misuse of the Ishihara test for colour blindness *Brit. J. Physiol. Opt.* **20** 113–18

—— 1964 Comments on some colour vision tests and their use for selection *Aust. J. Optom.* **3** 56–65

—— 1970 The colour blind driver *Aust. J Optom.* **53** 261–9

Cole B L and Brown B 1966 Optimum intensity of red road-traffic signal lights for normal and protanopic observers *J. Opt. Soc. Am.* **56** 516–22

Cole B L, Henry G H and Nathan J 1965 Phenotypical variations of tritanopia *Vision Res.* **6** 301–13

Cole B L and Vingrys A J 1982 A survey and evaluation of lantern tests of color vision *Am. J. Optom.* **59** 346–74

Cole B L and Watkins R D 1967 Increment thresholds in tritanopia *Vision Res.* **7** 939–47

Collin H B 1966 Recognition of acquired colour defects using the panel D15 *Aust. J. Optom.* **49** 342–7

Collins M 1925 *Colour Blindness, with a Comparison of Different Methods of Testing* (London: Kegan Paul)

Collins W E 1959 The effects of deuteranomaly and deuteranopia upon the foveal luminosity curve *J. Psychol.* **48** 285–97

Color Aptitude Test 1964 *Brochure* from Fed. Socs. for Paint Technology, Philadelphia

Colquhoun H 1829 An account of two cases of insensibility of the eye to certain of the rays of colour *Glasgow Med. J.* **2** 12–21 (as cited by Sherman 1981)

Committee on Vision 1981 *Procedure for Testing Colour Vision* Working Group 41 Report (Washington, DC: National Academy Press)

Condit R, Dresnick G, Korth K, Mattson D and Syrjala S 1982 Hue discrimination loss and retinopathy severity in diabetes mellitus *ARVO Meeting, Abstracts Document* (St Louis: Mosby)

Conner J D and MacLeod D I A 1976 Rod photoreceptors detect rapid flicker *Science* **195** 698–9

Conners M M (1964) Effect of surround and stimulus luminance on the discrimination of hue *J. Opt. Soc. Am.* **54** 693–5

—— 1968 Luminance requirements for hue perception in small targets *J. Opt. Soc. Am.* **58** 258–63

Conners M M and Kinney J A 1962 Relative red–green sensitivity as a function of retinal position *J. Opt. Soc. Am.* **52** 81–4

Conover D W and Kraft C L 1958 The use of colour in coding displays *WADC Technical Report* 55–471 (Ohio: USAF Wright–Patterson Base)

Cook W A 1931 Ability of children in colour discrimination *Child Dev.* **2** 303–20

Copenhaver R M and Gunkel R D 1959 The spectral sensitivity of color defective subjects determined by electroretinography *Am. Med. Assoc. Arch. Ophthalmol.* **62** 55–68

Cornsweet T 1960 *J. Opt. Soc. Am.* **50** 507

Courtney G R and Heath G G 1971 Color vision deficiency in the mentally retarded *Am. J. Ment. Def.* **76** 48–52

Cowan A 1942 Abstract of discussion on paper Tiffin and Kuhn (1942) *Arch. Ophthalmol.* **28** 857

Cox B J 1971 Validity of a preschool colour vision test *Can. J. Optom.* **33** 22–4

Cozijnsen M and Pinckers A 1969 Oogheelkundige aspecten van digitoxine intoxicatie *Ned. Tijd. Geneesk.* **113** 1735–7

Crawford A 1951 A description of an anomaloscope *Br. J. Physiol. Opt.* **8** 173–5

—— 1955 The Dvorine pseudoisochromatic plates *Br. J. Psychol.* **XLVI** 139–43

Crawford B H 1949 The scotopic visibility function *Proc. Phys. Soc.* B **62** 321–34

Crawford B H, Granger G W and Weale R A 1968 *Techniques of Photostimulation in Biology* (Amsterdam: North-Holland)

Crescitelli F and Dartnall H J A 1953 Human visual purple *Nature* **172** 195-7

Critchley M 1965 Acquired anomalies of colour perception of central origin *Brain* **88** 711–24

Crone R A 1955 Clinical study of colour vision *Br. J Ophthalmol.* **39** 170–3

—— 1956 Combined forms of congenital colour defects *Br. J. Ophthalmol.* **40** 462–72

—— 1959 Spectral sensitivity in colour-defective subjects and heterozygous carriers *Am. J. Ophthalmol.* **48** 231–8

—— 1961 Quantitative diagnosis of defective colour vision *Am. J. Ophthalmol.* **51** 298–305

—— 1968 Incidence of known and unknown colour vision defects *Ophthalmologica* **155** 37–55

Cruz-Coke R 1964 Colour blindness and cirrhosis of the liver *The Lancet* **ii** 1064

—— 1970 *Colour Blindness—an Evolutionary Approach* (Springfield, IL: Thomas)

—— 1972 Defective colour vision and alcoholism *Mod. Probl. Ophthalmol.* **11** 174–7

Cutting W C 1972 *Handbook of Pharmacology: the Actions and Uses of Drugs* 5th edn (New York: Appleton–Century–Crofts)

Dain S 1971 A new colour vision test *PhD Thesis, The City University, London*

—— 1974 The Lovibond colour vision analyser *Mod. Probl. Ophthalmol.* **13** 79–82

Dain S J, Bard D and Sue G 1980b New targets for clinical colour perimetry *Colour Vision Deficiencies V* ed G Verriest (Bristol: Adam Hilger) pp 217–20

Dain S J, Strange G and Boyd R 1980a A solid state anomaloscope *Colour Vision Deficiencies V* ed G Verriest (Bristol: Adam Hilger) pp 181–3

Dalton J 1798 Extraordinary facts relating to the vision of colours with observations *Mem. Lit. Phil. Soc. Lond.* **5** 28–45

Dartnall H J A 1953 The interpretation of spectral sensitivity curves *Br. Med. Bull.* **9** 24–30

—— 1957 *The Visual Pigments* (London: Methuen)

—— 1972 *Handbook of Sensory Physiology* vol VII/I Photochemistry of vision (Berlin: Springer)

Dartnall H J A, Bowmaker J K and Mollon J D 1982 Microspectrophotometry of human photoreceptors paper given at *International Conference on Colour Vision, Cambridge, UK, August 23–27 1982* (Proceedings published by Academic Press 1983)

Davidoff J B 1976a Hemispheric difference in hue discrimination *Mod. Probl. Ophthalmol.* **17** 353–6

—— 1976b Hemispheric sensitivity differences in the perception of colour *Q. J. Exp. Psychol.* **28** 387–94

Davis R and Gibson K S 1931 Filters for the reproduction of sunlight and daylight and the determination of color temperature *Misc. Publ. Bur. Standards* no 114 M114 (Washington, DC: US Govt. Printing Office)

Davis R, Gibson K S and Haupt G W 1953 Spectral energy distribution of the International Commission on illumination light sources A, B and C *J. Opt. Soc. Am.* **43** 172–6

Davson H 1976 *The Eye* vol 2B The photobiology of vision (New York: Academic)

—— 1980 *Physiology of the Eye* 4th edn (Edinburgh: Churchill Livingstone)

Departmental Committee on Sight Tests 1912 *Report* (London: HMSO)

van Dijk B W and Spekreijse H 1982 Ethambutol affects selectively nonlinear color opponency in vertebrate retina *International Conference on Colour Vision, Cambridge, UK, 23–27 August 1982* (Proceedings published by Academic Press 1983)

Dimmick F L 1956 Color Aptitude Test *J. Opt. Soc. Am.* **46** 389

Dittrich H and Neubauer O 1967 Störungen des Farbsehens beï Leberkrankheiten *Münch Med. Wschr.* **109** 2690

Dobson V 1976 Spectral sensitivity of the two month infant as measured by VECP *Vision Res.* **16** 367–74

Dodt E, Vanlith G H M and Schmidt B 1967 Electroretinographic evaluation of the photopic malfunction in a totally colour blind *Vision Res.* **7** 231–41

Donaldson R 1935 A trichromatic colorimeter *Proc. Phys. Soc.* **47** 1068

Dow B M 1974 Functional classes of cells and their laminar distribution in monkey visual cortex *J. Neurophysiol.* **37** 927–45

Dow B M and Gouras P 1973 Colour and spatial specificity of single units in the rhesus monkey striate cortex *J. Neurophysiol.* **36** 79–100

Dowling J E 1960 Chemistry of visual adaptation in the rat *Nature* **188** 114–18

Dowling J E and Boycott B B 1966 Organization of the primate retina: electron microscopy *Proc. R. Soc.* B **166** 80–111

Dreher B, Fukada Y and Rodieck R W 1976 Identification, classification and anatomical segregation of cells with X-like and Y-like properties in the lateral geniculate-nucleus of old-world Primates *J. Physiol.* **258** 433–52

Dreher E 1911 Methodische Untersuchung der Farbentonänderungen homogener Lichter bei zunehmend indirektem Sehen und veränderter Intensität *Z. Sinnesphysiol.* **46** 1–82

Dreyer V 1969 Occupational possibilities of colour defectives *Acta Ophthalmol.* **47** 523–34

Dronamraju K R and Meera Khan D 1963 Frequency of colour blindness in Andra Pradesh *Ann. Hum. Genet.* **29** 77

Drum B A 1976 Chromatic saturation derived from increment thresholds for white and coloured targets *Mod. Probl. Ophthalmol.* **17** 79–85

Dubois-Poulsen A 1952 *Le Champ Visuel* (Paris: Masson et Cie)

—— 1972 Acquired dyschromatopsias *Mod. Probl. Ophthalmol.* **11** 84–93

—— 1981 Colour vision in brain lesions *Colour Vision Deficiencies VI* ed G Verriest (The Hague: Junk) pp 429–40

Dubois-Poulsen A, Magis G, de Ajuria Guerra J and Hecaen H 1952 Les conséquences visuelles de la lobectomie chez l'homme *Ann. Ocul.* **195** 305–47

Duke-Elder W S 1932 *Textbook of Ophthalmology* vol 1 (London: Kimpton) p 970

—— 1949 *Textbook of Ophthalmology* vol IV (London: Kimpton)

—— 1963 *System of Ophthalmology* vol III pt I (London: Kimpton)

—— 1964 *System of Ophthalmology* vol III pt II (London: Kimpton) p 667

—— 1968 *System of Ophthalmology* vol IV (London: Kimpton) p 639

—— 1970 *System of Ophthalmology* vol V (London: Kimpton) pp 420–9, 602–3

—— 1971 *System of Ophthalmology* vol XII (London: Kimpton)

Duke-Elder W S and Cook C 1963 in *System of Ophthalmology* vol III pt I Embryology (London: Kimpton)

Duke-Elder W S and Weale R A 1968 in *System of Ophthalmology* vol IV The physiology of the eye and of vision (London: Kimpton) p 637

Dutta P C 1966 Variability and regional differences of colour blindness in India *Hum. Genet.* **2** 204–6

Dvorine I 1963 Quantitative classification of the colour blind *J. Gen. Psychol.* **68** 255–65

Ebrey T G and Honig B 1977 New wavelength dependent visual pigment nomograms *Vision Res.* **17** 147–51

Edridge-Green F W 1891 *Colour Blindness and Colour Perception* (London: Kegan, Paul, Trench and Trubner)

—— 1911 Accidents which have occurred through colour blindness *The Lancet* **181** 879–80

—— 1920 *The Physiology of Vision* (London: Bell)

—— 1933 *Science and Pseudo-Science* (London: Bale and Danielson)

Eisner A and MacLeod D I A 1980 Blue-sensitive cones do not contribute to luminance *J. Opt. Soc Am.* **70** 121–3

Ellis D 1979 Visual handicaps of mentally defective people *Am. J. Ment. Def.* **83** 497–511

Engelbrecht K 1953 *Stillingsche Tafeln* (Leipzig: Thieme)

Engelking E 1925 Die tritanomalie, ein bisher unbekannter Typus anomaler Trichromaise *Graefes Arch. Ophthalmol.* **116** 196–244

Engelking E and Eckstein A 1920 Physiologische Bestimmung der Musterfarben für die klinische Perimetrie *Klin. Monatsbl. Augenheilkd.* **64** 88–106

Enoch J 1972 The two-colour threshold technique of Stiles and derived component-colour mechanism in *Handbook of Sensory Physiology* vol VII/4 (Berlin: Springer) pp 537–67

Enoch J and Stiles W S 1961 The colour change of monochromatic light with retinal angle of incidence *Optica Acta* **8** 329–58

Enroth-Cugell C and Robson J G 1966 The contrast sensitivity of retinal ganglion cells in the cat *J. Physiol.* **187** 517–52

Ensell J S 1978 Experience with the Lovibond colour vision analyser *Color Res. Appl.* **3** 11–15

Estevez O 1979 On the fundamental data base of normal and dichromatic colour vision *Thesis* University of Amsterdam

—— 1982 A better Colorimetric Standard Observer for colour-vision studies: the Stiles and Burch 2° colour-matching functions *Col. Res. Appl.* **7** 131–4

Estevez O and associates 1983 *The O.S.C.A.R. test* Unpublished communication describing device, issued by Keeler Instruments, London

Estevez O and Cavonius C R 1977 Human colour perception and Stiles' π mechanisms *Vision Res.* **17** 417–22

Estevez O and Spekreijse H 1974 A spectral compensation method for determining the flicker characteristics of the human colour mechanisms *Vision. Res.* **14** 823–30

Estevez O, Spekreijse H, van den Berg T J T P and Cavonius C R 1975 The spectral sensitivities of isolated colour mechanisms determined from contrast evoked potential measurements *Vision Res.* **15** 1205–12

Everson R W 1973 Spectral transmission of a new aviation and skiing filter *Am. J. Optom.* **50** 413–15

Falls H F 1960 Discussion following paper by Alpen, Falls and Lee *Am. J. Ophthalmol.* **50** 1012

Falls H F, Wolter J P and Alpern M 1965 Typical total monochromacy – a histological and psychophysical study *Arch. Ophthalmol.* **74** 610–16

Fankhauser F, Enoch J and Cibis P 1961 Receptor orientation in retinal pathology *Am. J. Ophthalmol.* **52** 767–83

Farnsworth D 1943 The Farnsworth–Munsell 100 Hue and dichotomous tests for colour vision *J. Opt. Soc. Am.* **33** 568–78

—— 1945 The effect of coloured lenses upon color discrimination *Color Vision Report 9* (New London, CT: Naval Research Laboratory)

—— 1947 *The Farnsworth Dichotomous test for colour blindness—panel D15* (New York: Psychological Corporation)

—— 1955a *A polychromatic plate for detecting congenital tritanomalous colour vision* (New London, CT: Naval Research Laboratory)

—— 1955b Tritanomalous vision as a threshold function *Die Farbe* **4** H/4/6 185–93

—— 1957a Testing for colour deficiency in industry *Am. Med. Assoc. Arch. Ind. Health* **16** 100–3

—— 1957b *The Farnsworth–Munsell 100 Hue Test Manual* (Baltimore: Munsell Colour Corporation)

—— 1961 Let's look at those isochromatic lines again *Vision Res.* **1** 1–5

Farnsworth D and Foreman P 1941 Development and trial of New London Navy lantern as a selection test for serviceable colour vision *Report no 12, Medical Research Dept.* (New London, CT: Naval Research Laboratory)

Farnsworth D and Reed J D 1948 Effect of illuminants on scores on P.I.C tests *C. V. Report 4* BuMed X-261 (New London, CT: Naval Research Laboratory)

Farnsworth D, Sperling R, Kimble P 1949 A battery of pass–fail tests for detecting degree of colour deficiency *Report 147* (New London, CT: Naval Research Laboratory)

Favre A 1878 Daltonisme chez les employés de chemin de fer *Lyon Médical* no 19 6–20

Ferguson-Smith M A 1973 *Medical Genetics* ed V A McKusick and R Clairborne (New York: H P Publishing)

Feris 1876 Médicin de première classe de la marine du Daltonisme dans les rapports avec la navigation *Arch. Med. Nav.* **25** 270–99

Ferree C E and Rand G 1919 Chromatic thresholds of sensation from centre to periphery of the retina and their bearing on colour theory *Psychol. Rev.* **26** 16–41, 150–63

—— 1924 Effect of brightness of pre-exposure and surrounding field of breadth and shape of the colour fields for stimuli of different sizes *Am. J. Ophthalmol.* **7** 843–50

Ferree C E, Rand G and Sloan L L 1931 Selective methods for the detection of Bjerrum and other scotomas *Arch. Ophthalmol.* **5** 224–60

Fetter M C 1963 Colorimetric tests read by colour blind people *Am. J. Med. Technol.* **29** 349–55

Fiat G 1967 Colour vision defects in the population of North Thailand *Humangenetik* **3** 328

Fick A 1879 Die Lehre von der Lichtempfindung *Handbuch der Physiologie* vol III ed L Harmann (Leipzig: Vogel) pp 139–40

Fincham E F 1951 The accommodation reflex and its stimulus *Br. J. Ophthalmol.* **35** 381–93

—— 1953a Defects of the colour sense mechanism as indicated by the accommodation reflex *J. Physiol.* **121** 570–80

—— 1953b The accommodation reflex in colour-blind subjects *Colloq. Probl. Opt. Vision* **II** 1–5 (Madrid: Barmejo)

Fischer F P, Bouman K A and Doesschate J 1951 A case of tritanopy *Doc. Ophthalmol.* **5** 73–87

Fishman G A 1971 Techniques, merits and limitations of basic tests for colour defectiveness *Surv. Ophthalmol.* **14** 370–3

Fishman G A, Krill A and Fishman M 1974 Acquired colour defects in patients with open angle glaucoma and ocular hypertension *Mod. Probl. Ophthalmol.* **13** 335–8

Fletcher R J 1961 *Ophthalmics in Industry* (London: Hatton) pp 26,96

—— 1971 *A Portable Filter Anomaloscope* unpublished (described in a special study report by J Alderton, 1972, The City University, London)

—— 1972 A modified D-15 test *Mod. Probl. Ophthalmol.* **11** 22–4

—— 1975 *The City University Colour Vison Test* (London: Keeler Instruments)

—— 1976 Multi-colour tests for abnormal colour vision *Mod. Probl. Ophthalmol.* **17** 202–3

—— 1978a Confusion spot displays *Mod. Probl. Ophthalmol.* **19** 95–6

—— 1978b Recent experiences with the City spot test *Mod. Probl. Ophthalmol.* **19** 142–3

—— 1979 Investigating juvenile Daltonism *The Optician* **177** 9–14

—— 1980a *The City University Colour Vision Test* 2nd edn (London: Keeler Instruments)

—— 1980b Second Edition of the City University test *Colour Vision Deficiencies V* ed G Verriest (Bristol: Adam Hilger) pp 195–6

—— 1980c The prescription of filters for Daltonism *Ophthal. Opt.* **20** 334–40

—— 1980d Testing the Daltonic child *Colour Vision Deficiencies V* ed G Verriest (Bristol: Adam Hilger) pp 204–6

—— 1981 Colour test decisions in practice *The Optician* **181** 17–18

—— 1984 *The Fletcher–Hamblin Simplified Colour Vision Test* (London: Hamblin Instruments)

Fletcher R J and Nisted M 1963 A study of coloured contact lenses *Opthal. Opt.* **3** 1151

Forshaw C R 1954 A new form of anomaloscope *J. Sci. Instrum.* **31** 16–17

Foulds W S, Chisholm I A and Bronte-Stewart J M 1974 Effects of raised intra-ocular pressure on hue discrimination *Mod. Probl. Ophthalmol.* **13** 328–34

Foulds W S, Chisholm I A, Bronte-Stewart J M and Reid H C R 1970 The investigation and theory of the toxic amblyopias *Trans. Ophthalmol. Soc. UK* **90** 739–72

Franceschetti A 1928 Die Bedeutung der Einstellungstreite am Anomaloskop für die Diagnose der einzelnen Typen der Farbensinnstörungen *Schweiz. Med. Wochenschr.* **52** 1273–8

Franceschetti A and Klein D 1957 Two families with parents of different types of red-green blindness *Acta Genet.* **7** 255–9

Franceschetti M 1939 *Bull. Mem. Soc. Fr. d'Ophthalmol.* **52** 135

Francois J, de Rouck A and Cambie E 1972 Retinal and optic evaluation in quinine poisoning *Ann. Ophthalmol.* **4** 177–85

—— 1974 Dégénérescence hyadoidéo-tapéto-retinieme de Goldmann-Favre *Ophthalmologica* **168** 81–96

Francois J, de Rouck A and Verriest G 1963b L'électroretinographie dans les dyschromatopsies et dans l'achromatopsie *Ophthalmologica* **146** 87–100

Francois J and Verriest G 1957a Les dyschromatopsies acquises *Ann. Ocul.* **190** 713–46, 812–59, 893–943

—— 1957b Classification et symptomologie des dyschromatopsies acquises *Bull. Soc. Belg. Ophthalmol.* **116** 351-92

—— 1959 Les dyschromatopsies acquises dans le glaucome primaire *Ann. Ocul.* **192** 191–9

—— 1961a On acquired deficiency of colour vision with special reference to its detection and classification by means of the tests of Farnsworth *Vision Res.* **1** 201–19

—— 1961b Functional abnormalities of the retina in heredity in *Ophthalmology* ed J Francois (St Louis: Mosby) pp 396–440

—— 1967 La discrimination chromatique dans l'amblyopie strabique *Doc. Ophthalmol.* **23** 318–31

—— 1968 Nouvelles observations de déficiences acquises de la discrimination chromatique *Ann. Ocul.* **201** 1097–114

Francois J, Verriest G, Francois P and Asseman R 1961 Etude comparative des dyschromatopsies acquises associées aux différents types d'atrophie optique hérédo-familiale *Ann. Ocul* **194** 217–35

Francois J, Verriest G and Israel A 1964 La périmétrie statique colorée en pathologie oculaire *Bull. Soc. Belg. Ophthalmol.* **138** 512–21

—— 1966b Périmétrie statique colorée effectuée à l'aide de l'appareil de Goldmann. Résultats obtenus en pathologie oculaire *Ann. Ocul.* **199** 113–54

Francois J, Verriest G, Matton-Van Leuven M T, de Rouck A and Manavian D 1966a Atypical achromatopsia of sex-linked recessive inheritance *Am. J. Ophthalmol.* **61** 1101–8

Francois J, Verriest G and Metsala P 1963a La discrimination chromatique latérale dans les dyschromatopsies acquises *Bull. Soc. Belg. Ophthalmol.* **134** 312–16

Francois J, Verriest G, Mortier, V and Vanderdonck R 1957a De la fréquence des dyschromatopsies congenitales chez l'homme *Ann. Ocul.* **190** 5–16

Francois J, Verriest G and de Rouck A 1953 La maladie d'Oguchi *Ophthalmologica* **131** (Jan)

—— 1955 L'achromatopsie congénitale *Doc. Ophthalmol.* **9** 338–424

—— 1957b Les Fonctions Visuelles dans le glaucome congénitale *Ann. Ocul.* **189** 81–107

—— 1960 New electroretinographic findings obtained in congenital forms of dyschromatopsia *Br. J. Ophthalmol.* **44** 430–5

Fraser Roberts J A 1970 *Introduction to Medical Genetics* 5th edn (Oxford: Oxford University Press)

Fraunfelder F T 1976 *Drug-induced Ocular Side Effects and Drug Interactions* (Philadelphia: Lea and Febiger)

Frey R G 1958 Welche pseudoisochromatischen Tafeln sind für die Praxis ain besten Geeignet? *Grafes Arch. Ophthalmol.* **160** 301–20

Friedmann A I 1969 The early detection of chloroquine retinopathy with the Friedmann visual field analyser *Ophthalmol. Addit.* **158** 583–91

Frisen L 1973 A color confrontation test for the central visual field *Arch. Ophthalmol.* **89** 3–9

Frisen L and Hedin A 1972 Illumination box for colour vision tests *Acta Ophthalmol.* **50** 520–4

Frisen L and Kalm H 1981 Sahlgren's saturation test for detecting and grading acquired dyschromatopsia *Am. J. Ophthalmol.* **92** 252–8

Frome F S, Piantianida T P and Kelly D H 1982 Psychophysical evidence for more than two kinds of cone in dichromatic colour blindness *Science* **215** 417–18

Fry G A 1944 A quantitative formulation of color mixture and chroma discrimination data for dichromats *J. Opt. Soc. Am.* **34** 159–69

—— 1981 Chromatic adaptation and the fundamental stimuli *Am. J. Optom.* **58** 125–35

Fuortes M G F, Schwartz E A and Simon E J 1973 Colour dependence of cone responses in the turtle retina *J. Physiol* **234** 199–216

Fuortes M G F and Simon E J 1974 Interactions leading to horizontal cell response in the turtle retina *J. Physiol.* **240** 177–98

Gallagher J R and Gallagher C D 1964 Colour vision screening of preschool and first grade children *Arch. Ophthalmol.* **72** 200–11

Ganter H 1955 Farbenfehlsichtige im Strassenverkehr *Zbl. Verkehrsmed.* **1** 7–18

—— 1956 Strassenverkehrsunfälle durch Farbenfehlsichtige *Zbl. Verkehrsmed.* **1/2** 273–4

Gardiner P A 1972 The Guy's colour vision test for young children (London: Keeler Instruments)

Gaydon A G 1938 Colour sensations produced by ultraviolet light *Proc. Phys. Soc.* **50** 714–20

Gayle V S 1978 Colour vision tests and variation of illuminants *Project Report* The City University, London

Gibbons H L and Lewis M F 1969 Colour signals and general aviation *Aerosp. Med.* **40** 668–9

Gibbons S 1974 *Stamp Colour Key* (London: Stanley Gibbons)

Gibson H C, Smith D M and Alpern M 1965 π_5 specificity in digitoxin toxicity *Arch. Ophthalmol.* **74** 154–8

Gibson I M 1962 Visual mechanism in a cone monochromat *J. Physiol.* **161** 10

Gibson K S 1940 Spectral luminosity factors *J. Opt. Soc. Am.* **30** 51–61

Gibson K S and Tyndall E P T 1923 Visibility of radiant energy *Br. Soc. Sci. Pap.* **19** 131

Gilbert J G 1957 Age changes in colour matching *J. Gerontol.* **12** 210–15

Gilbert M 1947 Colour perception in extra-foveal vision *PhD Thesis* University of London

—— 1950 Colour perception in parafoveal vision *Proc. Phys. Soc.* B **63** 83–9

Giles G H 1934 Clinical application of colour vision tests *The Refractionist* **23** 40–1, 62–3, 78–80, 102–4

—— 1950a Colour vision, some recent trends in practice *Br. J. Physiol. Opt.* **7** 90–95

—— 1950b Some aspects of colour vision with special reference to individual normal and abnormal monocular differences *Br. J. Physiol. Opt.* **7** 7–15

—— 1954 The clinical testing of colour vision *Br. J. Physiol. Opt.* **11** 1–7

—— 1960 *The Principles and Practice of Refraction* (London: Hammond) p. 447

Gillespie L, Terry L L and Sjoerdsma A 1959 The application of a monamime oxidase inhibitor to the treatment of primary hypertension. *Am. Heart J.* **58** 1–12

Gintle H E 1878 *Zeitung des Club österreichischen Eisenbahn-Beamten* **1** 93–103

Glansholm A, Hedin A and Tengroth B 1974 Pupillography in colour normals and colour defectives *Mod. Probl. Ophthalmol.* **13** 185–9

Glenn J J and Killian J T 1940 Trichromatic analysis of the Munsell Book of Color *J. Opt. Soc. Am.* **30** 609–16

Glickstein M and Heath G G 1975 Receptors in the monochromatic eye *Vision Res.* **15** 633–6

von Goethe J W 1810 *Farbenlehre* see Eastlake's 1970 version (Cambridge, MA: MIT Press)

Gordon J and Abramov I 1977 Colour vision in the peripheral retina. II Hue and saturation *J. Opt. Soc. Am.* **67** 202–7

Göthlin G F 1941 *Acta Ophthalmol.* **19** 202

—— 1943 The fundamental colour sensation in man's colour sense *K. Sven. Vet. Handl.* **20** 1–76

Gouras P 1968a Identification of cone mechanisms in monkey ganglion cells *J. Physiol.* **199** 537–47

—— 1968b Cone receptive field organisation of monkey ganglion cells *Fed. Proc* **27** 637

—— 1970 Electroretinography: some basic principles *Invest. Ophthalmol.* **9** 557

Gouras P and Kruger J 1979 Responses of cells in foveal visual cortex of the monkey to pure colour contrast *J. Neurophysiol.* **42** 850–60

Gouras P and Zrenner E 1979 Enhancement of luminance flicker by colour-opponent mechanisms *Science* **205** 587

—— 1981a Color vision: a review from a neurophysiological perspective *Progress in Sensory Physiology* vol 1 ed D Ottoson (Berlin: Springer) pp 139–79

—— 1981b Colour coding in primate retina *Vision Res.* **21** 1591–8

von Grafe A 1856 Uber die Untersuchung des Gesichtfeldes bei Amblyopischen Affecktionen *Grafes Arch. Ophthalmol.* **2** 258–98

Graham B V 1972 Colour vision in the peripheral visual field *Phd Thesis* Indiana University, USA (published as Xerox Microfilm, Ann Arbor, Michigan)

Graham B V, Turner M E, Holland R and Bradley E L 1976 Wavelength discrimination derived from color naming *Vision Res.* **16** 559–62

Graham C H 1965 *Vision and Visual Perception* (New York: Wiley)

Graham C H and Hsia Y 1958a The spectral luminosity curve for a dichromatic eye and a normal eye in the same person *Proc. Natl. Acad. Sci USA* **44** 46–9

—— 1958b Some visual functions of a unilaterally dichromatic subject *Visual Problems of Colour* vol 1 (London: HMSO) pp279–95

—— 1958c Color defect and color theory *Science* **127** 675–82

Graham C H, Sperling H G, Hsia Y and Coulson A H 1961 The determination of some visual functions of a unilaterally color-blind subject: methods and results *J. Psychol.* **51** 3–32

Gramberg-Danielson B 1961 Significance of normal colour vision in traffic *Klin. Monatsbl. Augenheilkd.* **137** 811

Granit R 1933 The components of the retinal action potential in mammals and their relation to the discharge in the optic nerve *J. Physiol.* **77** 207

—— 1947 *Sensory Mechanisms of the Retina* (London: Oxford University Press)

Granit R and Wrede C M 1937 The electrical responses of light-adapted frog's eyes to monochromatic stimuli *J. Physiol.* **89** 239–56

Grant W M 1974 *Toxicology of the Eye* 2nd edn (Springfield, IL: Thomas)

Granville W C, Nickerson D and Foss C E 1943 Trichromatic specifications for intermediate and special colors of the Munsell system *J. Opt. Soc. Am.* **33** 376–85

Green B F and Anderson L K 1956 Colour coding in a visual search task *J. Exp. Psychol.* **51** 19–24

Green D G 1968 Contrast sensitivity of the colour mechanisms of the human eye *J. Physiol.* **196** 415–29

—— 1969 Sinusoidal flicker characteristics of the colour sensitive mechanisms of the eye *Vision Res.* **9** 591–600

—— 1972 Visual acuity in the blue cone monochromat *J. Physiol.* **222** 419–21

Green E L and Sloan L L 1945 Two criteria for the selection of colour vision test plates *J. Opt. Soc. Am.* **35** 723–30

Green G J and Lessell S 1977 Acquired cerebral dyschromatopsia *Arch. Ophthalmol.* **95** 121–8

Greenaway F 1966 *John Dalton and the Atom* (London: Heinemann)

Grether W F 1939 Colour vision and colour blindness in monkeys *Comp. Psychol. Monogr.* **15** 1–38

—— 1940 Chimpanzee colour vision I — Hue discrimination of three spectral points *J. Comp. Psychol.* **29** 167–77

Greve E L, Verduin W M and Ledeboer M 1974 Two-colour threshold in static perimetry *Mod. Probl. Ophthalmol.* **13** 113–18

Grey Walter W 1956 Colour illusions and aberrations during stimulation by flickering light *Nature* **177** 710

Griffin J F and Wray S H 1978 Acquired color vision defects in retrobulbar optic neuritis *Am. J. Ophthalmol.* **86** 193–201

Grum F and Bartleson C J 1980 *Optical Radiation Measurements* vol 2 Color measurement (New York: Academic)

Grützner P 1961 Typische erworbene Farbensinnstörungen bei heredodegenerativen Maculaleiden *Graefes Arch. Ophthalmol.* **163** 99–116

—— 1963 Über Diagnose und Funktionsstörungen bei der Infantilen, dominant vererbten Opticusatrophie bericht uber die 65 zusammenkunft der Deutschen *Ophthalmol. Ges. Heidelberg, Munich* 268–73

—— 1966 Über erworbene Farbeninnstörungen bei Sehnervenerkrankungen *Graefes Arch. Klin. Exp. Ophthalmol.* **169** 366–84

—— 1969a Acquired colour vision defects secondary to retinal drug toxicity *Ophthalmol. Addit.* **158** 592–604

—— 1969b Acquired colour vision defects from drug intoxication of the retina *Colour 69* (Göttingen: Musterschmidt) (published 1970)

—— 1970 Funktionsstorungen bei Digitalis intoxikation *Klin. Monatsbl. Augenheilkd.* **156** 260

—— 1971 Über Farbenanomalie, Farbenamblyopie und Farbenasthenopie *Klin. Monatsbl. Augenheilkd.* **158** 89–96

—— 1972 *Handbook of Sensory Physiology* vol VII/4 Acquired color vision defects (Berlin: Springer) pp 643–59

Grützner P and Schleicher S 1972 Acquired color vision defects in glaucoma patients *Mod. Probl. Ophthalmol.* **11** 136–40

Guild J 1925/26 A trichromatic colorimeter suitable for standardisation work *Trans. Opt. Soc.* **27** 106–28

—— 1926 A criticism of the monochromatic-plus-white method of colorimetry *Trans. Opt. Soc.* **27** (proof) 1–6

—— 1931 The colorimetric properties of the spectrum *Phil. Trans. R. Soc.* A **230** 149–87

Guilino G and Wieczorek H-L 1976 A new screening method for detecting colour vision deficiencies *Mod. Probl. Ophthalmol.* **17** 204–10

Gullstrand A 1906 Die Farbe der Macula centralis Retinae *Graefes Arch. Ophthalmol.* **62** 1–72

Gutemann A 1907 Eigene Erfahrungen eines Farbenschwachen auf Binnengewässern und auf See *Hansa Dtsch. Naut. Z.* **44** no 15 April 186–9

Hache J Cl, Constantinides G and Turut P 1972 Les aspects évolutifs de la choroidite serëuse centrale *Bull. Soc. Ophthalmol. Fr.* **72** 257–60

Hager G 1961 The importance of visual defects for accident risks in road traffic *Ber. Dtsch. Ophthalmol. Ges.* **64** 572

—— 1963 Das Sehorgan und das unfallgeschehen im Strassenverkehr *Klin. Monatsbl. Augenheilkd.* **142** 427–33

Hallden U 1974 The source of light of the anomaloscope with a nomogram to calculate the green–red quotient *Acta. Ophthalmol.* **52** 260–5

Hanaoka T and Fujimoto K 1957 Absorption spectra of a single cone in carp retina *Jpn. J. Physiol.* **7** 276–85

Hansen E 1963 Factors causing uncertainty when conducting colour discrimination tests *Ann. Inst. Barraquer* **4** 250–92

—— 1972 Colour vision defect after cranial trauma *Mod. Probl. Ophthalmol.* **11** 160–4

—— 1974a Chromatic adaptation in the Goldmann perimeter. Evaluation of congenital colour vision defects *Phys. Novegica* **7** 207–10

—— 1974b Chloroquine retinopathy evaluated with colour perimetry *Ann. Theor. Clin. Ophthalmol.* **25** 323–31

—— 1974c The photoreceptors in cone dystrophies *Mod. Prob. Ophthalmol.* **13** 318–27

—— 1974d The colour receptors studied by increment threshold measurements during chromatic adaptation in the Goldmann perimeter *Acta Ophthalmol.* **52** 490–500

—— 1976a Examination of colour vision by use of induced contrast colours *Acta. Ophthalmol.* **54** 611–21

—— 1976b The value of tissue paper contrast tests *Mod. Probl. Ophthalmol.* **17** 179–84

—— 1978 Noen data om Anders Daae og hans farvesansprøve *Nord. Med. Årsbok* 113–20

—— 1979 Selective chromatic adaptation studies with special reference to a method combining Stiles' two-colour threshold technique and static colour perimetry *Thesis* Oslo

Hansen E and Seim T 1978 Calibration of the Goldmann perimeter and accessories used in specific quantitative perimetry *Acta Ophthalmol.* **56** 241–51

Hård A and Sivik L 1981 NCS – Natural Color System: a Swedish standard for color notation *Color Res. Appl.* **6** 129–38

Hardy L H, Rand G and Rittler M C 1946 Effect of quality of illumination on results of the Ishihara test *J. Opt. Soc. Am.* **36** 86–94

—— 1954 The HRR polychromatic plates *J. Opt. Soc. Am.* **44** 509–23

Harosi F I 1982 Recent results from single cell microspectrophotometry: cone pigments in frog, fish and monkey *Color Res. Appl.* **7** 136–41

Harrington D 1971 *The Visual Fields* (St Louis: Mosby)

Harris M G and Cabrera C R 1976 Effect of tinted contact lenses in color vision *Am. J. Optom.* **53** 145–8

Harrison R, Hoefnagel D and Hayward J N 1960 Congenital total colour blindness; a clinicopathological report *Arch. Ophthalmol.* **64** 685–92

Hartline H K 1938 The response of single optic nerve fibres of the vertebrate eye to illumination of the retina *Am. J. Physiol.* **121** 400–15

—— 1940 The receptive fields of optic nerve fibres *Am. J. Physiol.* **130** 690

Hartline H K, Wagner H G and Ratliff F 1956 Inhibition in the eye of *Limulus J. Gen. Physiol.* **39** 651–73

Hartridge H 1945a The change from trichromatic to dichromatic vision in the human retina *Nature* **155** 657–62

—— 1945b Colour vision of the fovea centralis *Nature* **155** 391–2

—— 1947 Some fatigue effects on the human retina produced by using coloured lights *Nature* **160** 538

—— 1950 *Recent Advances in the Physiology of Vision* (London: Churchill)

Hartung H 1926 Über drei Familiäre fälle von Tritanomalie *Klin. Monatsbl. Augenheilkd.* **76** 229–40

Harvey G 1826 On an anomalous case of vision with regard to colour *Trans. R. Soc. Edinburgh* **10** 253–62

Haughey A and Haughey A E 1976 A study of colour vision defects in a valley population in the West of Scotland *Mod. Probl. Ophthalmol.* **15** 512

Hayhoe M M and MacLeod D I A 1976 A single anomalous pigment *J. Opt. Soc. Am.* **66** 276–7

Haynes H, White B L and Held R 1965 Visual accommodation in human infants *Science* **148** 528–30

Heath G G 1956 Accommodative responses in totally color blind observers *Am. J. Optom.* **33** 457–65

—— 1958 Luminosity curves of normal and dichromatic observers *Science* **128** 775–6

—— 1960 Luminosity losses in deuteranopes *Science* **131**

Health G G and Schmidt I 1959 Signal color recognition by color defective observers *Am. J. Optom.* **36** 421–37

Heaton J M, McCormick A J A and Freeman A G 1958 Tobacco amblyopia *The Lancet* **2** 286

Hecht S 1930 The development of Thomas Young's theory of color vision *J. Opt. Soc. Am.* **20** 231–70

Hecht S and Hsia Y 1947 Colourblind vision: I Luminosity losses in the spectrum for dichromats *J. Gen. Physiol.* **31** 141–52

Hecht S and Shlaer S 1937 The color vision of dichromats *J. Gen. Physiol.* **20** 57–93

Hecht S, Shlaer S, Smith E, Haig C and Peskin J C 1948 The visual functions of the complete color blind *J. Gen. Physiol.* **31** 459–72

Hedin A 1974 A study of the new series of Bostrom–Kugelberg pseudoisochromatic plates *Mod. Probl. Ophthalmol.* **13** 64–6

—— 1979 Color in peripheral vision in *Topics in Neuro Ophthalmology* ed H S Thompson (Baltimore: Williams and Wilkins) pp 46–66

Hedin A and Glansholm A 1976 Pupillary spectral sensitivity in normals and colour defectives *Mod. Probl. Ophthalmol.* **17** 231–6

Heine L 1931 *The Lancet* **1** 631

Heinsius E 1972 Effects of the beginning of macula degeneration on colour vision tests *Mod. Probl. Ophthalmol.* **11** 106–10

von Helmholtz H 1852 On the theory of compound colours *Phil. Mag.* **4** 519

—— 1866 *Handbuch der Physiologischen Optik* 1st edn (Leipzig: Voss)

—— 1924 *Works* (English edition) vol II (New York: Optical Society of America)

Helve J 1972 A comparative study of several diagnostic tests of colour vision used for measuring types and degrees of congenital red-green defects *Acta. Ophthalmol. Suppl.* **115** 1–64

Helve J and Krause J 1972 The influence of age on performance in the panel D-15 colour vision test *Acta Ophthalmol.* **50** 896–900

Henderson S T 1970 *Daylight and its Spectrum* (London: Adam Hilger)

Henry G H, Cole B L and Nathan J 1964 The inheritance of congenital tritanopia with the report of an extensive pedigree *Ann. Hum. Genet.* **27** 219–31

Hering E 1878 *Outlines of a Theory of the Light Sense* trans L M Hurvich and D Jameson (1964) (Cambridge, Ma: Harvard University Press)
—— 1891 Untersuchung eines total Farbenblinden *Pflugers Arch. Ges. Physiol.* **49** 563–603

Herschel J F W 1833 Personal communication from Sir J F W Herschel to John Dalton, May 20 1833. Quoted by Sherman 1981 as being cited in Henry W C 1854 *Memoirs of the Life and Scientific Researches of John Dalton* (London: Cavendish Society)
—— 1845 Article on light *Encycl. Metrop.* **III** 432–5 (as cited by Sherman 1981)

Hertel E 1939 *Farben proben zur Prüfung des Farbensinnes 20 neu bearbeitete auflage der Stilling'schen Tafeln* (Leipzig: Thieme)

von Hess C 1889 Über den Farbensinn bei indirektem Sehen *Graefes Arch. Ophthalmol.* **35** 1–42

Higgins K E, Moskowitz-Cook A and Knoblauch K 1978 Color vision testing—an alternative source of illuminant C *Mod. Probl. Ophthalmol.* **19** 113–21

Hill A R 1980 Decision uncertainty for a homozygous or heterozygous female *Colour Vision Deficiencies V* ed G Verriest (Bristol: Adam Hilger) pp 261–7

Hill A R, Connolly J E and Dundas J 1978a The performance of ten colour vision tests at three illumination levels *Mod. Probl. Ophthalmol.* **19** 64–6
—— An evaluation of The City colour vision test *Mod. Probl. Ophthalmol.* **19** 136–41

Hill A R, Heron G, Lloyd M and Lowther P 1981 An evaluation of some colour vision tests for children *6th Symp. IRGCVD, Berlin, 1981* (Junk: The Hague) pp 183–7

Hillmann B, Connolly K and Farnsworth D 1954 Colour perception of small stimuli with central vision *Report 257, Medical Research Dept.* (New London, CT: Naval Research Laboratory)

Hilz R L, Huppman G and Cavonius C R 1974 Influence of luminance contrast on hue discrimination *J. Opt. Soc. Am.* **64** 763–6

von Hippel A 1880 Ein Fall von einseifiger congenitaler Roth-Grun-blindheit bei normalem Farbensinn des anderen Auges *Graefes Arch. Ophthalmol.* **26** 176–85

Hofman H 1975 The recognition of colours of light signals in photopic and scotopic vision *1975 CIE Conference, London*

Holmes J G 1981 Personal communications to J Voke

Holmes J G and Wright W D 1982 A new colour-perception lantern *Color Res. Appl.* **7** 82–8

Holmgren F 1877 Colour blindness and its relation to accidents by rail and sea *Smithsonian Rep.* (1877) 131–95
—— 1878 *De la Cécité des Couleurs dans ses Rapports avec les Chemins de Fer et la Marine Traduit du Suédois avec l'autorisation de l'Auteur* (Stockholm)
—— 1881 How do the colour blind see the different colours? (Communicated by Pole W) *Proc. R. Soc.* 302–6

Holth S 1928 An indelible test for central colour scotoma *Br. J. Ophthalmol.* **12** 309–13

Hood D and Finkelstein M 1982 A case for the revision of physiological models of color vision *International Conference on Colour Vision, Cambridge, UK, 23–27 August 1982* (Proceedings published by Academic Press 1983)

Hoogerheide J 1959 Considerations about the acceptability of mild colour defective trichromats in flying personnel *Aeromed. Acta Soesterberg* **7** 17–23

Hope A 1972 Mosaic spectacles excite colour blind eyes *New Sci.* **56** 705

Horner J F 1876 Die erblichkeit des Daltonismus *Amtl. Ber. Verwalt. Med. Kant. Zurich* 208–11

Horner R G and Purslow E T 1947 Dependence of anomaloscope matching on viewing distance or field size *Nature* **160** 23–4, **161** 484

Hough E A and Ruddock K H 1969 The parafoveal visual response of a tritanope and an interpretation of the V_λ sensitivity function of mesopic vision *Vision Res.* **9** 935–46

Houston R A 1932 A new trichromatic colorimeter *Trans. Opt. Soc.* **33**

Hsia Y and Graham C H 1957 Spectral luminosity curve of protanopic, deuteranopic and normal subjects *Proc. Natl. Acad. Sci. USA* **43** 1011–19

Hubel D H and Wiesel T N 1960 Receptive fields of optic nerve fibres in the spider monkey *J. Physiol.* **154** 572–80

—— 1962 Receptive fields, binocular interaction and functional architecture in the cat's visual cortex *J. Physiol.* **160** 106–54

—— 1968 Receptive fields and functional architecture of monkey striate cortex *J. Physiol.* **195** 215–43

Huddart J G 1777 An account of persons who could not distinguish colours *Phil. Trans. R. Soc.* **67** 260–5

Hukami K 1959 Results of a color discrimination test for chorioretinitis centralis serosa (Masuda) *Folia Ophthalmol. Jpn.* **10** 675–8

—— 1967 Relationship between the results of color discrimination test and of lantern test. *Folia Ophthalmol. Jpn.* **18** 263–6

Hunt R W G 1954 A visual tricolorimeter using the CIE stimuli X, Y, and Z *J. Sci. Instrum.* **31** 122–4

—— 1957 *The Reproduction of Colour* (London: Fountain Press)

—— 1967 The strange journey from retina to brain. A Fleming memorial lecture *J.R. Telev. Soc.* (Summer)

—— 1980 Color terms, symbols and their usage *Optical Radiation Measurement* vol II ed F Grum and C J Bartleson (London: Academic) p 13

—— 1982 A model of colour vision for predicting colour appearance *Color Res. Appl.* **7** 95–112

Hurvich L M 1972 Color vision deficiencies *Handbook of Sensory Physiology* vol XVII/4 ed D Jameson and L M Hurvich (Berlin: Springer) pp 582–624

—— 1977/8 Two decades of opponent processes *Color 77* (Bristol: Adam Hilger) pp 33–62

Hurvich L M and Jameson D 1955 Brightness, saturation and hue in normal and dichromatic vision *J. Opt. Soc. Am.* **45** 602–16

—— 1957 An opponent-process theory of color vision *Psychol. Rev.* **64** 384–90

—— 1959 Perceived colour and its dependence on focal surrounding and preceding stimulus variables *J. Opt. Soc. Am.* **49** 890–8

—— 1961 *Neurophysiology and Psychophysics* ed R Jung and H Kornhuber (Berlin: Springer)

—— 1962 Color theory and abnormal color vision *Docum. Ophthalmol.* **16** 409–42

—— 1964 Does anomalous vision imply color weakness? *Psychonom. Sci.* **1** 11–12

—— 1974 Evaluation of single pigment shifts in anomalous colour vision *Mod. Probl. Ophthalmol.* **13** 200–9

IALA 1968 Recommendations for the standardization of lighted aids to navigation *Bulletin* no 37 (July) (Paris: IALA)

—— 1976 Maritime buoyage systems—System A *Bulletin Suppl.* no 6 (Paris: IALA)

—— 1977 Recommendations for the colours of light signals on aids to navigation *Bulletin* no 72 1977/4 (Paris: IALA)

—— 1980 Recommendations for the surface colours used as visual signals on aids to navigation *Bulletin* no 84 1980/4 (Paris: IALA)

Ichikawa H *et al* 1978 *Standard Pseudoisochromatic Plates* (Tokyo: Igaku-Shoin)

Ichikawa H and Majima A 1974a Genealogical studies on interesting families of defective colour vision discovered by a mass examination in Japan and Formosa *Mod. Prob. Ophthalmol.* **13** 265–71

—— 1974b Studies on genetic carriers of protan and deutan types of defective color vision *Trans. 4th Congr. Eur. Soc. Ophthalmol.* **II** 342–7 (582–7)

Ichikawa H, Tanabe S, Hukami K 1983 New pseudoisochromatic plates for acquired colour defects *7th Int. Symp. IRGCVD, Geneva, 1983* (The Hague: Junk)

ICO (International Council of Ophthalmology) 1979 *Perimetric Standards and Glossary* (The Hague: Junk)

Iinuma I and Handa Y 1976 A consideration of the racial incidence of congenital dyschromats in males and females *Mod. Probl. Ophthalmol.* **17** 151–7

Ikeda H and Ripps H 1966 The electroretinogram of a cone monochromat *Arch. Ophthalmol.* **75** 513–17

Ikeda M, Hukami K and Urakubo M 1974 Separation of carriers of colour vision defect with flicker photometry *Acta Chromatica* **2** 217–21

Ikeda M, Uetsuki T and Stiles W S 1970 Interrelations among Stiles mechanisms *J. Opt. Soc. Am.* **60** 406–15

Imaizumi K 1966 The clinical application of EOG *Clinical Retinography* (Oxford: Pergamon) pp 311–26

Imamura T 1964 *The Principle of Training for Color Blindness* (Osaka: Hayakawa Company)

Ingling C R and Drum B A 1973 Retinal receptive fields: correlations between psychophysics and electrophysiology *Vision Res.* **13** 1151–63

Ingling C R and Martinez E 1982 The spectral sensitivity of the R–G achromatic channel *ARVO Meeting Abstracts Document* (St Louis: Mosby) (supplement to *Invest. Ophthalmol.*)

Ingling C R, Scheibner H M O and Boynton R M 1970 Color naming of small foveal fields *Vision Res.* **10** 501–11

Inui T, Nimura O and Kani K 1981 Retinal sensitivity and spatial summation in the foveal and parafoveal regions *J. Opt. Soc. Am.* **71** 151

Ishak I G H 1952 The photopic luminosity curve for a group of fifteen Egyptian trichromats *J. Opt. Soc. Am.* **42** 529–34

Isobe K 1955 Maxwell's spot and local differences of colour response in the human retina *Jpn. J. Psychol.* **5** 1

Israel A and Verriest G 1972 Comparison in the central visual field of normal and amblyopic eyes of the increment thresholds for lights of short and long wavelengths *Mod. Probl. Ophthalmol.* **11** 76–81

Jacobs G H 1982 *Comparative Colour Vision* (London: Academic)

Jacobs G H, Bowmaker J K and Mollon J D 1982 Colour vision variations in monkeys. Behavioural and microspectrophotometric measurements on the same individuals *Docum. Ophthalmol. Proc.* Series 33 269–80

Jaeger W 1951 Gibt es kombinationsformen der verschiedenen Typen angeborener Farbensinnstörung? *Graefes Arch. Ophthalmol.* **151** 229–48

REFERENCES

—— 1955 Tritoformen angeborener und erworbener Farbensinnstörungen *Die Farbe* **4** 197–215

—— 1956 Defective colour-vision caused by eye diseases *Trans. Ophthalmol. Soc.* **LXXVI** 477–89

—— 1972 Genetics of congenital colour deficiencies *Handbook of Sensory Physiology* vol VII/4 (Berlin: Springer) pp 626–42

Jaeger W, Früh D and Laver H J 1972 Types of acquired colour deficiencies caused by autosomal-dominant infantile optic atrophy *Mod. Probl. Ophthalmol.* **11** 145–7

Jaeger W and Grützner P 1963 Erworbene Farbensinnstörungen Entwicklung und Fortschritt in *Augenheilkunde, Fortbildungskurs für Augeniarzte, Hamburg 1962* (Stuttgart: Enke)

Jaeger W and Kroker W 1952 Über das Verhalten der Protanopen und Deuteranopen bei groben Reizflachen *Klin. Monatsbl. Augenheilkd* **121** 445–9

Jaeger W and Laver H J 1976 Non-allelic compounds of protan and deutan deficiencies *Mod. Probl. Ophthalmol.* **17** 121–30

Jameson D and Hurvich L 1953 Spectral sensitivity of the fovea. II Dependence on chromatic adaptation *J. Opt. Soc. Am.* **43** 552–9

—— 1955 Some quantitative aspects of an opponent colors theory. I Chromatic responses and spectral saturation *J. Opt. Soc. Am.* **45** 546–52

—— 1956a Some quantitative aspects of an opponent-colours theory. III Changes in brightness, saturation and hue with chromatic adaptation *J. Opt. Soc. Am.* **46** 405–15

—— 1956b Theoretical analysis of anomalous trichromatic colour vision *J. Opt. Soc. Am.* **46** 1075–89

—— 1959 Perceived colour and its dependence on focal, surrounding and preceding stimulus variables *J. Opt. Soc. Am.* **49** 890–8

—— 1964 Theory of brightness and colour contrast in human vision *Vision Res.* **4** 135–54

—— 1968 Opponent-response functions related to measured cone photo-pigments *J. Opt. Soc. Am.* **58** 429–30

—— (eds) 1972 Visual psychophysics *Handbook of Sensory Physiology* vol VII/4 (Berlin: Springer)

Jameson D, Hurvich L M and Varner D 1982 Discrimination mechanisms in colour deficient systems *Docum. Ophthalmol. Proc.* Series 33 95–301

Jeffries B J 1883 *Colour Blindness — its Dangers and Detection* (Cambridge, MA: Riverside Press) (2nd edn cited as Boston 1879)

Jennings J E 1896/1905 *Color-Vision and Color-Blindness* (Philadelphia: Davis) 2nd edn 1905

Johnson M A and Massof R W 1982a The effect of stimulus size on chromatic thresholds in the peripheral retina *Docum. Ophthalmol. Proc.* Series 33 15–18

—— 1982b Changes in the spatial characteristics of chromatic mechanisms with eccentricity *ARVO Meeting, Abstracts Document* (St Louis: Mosby)

Jolly V G 1954 The selection of colour matchers *J. Oil Colour Chem. Assoc.* **37** 414, 666–9

Jones D J 1970 The City University, London, Dept. Optometry and Visual Science *Third Year Project*

Jones L A and Lowry E M 1926 *J. Opt. Soc. Am.* **13** 25

Jones M R 1962 Colour coding *Hum. Factors* **4** 355–65

Jordinson F 1961 Further tests for colour vision *J. Soc. Dyers Colour.* **77** 550–4

Judd D B 1932 Chromaticity sensibility to stimulus differences *J. Opt. Soc. Am.* **22** 72–108

—— 1935 A Maxwell triangle yielding uniform chromaticity scales *J. Opt. Soc. Am* **25** 24–35

—— 1940 Hue, saturation and lightness of surface colours with chromatic illumination *J. Res. NBS* **24** 293–333

—— 1943 Facts of color-blindness *J. Opt. Soc. Am.* **33** 294–307

—— 1944 Standard response function for protanopic and deuteranopic vision *J. Res NBS* **33** 407–37

—— 1945 Standard response functions for protanopic and deuteranopic vision *J. Opt. Soc. Am.* **35** 199–221

—— 1947 *Docum. Ophthalmol.* **3** 251

—— 1948 Color perception of deuteranopic and protanopic observers *J. Res. NBS* **41** 247–71

—— 1952 *Standard Filters for Electronic Equipment* (University of Michigan: Armed-Forces National Research Council Vision Committee)

—— 1953 Entoptic color-perceptions of the macular pigment by observers of normal and colour-defective vision according to a three-components theory *Colloq. Probl. Opt. Vision* (Madrid: Barmejo) pp 197–217

—— 1961a Blue-glass filters to approximate the blackbody at 6500 K *Die Farbe* **10** 31–6

—— 1961b Maxwell and modern colorimetry *J. Photogr. Sci.* **9** 341–52

—— 1973 Colour in visual signalling in colour vision *Symp. Proc. Committee on Vision, National Research Council* (Washington, DC: National Academy of Science)

Judd D B, MacAdam D L and Wyszecki G W 1964 Spectral distribution of typical daylight as a function of correlated color temperature *J. Opt. Soc. Am.* **54** 1031

Judd D B and Wyszecki G 1975 *Colour in Business, Science and Industry* 3rd edn (New York: Wiley)

Kaiser P K 1968 Colour names of very small fields varying in duration and luminance *J. Opt. Soc. Am.* **58** 849–52

Kaiser P K and Hemmendinger H 1980 The Color Rule: a device for color-vision testing *Color Res. Appl.* **5** 65–71

Kalmus H 1955 The familial distribution of congenital tritanopia, with some remarks on some similar conditions *Ann. Hum. Genet.* **20** 33–56

—— 1962 Distance and sequence of the loci for protan and deutan defects and for G6PD deficiency *Nature* **194** 215

—— 1965 *Diagnosis and Genetics of Defective Colour Vision* (London: Pergamon)

—— 1971 Observations with Ishihara charts at low colour temperatures low light and limited exposure time *Vision Res.* **11** 1487–90

—— 1972a Metameric 'Colour Rule' matches of normal, colour deviant, cataractic and aphakic observers *Ann. Hum. Genet.* **36** 109–18

—— 1972b Pure (unique) green and a neutral zone in the spectrum of colour defectives *Ann. Hum. Genet* **35** 375–7

—— 1979 Dependence of colour naming and monochromator setting on the direction of preceding changes in wavelength *Br. J. Physiol. Opt.* **33** 1–9

Kalmus H, Amir A, Levine O, Barak E and Goldschmidt E 1961 The frequency of inherited defects of colour vision in some Israeli populations *Ann. Hum. Genet.* **25** 51–5

Kalmus H and Case R M 1972 The distribution of the spectral locus of 'unique green' in samples of normal trichromats *Ann. Hum. Genet.* **35** 369–74

Kalmus H, Luke I and Seedburgh D 1974 Impairment of colour vision in patients with ocular hypertension and glaucoma *Br. J. Ophthalmol.* **58** 922–6

Kaneko A 1971 Physiological studies of single retinal cells and their morphological identification *Vision Res. Suppl.* **3** 17–26

—— 1973 Receptive field organisation of bipolar and amacrine cells in the goldfish *J. Physiol.* **235** 133–54

Keating E G 1979 Rudimentary colour vision in the monkey after removal of striate and preoccipital cortex *Brain Res.* **179** 379–84

Kelly D H 1973 Lateral inhibition in human colour mechanisms *J. Physiol.* **228** 55–72

—— 1974 Spatio-temporal frequency characteristics of color vision mechanisms *J. Opt. Soc. Am.* **64** 983–90

Kelly L L 1958 Observer differences in colour-mixture functions studied by means of a pair of metameric greys *J. Res NBS* **60** 97–103

Kemmetmüller H, Keck G and Cabaj A 1980 L'effect des lentilles de contact teintées sur le sens chromatique *Contactologica* **2** 273–80

Kerr L 1974 Detection and identification of monochromatic stimuli under chromatic contrast *Vision Res.* **14** 1095–105

Kessler I 1977 What can be done for the colour blind? *Ann. Ophthalmol.* **9** 431–3

Kherumian R and Pickford R W 1959 *Héredité et Fréquences des Anomalies Congenitales du Sens Chromatic* (Paris: Vigot Frères)

King-Smith P E 1973a The optical density of erythrolabe determined by retinal densitometry using the self-screening method *J. Physiol.* **230** 535–49

—— 1973b The optical density of erythrolabe determined by a new method *J. Physiol* **230** 551–60

—— 1975 Visual detection analysed in terms of luminance and chromatic signals *Nature* **255** 1, 69–70

King-Smith P E and Carden D 1976 Luminance and opponent colour contributions to visual detection and adaptation and to temporal and spatial integration *J. Opt. Soc. Am.* **66** 709–17

Kinnear P R 1965 The risk of colour vision deterioration in young diabetics *Proc. Int. Colour Meeting, Lucerne* (Göttingen: Musterschmidt)

—— 1970 Proposals for scoring and assessing the 100 Hue test *Vision Res.* **10** 423–33

—— 1974 Luminosity curves of anomalous subjects *Mod. Probl. Ophthalmol.* **13** 180–4

Kinnear P R, Aspinall P A and Lakowski R 1972 The diabetic eye and colour vision *Trans. Ophthalmol. Soc. UK* **92** 69–78

Kinney J A 1962 Factors affecting induced colour *Vision Res.* **2** 503–25

—— 1967a Induced colours seen by a deuteranope *J. Opt. Soc. Am.* **57** 1149–54

—— 1967b Colour induction using flashes *Vision Res.* **7** 299–318

Kinney J A and McKay C L 1974 Test of colour defective vision using the visual evoked response *J. Opt. Soc. Am.* **64** 1244–50

Kinney J A, Paulson H and Beare A 1979 The ability of colour defectives to judge signal lights at sea *J. Opt. Soc. Am.* **69** 106–13

Kitahara H 1980 Measurement of colour mechanisms on extrafoveal retina *Acta. Soc. Ophthalmol. Jpn.* **84** 1603–11

Kittel V 1957 *Ophthalmol. Ges.* **61** 251–4

Kleefeld M G 1960 Les filtres dichromique *Bull. Soc. Belg. Ophthalmol.* **125** 935–50

Klein R and Nordmann J 1948 L'éxagération du contraste des couleurs chez les dyschromatopes *Bull. Soc. Ophthalmol. Paris* 453–5

Klingaman R L 1979 A comparison of psychophysical and VECP increment threshold functions of a rod monochromat *Invest. Ophthalmol.* **16** 870–3

Klingaman R L and Moskowitz-Cook A 1979 Assessment of the visual acuity of human color mechanisms with the visually evoked cortical potential *Invest. Ophthalmol.* 1273-7

Knowles A and Dartnall H J A 1977 *The Eye* ed H Davson (London: Academic) ch 3

Kogure S 1980 Saturation sensitivity *Acta. Soc. Ophthalmol. Jpn.* **84** 537–52

Kok van Alphen C C 1960 A family with the dominant infantile form of optical atrophy *Acta. Ophthalmol.* **38** 675–85

Kolb H 1970 Organisation of the outer plexiform layer of the primate retina: electron microscopy of Golgi-impregnated neurones *Phil. Trans. R. Soc.* B **258** 261–83

Koliopoulos J, Iordanides P, Palimeris G and Chimonidou E 1976 Data concerning colour vision deficiencies amongst 29 985 young Greeks *Mod. Probl. Ophthalmol.* **17** 161–4

Koliopoulos J and Palimeris G 1972 On acquired colour vision disturbances during treatment with ethambutol and indomethacin *Mod. Probl. Ophthalmol.* **11** 178–84

Kollner H 1912 *Stroungen des Farbensinnes ihre Klinische Bedeutung und ihre Diagnose* (Berlin: Karger)

König A 1884 Zur Kenntniss dichromatischer Farbensysteme *Ann. Phys. Chem.* **22** 567–78

—— 1894 see König and Köttgen 1894

—— 1897 Die Abhängigkeit der Sehschärfe von der Beleuchtungsintensität *Sitzungsber. Akad. Wiss. Berlin* 559

—— 1903 in *Gesammelte Abhandlungen zur Physiologischen Optik* Leipzig p 23

König A and Dieterici C 1884 Ueber die Empfindlichkeit des normalen Auges für Wellenlangenunterschiede des Lichtes *Ann. Phys. Chem.* **22** 579–89

—— 1893 Die Grundempfindungen in normalen und anomalen Farben Systemen und ihre intensitats-verteilung im Spektrum *Z. Psychol. Physiol. Sinnesorg.* **4** 241–347

König A and Köttgen E 1894 Ueber den menschlichen Sehpurpur und seine Bedutung für das sehen *Sitzungsber. Akad. Wiss. Berlin* 577–98

Kornerup A 1968 *AIMS — Advanced Ink Mixing System* (Copenhagen: Danish Paint and Ink Research Laboratory)

Kratter F E 1957 Color-blindness in relation to normal and defective intelligence *Am. J. Ment. Def.* **62** 436–41

Krause I B 1967 *Report* cited by D Ellis 1979

Krauskopf J and Mollon J D 1971 The independence of the temporal properties of the individual chromatic mechanisms in the human eye *J. Physiol.* **219** 611–23

Krauskopf J, Williams D R and Heeley D W 1981 Computer controlled color mixer with laser primaries *Vision Res.* **21** 951–3

Krekling S and Andersen P 1973 Visual performance of children in Norwegian special schools *Br. J. Physiol. Opt.* **28** 149–61

Kreyer V 1969 Occupational possibilities of colour deficiencies *Acta Ophthalmol.* **47** 823–34

von Kries J 1897 Uber Farbensysteme *Z. Psychol. Physiol. Sinnesorg.* **13** 241–324

—— 1911 Appendix *Helmholtz Physiological Optics* vol II p 410

—— 1924 Normal and anomalous colour systems *Helmholtz Physiological Optics* ed J P C Southall vol II (New York: Optical Society of America) pp 395–428

von Kries J and Küster F 1879 Uber angeborene Farbenblindheit *Arch. Physiol.* 513–24

Krill A E 1964 A technique for evaluating photopic and scotopic flicker function with one light intensity *Docum. Ophthalmol.* **XVIII** 452–61

—— 1977 Congenital colour vision defects *Krill's Hereditary Retinal and Choroidal Diseases* vol II Clinical Characteristics ed A E Krill and D P Archer (New York: Harper and Row) pp 355–90

Krill A E and Beutler E 1964 The red-light absolute threshold in heterozygote protan carriers *Invest. Ophthalmol.* **3** 107–18

Krill A E and Schneiderman A 1964 A hue discrimination defect in so-called normal carriers of color vision defects *Invest. Ophthalmol.* **3** 445–50

—— 1966 *Proc. 3rd Int. Symp. Clinical Electroretinography* (supplement to *Vision Res.*) (Oxford: Pergamon) pp 351–61

Krill A E, Smith V C and Pokorny J 1970 Similarities between congenital tritan defects and dominant optic nerve atrophy *J. Opt. Soc. Am.* **60** 1132–9

—— 1971 Further studies supporting the identity of congenital tritanopia and hereditary dominant optic atrophy *Invest. Ophthalmol.* **10** 457–65

Kruger J 1977 Stimulus dependent colour specificity of monkey lateral geniculate neurones *Exp. Brain Res.* **30** 297–311

Kruger J and Gouras P 1980 Spectral selectivity of cells and its dependence on slit length in monkey visual cortex *J. Neurophysiol.* **43** 1055–69

Kuffler S W 1953 Discharge patterns and functional organisaton of mammalian retina *J. Neurophysiol.* **16** 37–68

—— 1973 The single-cell approach in the visual system and the study of receptive fields *Invest. Ophthalmol.* **12** 794

Kugelberg I 1972 *Tabulae Pseudoisochromaticae B-K* 2nd edn (Stockholm: Nordiska Bokhandeln)

Kühne W 1878 Zur Photochemie der Netzhaut *Unters. Physiol. Inst. Univ. Heidelberg* **1** 103

Kuypers L and Evens L 1970 Colour blindness among 10 000 recruits *Acta Belg. Arte Med. Pharm. Mil.* **3–4** 323–31

Lagerlöf O 1980 Drug-induced colour vision deficiency *Colour Vision Deficiencies V* ed G Verriest (Bristol: Adam Hilger) pp 317–19

—— 1983 Pseudoisochromatic charts in acquired dyschromatopsia *7th Int. Symp. IRGCVD, Geneva, June 1983* (The Hague: Junk)

Lakowski R 1958 Age and colour vision *Adv. Sci.* **15** 231–6

—— 1962 Is the deterioration of colour discrimination with age due to lens or retina changes? *Die Farbe* **11** 1–6

—— 1965a Colour vision and ageing *Bull. Br. Psychol. Soc.* **18** 61

—— 1965b Colorimetric and photometric data for the 10th edition of the Ishihara plates *Br. J. Physiol. Opt.* **22** 195–207

—— 1966 A critical evaluation of colour vision tests *Br. J. Physiol. Opt.* **23** 186–209

—— 1968 The Farnsworth-Munsell 100 Hue test *Ophthal. Opt.* **81** 862–72

—— 1969a Psychological variables in colour vision testing *Color 69* (Göttingen: Musterschmidt) (1970) pp 79–86

—— 1969b Theory and practice of colour vision testing *Br. J. Ind. Med.* **26** part I 173–89, part II 265–88

—— 1971 Calibration, validation and population norms for the Pickford–Nicolson anomaloscope *Br. J. Physiol. Opt.* **26** 166–82

—— 1972 The P-N Anomaloscope as a test for acquired dyschromatopsias *Mod. Probl. Ophthalmol.* **11** 25–33

—— 1974 Effects of age on the 100-Hue scores of red–green deficient subjects *Mod. Probl. Ophthalmol.* **13** 124–9

—— 1976 Objective analysis of the Stilling Tables *Mod. Probl. Ophthalmol.* **17** 166–71

—— 1978a Colour vision in glaucomatous eyes *Reg. Symp. IRGCVD, Sept. 5–6 1978, Dresden*

—— 1978b *Colorimetric data for Dr Velhagen's 'Tafeln zur prufung des Farbensinnes'* 26th edn

Lakowski R and Aspinall P A 1972 Transformation of arbitrary anomaloscope data to the CIE system of specifications *Optica Acta* **19** 399–402

Lakowski R, Aspinall P A and Kinnear P R 1972 Association between colour vision losses and diabetes mellitus *Ophthalmol. Res.* **4** 145–59

Lakowski R and Drance 1979 Acquired dyschromatopsias: The earliest functional losses in glaucoma *Docum. Ophthalmol. Proc.* Series 19 159–65

Lakowski R and Kinnear P R 1974 Diagnosis of congenital red–green anomalies in patients with clinical conditions *Mod. Probl. Ophthalmol.* **13** 369–74

Lakowski R and Morton A 1977 The effect of oral contraceptives on colour vision in diabetic women *Can. J. Ophthalmol.* **12** 89–97

—— 1978 Acquired colour losses and oral contraceptives *Mod. Probl. Ophthalmol.* **19** 314–18

Lakowski R and Oliver K 1974 Effect of pupil diameter on colour vision test performance *Mod. Probl. Ophthalmol.* **13** 307–11

Lakowski R and Tansley B W 1974 Energy modification of the Pickford–Nicolson anomaloscope *Mod. Probl. Ophthalmol.* **13** 42–6

Lampe J M, Doster M and Beal B 1973 Summary of a three year study of academic and school achievement between colour defective and normal primary age pupils *J. Sch. Health* **43** 309–11

Land E H 1965 The retinex *CIBA Found. Symp.* (London: Churchill) pp 217–23

de Lange H 1952 Experiments on flicker and some calculations on an electrical analogue of the foveal system *Physica* **18** 935

—— 1954 Relationship between critical flicker frequency and a set of low-frequency characteristics of the eye *J. Opt. Soc. Am.* **44** 350–89

—— 1957 Attenuation characteristics and phase shift characteristics of the human foveal-cortex system in relation to flicker-fusion phenomena *Thesis* Technische Hogeschool, Delft, Netherlands

Lanthony P 1975 *New Colour Test—Selon Munsell* (Paris: Luneau)

—— 1977 Etude clinique du City University colour vision test *Bull. Soc. Ophthalmol.* **77** 379–82

—— 1980 Boynton's colour naming method in acquired dyschromatopsias *Colour Vision Deficiencies V* ed G Verriest (Bristol: Adam Hilger) pp 278–9

Lanthony P and Dubois-Poulsen A 1973 Le Farnsworth 15 désaturé *Bull. Soc. Ophthalmol.* **73** 861

Larimer J, Krantz D H and Cicerone C M 1974 Opponent-process additivity. I red–green equilibria *Vision Res.* **14** 1127–40

—— 1975 Opponent-process additivity. II yellow–blue equilibria *Vision Res.* **15** 723–31

Laroche J and Laroche C 1970 Action de quelque antibiotiques sur la vision des couleurs *Ann. Pharm. Fr.* **28** 33–41

Larsen H 1921 Demonstration mikroskopischer Präparate von einem Monochromatism *Klin. Monatsbl. Augenheilkd.* **67** 301

Laxar K 1967 Performance of the Farnsworth lantern test as related to type and degree of colour vision defect *Mil. Med.* **132** 726–31

Leber T 1873 Über die Theorie der Farbenblindheit und über die Art und Weise, wie gewisse, der Untersuchung von Farbenblinden entnommenen Einwände gegen die Young–Helmholtz'sche Theorie sich mit derselben vereinigen lassen *Klin. Monatsbl. Augenheilkd.* **II** 467–73

Lee G B 1969 Luminosity curve differences among subjects with normal colour vision *J. Opt. Soc. Am.* **56** 1451

Lee G B and Virsu V 1982 A quantitative description of responses to coloured stimuli of cells in the macaque lateral geniculate nucleus *International Conference on Colour Vision, Cambridge UK, August 23–27 1982* (Proceedings published by Academic Press 1983)

Le Grand Y 1968 *Light, Colour and Vision* 2nd edn (London: Chapman and Hall)

—— 1969 Photopigments des cônes humains *Docum. Ophthalmol.* **26** 257–63

—— 1970 Les pigments des cônes chromatiques *Die Farbe* **19** 15–22

—— 1972a About the photopigments of colour vision *Mod. Probl. Ophthalmol.* **11** 186–92

—— 1972b Spectral luminosity *Handbook of Sensory Physiology* vol III/4 (Berlin: Springer)

Legras M and Coscas G 1972 Oedematous maculopathies and colour sense *Mod. Probl. Ophthalmol.* **11** 111–16

Lennox-Buchtal M A 1962 Single units in monkey cortex with narrow spectral responsiveness *Vision Res.* **2** 1

Leventhal A G, Rodieck R W and Dreher B 1981 Retinal ganglion cell classes in old-world monkey morphology and central projections *Science* **213** 1139–42

Lewis S D and Mandelbaum J 1943 Achromatopsia: Report of three cases *Arch. Ophthalmol.* **30** 225–31

Li C C 1955 *Population Genetics* (Chicago: University of Chicago Press)

Liebman P A 1972 Microspectrophotometry of photoreceptors *Handbook of Sensory Physiology* vol VII/1 ed H J A Dartnall (Berlin: Springer)

Linksz A 1966 The Farnworth Panel D-15 test *Am. J. Ophthalmol.* **62** 27–37

Long W F and Woo G C S 1980 Measuring light levels with photographic meters *Am. J. Optom.* 51–5

Lort M 1778 An account of a remarkable imperfection of sight in a letter from

J Scott to Rev Mr Whisson of Trinity College Cambridge *Phil. Trans. R. Soc.* **68** 611–14

Luckiesh M and Moss F K 1933 Visual acuity and sodium vapour light *J. Franklin Inst.* **215** 401–10

Ludvigh E and McCarthy E F 1938 Absorption of the visible light by the refractive media of the human eye *Arch. Ophthalmol.* **20** 37–51

Lyle W M 1974 Diseases and conditions (Part II of drugs and conditions which may affect color vision) *J. Am. Optom. Assoc.* **45** 173–82

Lyon M F 1962 Sex chromatin and gene action in the mammalian X-chromosome *Am. J. Hum. Genet.* **14** 135–48

Lyons J G 1958 Two unusual cases of colour blindness *The Optician* **136** 341–3

Lythgoe R J 1931 Dark adaptation and the peripheral colour sensations of normal subjects *Br. J. Ophthalmol.* **15** 193–210

MacAdam D L 1942 Visual sensitivities to colour difference in daylight *J. Opt. Soc. Am.* **32** 247–74

—— 1966 Colour science and colour photography *J. Photogr. Sci.* **14** 229–50

—— 1979 Judd's contributions to color metrics and evaluation of color differences *Color Res. Appl.* **4** 177–93

McCree K J 1960a Small field tritanopia and the effects of voluntary fixation *Optica Acta* **7** 317–23

—— 1960b Colour confusion caused by voluntary fixation *Optica Acta* **7** 281

Mace J and Nicati W 1879 Recherches sue le daltonisme *C.R. Acad. Sci. Paris* **89** 716–18

MacGillivray A M 1947 Defective colour vision and its practical importance *Med. Pract.* **218** 435–8

McKay Taylor C O 1978 *The Optician* **175** 9

McKee S P and Westheimer G 1970 Specificity of cone mechanisms in lateral interaction *J. Physiol.* **206** 117–28

McKeon W M and Wright W D 1940 The characterisics of protanomalous vision *Proc. Phys. Soc.* **52** 464–79 (note correction stated in Wright 1946 p 5)

McKusick V A 1962 On the X chromosome of man *Q. Rev. Biol.* **37** 69–175

McLaren K 1960 A simple test for detecting differences among colour matchers *J. Soc. Dyers Colour.* **76** 434–5

—— 1966 Defective colour vision. II its diagnosis *J. Soc. Dyers Colour.* **82** 382–7

MacLeod D I A and Hayhoe M M 1974 Three pigments in normal and anomalous color vision *J. Opt. Soc. Am.* **64** 92–6

MacLeod D I A and Lennie P 1974 A unilateral defect resembling deuteranopia *Mod. Prob. Ophthalmol.* **13** 130–4

MacLeod D I A and Webster M A 1982 Factors underlying individual differences in colour matching *International Conference on Colour Vision, Cambridge, UK, August 23–27 1982* (Proceedings published by Academic Press 1983)

MacNichol E F Jr 1964 Three-pigment colour vision *Sci. Am.* **211** 48–56

MacNichol E F, Levine J S, Lipitz L E, Mansfield R J W and Collins B A 1982 Microspectrophotometry of visual pigments in primate photoreceptors *International Conference on Colour Vision, Cambridge, UK, August 23–27 1982* (Proceedings published by Academic Press 1983)

MacNichol E F and Svaetichin G 1958 Electrical responses from the isolated retinas of fishes *Am. J. Ophthalmol.* **46** 26–40

Maggiore Prof (undated) *Instrument for the Colour Blindness Detection* (Firenze: Sbisa)

Maione M and Carta F 1972 The visibility of the Goldmann's perimeter coloured targets in the ageing and in retrobulbar neuritis *Mod. Prob. Ophthalmol.* **11** 72–5

Maione M, Carta F, Barberini E and Scocciant L 1976 Achromatic isopters on coloured backgrounds in some acquired colour vision deficiencies *Mod. Probl. Ophthalmol.* **17** 86–93

Maione M, Moreland J D, Carta F, Barberini E and Lettieri S 1978 Further observations on the extra-macular chromatic mechanism *Mod. Probl. Ophthalmol.* **19** 258–65

Majima A 1972a Lantern test *Folia Ophthalmol. Jpn.* **23** 83–7

—— 1972b Diagnostic criteria of defective color vision *Folia Ophthalmol. Jpn.* **23** 170–5

Malpeli J G and Schiller P H 1978 Lack of blue off-centre cells in the visual system of the monkey *Brain Res.* **141** 385–9

Mandola J 1969 The role of colour anomalies in elementry school achievement *J. Sch. Health* **39** 633–6

Mann I and Turner C 1956 Colour vision in native races of Australasia *Am. J. Ophthalmol.* **41** 797–800

Marc R E 1980 Retinal colour channels and their neurotransmitters *Colour Vision Deficiencies V* ed G Verriest (Bristol: Adam Hilger) pp 15–29

—— 1982 Retinal neurotransmitters; morphology and colour coding *Color Res. Appl.* **7** 155–8

Marc R E and Lam D M K 1981a Glycinergic pathways in the goldfish retina *J. Neurosci.* **1** 152–65

—— 1981b Uptake of aspartic and glutamic acid by photoreceptors in the goldfish retina *Proc. Natl. Acad. Sci. USA* **78** 7185–9

Marc R E and Sperling H G 1977 Chromatic organisation of primate cones *Science* **196** 454–6

Marc R E, Stell W K, Bok D and Lam D M K 1978 GABA-ergic pathways in the goldfish retina *J. Comp. Neurol.* **182** 221–45

Marks W B, Dobelle W H and MacNichol E F 1964 Visual pigments of single primate cones *Science* **143** 1181–3

Marmion V J 1977 The results of a comparison between the 100 Hue test and static colour perimetry *Docum. Ophthalmol. Proc.* Series 14 473–4

—— 1978 The colour vision deficiency in open angle glaucoma *Mod. Probl. Ophthalmol.* **19** 305–7

Marré M 1969 *Eine quantitative Analyse erworbener Farbensehstörungen* (Magdeburg: Habilitationsschrift)

—— 1971 Clinical examination of the three colour vision mechanisms in acquired colour vision defects *Mod. Probl. Ophthalmol.* **11** 224–7

—— 1973 The investigation of acquired colour vision deficiencies *Colour 73* (London: Adam Hilger) pp 99–135

Marré M and Marré E 1972 The influence of the three colour vision mechanisms on the spectral sensitivity of the fovea *Mod. Probl. Ophthalmol.* **11** 219–23

—— 1978 Colour vision in squint amblyopia *Mod. Probl. Ophthalmol.* **19** 308–13

—— 1982a The blue mechanism in diseased eyes with eccentric fixation *Docum. Ophthalmol. Proc.* Series 33 133–8

—— 1982b 'Eccentrisation' and 'scopotisation' in acquired color vision defects *Docum. Ophthalmol. Proc.* Series 33 373–8

Marré M, Nemetz U and Neubauer O 1974 Colour vision and the Pill *Mod. Probl. Ophthalmol.* **13** 345–8

Marrocco R T 1976 Sustained and transient cells in monkey lateral geniculate nucleus: condition velocities and response properties *J. Neurophysiol.* **39** 340–53

Martin L C 1939 A standardised lantern for testing colour vision *Br. J. Ophthalmol.* **23** 1–20

Martin L C and Pearse R W B 1947 The comparative visual acuity and ease of reading in white and coloured light *Br. J. Ophthalmol.* **31** 129–44

Martin L C, Warburton, F L and Morgan W J 1932 Some recent experiments on the sensitiveness of the eye to differences in the saturation of colours *Physical Society Discussion on Vision* pp 92–100

—— 1933 Determination of the sensitiveness of the eye to differences in the saturation of colours *MRC Report* 188 (London: HMSO)

Martin N J 1976 Effects of temperature on the properties of glass colour filters used in signalling equipment *Light. Res. Technol.* **8** 146–50

Massof R W and Bailey J E 1976 Achromatic points in protanopes and deuteranopes *Vision Res.* **16** 53–7

Massof R W and Bird J F 1978 A general zone theory of colour and brightness vision. 1. Basic formulation *J. Opt. Soc. Am.* **68** 1465–71

Massof R W and Guth S L 1976 Central and peripheral achromatic points in protanopes and deuteranopes *Mod. Probl. Ophthalmol.* **17** 75–8

Maxwell J C 1854 (?1857) On colour as perceived by the eye, with remarks on colour-blindness *Trans. R. Soc. Edinburgh* **XXI** 275–98

—— 1855a On the theory of colours in relation to colour blindness *Trans. Scott. Soc. Arts* **4** part III

—— 1855b *Letter to G Wilson* (27.7.85) quoted in Wilson G (1855) *Researches on Colour-Blindness* (Edinburgh: Sutherland and Knox)

—— 1860 On the theory of compound colours, and the relations of the colours of the spectrum *Phil. Trans. R. Soc.* **150** 57–84

—— 1890a On the theory of colours in relation to colour blindness *Trans. R. Scott. Soc. Arts.* 1855a **4** part III in W D Niven (ed) *Scientific Papers* vol 1 (London: Cambridge University Press) pp 119–25

—— 1890b Experiments on colour, as percieved by the eye, with remarks on colour blindness *Trans. R. Soc. Edinburgh* 1855b **21** part 2 in W D Niven (ed) *Scientific Papers* vol 1 (London: Cambridge University Press) pp 126–54

—— 1890c On the theory of compound colours and the relations of colours on the spectrum *Phil. Trans R. Soc.* 1860 **150** in W D Niven (ed) *Scientific Papers* vol 1 (London: Cambridge University Press) pp 410–44

—— 1890d On the theory of the three primary colours *Lecture at the Royal Institution of Great Britain May 17 1861* in W D Niven (ed) *Scientific Papers* vol 1 (London: Cambridge University Press) pp 445–50

Medical Research Council 1965 Spectral requirements of light sources for clinical purposes *Memo 43* (London: HMSO)

Mehaffey L and Berson E L 1974 Cone mechanisms in the electroretinogram of the cynomolgus monkey *Invest. Ophthalmol.* **13** 266

Meitner H J 1941 Übereinen eigenartigen Fall von Anomalie des Blaiusinnes *Klin. Monatsbl. Augenheilkd.* **106** 293–301

Mellerio J 1971 Light absorption and scatter in the human lens *Vision Res.* **11** 129–41

van de Meredonk S and Went L N 1980. Two cases of inherited deutan and tritan disturbances in the same person and a study of their families *Colour Vision Deficiences V* ed G Verriest (Bristol: Adam Hilger) pp 268–72

Metzstein F 1982 Colour vision testing in glaucoma testing *Bateman Rev.* **41** 13–15

Meyer J J, Korol S, Gramoni R and Tupling R 1978 Psychophysical flicker thresholds and ERG flicker responses in congenital and acquired colour vision deficiencies *Mod. Probl. Ophthalmol.* **19** 33–49

Michael C R 1978 Colour vision mechanisms in monkey striate cortex; dual opponent cells with concentric receptive fields *J. Neurophysiol.* **41** 572–88

Miles W R 1948 A functional analysis of regional differences in the human fovea *Science* **108** 683

—— 1954 Comparison of functional and structural areas in human fovea. I. Method of entoptic plotting *J. Neurophysiol.* **17** 22–38

Miller S S 1972 Psychophysical estimates of visual pigment densities in red-green dichromats *J. Physiol.* **223** 89–107

Milliken A B 1978 Quadrantic differences in saturation discrimination *MSc Project* The City University, London

Millodot M and Lamont A 1974 Colour vision deficiencies in French Canadian school children *Can. J. Public Health* **65** 461–2

Mitchell D E and Rushton W A H 1971a Visual pigments in dichromats *Vision Res.* **11** 1033–44

—— 1971b The red-green pigments of normal vision *Vision Res.* **11** 1045–56

Mollon J D and Polden P G 1975 Colour illusion and evidence for interaction between cone mechanisms *Nature* **258** 421–2

—— 1976 Absence of transient tritanopia after adaptation to very intensive yellow light *Nature* **259** 570–2

—— 1977 An anomaly in the response of the eye to light of short wavelengths *Phil. Trans. R. Soc.* B **278** 207–40

de Monasterio F M 1978 Properties of concentrically organized X and Y ganglion cells of macaque retinae *J. Neurophysiol.* **41** 1394–417

de Monasterio F M and Gouras P 1975 Functional properties of ganglion cells of the rhesus monkey retina *J. Physiol.* **251** 167–95

de Monasterio F M, Gouras P and Tolhurst D J 1975a Trichromatic colour opponency in ganglion cells of the rhesus monkey retina *J. Physiol.* **251** 197–216

—— 1975b Concealed colour opponency in ganglion cells of the rhesus monkey retina *J. Physiol.* **251** 217–29

de Monasterio F M and Schein S J 1982 Spectral bandwidths of colour-opponent cells of geniculocortical pathway of macaque monkeys *J. Neurophysiol.* **47** 214–24

Moreland J D 1955 The perception of colour by extrafoveal and peripheral vision *PhD Thesis* University of London

—— 1972a The effect of inert ocular pigments on anomaloscope matches and its reduction *Mod. Probl. Ophthalmol.* **11** 12–18

—— 1972b Inert pigments and the variability of anomaloscope matches *Am. J. Optom.* **49** 735–41

—— 1972c Peripheral colour vision in *Handbook of Sensory Physiology* vol VII/4 Visual psychophysics ed D Jameson and L M Hurvich (Berlin: Springer)

—— 1974 Calibration problems with the Nagel anomaloscope *Mod. Probl. Ophthalmol.* **13** 14–18

—— 1978 Temporal variations in anomaloscope equations *Mod. Probl. Ophthalmol.* **19** 167–72

Moreland J D and Cruz A C 1955 Peripheral small field tritanomaly *Die Farbe* **4** 241–5

—— 1959 Colour perception with the peripheral retina *Optica Acta* **6** 117–51

Moreland J D and Kerr J 1978 Optimisation of stimuli for trit-anomaloscopy *Mod. Probl. Ophthalmol.* **19** 162–6

Moreland J D, Kogan D and Smith S S 1976 Optokinetic nystagmus an objective indicator of defective colour vision *Mod. Probl. Ophthalmol.* **17** 220–30

Moreland J D, Maione M, Carta F, Barberini E and Scoccianti L 1977 The clinical assessment of the chromatic mechanism of the retinal periphery *2nd Int. Vis. Field Symp.* (Docum. Ophthalmol. Proc.) (The Hague: Junk) pp 413–21

Moreland J D, Maione M, Carta F and Scoccianti L 1978 Acquired 'tritan' deficiencies in macular pathology *Mod. Probl. Ophthalmol.* **19** 270–5

Moreland J D and Smith S S 1974 Color vision and optokinetic nystagmus *J. Opt. Soc. Am.* **64** 1387

Moreland J D and Young W B 1974 A new anomaloscope employing interference filters *Mod. Probl. Ophthalmol.* **13** 47–55

Morgan T H 1910a The method of inheritance of two sex-linked characters in the same animal *Proc.Soc. Exp. Biol. Med.* **8** 17–19

—— 1910b Sex-limited inheritance in *Drosophila Science* **32** 120–2

Moskowitz-Cook A 1979 The development of photopic spectral sensitivity in human infants *Vision Res.* **19** 1133–42

Motokawa K and Isobe K 1955 Spectral response curves and hue discrimination in normal and color-defective subjects *J. Opt. Soc. Am.* **45** 79–88

Motokawa K, Taira N and Okuda J 1962 Spectral responses of single units in the primate visual cortex *Tohoku J. Exp. Med.* **78** 320–37

Mount and Thomas 1968 Relation of spatially induced brightness changes to test and inducing wavelengths *J. Opt. Soc. Am.* **58** 23–7

Mullen K T 1982 The effect of spatial frequency on opponent colour contributions to modulation thresholds *ARVO Meeting, Abstracts Document* (St Louis: Mosby)

Müller G E 1924 *Darstellung und Erklärung der verschiedenen Typen der Farbenblindheit* (Göttingen: Vandenhoeck und Ruprecht)

Murakami M and Pak W 1970 Intracellularly recorded early receptor potential of vertebrate photoreceptors *Vision Res.* **10** 865–975

Murakami M, Shimoda Y and Nakatani K 1978 Effect of GABA on neural activities in the distal retina of the carp *Sens. Process.* **2** 334–8

Murray E 1943 Evolution of colour vision tests *J. Opt. Soc. Am.* **33** 316–34

—— 1945 A reply to Dvorine's comments *Am. J. Psychol.* **48** 399–402

—— 1954 Review of Walls G L and Mathews R W (1952) publication *Am. J. Psychol.* **67** 182–8

Murray G C 1968 Visual pigment multiplicity in cones of the primate retina *Thesis* Johns Hopkins University, Maryland, USA

Murray H D 1952 *Colour in Theory and Practice* (London: Chapman and Hall)

Nagel W A 1904 Die diagnose der anomalen trichromatischen Systeme *Klin. Monatsbl. Augenheilkd.* **42** 366–70

—— 1907a Ueber die Gefahren der Farbenblindheit im Eisenbahn und Marinedienst *Centralbl. Prakt. Augenheilkd.* 206–7

—— 1907b Zwei Apparate für die augenärztliche Funktionsprüfung: Adaptometer und kleines Spektral–photometer(Anomaloskop) *J. W.S. Z. Augenheilkd.* **17** 201

—— 1907c Farbenblindheit und Verkerssicherheit zur See und auf der Eisenbahn *Rundschan Taglichen* (July 23rd)

—— 1907d Neue Erfahrungen über das Farbensehen der Dichromaten auf grossem Felde *Z. Sinnesphysiol.* **41** 319–37

Nagy A L 1982 Homogeneity of large-field colour matches in congenital red–green colour deficiency *J. Opt. Soc. Am.* **72** 571–7

Nagy A L, Macleod D I A, Heyneman N E and Eisner A 1981 Four cone pigments in women heterozygous for colour deficiency *J. Opt. Soc. Am.* **71** 719–22

Nagy A L and Zacks J L 1977 The effects of psychophysical procedure and stimulus duration in the measurement of Bezold–Brücke hue shifts *Vision Res.* **17** 193–200

Naka K I and Rushton W A H 1966 An attempt to analyse colour reception by electrophysiology *J. Physiol.* **185** 556–86

—— 1967 The generation and spread of S-potentials in fish (Cyprinidae) *J. Physiol.* **192** 437–61

Nathan J, Henry C H and Cole B L 1963 Recognition of road traffic signals by persons with normal and defective colour vision *J. Aust. Road Res. Board* **1/7**

—— Recognition of coloured road traffic light signals by normal and colour vision defective observers *J. Opt. Soc. Am.* **54** 1041–5

de Nazaré Trinadade Marques M 1977 X chromosome and color blindness *Ophthalmologica* **175** 305–8

Nelson J H 1938 Anomalous trichromatism and its relation to normal trichromatism *Proc. Phys. Soc.* **50** 661–97

Nettleship E 1908 Three new pedigrees of eye disease *Trans. Ophthalmol. Soc.* **XXVIII** 220–49

—— 1913 *On Cases of Accidents to Shipping and on Railways Due to Defects of Sight* (London: Adlard, Bartholomew)

Neubauer O 1973 Farbensinnstörungen bei längerer verwendung von Ouvulations-Hemmern *Klin. Monatsbl. Augenheilkd.* **162** 803

Neubauer O and Harrer S 1976 Tritan defects found by using Velhagen's pseudoisochromatic plates 24th edition *Mod. Probl. Ophthalmol.* **17** 172–4

Neubauer O, Harrer S, Marré M and Verriest G 1978 Colour vision deficiencies in road traffic *Mod. Probl. Ophthalmol.* **19** 77–81

Neubert F R 1947 Colour vision in the consulting room *Br.J. Ophthalmol.* **31** 275–88

Neuhann T, Kalmus H and Jaeger W 1976 Ophthalmological findings in the tritans, described by Wright and Kalmus *Mod. Probl. Ophthalmol.* **17** 135–42 (Basel: Karger)

Newton I 1672 New theory about light and colours *Phil. Trans. R. Soc.* **80** 3075–87

—— 1704 *Opticks* 1st edn (London: W Innys)

—— 1730 *Opticks* reprint from 4th edn (London: Bell, 1931)

Nickerson D 1940 History of Munsell color system and its scientific application *J. Opt. Soc. Am.* **30** 575

—— 1946 A handbook on the method of disk colorimetry *US Dept. Agric. Misc. Publ. 580*

—— 1969 History of the Munsell color system *Color Eng.* **7** 42–51

Nickerson D and Granville W C 1940 Hue sensibility *J. Opt. Soc. Am.* **30** 159–62

Nimeroff I 1970 Deuteranopic convergence point *J. Opt. Soc. Am.* **60** 966–9

Nolte W 1962 Bestimmung achromatischer Schwellen für verschiedene Spektrallichter *Inaugural Dissertation* Tübingen

Norden L C, Amos J F and Newcomb R D 1978 Cone-rod dystrophy: a case report *Am. J. Optom.* **55** 824–35

Norman L G 1948 Industrial aspects of defective colour vision *Proc. 9th Int. Conf. Ind. Med.* pp 1054–63

—— 1960 Medical aspects of road safety *The Lancet* **1** 1039

van Norren D and Vos J J 1974 Spectral transmission of the human ocular media *Vision Res.* **14** 1237–44

van Norren D and Went L N 1981 New test for the evaluation of tritan defects evaluated in two surveys *Vision Res.* **21** 1303–6

Nuel 1879 *Ann Ocul.* **LXXXII** 64–72

Nunn B J and Baylor D A 1982a Visual transduction in retinal rods of the monkey *Macaca fasicularis Nature* **299** 726–8

—— 1982b Transduction in single primate photoreceptors studied by electro-physiological recording *International Confrence on Colour Vision, Cambridge, UK, August 23–27 1982* (Proceedings published by Academic Press 1983)

Obstfeld H 1980 Human colour vision mechanisms along four cardinal meridians *Colour Vision Deficiences V* ed G Verriest (Bristol: Adam Hilger) pp 211–16

O'Connor N 1957 Imbecility and color-blindness *Am. J Ment. Def.* **62** 83–7

Ohara H and Akutsu S 1956 *J. Clin. Ophthalmol. Jpn.* **10** 1173–8

Ohta Y 1972 Central scotometric plates utilising color confusion resulting from an acquired anomalous color vision *Mod. Probl. Ophthalmol.* **11** 40–8

Ohta Y *et al* 1978a Clinical analysis of colour vision deficiencies with the City University test *Mod. Probl. Ophthalmol.* **19** 126–30

Ohta Y, Izutsu Y, Miyamoto T and Shimizu K 1980 An experimental anomaloscope based on the interference filter system and test results *Colour Vision Deficiencies V* ed G Verriest (Bristol: Adam Hilger) pp 184–8

Ohta Y, Kogure S and Yamaguchi T 1978b Clinical experiences with the Lovibond colour vision analyser *Mod. Probl. Ophthalmol.* **19** 145–9

Ohtani K, Ohta Y, Kogure S, Kato H, Shimizu K and Seki T 1974 Farnsworth tritan plate *J. Clin Ophthalmol.* **28** 25–30

Oliver C A 1893 A series of wools for the ready detection of colour blindness *Med. News* Sept. 2

Oloff H 1935 Ueberangeborene blauanomale Trichromasie (Tritanomalie) *Klin. Monatsbl. Augenheilkd.* **94** 11–20

Onley J W, Klingberg, C L, Dainoff M J and Rollman G B 1963 Quantitative estimates of saturation *J. Opt. Soc. Am.* **53** 487–93

Osbourne N N 1982 Chemical messages in the retina *Problems of Normal and Genetically Abnormal Retinas* ed R M Clayton *et al* (London: Academic)

Østerberg C 1935 Topography of the layer of rods and cones in the human retina *Acta Ophthalmol.* Suppl. 6

Ourgaud A G, Vola J L, Jayle C E and Baud C E 1972 A study on the influence of the illumination level and pupillary diameter on chromatic discrimination in glaucomatous patients *Mod. Probl. Ophthalmol.* **11** 141–4

Padmos P and Norren D V 1975 Cone systems interaction in single neurones of the lateral geniculate nucleus of the macaque *Vision Res.* **15** 617–19

Palmer G 1777 *Theory of Colours and Vision* (London) based on excerpt from *Sources of Colour Science* ed D MacAdam (Cambridge, MA: MIT Press, 1970)

—— 1786 *Théorie de la Lumière* (Paris) English excerpt from *Sources of Colour Science* ed D MacAdam (Cambridge, MA: MIT Press, 1970)

Parra F 1972 Attempts at a method of programmation of the 100 Hue test *Mod. Probl. Ophthalmol.* **11** 793–8

Parsons J J 1924 *An Introduction to the Study of Colour Vision* (London: Cambridge University Press)

Paulson H M 1973 Comparison of colour vision tests used by the armed forces *Colour Vision Symp. NRC* (Washington, DC: National Academy of Sciences) pp 34–64

Payne B F 1940 Glasses for the color blind motorist *Am. J. Ophthalmol.* **23** 566–7

Pearce O'D W 1934 *The Selection of Colour Workers* (London: Pitman)

Peeples D R and Teller D Y 1975 Colour vision and brightness discrimination in two-month-old human infants *Science* **189** 1102–3

—— 1978 White-adapted photopic spectral sensitivity in human infants *Vision Res.* **18** 49–53

Pedler C 1965 Rods and cones—a fresh approach *CIBA Found. Symp. on Colour Vision* (London: Churchill) pp 52–82

Perry N W, Childers D G and Dawson W W 1972 Reliability of the monochromatic VER *Vision Res.* **12** 357–8

Peters G A 1954 The new Dvorine colour perception test *Optom. Wkly.* **11** 1801–3

—— 1956 A colour blindness test for use in vocational guidance *Pers. Guid. J.* **34** 572–5

Petry H M, Donovan W J, Moore R K, Dixon W B and Riggs L A 1982 Changes in the human visually evoked cortical potential in response to chromatic modulation of a sinusoidal grating *Vision Res.* **22** 745–55

Phillips R A and Kondig W 1975 Recognition of traffic signals viewed through colored filters *J. Opt. Soc. Am.* **65** 1106–13

Piantanida T P 1974 A replacement model of X-linked recessive colour vision defects *Ann. Hum. Genet.* **37** 393–404

—— 1976 A portable filter anomaloscope *Opt. Eng.* **15** 325–7

Piantanida T P and Sperling H G 1973a Isolation of a third chromatic mechanism in the protanomalous observer *Vision Res.* **13** 2033–47

—— 1973b Isolation of a third chromatic mechanism in the deuteranomalous observer *Vision Res.* **13** 2049–58

Piccolino M, Neyton J and Gerschenfeld H M 1980 Synaptic mechanisms involved in responses of chromaticity horizontal cells of turtle retinae *Nature* **284** 58–60

Pickford R W 1949 A study of the Ishihara test for colour blindness *Br. J. Psychol.* **40** part 2 71–80

—— 1951 *Individual Differences in Colour Vision* (London: Routledge and Kegan Paul)

—— 1959 Some heterozygous manifestations of colour blindness *Br. J. Physiol. Opt.* **16** 83–95

—— 1962 Compound hemizygotes for red–green colour vision defects *Vision Res.* **2** 245–52

—— 1964 The genetics of colour blindness *Br. J. Physiol. Opt.* **21** 39–47

—— 1965 The genetics of colour blindness *CIBA Found. Symp. on Colour Vision* (London: Churchill) pp 228–44

—— 1967a Variability and consistency in the manifestation of red–green colour vision defects *Vision Res.* **7** 65–77

—— 1967b Colour blindness: anomaloscope tests and physiological problems *Int. J. Neurol.* **6** 210–21

—— 1968 A pedigree showing variability in deuteranomaly *Vision Res.* **8** 469–74

Pickford R W and Lakowski R 1960 The Pickford–Nicolson anomaloscope *Br. J. Physiol. Opt.* **17** 131–50

Pickford R W, Pickford R, Bose J, Joardar B S, Nag P K, Ray G G and Sen R N 1980 Incomplete achromatopsia in Bishnupur: genetics and heterozygote problems *Colour Vision Deficiencies V* ed G Verriest (Bristol: Adam Hilger) pp 251–5

Piéron H 1932 Les temps de reaction auchroma en excitation isolumineuse *C. R. Soc. Biol.* **111** 380–2

—— 1939 La dissociation da l'adaptation lumineux et de l'adaptation chromatiqûe *Ann. Psychol.* **40** 1–29

Pinckers A 1971 Combined Panel D15 and 100 Hue recording *Ophthalmologica* **163** 232–4

—— 1972a An analysis of colour vision in 314 patients *Mod. Probl. Ophthalmol.* **11** 94–7

—— 1972b Achromatopsie congenitale *Ann. Ocul.* **205** 821–34

—— 1972c The Farnsworth tritan plate *Ophthalmologica* **164** 137–42

—— 1975 Toxische neuropathie van de nervus opticus *Medikon* **4** 18–27

—— 1980a Colour vision and age *Ophthalmologica* **181** 23–30

—— 1980b Tokyo Medical College test in acquired dychromatopsia *Ophthalmologica* **181** 7–12

Pinckers A, Pokorny J, Smith V C and Verriest G 1979 in *Congenital and Acquired Colour Vision Defects* ed J Pokorny and V C Smith (London: Grune-Stratton) ch 4

Pitt F H G 1935 Characteristics of dichromatic vision *MRC Report* 14 (London: HMSO)

—— 1942 Colour blindness and its importance in relation to industry *Proc. Phys. Soc.* **54** 219

—— 1944a The nature of normal trichromatic and dichromatic vision *Proc. R. Soc.* B **132** 101–17

—— 1944b Monochromatism *Nature* **154** 466–8

—— 1949 Some aspects of anomalous vision *Docum. Ophthalmol.* **3** 307–17

—— 1961 Comments on Dean Farnsworth's paper 'Lets look at those isochromatic lines again' *Vision Res.* **1** 6–7

von Planta P 1928 Die Häufigkeit der angeborenen Farbensinns förungen bei Knaben Und Mädchen und ihre Festellung durg die üblichen klinischen Proben *Arch. Ophthalmol.* **120** 253

Platt J R 1943 A note on theoretical color-blindness frequencies *J. Opt. Soc. Am.* **33** 679

Plunkett E R 1976 *Handbook of Industrial Toxicology* (London: Heyden)

Poinoosawmy S, Nagasubramian S and Gloster J 1980 Colour vision in patients with chronic simple glaucoma and ocular hypertension *Br. J. Ophthalmol.* **64** 852–7

Pointer M R 1974 Colour discrimination as a function of observer adaptation *J. Opt. Soc. Am.* **64** 750–9

Pokorny J, Moreland J D and Smith V C 1975 Photopigments in anomalous trichromats *J. Opt. Soc. Am.* **65** 1522–4

Pokorny J and Smith V C 1977 Evaluation of single pigment shift model of anomalous trichromacy *J. Opt. Soc. Am.* **67** 1196–209

—— (eds) 1979 *Congenital and Acquired Colour Vision Defects* (London: Grune-Stratton)

—— 1981 A variant of red–green color defect *Vision Res.* **21** 311–18

Pokorny J, Smith V C and Ernst J T 1980 Macular color vision defects: specialized psychophysical testing in acquired and hereditary chorioretinal diseases *Int. Ophthalmol. Clin.* **20** 53–81

Pokorny J, Smith V C and Katz I 1973 Derivation of photopigment absorption spectra in anomalous trichromats *J. Opt. Soc. Am.* **63** 232–7

Pokorny J, Smith V C and Lund D 1978 Technical characteristics of 'color-test glasses' *Mod. Probl. Ophthalmol.* **19** 110–12

Polyak S L 1941 *The Retina* (Chicago: Chicago University Press)

—— 1957 *The Vertebrate Visual System* (Chicago: Chicago University Press)

Ponte F and Anastasi M 1978 Electroretinography as a diagnostic test in colour vision deficiencies *Mod. Probl. Ophthalmol.* **19** 29–32

Posada-Armiga 1885 cited by Jennings 1895

Post R H 1962 Population differences in red and green colour vision deficiencies: a view and a query on selection relaxation *Eugen. Q.* **9** 131–46

Pugh E N Jr 1976 The nature of the π_1 colour mechanism of W S Stiles *J. Physiol.* **257** 713–47

Pugh E N Jr and Siegel C 1978 Evaluation of the candidacy of the π mechanisms of Stiles for color-matching fundamentals *Vision Res.* **18** 317–30

Pulos E, Teller D Y and Buck S L 1980 Infant colour vision: A search for short wavelength sensitive mechanisms by means of chromatic adaptation *Vision Res.* **20** 485–93

Purdy D McL 1931a Spectral hue as a function of intensity *Am. J. Psychol.* **43** 541–59

—— 1931b *Am. J. Psychol.* **43** 541

—— 1937 The Bezold–Brücke phenomenon and contours for hue contrast *Am. J. Psychol.* **49** 313–15

Purkinje J 1825 Beobachtungen und Versuche zur Physiologie der Sinne. Neue Beiträge zur Kenntniss des Sehens in *Subjectiver Hinsicht* (Berlin: Reimer)

Rayleigh Lord 1881 Experiments on colour *Nature* **25** 64–6

Reddy V and Vijayalaxmi 1977 Colour vision in Vitamin A deficiency *Br. Med. J.* **1** 81

Regan D 1977 Evoked potential indications of the processing of pattern, colour and depth information *Visual Evoked Potentials in Man; New Developments* ed J Desmedt (Oxford: Clarendon) pp 234–46

Regan D and Spekreijse H 1974 Evoked potential indications of colour blindness *Vision Res.* **14** 89–95

Regan D and Tyler C W 1971 Wavelength-modulated light generator *Vision Res.* **11** 43–56

Reynolds R E, White R M and Hilgendorf R L 1972 Detection and recognition of colored signal lights *Hum. Factors* **14** 227–36

Richards O W 1975 Color vision—marked improvement age 10 to 14 *Opt. J. Rev. Optom.* **112** 7–12

Richards O W and Grolman B 1962 Avoid tinted contact lenses when driving at night *J. Am. Optom. Assoc.* **34** 53–5

Richards O W, Tack T O and Thome C 1971 Fluorescent lights for color vision testing *Am. J. Optom.* **48** 747–53

Richards W 1966 Opponent-process solutions for uniform Munsell spacing *J. Opt. Soc. Am.* **56** 1110–20

—— 1967 Differences among color normals; I and II *J. Opt. Soc. Am.* **57** 1047

Richards W and Luria S M 1968 Recovery and spectral sensitivity curves for color-anomalous observers *Vision Res.* **8** 929–38

Richer S and Adams A J 1984a Development of quantitative tools for filter-aided dichromats *Am. J. Optom. Physiol. Opt.* **61** 246–55

—— 1984b An experimental test of filter-aided dichromatic color discrimination *Am. J. Optom. Physiol. Opt.* **61** 256–64

Richter K 1980 Cube-root color spaces and chromatic adaptation *Color Res. Appl.* **5** 25–43

Richter M 1948 Vom gegenwärtigen Stand der Farbenlehre *Z. Wiss. Photogr.* **43** 209–37

—— 1950 Recent developments in the examination of the colour sense *Klin. Monatsbl. Augenheilkd.* **119** 561–75 (English transl. *Br. J. Physiol. Opt.* (1963) **10** 155–67)

—— 1954 Colour vision tests in the colour industries (in German) *Die Farbe* **3**(516) 175–192

von Ricklefs G and Wende D 1966 Can a colour-blind person become a Radio or T.V. technician? (in German) *Klin. Monatsbl. Augenheilkd.* **148** 148–50

Riddell W J B 1948 Discussion on colour vision in industry *Proc. R. Soc. Med.* **42** 145–50

Riggs L A and Sternheim C E 1969 Human retinal potentials evoked by changes in the wavelength of the stimulating light *J. Opt. Soc. Am.* **59** 635–40

Riggs L A and Wooten B R 1972 Electrical measures and psychophysical data on human vision *Handbook of Sensory Physiology* vol VII/4 (Berlin: Springer) pp 690–731

Rivers W H R 1901a The colour vision of the natives of Upper Egypt *J. R. Anthropol. Inst.* **31** 229

—— 1901b Colour vision *Rep. Camb. Anthropol. Exped. Torres Straits* **II**

—— 1905 Observations of the senses of the Todas *Br. J. Psychol.* **1** 326–39

Rivron B D 1959-60 Personal communications

Roberts C 1884 *The Detection of Colour Blindness and Imperfect Eyesight* 2nd edn (London: Churchill)

Roberts D F 1967 Colour blindness in the Niger delta *Eugen. Q.* **14** 7–13

Robinson C C 1967 Colour in traffic control *Traffic Eng.* **37** 25–9

Rogers E 1982 A cure for the colorblind *Pence Happiness and Prosperity Mag.* **13** 12–25

Romeskie M 1976 Chromatic opponent-response functions of anomalous trichromats *PhD Thesis* Brown University, Providence RI, USA

—— 1978a Chromatic opponent-response functions of anomalous trichromats *Vision Res.* **18** 1521–32

—— 1978b Opponent-colours theory and colour blindness *Frontiers in Visual Science* (New York: Springer) pp 178–87

Romeskie M and Yager D 1978 Psychophysical measures and theoretical analysis of dichromatic opponent-responce functions *Mod. Probl. Ophthalmol.* **19** 212–17

Ronchi L, Barca L and Vaccari G 1978 *Personal communication*

Rosencrantz C 1926 Über die Unterschiedsempfindlichkeit für Farbentöne bei anomalen Trichromaten *Z. Psychol.* **58** 5–27

Rosenstock H B and Swick D A 1974 Color discrimination for the color blind *Aerosp. Med.* **145** 1194–7

Roth A 1966a Etude clinique du sens chromatique 'central' dans l'amblyopie fonctionnelle *Docum. Ophthalmol.* **20** 631–5

—— 1966b Le test 28 hue selon Farnsworth *Bull Soc. Ophthalmol.* **66** 321

—— 1968 Le sens chromatique dans l'amblyopie fonctionnelle *Docum. Ophthalmol.* **24** 113–200

Roth A, Renaud J C and Vienot J C 1978 Advances in the realization of a direct observation anomaloscope *Mod. Probl. Ophthalmol.* **19** 173–6

Royal Horticultural Society 1966 *R.H.S. Colour Chart* (London: Royal Horticultural Society)

Royal Society 1892 *Report of the Committee on Colour-vision Appointed by the Council of the Royal Society on March 20th 1890* (London: Royal Society)

Rubin M L 1961 Spectral hue loci of normal and anomalous trichromates *Am. J. Ophthalmol.* **52** 166–72

Ruddock K H 1963 Evidence for macular pigmentation from colour matching data *Vision Res.* **3** 417–29

—— 1964 The effect of age upon colour matching and colour discrimination *PhD Thesis* London University

—— 1965a The effect of age upon colour vision 1. Response in the receptoral system of the human eye *Vision Res.* **5** 37–45

—— 1965b The effect of age upon colour vision II. Changes with age in light transmission of the ocular media *Vision Res.* **5** 47–58

—— 1966 Integration processes in colour vision *Die Farbe* **15** 63–72

—— 1969 Colour vision under small field conditions *Optica Acta* **16** (3) 391–8

—— 1971 Parafoveal colour vision responses of four dichromats *Vision Res.* **11** 143–56

Ruddock K H and Bender B G 1972 On an observer with anomalous colour vision *Mod. Probl. Ophthalmol.* **11** 199–204

Ruddock K H and Burton G 1972 The organisation of human colour vision at the central fovea *Vision Res.* **12** 1763–9

Ruddock K H and Naghshineh S 1974 Mechanisms of red-green anomalous trichromacy: hypothetical analysis *Mod. Probl. Ophthalmol.* **13** 210–14

Rushton W A H 1952 Apparatus for analysing the light reflected from the eye of the cat *J. Physiol.* **117** 47–8

—— 1955 Foveal photopigments in the normal and colour blind *J. Physiol.* **129** 41–2

—— 1956 The rhodopsin density in the human rods *J. Physiol.* **134** 30–46

—— 1958 Visual pigments in the colour blind *Nature* **182** 690–2

—— 1963a A cone pigment in the protanope *J. Physiol.* **168** 345–59

—— 1963b The density of chlorolabe in the foveal cones of the protanope *J. Physiol.* **168** 360–73

—— 1963c Cone pigment kinetics in the protanope *J. Physiol.* **168** 374–88

—— 1964 Color blindness and cone pigments *Am. J. Optom.* **41** 265–82

—— 1965a A foveal pigments in the deuteranope *J. Physiol.* **176** 24–37

—— 1965b Cone pigment kinetics in the deuteranope *J. Physiol.* **176** 38–45

—— 1965c Stray light and the measurement of mixed pigments in the retina *J. Physiol.* **176** 46–55

—— 1970 Pigments in anomalous colour vision *Br. Med. Bull.* **26** 179–81

—— 1972a Pigments and signals in colour vision *J. Physiol.* **220** 1–31

—— 1972b Visual pigments in man in *Handbook of Sensory Physiology* vol VII/1 (Berlin: Springer)

Rushton W A H and Baker H D 1964 Red/green sensitivity in normal vision *Vision Res.* **4** 75–85

Rushton W A H, Campbell F W, Hagins W and Brindley G S 1953 The bleaching regeneration of rhodopsin in the living eye of the albino rabbit and of man *Optica Acta* **1** 182–90

Rushton W A H, Powell D S and White K D 1973a Anomalous pigments in the eyes of the red-green colour blind *Nature* **243** 167–8

—— 1973b Exchange thresholds in dichromats *Vision Res.* **13** (ii) 1993–2002

—— 1973c The spectral sensitivity of 'red' and 'green' cones in the normal eye *Vision Res.* **13** 2003–15

—— 1973d Pigments in anomalous trichromats *Vision Res.* **13** 2017–31

Ryabushkina Z P 1970 Eye lesions in the hydrolytic industry *Vest. Oftalmol.* **3** 76

Sachs E 1929 Die Aktionsstrome des menschlichen Auges, ihre Beziehung zu Reiz und Empfindung *Klin. Wochenschr.* **8** 136–7

Sachsenweger R and Nothaas E 1961 Eine Analyse von 4011 verkehrsunfällen aus augenärztlicker Sicht *Dtsch. Gesund.* **16** 868–72

Safir A, Hyams L and Philpot J 1971 The retinal direction effect: a model based on the Gaussian distribution of cone receptor orientations *Vision Res.* **11** 819–33

Said F S and Weale R A 1959 The variation with age of the spectral transmissivity of the living human crystalline lens *Gerontologia* **3** 213–31

Sakuma Y 1973 Studies on colour vision anomalies in subjects with alcoholism *Ann. Ophthalmol.* **5** 1277–92

Salvia J A and Shugerts J 1970 Colour related behaviour of mentally retarded children with colour blindness and normal colour vision *Except. Child.* **37** 37–8

Salvia J A and Ysseldyke J E 1971 Validity and reliability of the red–green HRR P.I.C. plates with mentally retarded children *Percept. Mot. Skills* **33** 1071–4

—— 1972 Criterion validity of four tests for red–green colour blindness *Am. J. Ment. Def.* **76** 418–22

Saraux H, Labret R and Biais B 1966 Aspects actuels de la névrite optique de l'ethylique *Ann. Ocul.* **199** 943

Sarwar M 1961 Colour vision and defects in the posterior chamber *Br. J. Physiol. Opt.* **18** 171–3

Sassoon R F and Wise J B 1970 Diagnosis of colour vision defects in very young children *The Lancet* **ii** 419–20

Sassoon R F, Wise J B and Watson J J 1970 Alcoholism and colour vision; are there familial links? *The Lancet* **ii** 367

Sastri V D P and Das S R 1968 Typical spectral distributions and color for tropical daylight *J. Opt. Soc. Am.* **58** 391–8

Saunders J E 1976 A red-green anomaloscope using light emitting diodes *Vision Res.* **16** 871–4

Sax N I 1975 *Dangerous Properties of Industrial Materials* (New York: Van Nostrand Reinhold)

Schade O H 1956 Optical and photoelectric analog of the eye *J. Opt. Soc. Am.* **46** 721–39

Schaller M J 1975 Chromatic vision in human infants; conditioned operant fixation to 'hues' of varying intensity *Bull. Psychonom. Soc.* **6** 39–42

Scheibner H 1973 Colour matcher of anomalous trichromats *Mod. Probl. Ophthalmol.* **13** 190–5

Scheibner H and Boynton R M 1968 Residual red-green discrimination in dichromats *J. Opt. Soc. Am.* **58** 1151–8

Scheibner H and Thranberend C 1974 Colour vision in a case of neuritis retrobulbaris *Mod. Probl. Opthalmol.* **13** 329–44

Schein J D and Salvia J A 1969 Colour blindness in mentally retarded children *Except. Child.* **35** 609–13

Schein S J, Marrocco R T and de Monasterio F M 1982 Is there a high concentration of colour-selective cells in Area V4 of Monkey visual cortex? *J. Neurophysiol.* **47** 193–213

Schepens C L 1946 Is tobacco amblyopia a deficiency disease? *Trans. Ophthalmol. Soc. UK* **66** 309–31

Schiøtz H 1925 Blink-farvelykten *Nor. Mag. Laegevidensk. (Saertr. av)* June 549–52

Schlaegel T F 1977 *Ocular Histoplasmosis* (New York: Grune and Stratton)

Schläfer R and Jaeger W 1956 Demonstration eines Projektionsanomaloscopes *Ber. Dtsch. Ophthalmol. Ges.* **LX** 288–91

Schmidt I 1934 Ueber manifeste Heterozygotie bei konductorinen für Farbenninnstörungen *Klin. Monatsbl. Augenheilkd.* **92** 456–67

—— 1936 Ergebnis einer Massenunters chung des Farbensinns mit dem Anomaloskop *Z. Bahnharzt* **2**

—— 1952 Effect of illumination in testing colour vision with P.I.C. plates *J. Opt. Soc. Am.* **42** 951–5

—— 1955a A sign of manifest heterozygosity in carriers of color deficiency *Am. J. Optom.* **32** 404–8

—— 1955b Some problems related to testing colour vision with the Nagel anomaloscope *J. Opt. Soc. Am.* **45** 514–22

—— 1970 On congenital tritanomaly *Vision Res.* **10** 717–43

—— 1972 Comments on the X-Chrom lens *J. Am. Optom. Assoc.* **43** 199–201

—— 1973 On acquired colour deficiencies *Optom. Wkly.* June 24–8

—— 1974 Brightness matches on the Nagel anomaloscope *Mod. Probl. Ophthalmol.* **13** 19–25

—— 1976 Visual aids for correction of red–green colour deficiencies *Can. J. Optom.* **38** 38–47 (reprinted 1977 *The Optician* **173** 7–35)

Schmidt I and Bingel A G A 1953 Effects of oxygen deficiency and various other factors on color saturation thresholds *U.S.A.F. Sch. Avigt. Med. Project Rep.* 21–31–002

Schmidt I and Fleck H 1952 Rotator and illuminator for pseudoisochromatic plates *Arch. Ophthalmol.* **48** 75–82

Schneider P 1968 Perte de vision par traumatisme crânien fermé *Ophthalmologica* **156** 377–84

Schofield R K 1939 The Lovibond tintometer adapted *J. Sci. Instrum.* **XVI** 74–80

Schrödinger E 1920 Grundlinien einer Theorie der Farbenmetrik im Tagesschen *Ann. Phys., Lpz.* **63** 481

Schultz M 1866 Zur Anatomie und Physiologie der Retina *Arch. Mikrosk. Anat. Entwicklungsmech.* **2** 175–286

Schuster A 1890 Experiments with Lord Rayleigh's colour box *Proc. R. Soc.* **48** 140–9

Schuurmans R P and Zrenner E 1981 Responses of the blue sensitive cone system from the visual cortex and the arterially perfused eye in cat and monkey *Vision Res.* **21** 1611–15

The Science of Colour 1953 (Washington, DC: Optical Society of America)

Scott J and Whisson 1778 An account of a remarkable imperfection of sight *Phil. Trans. R. Soc.* **68** 611–14

Scott J G 1941 Hereditary optic atrophy with dominant transmission and early onset *Br. J. Ophthalmol.* **25** 461–79

Scott P and Williams K G 1958 The latency of visual perception in man to very brief coloured flashes of light *Report* 50/; 15 (1958) (Weybridge, Surrey: Vickers Group Research Establishment)

Seebeck A 1837 Ueber den bei manchen Personen vorkommenden Mangel an Farbensinn *Pogg. Ann. Phys. Chem.* **42** 177–233

Serra A and Mascia C 1978 Color discrimination of 57 congenital color defectives under different illuminants *Attiv. Fond. G. Ronchi* **6** 895–906

Shaxby J H 1944 A simple form of the Nagel anomaloscope *J. Sci. Instrum.* **22** 15–16

Sherman P D 1981 *Colour Vision in the Nineteenth Century* (Bristol: Adam Hilger)

Shipley T, Jones R W and Fry A 1968 *Vision Res.* **8** 409–31

Shirley A W 1966 A new design for a colorimeter *Br. J. Physiol. Opt.* **23** 254–7

Shlaer S, Smith E L and Chase A M 1942 Visual acuity and illumination in different spectral regions *J. Gen. Physiol.* **25** 553

Siegel I M and Smith B F 1961 Acquired cone dysfunction *Arch. Ophthalmol.* **77** 8–13

Siegel M H 1965 Colour discrimination as a function of exposure time *J. Opt. Soc. Am.* **55** 566–8

Siegel M H and Dimmick F L 1962a The discrimination of color: Comparison of 3 psychophysical methods and sensitivity as a function of spectral wavelength 510 to 630 nm *Report* 389 (New London, CT: US Naval Medical Research Lab)

Siegel M H and Dimmick M H 1962b Discrimination of colour II sensitivity as a function of spectral wavelength 510–630 nm *J. Opt. Soc Am.* **52** 1017–74

Siegel M H and Siegel A 1972 Hue discrimination as a function of stimulus luminance *Percept. Psychophys.* **12** 295–9

Sjöstrand F S 1965 The synaptology of the retina *CIBA Found. Symp. on Colour Vision* (London: Churchill) pp 110–44

Skeller E 1954 *Anthropological and Ophthalmological Studies on the Angmagsalik Eskimos* (Copenhagen)

Sloan L L 1942 The use of P.I.C. charts in detecting central scotomas due to lesions in the conducting paths *Am. J. Ophthalmol.* **25** 1352–6

—— 1944 A quantitative test for measuring degree of red–green color deficiency *Arch. Ophthalmol.* **27** 941–7

—— 1945 Improved screening test for red–green color deficiency composed of available P.I.C plates *J. Opt. Soc. Am.* **35** 761–6

—— 1947 Rate of dark adaptation and regional threshold gradient of dark adapted eye: physiologic and clinical studies *Am. J. Ophthalmol.* **30** 705–20

—— 1950 Comparison of the Nagel anomaloscope and a dichroic filter anomaloscope *J. Opt. Soc. Am.* **40** 41–7

—— 1954 Congenital achromatopsia: a report of nineteen cases *J. Opt. Soc. Am.* **44** 117–28

—— 1958 The photopic retinal receptors of the typical achromat *Am. J. Ophthalmol.* **46** 81–6

—— 1961 Evaluation of the T.M.C color vision test *Am. J. Ophthalmol.* **52** 650–9

—— 1963 Testing for deficient colour perception in children *Int. Ophthalmol. Clin.* **3**(4) 697–705

Sloan L L and Altman A 1951 Evaluation of the HRR plates for measuring degree of red–green deficiency *Wilmer Ophthalmol. Inst. Tech. Rep. Proj.* N60NR 24307

Sloan L L and Feiock K 1972 Selective impairment of cone vision *Mod. Probl. Ophthalmol.* **11** 50–62

Sloan L L and Gilger A P 1947 Visual effects of tridione *Am. J. Ophthalmol.* **30** 1387–405

Sloan L L and Habel A 1955a Color signal systems for the red–green color blind. An experimental test of the 3-color signal system proposed by Judd *J. Opt. Soc. Am.* **45** 592–8

—— 1955b Recognition of red and green point sources by color-deficient observers *J. Opt. Soc. Am.* **45** 599–601

—— 1956 Tests for color deficiency based on the P.I.C principle—a comparative study *Arch. Ophthalmol.* **55** 229–39

Sloan L L and Newhall S M 1942 Comparison of cases of atypical and typical achromatopsia *Am. J. Ophthalmol.* **25** 945–61

Sloan L L and Wollach L 1948a Comparison of tests for red–green color deficiency *J. Aviat. Med.* **19** 447–55

—— 1948b A case of unilateral deuteranopia *J. Opt. Soc. Am.* **38** 502–9

Smith D P 1971 Derivation of wavelength discrimination from colour naming data *Vision Res.* **11** 739–42

—— 1972a Diagnostic criteria in dominantly inherited juvenile optic atrophy *Am. J. Optom.* **49** 183–200

—— 1972b The assessment of acquired dyschromatopsia and clinical investigation of the acquired tritan defect in dominantly inherited juvenile atrophy *Am. J. Optom.* **49** 574–88

—— 1973 Color naming and hue discrimination in congenital tritanopia and tritanomaly *Vision Res.* **13** 209–18

Smith D P, Cole B L and Isaacs A 1973c Congenital tritanopia without neuroretinal disease *Invest. Ophthalmol.* **12** 608–17

Smith J W 1971 Colour vision in alcoholics *Am. Nat. Council on Alcoholism Conf., Anaheim, California, April 1971* (unpublished)

Smith S 1962 Colour coding and visual search *J. Exp. Psychol.* **64** 434–40

Smith V C, Burns S A and Pokorny J 1982 Colorimetric evaluation of urine-sugar tests used by diabetic patients *Docum. Ophthalmol. Proc.* Series 33 345–54

Smith V C and Pokorny J 1972 Spectral sensitivity of color-blind observers and the cone photopigments *Vision Res.* **12** 2059–71

—— 1973 Letter to editor: Autosomal dominant tritanopia *Invest. Ophthalmol.* **13** 706–7

—— 1975 Spectral sensitivity of the foveal cone photopigments between 400 and 500 nm *Vision Res.* **15** 161–72

—— 1977 Large field trichromacy in protanopes and deuteranopes *J. Opt. Soc. Am.* **67** 213–20

—— 1980 Cone dysfunction syndromes defined by colour vision *Colour Vision Deficiencies V* ed G Verriest (Bristol: Adam Hilger) pp 69–82

Smith V C, Pokorny J and Diddie K R 1978b Colour matching and Stiles-Crawford Effect in central serous choroidopathy *Mod. Probl. Ophthalmol.* **19** 284–95

Smith V C, Pokorny J and Newell F W 1978a Autosomal recessive incomplete achromatopsia with protan luminosity *Ophthalmologica* **177** 197–207

—— 1979 Autosomal recessive incomplete achromatopsia with deutan luminosity *Am. J. Ophthalmol.* **87** 393–402

Smith V C, Pokorny J and Starr S J 1976 Variability of colour mixture data. I. Interobserver variability in the unit coordinates *Vision Res.* **16** 1087–94

Smith V C, Pokorny J and Swartley R 1973b Continuous hue estimation of brief flashes by deuteranomalous observers *Am. J. Psychol.* **86** 115–31

Smith V C, Pokorny J and Zaidi Q 1982 Evaluation of colour matching functions *International Conference on Colour Vision, Cambridge, UK, August 23–27 1982* (Proceedings published by Academic Press 1983)

Sokol S 1978 Measurement of infant visual acuity from pattern reversal evoked potentials *Vision Res.* **18** 33–9

Solant D Y and Best C H 1943 The Royal Canadian Navy colour vision test lantern *Can. Med. Assoc. J.* **48** 18–21

Sorsby A 1970 *Ophthalmic Genetics* 2nd edn (London: Butterworths)

Spencer J A 1967 An investigation of Maxwell's spot *Br. J. Physiol. Opt.* **24** 103–47

Sperling H G 1973 Comparison of isolated red and green sensitive cone spectra obtained by microspectrophotometry and by psychophysical thresholds *Colour 73* (London: Adam Hilger) pp 261–2

—— 1980 Blue receptor distribution in primates from intense light and histochemical studies *Colour Vision Deficiencies V* ed G Verriest (Bristol: Adam Hilger) pp 30–44

Sperling H G and Harwerth R S 1971 Red–green cone interactions in the increment-threshold spectral sensitivity of primates *Science* **172** 180–4

Sperling H G and Hsia Y 1957 Some comparisons among spectral sensitivity data obtained in different retinal locations and with two sizes of foveal stimulus *J. Opt. Soc. Am.* **47** 707–13

Spivey B E 1965 The X-linked recessive inheritance of atypical monochromatism *Arch. Ophthalmol.* **74** 327–33

Spivey B E, Pearlman J F and Burian H M 1964 Electroretinographic findings (including flicker) in carriers of congenital X-linked achromatopsia *Docum. Ophthalmol.* **18** 367–75

Stabell B and Stabell U 1976 Rod and cone contributions to peripheral colour vision *Vision Res.* **16** 1099–104

—— 1977 The chromaticity coordinates for spectrum colour of extrafoveal cones *Vision Res.* **17** 1091–4

—— 1979 Rod and cone contributions to change in the hue with eccentricity *Vision Res.* **19** 1121–5

—— 1980a Spectral sensitivity in the far perception retina *J. Opt. Soc. Am.* **70** 959–63

—— 1981 Absolute spectral sensitivity at different eccentricities *J. Opt. Soc. Am.* **71** 836–40

Stabell U and Stabell B 1975 The effect of rod activity on colour matching functions *Vision Res.* **15** 1119–23

—— 1980b Variation in density of macular pigmentation and in short-wave cone sensitivity with eccentricity *J. Opt. Soc. Am.* **70** 706–11

Starr S J 1977 Effect of luminance and wavelength on the Stiles–Crawford effect in dichromats *PhD Thesis* University of Chicago, USA

Starr S J and Fishman G 1982 Visual function in a carrier of blue cone mono-chromacy *ARVO Meeting, Abstracts Document* (St Louis: Mosby)

Steen J A and Lewis M F 1972 Color defective vision and day and night recognition of aviation color signal light flashes *Aerosp. Med.* **43** 34–6

Stephenson S 1898 *Trans. Ophthalmol. Soc. UK* **18** 55

Stiles W S 1937 The luminous efficiency of monochromatic rays entering the eye pupil at different points and a new colour effect *Proc. R. Soc.* B **123** 90–118

—— 1946a A modified Helmholtz line-element in brightness-colour space *Proc. Phys. Soc.* **58** 41–65

—— 1946b Separation of the 'blue' and 'green' mechanisms of foveal vision by measurements of increment thresholds. *Proc. R. Soc.* B **133** 418–34

—— 1949a Increment thresholds and the mechanisms of colour vision *Docum. Ophthalmol.* **3** 138–63

—— 1949b Investigations of the scotopic and trichromatic mechanisms of vision by the two-colour threshold technique *Rev. Opt.* **28** 215–37

—— 1953 Further studies of visual mechanisms by the two-colour threshold technique *Colloq. Probl. Opt. Vision* **I** 65–103 (Madrid: Barmejo)

—— 1958 The average colour matching functions for a large matching field *Visual Probl. Colour* **1** 201 (London: HMSO)

—— 1959 Colour vision: the approach through increment threshold sensitivity *Proc. Natl. Acad. Sci. USA* **45** 100–14

—— 1978 *Mechanisms of Colour Vision* (London: Academic)

Stiles W S and Burch J M 1955 Interim Report to the Commission International de l'Éclairage, Zurich, 1955, on the National Physical Laboratory's investigation of colour-matching *Optica Acta* **2** 168

—— 1959 N.P.L. colour-matching investigation: Final report *Optica Acta* **6** 1–26

Stiles W S and Crawford B H 1933a The luminous efficiency of rays entering the eye pupil at different points *Proc. R. Soc.* B **112** 428–50

—— 1933b The liminal brightness increment as a function of wavelength for different conditions of the foveal and parafoveal retina *Proc. R. Soc.* B **113** 496–530

Stilling J 1909 *Pseudo-isochromatische Tafeln zu Prüfung des Farbensinnes* (Leipzig: Thieme)

Stocker H, Wolf A and Scheibner H 1979 Erythrolabe and deuteranomaly *Dtsch. Ophthalmol. Ges.* 409–14

Stonebridge E H 1968 Practical help for colour defectives *Ophthal. Opt.* **68** 62

Svaetichin G 1953 The cone action potential *Acta Physiol. Scand.* **29** Suppl. 106 565–99

—— 1956 Spectral response curves from single cones *Acta. Physiol. Scand.* **39** Suppl. 134 17–46

Svaetichin G, Negishi K and Fatehchand R 1965 Cellular mechanisms of a Young-Hering visual system *CIBA Found. Symp. on Colour Vision* (London: Churchill) pp 178–203

Swinson R P 1972 Colour vision defects in alcoholism *Br. J. Physiol. Opt.* **27** 43–50

Szentagothai J 1973 Synaptology of the visual cortex *Handbook of Sensory Physiology* vol VII/3 Central Processing of Visual Information part B ed B Jung (Berlin: Springer)

Takahama K and Sobagaki H 1981 Formulation of a nonlinear model of chromatic adaptation *Color Res. Appl.* **6** 161–71

Talbot S A and Marshall W H 1941 Physiological studies on neural mechanism of visual localization and discrimination *Am. J. Ophthalmol.* **24** 1255–64

Tann J 1967 *Gloucestershire Woollen Mills* (Newton Abbot, Devon: David and Charles)

Tansley B W and Glushko R J 1978 Spectral sensitivity of long-wavelength-sensitive photoreceptors in dichromats determined by elimination of border percepts *Vision Res.* **18** 699–706

Tansley B W, Robertson A W and Maughan K E 1982 Chromatic and achromatic border perception *International Conference on Colour Vision, Cambridge, UK, August 23–27 1982* (Proceedings published by Academic Press 1983)

Tate G W and Lynn J R 1977 *Principles of Quantitative Perimetry: Testing and Interpreting the Visual Field* (New York: Grune and Stratton)

Taylor S P 1982 The X-chrom lens—a case study *Ophthalmol. Physiol. Opt.* **2** 165–70

Taylor W O G 1970 Screening red–green blindness *Ann. Ophthalmol.* **2** 184–92

—— 1971 Effects on employment of defects in colour vision *Br. J. Ophthalmol.* **55** 753–60

—— 1972 Achromatism, an unlooked-for hazard with urine self-testing in diabetes *Trans. Ophthalmol. Soc. UK* **92** 95–9

—— 1974 Problems in performance and interpretation of Farnsworth's 100 Hue test *Mod. Probl. Ophthalmol.* **13** 73–8

—— 1975a Practical problems of defective colour vision *Practitioner* **214** 654

—— 1975b Constructing your own P.I.C. test *Br. J. Physiol. Opt.* **30** 22–4

—— 1977 Of divers colours *Trans. Ophthalmol. Soc. UK* **97** 768–80

—— 1978 Visual disabilities of oculocutaneous albinism and their alleviation *Trans. Ophthalmol. Soc. UK* **98** 423–45

Taylor W O G and Donaldson G B 1976 Recent developments in Farnsworth's colour vision tests *Trans. Ophthalmol. Soc. UK* **96** 262–4

Teller D, Peeples D R and Sekel M 1978 Discrimination of chromatic from white by two month old human infants *Vision Res.* **18** 41–8

Terstiege H 1980 Colours of transilluminated traffic signs *Light. Res. Technol.* **12** 69–73

Thomas D L (undated *c*. 1975) Personal communication from Western Regional Hospital Board, Glasgow

Thompson D G *et al* 1979 Defective colour vision in diabetes: a hazard to management *Br. Med. J.* **1** 859–60

Thompson W 1894 A new wool test for the detection of color blindness *Med. News* Aug. 18

Thomson L C 1954 Sensations aroused by monochromatic stimuli and their prediction *Optica Acta* **1** 93–102

Thomson L C and Trezona P W 1951 The variation of hue discrimination with change of luminance level *J. Physiol.* **114** 98–106

Thomson L C and Wright W D 1947 The colour sensitivity of the retina within the central fovea of man *J. Physiol.* **105** 316–31

—— 1953 The convergence of the tritanopia confusion loci and the derivation of the fundamental response functions *J. Opt. Soc. Am.* **43** 890–4

Thornton W A 1974 Colour discrimination enhancement by the illuminant *Mod. Probl. Ophthalmol.* **13** 313–14

Thuline H C 1967 Inheritance of alcoholism *The Lancet* **i** 274

—— 1972 Colour blindness in children; the importance and feasibility of early recognition *Clin. Pediat.* **11** 295–9

Tiffin J and Kuhn N S 1942 Colour discrimination in industry *Arch. Ophthalmol.* **28** 951–9

Tomita T 1970 Electrical activity of vertebrate photoreceptors *Q. Rev. Biophys.* **3** 179–222

Topley H 1959 Sight testing for the Merchant Navy *Br. J. Physiol. Opt.* **16** 36–46

Towbin E J, Pickens W S and Doherty J E 1967 The effects of D.O. upon colour vision and the ERG *Clin. Res.* **15** 60

Toyoda J, Kujiraoka T and Fujimoto M 1982 The role of L-type horizontal cells in the opponent-color processes *Color Res. Appl.* **7** 152–4

Traquair H M 1948 *Introduction to Clinical Perimetry* (St Louis: Mosby)

Tredici T J, Mims J L and Culver J F 1972 History, rationale and verification of colour vision standards and testing in the U.S. Air Force *AGARD Conference Proceedings 99* (Neuilly sur Seine: AGARD)

Trendelenburg W and Meitner S H J 1941 *Klin. Monatsbl. Augenheilkd.* **104** 12–19

Trezona P W 1970 Rod participation in the 'blue' mechanism and its effect on colour matching *Vision Res.* **10** 317–32

—— 1976 Aspects of peripheral colour vision *Mod. Probl. Ophthalmol.* **17** 52–70

Trick G, Guth S L and Massof R 1976 Wavelength discrimination in protanopes and deuteranopes *Mod. Probl. Ophthalmol.* **17** 17–20

Troxler D 1804 Über das Verschwinden gegebener Gegenstände innerhalb unseres Gesichtskreises *Ophthalmol. Bibl.* **II** ed Himley and Schmidt (Jena) pp 51–3

Trusov M S 1972 About disturbances in colour vision in hypertonic disease and artherosclerosis *Oftalmol. Zh.* **1** 19

Tuberville D 1684 Some remarkable cases in physick relating chiefly to the eyes *Phil. Trans. R. Soc.* **14** 736 van der Tweel L H and Sperkreijse H 1973 Psychophysics and electrophysiology of a rod achromat *Proc. 10th ISCERG Symp.* (Junk: The Hague)

Uchikawa K and Ikeda M 1981 Temporal deterioration of wavelength discrimination with successive comparison method *Vision Res.* **21** 591–5

Ugarte G, Cruz-Coke R, Rivera L, Altschiller H and Mardones J 1970 Relationships of colour blindness to alcoholic liver damage *Pharmacology* **4** 308

Umazume K 1957 *Tokyo Medical College Plates* (Tokyo: Murakami Laboratories)

Umazume K and Matsuo H 1962 Tokyo Medical College test for colour vision *Die Farbe* **11** 1/6 45–7

Valberg A and Tansley B W 1977 Tritanopic purity-difference function to describe the properties of minimally distinct borders *J. Opt. Soc. Am.* **67** 1330–5

de Valois R L 1965 Analysis and coding of colour vision in the primate visual system *Cold Spring Harbour Symp. Quant. Biol.* **30** 567–9

de Valois R L, Abramov I and Jacobs G H 1966 Analysis of response patterns of lateral geniculate nucleus cells *J. Opt. Soc. Am.* **56** 966–77

de Valois R L, Abramov I and Mead W R 1967 Single cell analysis of wavelength discrimination at the lateral geniculate nucleus in the macaque *J. Neurophysiol.* **30** 415–33

Vanderdonck R and Verriest G 1960 Femme protanomale et hétérozygote mixte ayant deux fils déuteranopes *Biotypologie* **21** 110–20

Varner D 1981 Temporal characteristics of colour vision *PhD Dissertation* University of Pennyslvania

Varner D, Piantanida T P and Baker H D 1977 Spatio-temporal Rayleigh matches *Vision Res.* **17** 187–91

Velhagen K 1974 *Tafeln zur Prüfung des Farbensinnes* (Stuttgart: Thieme)

—— 1980 *Pflügertrident-Plates for Testing the Sense of Colour* (Pflügerhakens-Tafeln zur Prüfung des Farbensinnes) (Leipzig: VEB, Georg Thieme) (Obtainable from: V E B Gustav Fischer Verlag, Villengang 2, 69 Jena DDR, or Erich Bieber, Wilhemstrasse 4, 7000 Stuttgart BRD)

Verduyn-Lunel H F E and Crone R A 1974 Static perimetry with purely chromatic stimuli *Mod. Probl. Ophthalmol.* **13** 103–8

Verin P, Besme D, Yacoubi M and Moray S 1971 Toxicité oculaire de l'ethambutol *Arch. Ophthalmol.* **31** 669–86

Vernon P E and Straker R A 1943 Distribution of colour blindness in Great Britain *Nature* **152** 690

Verrey A 1926 Variation du sens chromatique dans les dyschromatopsias acquises *Arch. Ophthalmol.* **43** 612–23

Verriest G 1960a in discussion (p.949 of Kleefeld MG 1960)

—— 1960b (ed) *A Study of Achromatic Visual Functions in the Congenital Sensory Abnormalities of the Human Eye and in some Amphibians and Reptiles* (The Hague: Junk)

—— 1963 Further studies on acquired deficiency of colour discrimination *J. Opt. Soc. Am.* **53** 185–95

—— 1964 Les déficiencies acquises de la discrimination chromatique *Mem. Acad. R. Med. Belg.* **II/IV** 5

—— 1968 Etude comparative des efficiencies de quelques tests pour la reconnaissance des anomalies de la vision des couleurs *Arch. Malad. Prof. Med. Trav.* **29** 293–314

—— 1970 The spectral curve of relative luminous efficiency in acquired colour vision deficiency *Proc. Int. Colour Meet. Colour '69* vol 1 (Göttingen: Musterschmidt) pp 115–30

—— 1971 Les courbes spectrales photopiques d'efficacité lumineuse relative dans les deficiences congénitales de la vision de coleurs *Vision Res.* **11** 1407–34

—— 1974a The spectrum curve of relative luminous efficiency in different age groups of aphakic eyes *Mod. Probl. Ophthalmol.* **13** 314–17

—— 1974b Recent advances in the study of the acquired deficiencies of colour vision *Attiv. Fond. G. Ronchi* **XXIV** 1–80

—— 1981 Color vision test in children *Attiv. Fond. G. Ronchi* **XXXVI** 83–119

Verriest G and Bozzoni F 1965 Comparative value of the best known tests for the diagnosis of dyschromatopia *Attiv. Congr. Soc. Oftalmol. Ital.* **XLIX** 92–100

Verriest G, Buyssens A and Vanderdonck R 1963 Etude quantitative de l'effect q'exerce sur les resultats de quelques tests de la discrimination chromatique une diminution non selective du riveau d'un eclairage *C. Rev. Opt.* **42** 105–19

Verriest G and Caluwaerts M R 1978 An evaluation of three new colour vision tests *Mod. Probl. Ophthalmol.* **19** 131–5

Verriest G, Francq P and Piérart P 1980c Results of colour vision tests in alcoholic and in mentally disordered subjects *Ophthalmologica* **180** 247–56

Verriest G, Gandibleux M F and Piérart P 1981b Colour vision tests in children. III. Results of the Farnsworth–Munsell 100 hue test and of the Burnham–Clark colour memory test in colour normal children from 6 to 13 years of age *Attiv. Fond. G. Ronchi* **36** 106

Verriest G and Israel A 1965 Application du périmètre statique de Goldmann au relevé topographique des seuils différentiels de luminance pour de petits objects colorés projetés sur un fond blanc *Vision Res.* **5** 151–74, 341–59

Verriest G, van Laethem J and Uvijls A 1982 A new assessment of the normal ranges of the Farnsworth–Munsell 100 Hue test scores *Am. Ophthalmol.* **93** 634–42 (See also *Docum. Ophthal. Proc.* Series 33 199–208)

Verriest G and Metsala P 1963 Résultats en vision latérale de quelques tests de la discrimination chromatique maculaire *Rev. Opt.* **42** 391–400

Verriest G, Neubauer O, Marré M and Uvijls A 1980a New investigations into the relationships between congenital colour vision defects and road traffic safety *Colour Vision Deficiencies V* ed G Verriest (Bristol: Adam Hilger) pp 331–42

—— 1980b New investigations concerning the relationship between congenital colour vision defect and road traffic security *Int. Ophthalmol.* **2** 87–99

Verriest G, Padmos P and Greve E L 1974 Calibration of the Tübingen perimeter for colour perimetry *Mod. Probl. Ophthalmol.* **13** 109–12

Verriest G and Popescu M P 1974 'Umstimmung' of normal subjects at the Nagel anomaloscope *Mod. Probl. Ophthalmol.* **13** 26–30

Verriest G and Uvijls A 1977a Spectral increment thresholds on a white background in different age groups of normal subjects and in acquired ocular diseases *Docum. Ophthalmol.* **43** 217–48

—— 1977b Central and peripheral increment thresholds for white and spectral lights on a white background in different kinds of congenital defective colour vision *Attiv. Fond. G. Ronchi* **32** 213–54

—— 1978 Central and peripheral spectral increment thresholds on white backgrounds in acquired ocular diseases *Mod. Probl. Ophthalmol.* **19** 253

Verriest G, Uvijls A Gandibleux M F, Pierart O, Malfroidt A and de Coninck M R 1981a Colour vision tests in children *6th Symp. IRGCVD, Berlin, 1981*

Verriest G, Vandevyvere R and Vanderdonck R 1962 Nouvelles recherches se rapportant à l'influence du sexe et de l'age sur la discrimination chromatique, ainsi qu'à la signification du test 100 Hue *Rev. Opt. Théor. Instrum.* **41** 499–509

Vienot F 1982 Can variation in macular pigment account for the variation of colour matches with retinal position? *International Conference on Colour Vision, Cambridge, UK, 23–27 August 1982* (Proceedings published by Academic Press 1983)

Vierling O 1935 *Die Farbensinnprufung bei der deutschen Reichsbahn* (Melsungen: Bernecker)

Vilter V 1949 Recherches biometrique sur l'organization synaptique de la retinae humaine *C. R. Soc. Biol.* **143** 830–2

Vints 1975 The effect of lead on the visual organ *Vest. Oftalmol.* **1** 74–5

Vogt G 1973 Spectral mixture curves for protanopic, deuteranopic, protanomalous and deuteranomalous subjects *Mod. Probl. Ophthalmol.* **13** 196–9

Voke J 1974 Stiles–Crawford chromatic effect in congenital colour defective observers *Mod. Probl. Ophthalmol.* **13** 140–4

—— 1976a The industrial consequences of deficiencies of colour vision *PhD Thesis* The City University, London

—— 1976b The Wright-Holmes colour perception lantern *The Optician* **172** 10

—— 1978a Congenital Rod Monochromatism in a brother and sister *Mod. Probl. Ophthalmol.* **19** 236–8

—— 1978b Colour vision testing in the school medical service *The Optician* **176** 15–24

—— 1978c Colour vision defects—occupational significance and testing requirements *J. Soc. Occup. Med.* **28** 51–6

—— 1980a Colour blindness *Pract. Electron.* **16**(6)48–51

—— 1980b *Colour Vision Testing in Specific Industries and Professions* (London: Keeler)

—— 1982 Colour vision problems at work *Health. Saf. Work* **4** 27–8

—— 1984 But spinach is black *The Optician* **187** 35–6

Voke J and Voke P R 1980 Congenital dyschromatopsia in Saudi Arabia *Saudi Med. J.* **1**(4) 209–14

Voke-Fletcher J and Fletcher R J 1978 A case of tritanopia *Mod. Probl. Ophthalmol.* **19** 229–31

Vola J L, Carruel C and Cornu L 1980 Further studies on Stiles' technique in acquired dyschromatopsias *Colour Vision Deficiencies V* ed G Verriest (Bristol: Adam Hilger) pp 127–9

Vola J, Leid J, Leid V, Gastaud P and Saracco J B 1982 Advantages of the two colour threshold method in diabetics and its comparison with an arrangement test *Docum. Ophthalmol. Proc.* Series 33 405–11

Vola J, Leprince G, Langle D, Cornu L and Saracco J B 1978 Preliminary results on the clinical interpretation of Stiles 2 colour thresholds method *Mod. Probl. Ophthalmol.* **19** 266–9

Vola J L, Riss M, Jayle G E, Gosset A and Tassy A 1972 Acquired deficiency of colour vision in lateral homonymous hemianopia *Mod. Probl. Ophthalmol.* **11** 150–9

Volk D and Fry G A 1947 Effect of quality of illumination and distance of observation and performance in the Ishihara test *Am. J. Optom.* **24** 99–122

Völker-Dieben H J, Went L N and de Vries-de Mol E C 1974 Comparative colour vision ... in three families with dominant inherited juvenile optic atrophy *Mod. Probl. Ophthalmol.* **13** 277–81

Vos J J 1972 Literature review of human macular absorption in the visible spectrum and its consequences for the cone receptor primaries *Report* 1972–17 (Soesterberg: Institute for Perception TNO)

—— 1982 On the merits of model making in understanding colour-vision phenomena *Color Res. Appl.* **7** 69–77

Vos J J, Verkaik W and Boogaard J 1972 The significance of the T.M.C. and H.R.R. color-vision tests as to red–green defectiveness *Am. J. Optom.* **49** 847–59

Vos J J and Walraven P L 1970a On the derivation of the foveal receptor primaries *Vision Res.* **11** 799–818

—— 1970b An analytical description of the line-element in the zone fluctuation model of colour vision *Report* IZF 1970–5 (Soesterberg: Institute for Perception TNO)

de Vries H 1946 On the basic sensation curves of the three-color theory *J. Opt. Soc. Am.* **36** 121–7

—— 1948 *Physica* **14** 367

de Vries-de Mol E C 1977 A study on the hereditary colour vision deficiencies *Doctoral Thesis* University of Leiden, Netherlands

de Vries-de Mol E C and Walraven P L 1982 Analysis of the chromatic Stiles-Crawford effect of colour vision deficient observers *Docum. Ophthalmol. Proc. Series* **33** 303–10

de Vries-de Mol E C and Went L N 1971 Unilateral colour vision disturbance. A family study *Clin. Genet.* **II** 15–27

de Vries-de Mol E C, Went L N, van Norren D and Pols L C W 1978 Increment spectral sensitivity of hemizygotes and heterozygotes for different classes of colour vision *Mod. Probl. Ophthalmol.* **19** 224–8

Vukovich V 1952 Das ERG des Achromaten *Ophthalmologica* **124** 354–9

Waaler G H M 1927 Ueber die Erblichkeitsverhältnisse der verschiedenen Arten von angeborenen Rotgrün-blindheit *Z. Indukt. Abstamm. Vererbungsl.* **45** 279

—— 1967 The heredity of normal and defective colour vision *Avh. Nor. Vidensk.* **I** (9)

—— 1968 New facts in the genetics of colour vision *Avh. Nor. Vidensk.* **I** (11)

—— 1969 Studies in colour vision *Avh. Utgitt. Nor. Vidensk. Akad. Oslo* **I** Mat-Naturv. Kl. N.S. no 12

—— 1977 Genetics and physiology of colour vision *Reprint* 27 XI-XII (Oslo City Hospital)

Waardenburg P J 1963a Colour sense and dyschromatopsia *Genetics and Ophthalmology* vol 2 (Assen, Netherlands: Royal van Gorcum) pp 1425–566

—— 1963b Achromatopsia congenita *Genetics and Ophthalmology* vol 2 (Assen, Netherlands: Royal van Gorcum) pp 1695–718

Waardenburg P J, Franceschetti A and Klein D 1961 *Genetics and Ophthalmology* (Oxford: Blackwell)

Waddington M 1965 Colour blindness in young chidren *Educ. Res.* **7** 236–40

Wagner G and Boynton R M 1972 Comparison of four methods of heterochromatic photometry *J. Opt. Soc. Am.* **62** 1508–515

Wald G 1937 Photo-labile pigments of the chicken retina *Nature* **140** 545–6

—— 1945 Human vision and the spectrum *Science* **101** 653–8

—— 1949 The photochemistry of vision *Docum. Ophthalmol.* **3** 94–137

—— 1964 The receptors of human colour vision *Science* **145** 1007–16

—— 1965 Molecular and fine structure of receptors—introductory lecture *Proc. Int. Congr. Photobiol.* (Oxford: Blackwell) pp 133–44

—— 1966 Defective colour vision and its inheritance *Proc. Natl. Acad. Sci. USA* **55** 1347–63

—— 1967 Blue blindness in the normal fovea *J. Opt. Soc. Am.* **57** 1289–301

Wald G and Brown P K 1958 Human rhodopsin *Science* **127** 222–6

—— 1965 Human colour vision and colour blindness *Cold Spring Harbour Symp. Quant. Biol.* **30** 345–61

Wald G, Brown P K and Smith P H 1955 Iodopsin *J. Gen. Physiol.* **38** 625–31

Wald G and Burian H 1944 The dissociation of form vision and light perception in strabismic amblyopia *J. Ophthalmol.* **27** 950–63

Wald G and Steven D 1939 Experiments in human Vitamin A deficiency *Proc. Natl. Acad. Sci. USA* **25** 344–9

Walls G L 1942 *The Vertebrate Eye and its Adaptive Radiation* (New York: Hafner)

—— 1956 The G Palmer story *J. Hist. Med.* 66–96

—— 1959a Peculiar colour blindness in peculiar people *Arch. Ophthalmol.* **67** 41–60

—— 1959b How good is the H.R.R. test for color blindness? *Am. J. Optom.* **36** 169–93

—— 1964 Notes on four tritanopes *Vision Res.* **4** 3–16

Walls G L and Heath G G 1954 Typical total colour blindness reinterpreted *Acta Ophthalmol.* **32** 279–97

—— 1956 Neutral points in 138 protanopes and deuteranopes *J. Opt. Soc. Am.* **46** 640–9

Walls G L and Judd H D 1933 The intra-ocular colour-filters of vertebrates *Br. J. Ophthalmol.* **17** 641–75, 705–25

Walls G L and Mathews R W 1952 New means of studying color blindness and normal foveal color vision *Univ. Calif. Publ. Psychol.* **7** (1)

Walraven P L 1961 On the Bezold-Brücke phenomenon *J. Opt. Soc. Am.* **51** 1113–16

—— 1962 On the mechanisms of colour vision *Thesis* University of Utrecht, Netherlands

—— 1966 A zone theory of colour vision *Proc. Int. Color Meeting Lucerne 1965* (Berlin: Musterschmidt) pp 137–40

—— 1973 Theoretical models of the colour vision network *Colour 73* (London: Adam Hilger) pp 11–20

—— 1974 A closer look at the tritanopic convergence point *Vision Res.* **14** 1339–43

Walraven P L and Bouman M A 1960 Relation between directional sensitivity and spectral response curves in human colour vision. *J. Opt. Soc. Am.* **50** 780–4

—— 1966 Fluctuation theory of colour discrimination of normal trichromats *Vision Res.* **6** 569–86

—— 1972 in *Handbook of Sensory Physiology* ed L M Hurvich and D Jameson (Berlin: Springer)

Walraven P L and Leebeck H L 1958 The recognition by normals and colour defi-
  ́cient persons of coloured resistor codes under different levels of illumination (in
  Dutch) *Rapport* 17F 1958–4 (Soesterberg: Institute of Perception TNO)
——1960 The recognition of colour codes by normals and colour defectives at
  general illumination levels—an evaluation of the HRR plates *Am. J. Optom.* **37**
  82–92
——1962a Kunstmatig daglicht voor kleurenzientests *Rapport* IZF 1962–20
  (Soesterberg: Institute of Perception TNO)
——1962b The chromatic Stiles-Crawford effect of anomalous trichromats *J. Opt.
  Soc. Am.* **52** 836–7
Walsh F B and Hoyt W F 1969 *Clinical Neurophthalmology* vol 3 (Baltimore:
  Williams and Wilkins)
Walsh F B and Sloan L L 1936 Idiopathic flat detachment of the macula *Am. J.
  Ophthalmol.* **19** 195–208
Walton H 1877 Letter headed 'Colour Blindness' *The Times* January 3rd p 8
Warburton F L 1954 Variations in normal colour vision in relation to practical col-
  our matching *Proc. Phys. Soc.* B **67** 477–84
Wardrop J 1808 *The morbid anatomy of the eye* 1st edn (London) (cited by Sherman
  1981)
Wasserman G S 1978 *Color vision: an Historical Introduction* (Chichester: Wiley)
Watkins R D 1969a Foveal increment thresholds in normal and deutan observers
  Vision Res. **9** 1185–96
——1969b Foveal increment thresholds in protan observers *Vision Res.* **9** 1197–204
Watson W 1914 On anomalous trichromatic colour vision *Proc. R. Soc.* A **90** 443–8
Weale R A 1950 Foveal hue discrimination in the presence of a white surround
  Nature **166** 872
——1951a Hue discrimination in para-central parts of the human retina measured
  at different luminance levels *J. Physiol.* **113** 115–22
——1951b The foveal and para-central spectral sensitivities in man *J. Physiol.* **114**
  435–46
——1953a Some aspects of total colour blindness *Trans. Ophthalmol. Soc. UK* **73**
  241–9
——1953b Spectral sensitivity and wavelength discrimination in the peripheral
  retina *J. Physiol.* **119** 170–90
——1953c Cone monochromatism *J. Physiol.* **121** 548–69
——1953d Photochemical reactions in the living cat's retina *J. Physiol.* **122**
  322–31
——1955 Bleaching experiments on eyes of living guinea-pigs *J. Physiol.* **127**
  572–85
——1959 Photo-sensitive reactions in foveae of normal and cone-monochromatic
  observers *Optica Acta* **6** 158–73
——1960 *The Eye and its Function* (London: Hatton Press)
——1963 *The Ageing Eye* (London: Lewis)
Weder W 1975 Special prescription of spectacles in cases of achromatopsia *Klin.
  Monatsbl. Augenheilkd.* **166** 380–2
Weinke R E 1960 Refractive error and the green/red ratio *J. Opt. Soc. Am.* **50**
  341–2
Weiskrantz L 1963 Colour discrimination in a young monkey with striate cortex
  ablation *Neuropsychology* **1** 145–64

Weitzman D O and Kinney J A 1969 Effect of stimulus size, duration and retinal location on the appearance of colour *J. Opt. Soc. Am.* **59** 640–3

Weller R E and Kaas J H 1981 Cortical and subcortical connections of visual cortex in primates *Cortical Sensory Organisation* vol 2 Multiple visual areas ed C N Woolsey (Clifton, NJ: Humana Press) pp. 121–48

Went L N and de Vries-de Mol E C 1976 Genetics of colour vision *Mod. Probl. Ophthalmol.* **17** 96–107

Wentworth H A 1930 A quantitative study of achromatic and chromatic sensitivity from centre to periphery in the visual field *Psychol. Monogr.* **40** (3) no 183

Werbling F S and Dowling J E 1969 Organisation of retina of mudpuppy *J. Neurophysiol.* **32** 339–55

Westfall J A (ed) 1982 *Visual Cells in Evolution* (New York: Raven Press)

Whisson 1778 An account of a remarkable imperfection of sight in a letter from J. Scott to Rev. Whisson *Phil. Trans. R. Soc.* **68** 611–14

White C T, Katoka R W and Martin J I 1977 Colour-evoked potentials; development of a methodology for the analysis of the processes involved in colour vision *Visual Evoked Potentials in Man; New Developments* ed J Desmedt (Oxford: Clarendon) pp 250–72

Wiesel T N and Hubert D H 1966 Spatial and chromatic interactions in the lateral geniculate body of the rhesus monkey *J. Neurophysiol.* **29** 1115–56

Wildberger H G H and van Lith G H M 1976 Colour vision and visually evoked response (VECP) in the recovery period of optic neuritis *Mod. Probl. Ophthalmol.* **17** 320–4

van der Wildt G L and Bouman M A 1968 The dependence of Bezold–Brücke hue shift on spatial intensity distribution *Vision Res.* **8** 303–13

Williams C M 1976 Visual acuity and colour vision tests *Br. J. Physiol. Opt.* **31** 29–31

Williams C M and Leaver P K 1980 Visual acuity and colour vision in central serous retinopathy *Colour Vision Deficiencies V* ed G Verriest (Bristol: Adam Hilger) pp 306–9

Williams D R, MacLeod D I A and Hayhoe M M 1981 Punctate sensitivity of foveal blue-sensitive cones *Vision Res.* **21** 1357–75

Willis MP and Farnsworth D 1952 Comparative evaluation of anomaloscope *Med. Res. Lab. Report* 190 (New London, CT: Naval Research Laboratory)

Willmer E N 1944 Colour of small objects *Nature* **153** 774–5

——1946 *Retinal Structure and Colour Vision* (London: Cambridge University Press)

——1949 Colour vision in the central fovea *Docum. Ophthalmol.* **3** 194–213

——1950 Further observations on the properties of the central fovea in colour-blind and normal subjects *J. Physiol.* **110** 422–46

Willmer E N and Wright W D 1945 Colour sensitivity of the fovea centralis *Nature* **156** 119–21

Wilson E B 1911 The sex chromosomes *Arch. Mikrosk. Anat.* **77**(II) 249–71

Wilson G 1855 *Researches on Colour-Blindness* (Edingburgh: Sutherland and Knox; London: Simkin, Marshall)

Wilson R F 1960 *Colour in Industry Today* (London: Allen and Unwin)

Wolffin E 1924 Ueber angeborene/totale Farbenblindheit *Klin. Monatsbl. Augenheilkd.* **72** 1–8

Wooten B R, Fuld K and Spillman L 1975 Photopic spectral sensitivity of the peripheral retina *J. Opt. Soc. Am.* **65** 334–41

Wooten B R and Wald G 1973 Colour-vision mechanisms in the peripheral retinas of normal and dichromatic observers *J. Gen. Physiol.* **61** 125–45

Wright A A 1972 Construct a monochromator from a single interference filter *J. Exp. Anal. Behav.* **18** 61–3

Wright W D 1927/28 A trichromatic colorimeter with spectral primaries *Trans. Opt. Soc.* **29** 225–41

—— 1928/29 A re-determination of the trichromatice coefficients of the spectral colours *Trans. Opt. Soc.* **31** 141–59

—— 1929/30 A re-determination of the mixture curves of the spectrum *Trans. Opt. Soc.* **31** 201–11

—— 1941 The sensitivity of the eye to small colour differences *Proc. Phys. Soc.* **53** 93

—— 1943 The graphical representation of small colour differences *J. Opt. Soc. Am.* **33** 632

—— 1946 *Researches on Normal and Defective Colour Vision* (London: Henry Kimpton)

—— 1952 Characteristics of tritanopia *J. Opt. Soc. Am.* **42** 509–20

—— 1953 Defective colour vision *Br. Med. Bull.* **9** 36–40

—— 1957a *Report on Colour Vision Tests by a Committee Appointed by the International Council of Ophthalmology* (unpublished but privately circulated)

—— 1957b Diagnostic tests for colour vision *Ann. R. Coll. Surg. Engl.* **20** 177–91

—— 1964 *The Measurement of Colour* 3rd edn (London: Hilger and Watts) (4th edn 1969)

—— 1981 Why and how chromatic adaptation has been studied *Colour Res. Appl.* **6** 147–52

Wright W D and Pitt F H G 1934 Hue discrimination in normal colour vision *Proc. Phys. Soc.* **46** 459

—— 1937 The saturation discrimination of two trichromats *Proc. Phys. Soc.* **49** 329

Wyszecki G 1964 Re-analysis of the N.R.C field trials of colour-matching functions *J. Opt. Soc. Am.* **54** 710

—— 1970 Development of new C.I.E. Standard Sources for colorimetry *Die Farbe* **19** 43–76

Wyszecki G and Stiles W S 1967 *Colour Science* (New York: Wiley)

Yates T 1974 Chromatic information processing in the foveal projection of unanesthetized primate *Vision Res.* **14** 163–79

Yorke H C 1976 Use of the Rayleigh equation as an objective determinant of colour vision defects *Mod. Probl. Ophthalmol.* **17** 237–40

Young R S L, Fishman G A and Chen F 1980 Traumatically acquired colour vision defect *Invest. Ophthalmol.* **19** 545–9

Young R W 1981 A theory of central retinal disease *New Directions in Ophthalmic Research* ed M L Sears (New Haven, CT: Yale University Press) pp 237–70

Young T 1802a On the theory of light and colours *Phil. Trans.* **92** 12–48

—— 1802b An account of some cases of the production of colours not hitherto described *Phil. Trans.* **92** 387–97

—— 1807 *Lectures on Natural Philosophy and the Mechanical Arts* vol 2 (London: Johnson) pp 315–16 (quoted by Judd 1943 and Sherman 1981)

Young W B 1974 Some temporal aspects of colour vision and a pilot study with a new interference filter anomaloscope *MSc Thesis* University of Waterloo, Canada

Zanen J 1953 Introduction a l'étude des dyschromatopsies rétiniennes centrales acquises *Bull. Soc. Belg. Ophthalmol.* **103** 7–148

—— 1971 La méthode des seuils en pathologie oculaire *Conc. Ophthalmol. Acta* **2** 1750–6

—— 1972 The foveal spectral thresholds in acquired dyschromataposia *Mod. Probl. Ophthalmol.* **11** 213–18

Zehnder E 1971 Die Bewährung farbensinngestörter motor fahrzeuglenker im Verkehr *Schweiz. Med. Wochenschr.* **101** 530–7

Zeki S M 1971 Cortical projections from two prestriate areas in the monkey *Brain Res.* **19** 63–75

—— 1973 Colour coding in rhesus monkey prestriate cortex *Brain Res.* **53** 422–7

—— 1975 The functional organization of projections from striate to prestriate visual cortex in the rhesus monkey *Cold Spring Harbour. Symp. Quant. Biol.* **40** 591–600

—— 1977 Colour coding in the superior temporal sulcus of the rhesus monkey visual cortex *Proc. R. Soc.* B **197** 195–223

—— 1978a The cortical projections of foveal striate cortex in the rhesus monkey *J. Physiol.* **277** 227–44

—— 1978b The third visual complex of rhesus monkey peristriate cortex *J. Physiol.* **277** 245–72

—— 1978c Uniformity and diversity of structure and function in rhesus monkey prestriate visual cortex *J. Physiol.* **277** 273–90

—— 1978d Functional specialisation in the visual cortex of the rhesus monkey *Nature* **274** 423–8

—— 1980 The representation of colours in the cerebral cortex *Nature* **284** 412–18

Zeltzer H I 1971 Methods of improving color discrimination *US Patent* 3 586 423 (See also the X-Chrom lens *J. Am. Optom. Assoc.* **42** 933–9)

—— 1973 The X-Chrom contact lens and color deficiency *Opt. J. Rev. Optom.* **110** 15–19

Zrenner E 1982 Interactions between spectrally different cone mechanisms *Docum. Opthalmol. Proc.* Series 31 287–96

Zrenner E and Gouras P 1978 Retinal ganglion cells lose colour opponency at high flicker rates *Invest. Ophthalmol. Vis. Sci.* Suppl. **17** 130

—— 1979 Cone opponency in tonic ganglion cells and its variation with eccentricity in rhesus monkey retina. *Invest. Ophthalmol. Vis. Sci.* Suppl. **18** 77

—— 1981 Characteristics of the blue sensitive cone mechanism in primate retinal ganglion cells *Vision. Res.* **21** 1605–9

Zrenner E and Kojima M 1976 Vsually evoked cortical potential in dichromats *Mod. Probl. Ophthalmol.* **17** 241–6

Zrenner E and Krüger C J 1981 Ethambutol mainly affects the function of red/green-opponent neurones *Docum. Ophthalmol. Proc.* Series 27 13–25

Author index

Subject index